Organic Farming

Organic Farming

Nicolas Lampkin

FARMING PRESS

First published 1990
Reprinted with amendments, 1992 & 1994
Reprinted 1997, 1998 & 1999

Copyright © Nicolas Lampkin

A catalogue record for this book is available from the British Library

ISBN 0-85236-191-2

Published by Farming Press
Miller Freeman UK Ltd
Miller Freeman House
Sovereign Way
Tonbridge TN9 1RW

Distributed in North America
by Diamond Farm Enterprises,
Box 537, Bailey Settlement Road,
Alexandria Bay, NY13607, USA

Cover design by Mark Beesley
Phototypeset by Cambridge Phototypesetting, Cambridge
Printed and bound in Great Britain by Butler & Tanner Ltd, Frome and London

Contents

Colour plates 1–28 appear between pages 334 and 335

Preface to the First Printing

Organic farming has become an established part of the farming scene, but in spite of the considerable media attention there is still very little in the way of published information on the subject. This book is an attempt to remedy the situation.

It is aimed at a range of readers, from farmers farming organically or considering converting, to advisers, researchers and students who require not only detailed information but also access to source material.

I am conscious of the fact that it is a long book—a result of the increasing body of scientific evidence, especially from outside Britain, which underpins the organic farming approach, and the need to avoid another relatively lightweight text which does little more than give an introduction to the subject.

Covering such a wide range of material may well mean that not all parts of the book are relevant to an individual reader, but I hope that all who work in the agricultural industry will find something of interest in this book.

Organic farming is developing at such a pace that it is inevitable that some of the material contained in this book will have been superseded even by the time it is published. It is intended that any such new developments will be fully reflected in future editions.

NICOLAS LAMPKIN
Aberystwyth
July 1990

Preface to the Third Printing

Since this book was published in 1990, there have been a number of developments which have significantly changed the characteristics of organic farming in Europe and in the United Kingdom. It has not yet been possible to produce a fully revised version of the book to take account of all these changes. The intention of this preface is to draw attention to the most important developments and their impact on the organic sector. In addition, recent publications which provide further information on the current situation are listed separately at the end of each chapter, but have not been referred to in the text.

Legislation

Since 1991, organic crop production in the European Community has been legally defined by EC Regulation 2092/91 (EC, 1991—see Chapter 12). This legislation came into force on 1st January 1993. Since this date, all produce sold with an organic label, or a label indicating organic methods of production, has had to comply with this legislation. There have been some significant changes to organic standards and inspection procedures to accommodate the new legislative framework. For this reason, the Soil Association's standards published as an Appendix to this book have been completely revised, and the most recent (1993) version has been included in this impression. Readers are strongly advised to use the latest version of the relevant organic sector body's standards when dealing with certification issues.

The second impact of the legislation has been to bring about significant changes to the organisations carrying out inspections and certification. In the UK, the United Kingdom Register of Organic Food Standards (UKROFS) is now the body with overall responsibility for implementation of the EC Regulation. The Soil Association Marketing Co. Ltd., Organic Farmers and Growers Ltd, the Scottish Organic Producers Association, the Bio-Dynamic Agricultural Association and the Organic Food Federation have been approved by UKROFS (see Chapter 12 and Appendix 1). The Irish Organic Farmers and Growers Association

(IOFGA) has also been approved by UKROFS to continue certification activities in Northern Ireland.

Conversion subsidies

Since 1989, conversion subsidies have been introduced in several European countries, including Germany, Sweden, Denmark, Norway and parts of Switzerland (see Chapter 15). More recently, schemes have been introduced in Finland, the Czech Republic, Slovakia, Austria, and Switzerland as a whole. The impact of these schemes on the uptake of organic farming has been dramatic, particularly in the German-speaking countries (Lampkin and Padel, 1994—see Chapter 15).

In Germany, the use of the European Community's extensification policy to encourage conversion to organic farming resulted in over 10,000 farms on over 376,000 ha converting to organic farming in the period 1989/90 to 1992/3, with the result that over 2.4% of the German agricultural land area is now managed organically. In some western German states, 5% of the land area has been converted. In eastern Germany, where the subsidies have only been available for two years, the results have been even more dramatic, with 7.9% of farms on 3.6% of the land area converting to organic farming. In some districts more than 20% of the land area has been converted under this policy. In Austria, a similar policy has resulted in an increase from 1000 to 9000 farms on 135,000 ha, or more than 3% of the Austrian agricultural land area. In Europe as a whole, the land area managed organically and in conversion has increased from less than 200,000 ha in 1989 to nearly 900,000 ha in 1993. As might be expected, increases of this magnitude have had an impact on established organic producers and their market outlets, with downward pressure on price premiums, but they have also provided new opportunities for developing mainstream rather than niche markets. Despite these challenges and the recession of the early 1990s, many of the marketing initiatives described in Chapter 12 are continuing to thrive.

From 1993, the European Community Regulation 2078/92 (EC, 1992—see Chapter 15) will provide, for the first time, scope for support for established organic farms as well as farms in conversion to organic farming. As a result, the majority of European countries will have in place schemes to give financial support to organic farmers by 1994. The Ministry of Agriculture, Fisheries and Food in the United Kingdom has announced proposals to support conversion to organic farming (but not existing organic farmers) at an average rate of £50 per ha per year over five years. A reduced rate of 20% of the lowland rate applies to rough grazing in Scotland and Northern Ireland, and to the Less Favoured Areas in England and Wales.

Although full details of the schemes in each EC member state were not available at the time of writing, it is clear that the MAFF proposals fall far short of the assistance likely to be available elsewhere, with payment rates typically only one-third of the level in other EC countries. Further details of the organic aid scheme in England, Wales, Scotland and Northern Ireland can be obtained directly from MAFF and the respective Agriculture Departments.

The environmental impact of organic farming

Evidence on the environmental impact of organic farming continues to accumulate, with recent reports indicating:

- improved soil biological activity and physical characteristics on bio-dynamic farms in New Zealand (Reganold *et al.*, 1993—see Chapter 2);
- the potential of organic farming as an option for reducing nitrate leaching in Nitrate Sensitive Areas (EFRC, 1992a,b—see Chapter 3);
- beneficial impacts of organic farming on bird populations (Wilson, 1993—see Chapter 15).

Many other studies on the environmental impact of organic farming have also been reviewed in Greenpeace, 1992 and Redman *et al.*, 1993 (see Chapter 15).

The economics of organic farming

Despite the pressure on price premiums caused by the rapid expansion in the supply of organic foods in Europe, demand for organic foods remains strong (Mintel, 1993—see Chapter 12). However, predicted growth rates in the retail value of the organic food sector have been reduced relative to some of the reports published in 1990/91, which indicated that 5–10% of retail food sales might be organically produced by the year 2000 (Tate in Lampkin and Padel, 1994—see Chapter 13).

A recent survey of organic farming as a business in Great Britain (Murphy, 1992—see Chapter 13) painted a gloomy picture of the financial viability of organic farming. The Murphy report has been criticised on a number of grounds, not least the basis of the comparisons made between organic and conventional farms (Bateman, 1993; Lampkin and Bateman, 1993—see Chapter 13). In particular, very small, effectively part-time, farms were not treated separately from the larger full-time units, whereas farms of this size are excluded from the conventional comparison groups. The poor financial performance of organic farming reported in the Murphy study reflects the financial problems of small farms in general. This is a problem which organic farming on its own cannot be expected to solve, although in some cases access to a premium market may help. A large number of the smallest organic units have been forced out of business recently due to a combination of these difficult financial circumstances and the added burdens imposed by the EC legislation. Up-to-date costings of organic enterprises and typical whole farm systems are contained in a new organic farm management hand-book (UWA/EFRC, 1992—see Chapter 13).

Specialist, intensive arable farmers, in contrast to livestock farmers, have also found it difficult to justify conversion to organic production because of the relatively high restructuring costs and the lack of suffi-ciently high premiums to compensate for yield reductions (Lampkin, 1993a—see Chapter 14). This situation has been exacerbated by the Common Agricultural Policy reform of 1992, which introduced arable area payments linked to 15% set-aside of arable land. Although these payments were of benefit to existing organic producers, they represent an additional barrier to conversion to organic production now because of the reduction in area qualifying for these payments under the rotational constraints of an organic system (Lampkin, 1993b—see Chapter 14).

The impact of the changes in the legislative and policy support framework on the development of organic farming in Europe over the next few years is very difficult to predict. What is clear is that the increased acceptance of organic farming by farmers, politicians and legislators has brought with it a much greater need to take account of the economic, market and policy framework within which all farmers operate and which dictates the financial results of the vast majority of farms, whether conventional or organic. Good management, within the resource and other constraints on the farm business, remains the key factor determining whether or not an organic farming venture will succeed.

NICOLAS LAMPKIN
Aberystwyth
November, 1993

Acknowledgements

Grateful acknowledgements are due to the many farmers who contributed their practical experience to the writing of this book and to all those who have helped, through the organisation of conferences, seminars and other events, to make information on organic farming more widely available.

Special thanks are due to David Bateman, Philip Harris, Patrick Holden, Mark Measures, Richard Moore-Colyer, Jon Newton, Steve Parish, Colin Spedding, Victor Stewart, Ian Sturgess, Hartmut Vogtmann and Lawrence Woodward for their comments on the draft of this book. Any errors which still remain are probably my responsibility.

Permission to use the Soil Association's standards for organic food production as an appendix is also gratefully acknowledged, as is permission from the Development Board for Rural Wales to use a number of their photographs and unpublished material.

I am grateful for kind permission from various sources to reproduce copyright material in tables and figures. Full documentation is given either in source notes or references at ends of chapters. While attempts were made to secure permission for all material reproduced, we have been unable to trace some copyright holders. Acknowledgements will be given in later editions if the copyright holders contact Farming Press.

Finally, recognition must be given to Roger Smith and Julanne Arnold at Farming Press Books for their patience and forbearance during the book's long gestation period.

Additional contributions from

Charles Arden-Clarke   (Chapter 15)
Pam Best   (Chapter 9)
Will Best   (Chapters 10, 14)
Stewart Biggar   (Chapter 9)
Josef Finke   (Chapter 11)
Susan Fowler   (Chapters 3, 4)
David Frost   (Chapter 12)
Edward Goff   (Chapters 10, 14)
Jeremy Harding   (Chapters 11, 14)
Patrick Holden   (Chapters 10, 11, 12)
Herbert Koepf   (Appendix 2)
Gerry Minister   (Chapter 11)
Christopher Raymont   (Chapter 9)
Julian Rose   (Chapter 9)
Gareth Rowlands   (Chapters 9, 12)
Mariette Smit   (Appendices 4, 5)
Charlotte Stewart   (Chapter 12)
Christopher Stopes   (Chapters 5, 8, 10, 12)
Iain Tolhurst   (Chapter 11)
John Wakefield-Jones   (Chapter 11)
Lawrence Woodward   (Chapters 1, 6, 8, 12, 14)
Richard Young   (Chapters 9, 11, 12)
Rosamund Young   (Chapters 9, 11)

Foreword

The recent increase in interest in organic farming, not only in the U.K. but in many other countries as well, makes this a very timely book. Many consumers are increasingly concerned about the ways in which their food is produced and are prepared to pay more for products which result from methods that are perceived to be "better" for the environment, for farming, for animals and for people.

Many farmers would prefer to shift away from heavy use of agrochemicals and see their activities as being much broader than just production.

The UK Register of Organic Food Standards (UKROFS) has been established to ensure that those who want to purchase genuine "organically-produced" foods can identify them and that those who produce them are protected from unfair competition from those whose methods do not justify the label.

There is, however, considerable ignorance about the facts, both amongst consumers and producers and, indeed, the latter are the first to acknowledge the lack of information and to call for more research.

There is also a considerable lack of understanding on the part of the general public as to what organic farming involves. It is more often seen in negative terms, involving not using agrochemicals, and much less in positive terms, especially in its concern, for example, for animal welfare.

In these circumstances, any attempt to produce a balanced and comprehensive account of the subject is to be welcomed.

This book will be invaluable to all those who wish and need to know more about organic farming, its principles and its practices.

Its comprehensiveness makes it a very long book but it is consistently readable and clear. It contains a great deal of information but it avoids unnecessary jargon and should appeal to a wide audience.

It is not afraid to expose myths and lack of evidence and well reflects the growing maturity of the organic movement.

C. R. W. SPEDDING
Chairman of the Board of UKROFS
December 1989

Chapter 1

Organic Farming— Agriculture with a Future

Organic farming took on a new lease of life during the 1980s, not just in Britain but around the world. The problems of overproduction in the industrialised countries, underproduction in developing countries and the environmental impact of agriculture have concentrated minds and brought about a dramatic reassessment of the achievements of the post 1945 era. The effect can be seen not only in the range of policies which give greater weight to environmental considerations, but also in the growth of the organic movement and the market for organically produced food.

The number of organic farmers in Britain increased from fewer than 100 in 1980 to more than 700 in 1989; more farmers are converting all the time. Consumer interest developed even faster, for both health and environmental reasons, with most of the major multiples including Safeway, Sainsbury, Waitrose, Tesco, Gateway and most recently Marks and Spencer becoming involved. The organic food retail market, estimated at £50 million in 1989, looks set to expand by as much as 100% in 1990. In response to this growth, the Government established the United Kingdom Register of Organic Food Standards (UKROFS) in 1987 to provide nationally agreed standards; these were launched in May 1989. The European Commission introduced a regulation legally defining organically produced food in July 1991.

Organic farming is increasingly being recognised as a potential solution to many of the policy problems facing agriculture in both developed and developing countries. Denmark, Sweden, West Germany and the Canton of Berne in Switzerland have introduced schemes to support farmers financially during the critical conversion period; more European Community countries are set to do so under the new extensification legislation. Other countries like Israel and New Zealand have given considerable support to the development of export-oriented marketing strategies and research. The Government of Burkina-Faso in West Africa hosted the 1989 conference of the International Federation of Organic Agriculture Movements and is committed to following an ecological approach to the development of its own agricultural resources. Ecologically oriented projects are being developed in many other countries in the so-called Third World.

Research is taking place at universities and elsewhere, particularly in mainland Europe and in North America, on a scale which would have seemed unthinkable ten years ago. West Germany has three Professors of Organic Agriculture with their own University Departments. The Agricultural Development and Advisory Service in England and Wales has appointed a national coordinator and advisors in each region with responsibility for organic farming related work. Courses in organic farming are being offered at a number of agricultural colleges and by the Agricultural Training Board. The availability of information on organic farming is considerably improved as a result, with one exception: printed material containing the basic scientific and technical information to enable producers to successfully convert and manage their farms organically. It is hoped that this book will go some way towards meeting this need.

What is Organic Farming?

There are several problems which arise when presenting an explanation or definition of organic farming. Firstly, there are a number of misconceptions surrounding the topic which tend to a prejudicial view and divert attention from the main issues. Secondly, the nomenclature varies in different parts of the world, causing understandable confusion to the uninitiated observer. Thirdly, many existing practitioners believe that successful organic farming involves conceptual understanding as well as the employment of specific practical techniques.

These problems prevent the framing of a short, sharp, clear definition of organic farming. It has, therefore, become commonplace to define what it is by stating what it is not. Definitions and descriptions are frequently framed around negatives. What organic farmers do not do or use is summarised in the phrase that 'organic farming is farming without chemicals'. While such a definition has the advantage of being concise and clear, it is unfortunately untrue and misses out on several characteristics which are of fundamental importance.

This notion about the non-use of chemicals is one of four misconceptions referred to above as problematic. In that all material, living or dead, is composed of chemical compounds, then organic farming utilises chemicals. Chemicals, albeit naturally derived, are also directly used in fertilising, plant protection and livestock husbandry. However, organic farming is a system which seeks to avoid the direct and/or routine use of readily soluble chemicals and all biocides whether naturally occurring, nature identical, or not. Where it is necessary to use such materials or substances, then the least environmentally disruptive at both micro and macro levels are used.

The second misconception is that organic farming merely involves substituting 'organic' inputs for so-called 'agro-chemical' ones. A straight substitution of NPK (nitrogen, phosphorus and potassium) as mineral fertiliser by NPK as organic manure is likely to have the same—probably adverse—effect on plant quality, disease susceptibility and environmental pollution. Contrary to the dearly held ideas of organic 'traditionalists', there is nothing magical about muck even if it is pushed in a heap and called 'compost'. The misuse of organic materials, either by excess, by inappropriate timing, or by a mixture of both, will effectively short circuit or curtail the development and working of natural or biological cycles. This approach has rightly been called 'neoconventional' and is rooted in the assumption that the farmer should seek to dominate rather than work with nature and natural cycles.

Another mistaken idea about organic agriculture is that it is a return to farming as it was pre-1939. While there is a shared focus on what has been described as 'good, sound husbandry', involving balanced rotations, mixed farms and mechanical methods of weed control, modern organic farming seeks to develop upon increased understanding of such things as mycorrhizal associations, rhizobia and the rhizosphere, the turnover of organic matter and other areas of soil life, crop and animal husbandry that modern science has revealed. Organic farmers cannot be Luddites, setting aside the developments of the last 50 years. Indeed, it is more the case that modern agriscience has constrained itself by concentrating far too much upon agrochemical inputs and not enough on understanding and developing the inherent qualities to be found within biological sciences. The fact is that, while organic farming in Britain today generally has the same ley or mixed farming base that could be found 40 years ago, many of its techniques and practices are modern developments. Crucially, the approach and attitude of its most successful practitioners is profoundly different.

The fourth misconception is that organic farming requires a change of lifestyle on the part of the farmer. While it is true that organic agriculture has been passionately supported by people with radical views on other issues and by those holding minority opinions about such things as nutrition, it has never been the case that organic farmers are either part of the love and magic, beard and sandals brigade, or that they are excessively puritan. Such cheap jibes have been the stock in trade of agricultural commentators unable to face up to the real issues generated by the growth of interest in organic food and organic farming. They ceased to be funny a long time ago, are now merely tiresome and just will not do as a substitute for genuine discussion and debate.

Turning to the problem of nomenclature, it has been estimated that there are about 16 different names used throughout the world for what we call organic farming. Some of the better known ones are biological farming, regenerative farming, and sustainable farming. In some cases, there is little or no difference between them. For example, organic and biological, in the United Kingdom, mean the same thing and are interchangeable. The term 'biological' is more favoured throughout mainland Europe, whereas English-speaking Britain and the United States stick doggedly to 'organic'.

In other cases, however, the difference in name indicates a conceptual or philosophical difference. Some of these differences—for example, the terms 'regenerative' and 'sustainable' agriculture—are the result of pedantic academics filling up lecture time or library shelf space, and it would have been wiser for them not to have bothered. Others, such as biodynamic farming, are genuinely part of a whole philosophy that encompasses education, art, nutrition and religion, as well as agriculture.

Nonetheless, the principles and practices that lie behind these different names are essentially similar. They have been concisely expressed in the standards document of the International Federation of Organic Agriculture Movements (IFOAM) as:

- to produce food of high nutritional quality in sufficient quantity;
- to work with natural systems rather than seeking to dominate them;
- to encourage and enhance biological cycles within the farming system, involving microorganisms, soil flora and fauna, plants and animals;
- to maintain and increase the long-term fertility of soils;
- to use as far as possible renewable resources in locally organised agricultural systems;
- to work as much as possible within a closed system with regard to organic matter and nutrient elements;
- to give all livestock conditions of life that allow them to perform all aspects of their innate behaviour;
- to avoid all forms of pollution that may result from agricultural techniques;
- to maintain the genetic diversity of the agricultural system and its surroundings, including the protection of plant and wildlife habitats;
- to allow agricultural producers an adequate return and satisfaction from their work including a safe working environment;
- to consider the wider social and ecological impact of the farming system.

For organic farmers worldwide, these principles provide the basis for day-to-day farming practice. They directly give rise to the techniques of organic agriculture, such as composting; the use of wide rotations which utilise leys and green manures; the avoidance of soluble fertilisers; the prohibition of intensive livestock operations; the avoidance of antibiotic and hormone stimulants; the use of mechanical and thermal methods of weed control; the emphasis towards on-farm processing and direct sales to the consumer; and the use of extra labour when not strictly necessary, as a positive contribution to the farm and rural community.

The United States Department of Agriculture has framed a handy definition of organic farming which, although it misses out some important aspects, provides a description of the key practices:

> Organic farming is a production system which avoids or largely excludes the use of synthetically compounded fertilisers, pesticides, growth regulators and livestock feed additives. To the maximum extent feasible, organic farming systems rely on crop rotations, crop residues, animal manures, legumes, green manures, off-farm organic wastes, and aspects of biological pest control to maintain soil productivity and tilth, to supply plant nutrients and to control insects, weeds and other pests.

> The concept of the soil as a living system. . . that develops. . . the activities of beneficial organisms. . . is central to this definition.

This definition can be divided into three parts:

(1) What organic farmers do not do;
(2) What positive things they do instead;
(3) An indication of the underlying view of the soil as a living system that the farmer, in harmony with nature, should seek to develop.

This idea of the soil as a living system is part of a concept which maintains that there is an essential link between soil, plant, animal and man. Many people involved with organic agriculture believe that an understanding of this is the prerequisite for sustaining a successful organic farming system. This concept has been described as 'holistic', but it can be discussed in a less pretentious way.

Simplified, and put into a practical context, it is the recognition that—within agriculture, as within nature—everything affects everything else. One component cannot be changed or taken out of the farming or the natural system without positively or adversely affecting other things. For example, on an organic farm there is not one method of weed control or of supplying nitrogen. The ley, green manures and appropriate cultivations do both of these things, as well as their more obvious other functions.

Here indeed is the key to understanding what organic agriculture is about. It concentrates primarily on adjustments within the farm and farming system, in particular rotations and appropriate manure management and cultivations, to achieve an acceptable level of output. External inputs are generally adjuncts or supplements to this management of internal features. This theme will recur throughout the book and will become clear as it is discussed in both theory and practice, and as the components of organic agriculture are examined, individually and in detail.

Why Organic Farming?

Several factors have come together in recent years which highlight the necessity for a fundamental review and revision of agricultural policy in Britain and other European countries. The traditional goal of maximising output is being countered by widespread concern over the countryside and environment, and by the growing realisation that finite natural resources need to be more carefully managed. At the same time, subsidised overproduction in Europe has brought about unendurable financial strain and political embarrassment.

While increased productivity has resulted in European food self-sufficiency and the arrival of surplus, the real cost of support for the Common Agricultural Policy has increased by 28% since the mid 1970s as farm incomes in Britain have fallen by 50%. The current cost of the storage and export of the EC cereal surplus is of the order of £12.5 billion, equivalent to approximately £137 per year for every taxpayer in the United Kingdom.

Dramatic changes in farming practices have resulted in a loss of natural habitat and species: for example, a loss of one fifth of hedgerows and more than three quarters of wetland habitats. Environmental pollution of ground and surface water from agricultural sources is an increasing problem, with supplies in some parts of Britain at EC maximum permissible levels. The Soil Survey of England and Wales estimates that nearly 44% of UK arable land is at risk from erosion.

It is not really surprising, therefore, that increasingly people within and outside agriculture are questioning the desirability of its continuing in its present form. A recent NOP poll, for example, showed that almost 60% of those questioned felt that farmers should avoid using 'modern methods of farming', and more than 25% of respondents thought that farmers should be paid for changing from these methods.

The major questions and criticisms of current agricultural practice include the following:

- that it damages soil structure;
- that it damages the environment;
- that it creates potential health hazards in food;
- that it has brought about a reduction in food quality;
- that it is an energy-intensive system;
- that it involves intensive animal production systems which are ethically unacceptable;
- that it is economically costly to society, and increasingly so to the farmer.

It is not the intention here to write a definitive critique nor to repeat the propaganda of the organic movement. These questions and criticisms can be found in the farming press and in the scientific literature.

Conversely, organic agriculture has a positive contribution to make in some of these areas. It is absolutely dependent upon maintaining ecological balance and developing biological processes to their optimum. The preservation of soil structure, earthworms, microorganisms and larger insects is essential to the working of an organic system. Therefore, the protection of the soil and the environment is a fundamental 'must' for the organic farmer and not something that can be tacked on at the end if profits allow.

The potential health hazards of pesticide residues and nitrates resulting from conventional agriculture are now receiving some attention. However, there is growing scientific evidence about the positive quality aspects of organically produced food, for example higher dry matter and vitamin content and improved storage quality.

As it largely avoids the use of chemical inputs produced from finite resources and manufactured in an energy-intensive fashion, and uses little or no external inputs, organic agriculture is not a major drain on the earth's finite resources. Nor does it add to agriculture's hidden costs—those borne by society and not the farmer—such as coping with the problems of excess nitrates in water.

Organic farmers avoid the excesses of intensive animal production systems. This is especially so with pigs, poultry and the use of growth promoters.

It must be acknowledged that, like conventional farming, organic farming in Britain today is economically difficult. However, unlike conventional agriculture, it has not been blessed with extensive research and development, nor have organic farmers had the back-up of advisory services. Even so, some organic farms are out-performing the conventional average. It must surely be reasonable to assume that organic systems would be much more productive if just a fraction of the research effort that has gone into the 'chemical' approach were spent on developing the techniques of organic husbandry.

These aspects will be discussed in detail elsewhere in this book. There is, however, one other striking answer to the question 'Why organic farming?', and that is quite simply that people want to buy organic food. A study commissioned by the National Farmers' Union in England and Wales in 1988 showed that 28% of the people interviewed were 'definitely interested', and a further 23% 'possibly interested' in buying organic food. Other surveys have shown a larger percentage of consumer interest if organic food were more readily available, even to the extent of paying a price premium of 15% or more.

WHO ARE ORGANIC FARMERS?

Organic farmers consist of a wide range of people with often very different motives, from the hard-headed businessman or woman keen to exploit market opportunities to the self-sufficiency smallholder seeking the good life. Increasingly, however, the image of the organic farmer as either a 'gentleman' or 'hobby' farmer, or a 'brown bread and sandals' type, has become increasingly irrelevant, if it were ever true.

Admittedly, many of the people who started farming organically in the 1960s and '70s were newcomers to farming and faced many of the problems which any newcomer would face, let alone those trying to produce organically. But those early pioneers now have a wealth of experience and many are as hardheaded as any commercial farmer trying to make a go of things. More importantly, they have been supplemented in the 1980s by an increasing number of conventional farmers with existing practical experience who are better placed, both financially and practically, to make the change-over.

But organic farming is not suited to every farmer. It requires a commitment to make the system work, often to take risks where there is not a lot of information, and an eye for detail. There is no way that people could be forced to farm organically and do it successfully; the individual has to be sufficiently convinced and motivated to achieve the necessary level of management input.

THE ORGANIC MOVEMENT

Organic farmers, in order to succeed in an indifferent and occasionally hostile environment, have become particularly self-reliant, often learning by trial and error on their own holdings. But they have also learnt the need to cooperate and share experiences.

The Soil Association is the oldest of the organic organisations in Britain. Founded in 1946, it concentrates on researching and providing information on the links between the way food is produced and human health and environmental quality. It is perhaps best known, however, for its work with standards for organic food production which have become the most widely used and respected in Britain and which are used as the basis for this book.

In the early 1980s, two closely related producer organisations, the Organic Growers Association and British Organic Farmers, were formed specifically to improve the level of technical information available to commercial organic producers and conventional producers considering conversion. This is done through a quarterly magazine, *New Farmer and Grower*, a biennial conference at the Royal Agricultural College, Cirencester, and a series of farm walks and seminars.

Also in the early 1980s, Elm Farm Research Centre was established to carry out research and development work for organic producers. Elm Farm has established close links with the Universities and other research institutes in Britain and in mainland Europe. More recently, Elm Farm has joined with the Soil Association, British Organic Farmers and the Organic Growers Association to form an Organic Advisory Service using experienced organic practitioners to provide direct advice to farmers and growers.

Biodynamic producers (see Appendix 2) are represented in Britain by the Bio-Dynamic Agricultural Association, while leisure and amenity gardeners benefit considerably from the existence of the Henry Doubleday Research Association at the National Centre for Organic Gardening near Coventry. A range of other associated organisations, such as the McCarrison Society and the Farm and Food Society, as well as many of the conservation and environmental groups, cooperate closely with the organic movement.

There is also a number of marketing organisations, including producer cooperatives such as the Eastern Counties Organic Producers, Green Growers (Herefordshire), Organic Farmers and Growers Ltd, Organic Growers West Wales and Somerset Organic Producers. These help considerably to alleviate marketing problems, particularly for the smaller producers.

The number and diversity of organisations within the organic movement is reflected internationally, both physically and in terms of sometimes markedly differing opinions, but all are linked by the International Federation of Organic Agriculture Movements (IFOAM). IFOAM sets baseline standards to be adopted by national organisations and monitors these national standards to assist with international trade. In addition, IFOAM encourages the exchange of information and ideas through conferences, research seminars and its multi-lingual magazine, the IFOAM Bulletin.

The ideas behind organic farming have been around since the 1920s. They have evolved considerably and continue to evolve as new scientific research becomes available, while retaining the fundamental philosophical perspective of working with, not dominating, natural systems and respect for the environment which sustains us. The organic movement is clearly established and here to stay. It is heralding a change in agriculture which is occurring simultaneously in every developed agricultural nation in the world. Far from being a return to the past, organic farming is an agriculture for the future, our future.

Part One

The Principles of
Organic Farming

Chapter 2

The Living Soil

A DYING SOIL?

In 1862, Friedrich Albert Fallou wrote:

> There is nothing in the whole of nature which is more important
> than or deserves as much attention as the soil. Truly it is the soil
> which makes the world a friendly environment for mankind. It is
> the soil which nourishes and provides for the whole of nature; the
> whole of creation depends upon the soil which is the ultimate
> foundation of our existence.

There is no question that the fertility of the soil is crucial to the long
term sustainability of life on this planet, or that the soil has been very
generous to us. Crop yields have increased substantially over the last
50 years, and especially so in the last decade. Yet this remarkable
productivity increase has not been achieved without cost. A gradual
reduction in organic matter levels in the soil, especially in the intensively
cultivated arable areas, has been accompanied by a deterioration in soil
structure leaving the soil more prone to compaction and erosion.

The Strutt report, *Modern Farming and the Soil*, published by the
Ministry of Agriculture, Fisheries and Food (MAFF) in 1970, concluded
that 'some soils are now suffering from dangerously low organic matter
levels and could not be expected to sustain the farming systems
which have been imposed on them. ' The biological activity of the soil,
which depends on the availability of nutrients and energy supplied by
soil organic matter and crop and livestock residues, has declined
correspondingly. Given that much of the chemical weathering which
takes place in soils is the result of the activity of soil microorganisms,
then the ability of the soil to supply nutrients to the growing crop from
its own reserves is going to be diminished by a reduction in soil
biological activity.

While the MAFF report did not consider that the decline in organic
matter levels was of concern in terms of nutrient supply, presumably
because alternative (artificial) sources of nutrients are available, it is
clear that in the absence of fertiliser inputs the soil is unlikely to be able
to maintain the productivity levels which might otherwise be the case. In

13

fact, it can be argued that current levels of productivity can only be maintained by relying on ever increasing quantities of fertiliser inputs. We are witnessing a long-term decline in the inherent ability of the soil to grow crops without reliance on inputs from outside the system. Seen in this context, soil fertility as opposed to productivity is not improving. The consequences of this are very serious and we can no longer afford to ignore them.

This situation, described by Sir Albert Howard (1948) as a 'disease' in the wider sense of imbalance or disruption of a whole system rather than an individual organism, is manifested in many forms, one of the most important of which is soil erosion: the rapid loss of soil from agricultural lands.

Although a certain level of soil loss can be considered to be a natural, geological process, these losses are more or less compensated for by the formation of new soil from the bedrock. This process of soil 'denudation' occurs over long periods of time. In many natural ecosystems, well established climax vegetation provides a stable protection for the underlying soil, resulting in only minimal erosion. In some cases, the formation of new soils may even exceed soil losses, resulting in the development of deep soils. Agricultural activities, however, tend to leave the soil exposed and reduce the structural stability of the soil leading to what is termed 'accelerated erosion', where soil losses occur at a much faster rate than the formation of new soils. The long term effects of this process can result in the complete absence of soil over large areas of land.

At the time that Sir Albert Howard was writing, soil erosion was seen mainly in terms of the dust bowls of the United States and water erosion in some of the developing countries. Now we are seeing increasing evidence of it in Britain.

The process is not new. Throughout recorded history we have seen the effects of human activity on the soil. Civilisation after civilisation has mismanaged the soil and been forced to move elsewhere as a result of the devastating effects of soil erosion. The once productive lands of the Mediterranean were transformed into the mainly barren, soilless slopes which we see today. The BBC TV series *Far from Paradise* brought home the full seriousness of erosion around the world, both past and present (Seymour & Girardet, 1986). In the United States, water erosion is now thought to be as serious as, if not more serious than, the dust bowls of the 1930s (Sampson, 1981). In virtually every instance, human activity and inappropriate agricultural practices are the major causes.

In Britain, it has long been argued that there is no problem of soil erosion. Recently, however, a growing body of evidence has accumulated to suggest that erosion is not insignificant and that it should now be a cause for serious concern. The Government and the Ministry of Agriculture remain unconvinced; research into soil erosion and its causes in Britain has been cut back at a time when it should be greatly expanded. In December 1987, John Boardman of Brighton Polytechnic wrote in *The Guardian*:

> In October and November [1987], soil losses on the Downs were the highest ever recorded in Britain. In addition, damage to property due to soil-laden runoff from farming land topped £1m at one site alone... Soil loss on the Downs this autumn has frequently exceeded 30 t/ha and in some cases exceeded 100 tonnes. One field recorded the British record loss of 270 tonnes per hectare over 9 hectares. On soils that are already very thin for agricultural operations, being rarely more than 20 cm thick, a loss of 200 tonnes per hectare represents about 10% of the soil in the field.

A report prepared for the Soil Association (Hodges & Arden-Clarke, 1987) reviews the current evidence in detail. It also does not make comfortable reading. The report concludes that about 44% of our arable soils are now clearly at risk and that rates of soil loss are often greatly in excess of natural soil renewal rates (0.1–0.5 t/ha/year) and also of the 2 t/ha/year which, on many soils, can be considered as a tolerable erosion loss.

Erosion in Britain is primarily associated with the shift to arable crops on soils which should have remained covered by grassland, such as the

Plate 2.1   Soil erosion on recently drilled winter cereal fields at Rottingdean, East Sussex in 1987   (J. Boardman)

Downs. The increasing areas of maize production in Europe have also been identified as a major cause of the problem, as has the removal of field boundaries.

There is, however, considerable evidence that organic farming practices, including rotations with grass leys, green manuring and undersowing of crops such as maize, can make a significant contribution to reducing the risk of erosion. For example, Reganold *et al.* (1987) compared soils on two neighbouring farms, one conventional and one organic, and found that the soil loss since 1948 was 16 cm greater on the conventionally managed soils (Figure 2.1). Although on the organic farm 5 cm of topsoil had been lost over this period, the loss of 21 cm on the conventional farm represented one third of the total topsoil present.

Figure 2.1   Organically and conventionally farmed soil losses due to water erosion between 1948 and 1985.

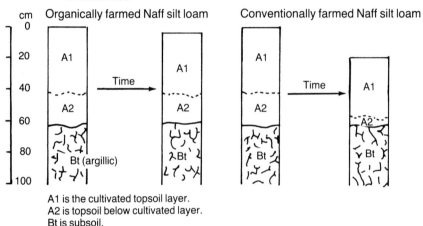

A1 is the cultivated topsoil layer.
A2 is topsoil below cultivated layer.
Bt is subsoil.

Source: Reganold *et al.* (1987).

Erosion is not the only illustration of soil 'disease', however. Desertification and salinisation/alkalisation also result from inappropriate agricultural techniques. Desertification is a direct result of human pressure on what is already a very fragile ecosystem in arid areas, and can also be a long-term consequence of deforestation and the removal of natural vegetation such as the tropical rainforests which formerly provided a protective cover for the soil. Salinisation and alkalisation of the soil result from the careless use of irrigation techniques in semi-arid but potentially fertile areas of the world. The current situation in the San Joaquin Valley in California is a disturbing example, where the build-up of selenium in the irrigated soils threatens the health of consumers and the livelihoods of many farmers.

The effect of all these consequences of the loss of soil fertility—erosion, desertification and salinisation/alkalisation—is to render the soil useless for the production of crops except in situations where all the necessary crop nutrients can be purchased or brought in from outside the area. If we are to continue to be able to produce food in the future, in the quantities which are required to feed a rapidly expanding human population and under conditions where manufactured inputs may not be readily or economically available, then action needs to be taken now by all farmers to reverse this trend. In order to do this, a much better understanding of how the soil works, in a biological as well as a mechanical sense, is needed so that we can learn to manage the soil in a way which is sustainable in the long term.

CREATING A HEALTHY SOIL

The fertility of the soil not only has implications in terms of soil degradation, it also has an important impact on the health of the crops, animals and human beings which derive their sustenance from it, a fact emphasised by many of the early organic farming protagonists including Eve Balfour (1975), Albert Howard (1943), Newman Turner (1951) and Friend Sykes (1946).

The crops grown today have evolved over millions of years to utilise the nutrients made available in the soil by the activity of soil micro-organisms. They have developed symbiotic relationships with the soil fauna and flora, the most notable examples being *Rhizobia* bacteria which live in nodules on the roots of legumes and fix nitrogen from the atmosphere, and the mycorrhizae, small fungal threads which penetrate plant roots and allow nutrients to be transferred directly from the soil into the plant's root system. The use of readily available, soluble mineral fertilisers by-passes or short circuits the biological processes to which the plants have become adapted and makes the nutrients directly available to the growing crop. This can result in luxury uptake of nutrients by the plant, and has consequences not only in terms of the health of the growing crop and its susceptibility to pests and diseases, but also in terms of the quality of the end product. Excessive levels of nitrates in leafy vegetables are a good example of the type of problem which can occur, but not the only one. Other examples are considered in the section on food quality in Chapter 15 as well as in a detailed review of the subject by Hodges and Scofield (1983).

The basis of healthy livestock, crops and human beings is a healthy soil, one which through its biological activity and inherent fertility can grow high quality crops and remain productive over long periods of time without the need to rely on large quantities of inputs from outside the system.

A starting point in the creation of a healthy soil must be to break away from the idea of the soil as merely a growing medium for crops, a place where crops can anchor their roots and pick up the nutrients they need while being fed from the fertiliser bag. It is well known that it is possible to grow high yields of crops hydroponically, without any soil at all, so long as all the necessary nutrients are supplied in a form acceptable to the growing plant. Even so, not only does the soil still provide the most practical place to grow the quantities of food which we need, it is also much more than just a mixture of rock rubble and mineral particles with more or less organic matter added. The soil is a living entity, an ecosystem containing a wide variety of different flora and fauna which fulfil myriad different roles.

The main components of the soil ecosystem can thus be categorised as living organisms, minerals, organic matter, water and air, all of which are required for its smooth and efficient functioning. The interactions between the living and non-living components of the soil are as important as the presence or absence of any one of them.

Nutrient and energy flows in the soil ecosystem

An ecosystem such as the soil depends on the cycling of nutrients and the availability of energy to enable that cycling to take place. The energy cannot be created within the system; it has to come from outside because, however efficient the system is, there will always be losses from it. Without an outside source, the system will eventually grind to a halt. From a global perspective, all the energy which ecosystems require comes from the sun. In the plant–soil system, the plants are the principal organisms which trap the sun's energy and store it in the form of organic (carbon) compounds. These plants are therefore referred to as 'producers'. Herbivorous animals and other higher life-forms which depend on plants for their sources of nutrients and energy are known as 'consumers'. Eventually, the complex organic compounds are broken down by other organisms called 'decomposers' into an inorganic state as and when the energy is required to mobilise nutrients.

On its own, the soil is not nearly as effective at capturing the sun's energy as plants are. The only 'producer' organisms present are the photosynthesising algae which can capture the sun's energy directly. To some extent, energy in the form of heat can be absorbed by the soil and the effects of this are obvious from seasonal growth patterns. The main source of energy, however, is crop residues and to a lesser extent animal wastes which are broken down by the soil organisms to release the necessary energy to allow the soil ecosystem to function.

The other main aspect of the soil ecosystem is the availability and cycling of nutrients. The minerals in the soil represent a massive store of nutrients, and together with the nutrients in the air, particularly nitrogen, are sufficient to support and maintain the production of large quantities of biomass or living material. However, the larger the quantity of living material being produced, the faster the cycling and decay process needs to take place to sustain the level of biomass production.

Over time, a natural ecosystem will adjust the total levels of biomass production and the rates of nutrient cycling to suit its particular environment. Eventually, it will reach a more-or-less stable situation which is known as a 'climax ecosystem'. The rain forests of the Amazon and central Africa, the redwood forests of the North American Pacific Coast and the prairies and savannah areas of North America and Africa are good examples of climax ecosystems which have evolved to cope with different climatic and environmental conditions.

Direct intervention by humans simplifies existing natural ecosystems and reduces the diversity and complexity which have evolved over long periods of time. Agriculture has the effect of bringing the development of an ecosystem back to square one, not only as far as the vegetation is concerned, but also in terms of the soil ecosystem and the complexity of biological activity in the soil. As a result, the soil system lacks the protection which is necessary to preserve it intact, unless very carefully managed.

The biological life in the soil

The soil contains an enormous number of different organisms, varying widely in both size and function. All of them, however, play an important role in the mobilisation of nutrients in the soil. At the lower end of the scale, there are the microorganisms such as algae, protozoa, fungae and bacteria. Further up the scale in terms of size come the nematodes, springtails, small arthropods and enchytraeid worms, while the largest organisms include earthworms, molluscs and the larger enchytraeids and arthropods. Some spend their whole lifecycle in the soil, others live in the soil only at certain stages. An indication of their relative abundance in terms of biomass and numbers can be obtained from Figure 2.2.

Each of these organisms has a specific role or function within the soil. Producers, those organisms which are able to use the sun's energy through photosynthesis to form complex carbon compounds, are less obvious in the soil than in other ecosystems. Only the algae are photosynthetic, although they can also grow in the absence of light if simple organic solutes are also available. The blue-green algae, which

Figure 2.2 The major kinds of animals found in the soil, showing the range of population densities likely to be encountered in a cool temperate grassland eco-system.

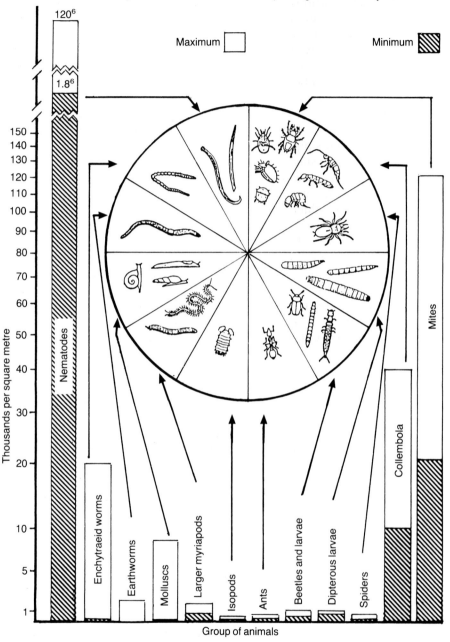

Thousands per square metre

Group of animals

Source: Kevan, D. K. McE. (1965) in *Ecology of Soil-borne Plant Pathogens: Prelude to Biological Control*, eds. K. F. Baker and W. C. Snyder. Copyright © 1965 The Regents of the University of California.

occur in great abundance under certain conditions, are also able to fix nitrogen from the atmosphere, while receiving their energy requirements directly from the sun. They prefer neutral—alkaline soils whereas the green algae are more common in acid soil.

The vast majority of soil organisms function as consumers and decomposers. Some, such as snails and slugs, utilise living vegetable matter as a source of both nutrients and energy. The nutrients which are obtained in this way are built up into more complex compounds than existed initially, and hence they can be regarded as consumers. Other organisms specialise in breaking down organic matter of both plant and animal origin into simple, inorganic forms. A whole range of organisms are involved at this stage, from earthworms to fungi and bacteria. These are the decomposers which utilise the energy available in the organic matter as a means of breaking down the material into its component parts. The end product of this demolition job are nutrients which once again can be taken up by plants, or which are released back to the atmosphere or leached into the water draining from the soil.

There are other organisms which do not feed directly on either dead or living organic matter, but which form close, symbiotic relationships with other living organisms. Their main feature is that they obtain the energy which they require in the form of carbon compounds in exchange for some other nutrient which they are able to provide. *Rhizobia* bacteria, which are able to fix nitrogen from the atmosphere, do so using energy obtained from the leguminous plants with which they form their symbiotic associations. In return, the plants benefit from the additional nitrogen which is made available. This contrasts with the blue-green algae, which capture their energy directly through photosynthesis. Another example of this kind of beneficial association is the mycorrhizae, fungi which penetrate plant roots to obtain carbon directly from the plant, but which also facilitate the transport of nutrients from the soil to the plant, in particular phosphates and to some extent larger molecules containing organically bound nutrients such as nitrogen.

Crops can therefore benefit considerably from a balanced soil eco-system, but they also have specific mechanical requirements of the soil, which include anchorage, the opportunity for deep, exploitive rooting to obtain nutrients, and adequate supplies of both oxygen and water. They are therefore highly dependent on the physical characteristics and structure of the soil in which they are growing.

The physical basis of the soil

The physical nature of the soil is determined by a combination of geological and chemical processes which have taken place over very long periods of time. The predominant particles, derived from the disintegration and physical comminution of primary igneous rocks, will have been subject over time to movement by wind, erosion and subsequent deposition, as well as the effect of glaciation in certain parts of the world. The consequence of these processes is that the mineral composition of the soil does not always directly reflect the underlying rock strata.

It is possible to distinguish them by sizes into categories such as gravel, sand, and silt. These particles are irregularly shaped and their origin differentiates them from the clays, which are formed by a process of chemical weathering and are platy in nature. The relative proportions of these particles in the soil determine the soil 'texture'. In Britain, the influence of the ice age and glaciation has resulted in soils which are characterised by a wide range of particle sizes, but which are remarkably constant in the relative proportions of silt and clay which they contain, with a silt to clay ratio of about 6:4 (Stewart et al., 1980).

The characteristic behaviour of the clay particles is very different from that of gravel, sand or silt. The latter are largely chemically inert; their significance is the effect which they have on water retention and drainage. Capillarity is the process by which water is raised upwards through the soil. The coarser particles, especially stones and gravel, are not able to raise water at all in this way, whereas the finer sands can raise water about 30 cm and silt as much as several metres. The problem with silts is that, when closely packed, they are so retentive of water that they cannot lose it by drainage. The only way in which the water can be removed is through evaporation, if at all.

Traditionally, agriculturalists have distinguished soils in terms of their ease of cultivation, although the descriptions often differ from the strict definitions used by soil scientists. For example, in many cases, a heavy or clay soil in common parlance would more accurately be defined as a silty loam or a silty clay loam, especially if they are inclined to be wet and difficult to manage.

Texture not only affects the ease of cultivation of a soil or its water and nutrient retention capacity; it also has a pronounced effect on soil temperature. Soils which hold more water, such as silts and clays, will be slower to warm up than sandy soils which are less retentive of water.

Soil structure

Soil structure refers to the way in which the soil particles are arranged in the soil matrix. The structure of the soil can be thought of in two ways; either as a mass of particles mixed together, or as a series of channels bounded by solid surfaces within which plant roots grow and the soil fauna and flora can be active, known collectively as the pore space. The importance of thinking in terms of pore space when discussing soil structure lies in the requirements of most soil organisms and plant roots for both air and water as well as nutrients.

The pore space can be divided into two size categories: macropore and micropore. The macropore space is the space which exists between large soil particles, such as sand, gravel, crop residues and soil aggregates, draining rapidly to admit air within about an hour of saturation. This is essential to provide an adequate level of aeration for crop roots and soil microorganisms. The micropore space, such as exists within soil aggregates or between silt and clay particles, is much smaller and holds water, which is important for moisture retention. The ideal soil structure would contain about 30% macropore and 70% micropore space (Stewart & Adams, 1968), enabling all the water reserves within aggregates to be fully exploited by root hairs and to drain rapidly, allowing roots access to air and facilitating their deep penetration.

Sandy soils have too high a proportion of macropore space and are not sufficiently moisture or nutrient retentive. They can be helped to some extent by the addition of organic matter, but they do not generally require much attention in terms of soil structure. Silts, on the other hand, consist almost entirely of micropore space and are likely to be continuously waterlogged. Even if some form of structure can be achieved, it is likely to disintegrate easily with resultant packing and slumping. Predominantly clay soils will crack as a result of continual drying and wetting, creating drainage channels and the macropore space necessary for crop growth. Careful management of clay soils, by cultivating only when soil moisture conditions are right, is needed to gain the full benefit of structure formation by such fracturing.

Most of the soils in Britain, however, are loams: soils with a variety of particle sizes and generally a high proportion of silt. If these are allowed to pack, they effectively become single particle size systems at risk from waterlogging. Structure can be maintained in these types of soil by a process of soil particle aggregation, or the formation of a crumb structure characteristic of a soil in good heart. To achieve this, a clay content of at least 10% is necessary, as well as an adequate return of fresh organic residues to the soil. In addition, lime is required to maintain pH levels and, crucially, an active microbial population to promote the decay of fresh organic matter residues and the production of an adequate and continuous supply of organic, structure-stabilising agents.

The reason for this is the way soil particle aggregates are held together. Clay, unlike gravel, sand or silt, is able, because of its negative charge, to hold nutrients such as calcium in the form of positively charged cations. Organic matter behaves in a similar way. Calcium ions, which are strongly charged, can form a bridge between negatively charged clay and/or organic particles, holding them together when they would normally be repelled. The effect is reinforced by the presence of organic resins or gums formed by soil microbial activity.

The role of earthworms
This is where the burrowing earthworms play a unique role. Their contribution, however, has often been underrated; the Strutt report (MAFF, 1970), for example, regarded soil crumbs and granules as being:

> . . . products of root action and the decomposition of fresh organic matter. They are more common in the topsoils than subsoils and are produced more readily by some plant root systems than others. The root systems of grasses vary, but usually excel in causing granulation.

The argument that soil crumb structure is improved in the presence of grass roots is certainly true, but it is difficult to see how grass roots on their own could bring about the mechanical combination of clay, lime and organic matter which is needed to form water-stable soil aggregates. This is where the earthworm comes in. By processing soil and organic residues through its gut, clay and organic matter are intimately mixed and coated with organic stabilising gums and lime secreted from a special gland within the digestive tract. The result, the wormcast, consists of just the type and size of water-stable soil aggregate which is needed to hold water while allowing the crop root hairs which penetrate it to obtain sufficient air and fully exploit the nutrient and moisture reserves contained within. Various estimates suggest that an active worm population can process as much as 40 tonnes of dry earth per hectare, or the equivalent of at least 0.5 cm of soil annually.

The elimination of worms, through the use of wormicides, acidification of the soil, the use of fertilisers such as ammonium sulphate, the use of certain herbicides and fungicides, inappropriate cultivation techniques (e.g. rotary cultivators), and the failure to retain sufficient organic residues in the soil, leads to progressive soil compaction and associated problems of drainage. Organic residues accumulate on the surface, leading, in a grassland situation, to increased surface rooting and matted pasture. In arable soils, the reduced water stability of the soil aggregates results in increased susceptibility to packing and to soil erosion by water.

The presence of burrowing earthworms is therefore an essential element of the management of loams in temperate climates. In addition to their direct mixing of soil constitutents, they also make a significant contribution to organic matter incorporation generally and to the enrichment of nutrients and humus in topsoil, bringing nutrients up from lower down the soil profile. The channels facilitate drainage and allow exploration of the deeper soil layers by plant roots, along with a concentrated and readily available supply of nutrients. They also provide channels for other small animals to move through the soil, increasing the rate at which surface organic residues are incorporated into the soil profile. This can help with the destruction of pathogens by incorporating crop residues which would otherwise become sources of infection. (See Colour plate 1.)

Earthworms cannot tolerate acidic soils with a pH much below 5.0, so that under these conditions, mineral and organic matter are not blended together by their activity. This phenomenon can be observed in coniferous forests and heathland, accounting to a large extent for the dichotomy between mull and mor soils and calcicole and calcifuge plant communities (Stewart, 1984), with consequent impact on the availability of cations such as aluminium and iron. (See Colour plate 2.)

Research by soil scientists such as Dr Stewart at Aberystwyth has clearly shown the benefits of encouraging earthworms in terms of soil structure formation, in particular with respect to the management of sports turfs and the rehabilitation of former open cast mining land (Stewart *et al.*, 1988). The farming practices favoured by organic producers, such as intensive use of organic residues, the inclusion in rotations of a fertility building phase usually based on a grass ley, and the avoidance of substances such as mineral fertilisers and pesticides which are potentially toxic to soil life, have been shown to have a beneficial effect on earthworm populations (e.g. Stewart & Salih, 1981 and Scullion & Ramshaw, 1987). In particular, top-dressing rather than direct incorporation of organic manures has been found to promote earthworm activity, but the evidence is not always conclusive. Low levels of urea use also seemed to have a beneficial effect. White clover, on the other hand, has been found to inhibit earthworm activity.

There is still a great deal to be learnt about the interactions between type and quantity of fertiliser use, crops grown, cultivation practices and earthworm activity. Large quantities of mineral nitrogen fertiliser can have a beneficial effect by increasing the supply of crop residues available to earthworms, as work at Rothamsted has shown. Work in Germany (e.g. Appel, 1980; Walthemathe, 1985; Bauchhenss & Herr, 1986; Dormann & Meier, 1986) comparing earthworm numbers on organic and conventional farms, however, has shown a significantly greater level of earthworm activity and soil structure formation on organically managed soils. It would therefore seem that the combination of practices used by organic farmers has a beneficial impact on earthworm populations, notwithstanding the possible benefits of fertilisers or disadvantages of white clover.

This is further supported by evidence from researchers such as Edwards and Lofty at Rothamsted (1982), which showed that straw removal, particularly by burning, and certain cultivation techniques could have an adverse effect on earthworms and other soil organisms. They also found in arable experiments at Rothamsted that the species of earthworms were more numerous in plots treated with organic fertilisers than in untreated plots. There was a strong positive correlation between the amounts of inorganic nitrogen applied and populations of earthworms, probably because of the increased production of crop roots and residues. Organic fertilisers applied to arable land, however, increased earthworm populations much more than inorganic nitrogenous fertilisers. Plots receiving both inorganic and organic nitrogen had the largest populations of earthworms. The effects of both inorganic and organic nitrogen were much less on earthworm populations in grassland than on those in arable crops. Organic fertilisers had a greater impact on populations of *Lumbricus terrestris*, which feed directly on surface organic matter, than on those of the *Allolobophora* species which feed below the surface, utilising both the organic matter and the microorganisms which also feed on the organic matter. Edwards and Lofty consider that the harmful effects of fertilisers reported elsewhere may be due to increased soil acidity, which has been shown to have a severe impact on earthworm activity. They also found that slurry may be toxic to earthworms, particularly in situations where toxic compounds have accumulated during anaerobic storage.

An important new threat to the earthworm in Britain comes not from agricultural practices, but from the New Zealand flatworm *Artioposthia triangulata*. The flatworm has been found in parts of Northern Ireland and Scotland and is believed to have entered Britain in the soil of imported ornamental plants. It attacks and eats earthworms, which appear to be its only food, and it has no natural enemies in Britain. This means that earthworm populations in some areas could be decimated, with consequent effects on soil structure and fertility. Scientists are currently investigating options for biological control, as chemical controls are likely to prove ineffective.

The role of other organisms
Emphasis has been placed here on the role of earthworms in soil structure formation, but it is important to recognise the contribution of other soil organisms as well as the multitude of processes which take place in a very specific part of the soil ecosystem, the rhizosphere. The rhizosphere is where the plant roots make contact with the soil and form an integral part of the soil ecosystem. Within the rhizosphere, a wide variety of microbial relationships exist, to a large extent associated

with the flow of nutrients and energy between the soil and the growing plant. These deserve special consideration and are given more detailed attention in Chapter 3.

Many microorganisms produce substances known as polymers, particularly polysaccharides, which provide a cementing action between soil particles. These microorganisms are encouraged by the growth of plant roots, feeding on the nutrients contained in the mucigel excreted by the plant at the root tip. The mucigel also contains polysaccharides produced by the plant itself. These effects are, however, more prominent with perennial crops such as grass than with annual crops. There is also some evidence that fungi such as the vesicular-arbuscular mycorrhizae also have a role in promoting soil aggregate stability. The mechanisms are not well understood, but it is possible that the fungal hyphae work in the same way as plant roots, mechanically pressing soil particles together. In addition, complex chemical reactions involving clays and organic acids produced by plants, bacteria and fungi, have been implicated.

Various fertilisers, pesticides and cultivation techniques have been associated with rapid decline in soil microbial activity. In particular, the use of fungicides has been shown to shift the balance of soil organisms from fungi to bacteria. Other pesticides and some fertilisers have been shown to be directly toxic to soil fauna. Straw burning, as has already been mentioned, deprives the soil ecosystem of an important source of energy, in addition to the direct impact of burning on soil life. The net effect of arable monocultures, straw burning and the use of biocides and mineral fertilisers must be a long-term decline in soil biological activity and hence soil structure and soil fertility.

On the other hand, various studies (e.g. Schroeder, 1980; Bolton, 1983; Fraser, 1984; Fraser et al., 1988; Dietz et al., 1986; Gehlen, 1987; Trolldenier, 1987; Helweg, 1988; Bourguignon, 1989; and Elmholt & Kjøller, 1989) have shown that organically managed soils have higher levels of microbial activity, as indicated by number and species diversity of fungi, rates of cellulose (straw) breakdown and by the activity of enzymes such as dehydrogenase, urease and phosphatase, all crucial for the release of nutrients to the growing crop. The study by Fraser comparing organic and conventional practices confirmed the concept that the soil physical environment, as related to soil and crop management systems, is also a primary determinant of microbial populations and activities; it also recognised the prime role of organic manuring using either livestock wastes or crop residues. Soil bacterial and fungal counts, dehydrogenase activity and microbial biomass were significantly greater in surface soil amended with manure. However, the same increases in soil biological properties were seen in soil planted to red clover or oats/clover. As observed by other researchers, these increases were a result of many factors in a changing soil microenvironment, including soil water content and organic carbon and nitrogen levels.

The crucial message that emerges from this consideration of the soil as a living entity is that fertilising and manuring need to be looked at not only in terms of crop requirements, but also in terms of the requirements of soil organisms and the soil ecosystem. This message has been succinctly put by Eve Balfour in *The Living Soil:* 'feed the soil and let the soil feed the plant'. In this way, a balanced relationship between soil and crops can evolve, ensuring that both are maintained in a healthy state.

Soil Assessment

Part of the key to the creation of a biologically active 'living' soil lies in the assessment of the soil. It is only in this way that the consequences of faulty cultivations, poor drainage or nutrient deficiencies can be un-covered and remedied. There are several different ways of approaching soil assessment, varying greatly in their complexity and applicability to practical farming situations. They range from direct, visual assessment of soil structure and root development to complex biochemical analyses which indicate the type and nature of biological activity in the soil.

Visual assessment is probably the most important as far as the farmer is concerned. A spadeful of soil carefully removed from the ground can often indicate drainage or compaction problems which may be improved by changes in cultivation practice and can also show layers of crop residues such as straw which have not been successfully incorporated into the soil and may be impeding root growth (MAFF/ADAS, 1983a, EFRC, 1984). This type of assessment also means that unnecessary cultivations can be avoided where there is no real need to correct soil structure problems (although cultivations may still be necessary for weed control).

Examining a spadeful of soil requires little effort and is a simple operation. Samples taken should be representative of a field, and it may be necessary to take several. Compaction may already be indicated by the presence of standing water in wet weather. If the soil is moist and not too stony, the pressure required to push the spade into the ground can also give a preliminary indication of compaction problems.

Once the spadeful of soil has been obtained, the assessment should concentrate on structure (friable soil, blocks, clods, plates), density, root penetration, distribution of organic manures etc. Where parcels of straw, platy structure, few coarse pores and roots growing horizontally are found, appropriate remedial action needs to be taken. Often it will be possible to detect the depth of previous cultivations or ploughpans. Any compaction is likely to exist just below that level. In some cases, it may be necessary to dig down into the subsoil to get a proper assessment of a compaction problem. Colour (e.g. gleying) and smell can also indicate areas where anaerobic conditions exist, although the actual cause of the problem may not be readily apparent on the basis of this information alone.

Soil texture or the physical composition of the soil plays an important role in determining soil structure and the way a soil behaves in response to management practices. Physical soil assessment does not necessarily require great precision and can involve a very rough assessment of soil texture by feel alone.

Moisten a soil sample until it glistens and knead it between the fingers and thumb until the aggregates are broken down and the soil grains are thoroughly wetted. Coarse sands are easily visible and the individual grains grate against each other. Fine sands are less easy to detect, unless they form more than 10% of the sample, when the characteristic gritty feel can be observed. Individual silt particles are too small to be detected by feel or sight, but their presence gives the soil a smooth, silky feel with only very slight stickiness, whereas clays are characteristically sticky and tenacious. This arises from the platy nature of the clays, with individual particles arranged in layers and sticking to each other in much the same way as slightly moistened plates of glass. The stickiness of clays can be affected by the type of clay and the organic matter content, declining as organic matter levels increase. High organic matter contents will also tend to make sandy soils feel more silty, as does finely divided calcium carbonate.

Other tests exist which can be carried out by the farmer without any special equipment and can give a greater level of accuracy. A more precise assessment is provided as part of some soil analysis services, giving percentages of sand, silt and clay. These details are important, because they not only influence the workability of soils, but also the availability of plant nutrients and the possible losses from the system.

Soil structure in terms of water-stable aggregates can be assessed in the laboratory, but a rough estimate can be obtained by shaking a sample of the larger particles with water. Stable aggregates should not disintegrate, and the water should remain clear if the soil is in good heart. This test is not appropriate, however, for sands, loamy sands and loams which have insufficient clay (less than 10%) to hold a structure.

Biological approaches to soil assessment involve determining the level of biological activity in the soil, particularly that of earthworms. The importance of earthworms in helping to incorporate organic matter and to create soil structure is so important that low numbers should be

Plate 2.2 (Left)
Taking a spade sample
for soil examination
(FIBL, Switzerland)

Plate 2.3 (Below)   *A*
spade sample showing
good soil structure with
deep root penetration and
even granulation
(G. Hasinger)

Plate 2.4 (Below)   *A spade sample showing poor soil structure and compaction with roots*
forced to grow horizontally   (G. Hasinger)

regarded as a cause for concern. High numbers, on the other hand, provide a good indication of a healthy, living soil which is in good condition. Earthworms react unfavourably to acid conditions, as well as to hostile climatic circumstances (drought, frost, waterlogging, etc.) and it is necessary to distinguish between the various factors which can affect earthworm populations and their species composition. The average number of worms in a biologically active soil is between 800,000 and 1.2 million per ha, or about 80–120 per square metre. Accurate counts can be made using suitable sampling techniques, but a good impression can be obtained from field evidence of abundant earthworm activity such as mole heaps, earthworm channels and casts, and the absence of a surface organic mat (see Colour plate 3).

An alternative approach to biological assessment of soil condition is an assessment of microbial activity in the soil either by culturing bacteria, for example, or biochemically by measuring enzyme activity and the rates at which certain gases are released. These biological and biochemical approaches to soil assessment are not extensively used, but it has been argued that they have particular relevance to organic systems which place special emphasis on high levels of biological activity in the soil to maintain its fertility and productivity. The main problem with the application of these tests is their complexity and cost, combined with considerable variability which makes interpretation difficult.

As far as pH values, organic matter levels and nutrient content of the soil are concerned, chemical analysis is widely used and is an appropriate tool. It is particularly important, in a farming system where few if any nutrients are being brought in from outside, that a constant watch is kept on soil reserves. A range of standard tests exists to try to estimate the available fractions of crucial nutrients like phosphate and potash, but often, particularly with phosphate, the tests give no indication of total reserves or the proportion of total reserves which are available to the crop. Microbial activity, and the acids which are excreted by soil microbes, play a major part in making nutrients like phosphate available to be taken up by plant roots; chemical analyses of the different phosphate fractions and their relative availability can provide a good indication of biological activity in the soil. In addition, information on certain minor elements such as iron can indicate compaction or waterlogging problems as well as fulfilling the usual role of pinpointing trace element deficiencies. The chemical analysis service provided by Elm Farm Research Centre in Britain takes some of these factors into account (EFRC, 1984).

SOIL MANAGEMENT

Soil management in an organic system has the twin aims of maintaining a healthy, biologically active soil environment and providing optimal conditions for plant growth. To achieve this, a detailed understanding of soil–crop interactions is required and, in particular, the effects which management practices have on the soil.

A well structured soil, consisting of a continuous network of pore spaces, is vital to allow drainage of water, free movement of air and the unrestricted development of roots. A soil which has become compacted will have a markedly reduced natural porosity which limits water and air movement and restricts root development. Nutrient movement through the soil is very slow; in most circumstances, plant roots need to grow towards new sources of nutrients. Particularly in organic farming systems, where the use of mineral fertilisers on the soil surface is avoided, restricted rooting will result in significant yield losses and leave the crop at risk from both lack of moisture in drought periods and waterlogging after periods of heavy rainfall. For optimal plant growth, therefore, conditions are required which permit deep, exploitive rooting in well drained, well aerated soils.

Although biological agents in the soil can play a major part in creating the necessary conditions for plant growth, direct intervention is often required to either assist the natural processes, or to repair damage caused by previous activities. All methods of intervention, however, have both desirable and undesirable consequences and these can vary considerably under different circumstances.

Soil management covers a wide range of activities, all of which are central to crop production, including the control of erosion by wind and water, the control of water loss in semi-arid areas, the removal of excess water by drainage, cultivation to provide suitable growing conditions for crops and making nutrients available to crop plants.

Soil erosion problems arise primarily because of inappropriate cropping activities, often under conditions where organic matter levels are low. Maintaining vegetation cover, minimal or reduced cultivations, and attention to the direction of crop rows on slopes (keeping to the contours as much as possible) are some of the techniques which can be adopted to reduce the risk of erosion. In particular, vegetation cover, whether it be 'weeds', green manures, crop residues or leys, lessens the impact of raindrops on the soil and thus reduces the disintegration of soil aggregates (see Colour plate 4).

Avoidance of water loss in semi-arid areas is not a topic of major concern in Britain, although some areas suffer from drought periods in the summer. Traditional cultivations can result in excessive moisture

Plate 2.5   Contour cropping to reduce the risk of soil erosion on a Pennsylvania low-input farm

loss, but more importantly, attention should be paid to maintaining a good soil structure which allows for deep root penetration and for plants to obtain water from further down the soil horizon as the surface layers dry out. The selection of deep-rooting crops, such as lucerne, or suitable grass species, is another way of ensuring that water reserves deeper down can be exploited during dry periods.

Crop nutrition and nutrient cycling are the subject of the next chapter. However, many activities such as organic manuring and liming have also to be seen in the context of feeding soil organisms and helping to maintain the free lime content of the soil, both of which play a crucial role in the biological formation and stabilisation of soil structure. Adequate levels of free lime are also necessary if mechanical structuring of soils is going to have anything other than short-term benefits. In the long run, mechanical structuring must be stabilised by biological means anyway. Direct physical intervention to assist the creation of optimal conditions for plant growth must, however, involve specific attention to drainage and cultivation practices.

Drainage

Drainage becomes necessary if the soil has too high a water table or if excess surface water cannot penetrate sufficiently rapidly to below the root zone of the crop. The significance of the height of the water table depends on soil type and the crop which is being grown. For example, a high water table is advantageous for peat soils as this slows down the rate of oxidation caused by cultivations. Crops vary in their rooting depths; a lower water table is desirable for trees than would be necessary for grass.

Reducing the level of the water table is largely a question of engineering and/or conventional drainage activities, but the general desirability of reducing the level of the water table is another matter. Environmental consequences and the damage which can be caused to wild flora populations and sensitive ecological habitats must also be considered. This is a very important issue which can no longer be ignored purely in the interests of increasing production. It is of particular importance in organic farming, precisely because of the emphasis which is placed on farming in a way which is sensitive and less destructive to the environment.

The problem of water which cannot penetrate sufficiently rapidly through the soil, on the other hand, is one where management practices are important. Inadequate drainage and waterlogging lead to reduced aeration and root penetration and limit the plant's ability to fully exploit nutrients in the soil. Within roots, physiological damage occurs because of the build-up of toxic metabolic products such as ethanol and the availability of energy for the uptake of nutrients becomes restricted. The range of soil microorganisms shifts in favour of those which are able to obtain oxygen from sources other than air, usually by the reduction of oxidised ions. This leads to lower levels of nitrogen in a form which plants can utilise, in some cases causing the characteristic yellowing of surface vegetation. In addition, the plants will suffer from the build up of toxic substance such as ferrous iron, hydrogen sulphide and methane in the root environment.

Improved drainage obviously carries with it several benefits, but does not necessarily result just from the installation of drainage pipes. Physical and biological changes in the soil are also important. On heavy land, where structure is achieved largely by cracking following alternate drying and wetting of clays, vigorous crops help create and stabilise fissures, and deeper penetration can help dry out lower layers more during drought, encouraging the cracking process. Soils warm up faster, resulting in higher microbiological activity, increased levels of nutrient availability and rates of nutrient cycling. At the same time, however, water is the main transport mechanism for products of mineral weathering and organic decomposition; very rapid downward movement could lead to excessive leaching.

Drainage can only take place through cracks and spaces between soil crumbs in medium and heavy soils, where capillary action in the small pore spaces would not otherwise allow water to escape. In sandy soils, the pore spaces are large enough so that they are usually free-draining

Figure 2.3   Poorly structured and well structured soil profiles.

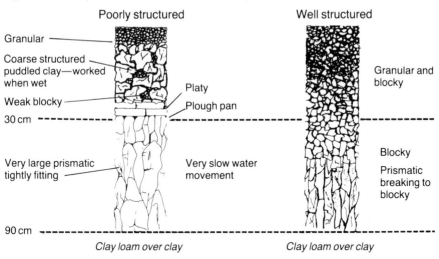

Poorly structured   Well structured

Granular

Coarse structured
puddled clay—worked
when wet   Granular and
                                        Platy   blocky
Weak blocky
                                        Plough pan
30 cm

Very large prismatic   Very slow water
tightly fitting   movement

                                                                Blocky

                                                                Prismatic
                                                                breaking to
                                                                blocky

90 cm

Clay loam over clay   Clay loam over clay

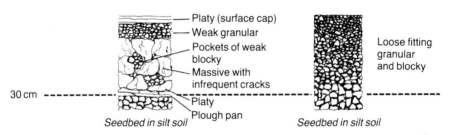

                        Platy (surface cap)
                        Weak granular
                        Pockets of weak
                        blocky   Loose fitting
                        Massive with   granular
                        infrequent cracks   and blocky
30 cm
                        Platy
Seedbed in silt soil   Plough pan   Seedbed in silt soil

Source: Davies *et al.* (1982).

anyway. The problem, therefore, is how to manage silts, clays and loams
in a way which encourages improved water penetration. Poor structure
(low crumb stability) can lead to the clogging up of pore spaces. The
activities of burrowing earthworms and other organisms in creating
stable aggregates is one important aspect of improving structure and
reducing the risk of water-logging. Earthworm channels, the spaces
occupied by dead roots and vertical fissures greatly assist the downward
movement of water.

Even if the structure of the topsoil is adequate, soil compaction
problems occur lower down because of faulty cultivations or soil types
which are generally difficult to manage. The smearing caused by
ploughs or rotary cultivators tends to seal off the vertical channels even if
there is no actual compaction. Compaction zones or ploughpans may
need to be broken up by subsoiling or mole drains to allow water to
penetrate.

Soil cultivation

Cultivation practices probably have the most significant impact on the soil of any agricultural activity. Given that soil, although ubiquitous, is a precious resource, the impact of cultivation techniques on the soil ecosystem deserves much greater attention than has been paid in the past. Any consideration of the appropriateness of cultivation practices depends primarily on what we are trying to achieve by them. The objectives and principles of soil cultivation can be summed up in the following list:

1. the production of suitable tilth or soil structure;
2. the control of soil moisture, aeration and temperature;
3. the destruction or control of weeds and soil pests (this can be achieved indirectly by the improvement of conditions for plant growth);
4. burying or clearing crop residues, and incorporation of manures in the soil;
5. remedying of compaction caused by previous activities.

All of these points are interlinked to a greater or lesser extent, but the central aim is the creation of optimal conditions for seed germination and growth of the *intended* crop. One aspect of this is the preparation of a suitable seedbed, with the right degree of tilth or soil structure. The actual requirements of different crops vary, and it is not always clear what the ideal tilth should be. It should also be realised that there are several factors involved in improving soil structure, including (as has already been mentioned), the presence of free lime, a loamy or clayey mixture, good natural or artificial drainage, high biological activity (especially earthworms), winter rather than spring cropping and early establishment of winter crops, minimising field traffic, low tyre pressures and strong drying and freezing cycles in heavy soils. Cultivation is only one of the range of factors and then only on some soils.

The traditional approach to cultivation is the mouldboard plough, which has been around for nearly 200 years. During this time, its function has not changed. The plough's most significant role in the preparation of ground for a new crop was the burial of weeds and debris, which remains the case today. To bury the crop residues, ploughing inverts the soil and leaves a clean, but exposed, surface. The porosity and aeration of the soil is increased, stimulating microbial oxidation of organic matter. Large quantities of nitrogen are released in a form which is available to crop plants, or which may be leached out of the soil if there is insufficient vegetation to utilise it. Ploughing carries with it the further danger on medium and heavy land that the soil can become compressed if it is too wet, creating compaction zones and ploughpans which impede drainage. Under very dry conditions, ploughing often only serves to turn up large clods which are very difficult to break down using secondary cultivations.

Although it can be argued that the plough has served farmers well, in particular with respect to weed control, the expansion of plough-based arable cropping since the 1940s has led to the increased exposure of the soil to the elements with consequent structural decline, increased erosion risk, compaction and waterlogging. These problems have been exacerbated by the impact of heavy machinery and reduced biological activity to repair structural damage.

In 1975, Sir Charles Pereira argued that 'the most logical approach to tillage should start from the observation that, when free of man's interference, most of the earth's soils are clothed with a vigorous system of vegetation which requires no uniform seed beds'. He urged that 'we should look more carefully at the mechanism by which soils and their plant cover interact to provide a tilth in which the seeds are germinated and plants established without any application of applied horsepower'.

He has not been alone in arguing that the traditional approach to cultivation, the use of the plough, is not as benign or essential as it might appear. Edward Faulkner in his book *The Ploughman's Folly* took up a similar theme, but it is also clear that the deleterious effects of ploughing were not so obvious under orthodox rotational farming as they became in the '50s and '60s as agriculture intensified.

The concern about several aspects of traditional cultivations, and the realisation that on many soils ploughing was not strictly necessary for purposes other than weed control, has led to the development of new techniques in conventional farming systems. Effective straw burning and modern herbicides have meant that cultivations are needed solely to provide satisfactory growing conditions for the next crop. The emphasis is now on reducing the depth of working within the limits of structure maintenance and satisfactory seed germination and establishment. The cultivator has been taken up as the primary tillage implement, particularly by heavy land farmers; little heavy land now sees the plough.

These new techniques come under the general category of reduced or minimal cultivations, the most extreme version of which is direct drilling. Shallow or reduced cultivations refer to a range of techniques where the soil is disturbed with tines or discs to less than 10 cm to achieve an 'acceptable seed bed'. The cultivators can mix the crop residues into the soil, but cannot produce a 'clean' field. This has implications for the design of seed drills which need to be able to cope with an increased amount of surface debris.

With direct drilling, the seed is sown directly into the stubble of the previous crop, without any other cultivations taking place. The direct drilling techniques as currently conceived, and to a lesser extent reduced cultivations, rely heavily on herbicides to kill existing vegetation and strawburning to clear crop debris from the surface. Since neither the use of herbicides nor strawburning are acceptable in organic systems, the applicability of current direct drilling techniques is limited. However, there is no reason why the concept should not be developed further within an organic farming context.

The other reduced or minimum cultivation practices are applicable in organic systems and they share with direct drilling a number of advantages over traditional deep cultivations such as ploughing. The most obvious in terms of the farm business are the substantial savings in time, labour and fuel use which they entail.

There are also significant advantages in terms of soil management and crop–soil interactions. Crop residues and stubble are left on or mixed into the surface of the soil. Like this, they will protect the soil from wind or water erosion and help maintain the permeability of the soil to rainwater. Evaporation of water from the soil is minimised in situations where surface moisture is critical. For these reasons, reduced cultivations have significant implications for areas of the world where erosion by water is a major problem, and have been widely adopted in parts of North America where they are referred to by the term 'conservation tillage'.

The concentration of organic matter in minimum tilled soils is near the surface, rather than spread throughout the plough layer, with a higher concentration of soil microorganisms and increased biological activity near the surface of the soil, which increases the porosity and stability of the surface tilth. Trials at Letcombe support this point, as illustrated by the data in Table 2.1.

Table 2.1 Effect of cultivation on the stability and organic matter content in the surface 2.5 cm of a calcareous clay (Evesham series) at Buckland, Oxon.

Treatment in previous four years	Stability measurement (T) for aggregates 1.4–2.8 mm	Organic matter content (%)
Direct drilled	0.560	5.83
Shallow-tined cults	0.574	5.83
Deep-tined cults	0.555	5.43
Ploughed	0.527	5.05
LSD P*=0.05	0.039	

* LSD = Least Significant Difference; P = Probability.
Source: Douglas, J. (1977) Rep. A.R.C. Letcombe, 1976, 40–8.

The data also shows a close correlation between aggregate stability and the level of organic matter in the surface layer of the soil. This may be caused by the direct contribution of root exudates and root residues to the stability of soil aggregates. In spite of this, the tilth can still deteriorate in wet winters.

When compared with ploughing, direct drilling creates a moister, cooler, less oxidative environment which enhances the immobilisation of nitrogen while slowing mineralisation rates, thereby increasing the size of many of the soil nitrogen pools. This can have the effect of later spring growth. The warming up of the soil may be further delayed by the surface mulch in these conditions.

Reduced tillage techniques result in less subsoil compaction than do traditional cultivations, with the particular advantage that the soil retains the basic advantages of an undisturbed topsoil: good vertical communication from topsoil to subsoil via cracks, old root and earthworm channels, thus permitting good drainage. Trials comparing traditional and reduced cultivations indicate that following repeated applications of water, ploughed land became wetter than the undisturbed land, and this was most obvious below a depth of about 10 cm. The cracks and channels may also contribute to increased aeration, but the evidence is not yet conclusive.

The overall impact on soil structure, however, is very dependent on soil type. Certain soils, such as loamy soils over limestone and calcareous clay soils are suitable for shallow cultivations or direct drilling without any need for occasional ploughing to mechanically restore structure in the top soil. Non-calcareous loams and sandy loams with no free lime, on the other hand, will require ploughing after several years of shallow cultivations or direct drilling to restore structure. With these types of soil, compaction can take place in the 5–20 cm layer of the topsoil. It may be that organic rotations with a grass/legume ley are able to overcome this problem without the need to resort to the plough (other than to turn in the ley). Light sandy and silty soils compact as a result of the settling of the soil; they need annual topsoil loosening by ploughing and are not usually suitable for direct drilling.

An important consideration is the effect which cultivations have on soil life. Earthworm populations are of particular interest because of their role in helping to maintain the structure and permeability of the soil. The trials at Letcombe have also looked at this issue and the available data suggests that earthworm populations are greater under direct drilling conditions than with ploughing (Table 2.2). This applied to deep-burrowing and surface-inhabiting species alike. The population differences were greater with spring-sown than with autumn-sown crops. The disruption caused to the soil by ploughing with deep cultivations generally creates conditions which do not encourage earthworm survival. At the same time, however, some of the herbicides currently in use can have a deleterious impact on earthworm populations. Shallower cultivations would be expected to favour earthworm survival.

Table 2.2 Earthworm populations in a clay soil (Evesham series) at Buckland, which had been direct drilled or ploughed in successive years 1973–6.

Crop and year of experiment	Date of sampling	Number of earthworms		Ratio DD:P
		Direct drilled	Ploughed	
Spring barley				
Year 1	1/10/73	145	110	1:3
Year 2	1/10/74	345	218	1:6
Year 3	20/10/75	231	98	2:4
Year 4	20/11/76	197	50	3:9
Winter wheat				
Year 4	26/11/76	152	95	1:6

Source: Ellis, F. B. and Barnes, B. T. (1977) Rep. A.R.C. Letcombe, 1976, 50.

Cultivations play a major part in weed control, particularly in organic systems where herbicides are not available. Careful consideration needs to be given to the effect of reduced cultivations on the weed problem. This subject is dealt with in greater detail in the chapter on weed control, but a few comments are appropriate at this stage. Shallow cultivations encourage weed seed germination which, apart from making cultivation aimed at weed control easier, also means that the soil is rapidly covered with growing plants, protecting the crumb structure from damage by rain drops.

This may not mean that the mouldboard plough can be dispensed with altogether. As a farm goes through the conversion to an organic system, one of the main priorities is an improvement in soil structure. This will usually give rise to increased weed growth and control is often a problem, especially during the early years after conversion. Reduced cultivations provide ideal conditions for perennial weed development and without the use of herbicides may result in poor control of annual weeds as well. In these situations, ploughing gives better control of annual broadleaved weeds, black grass and brome grasses, but wild oat control is better with reduced cultivations, because buried seed is protected from rotting, birds and rodents and remains viable for many years. Harrowing the crop in spring and subsequent inter-row hoeing will then be necessary. However, harrowing is often difficult on heavy land in spring and hoeing equipment may not be available. In such cases, shallow ploughing (10–12 cm) provides a good method of weed control. Many of these weed problems are also influenced by the rotation and may not actually be a problem whichever approach to cultivation is adopted.

Disease and pest control is another important issue. Again, the subject is dealt with in detail in a separate chapter, but a decision to adopt reduced cultivations can have implications for rotation design. ADAS trials in conventional systems found that trash-borne diseases could be more serious in unploughed fields, but the increase was seldom great enough to warrant abandoning a non-plough technique (MAFF/ADAS, 1983b). These results were influenced by the use of straw-burning as a means of disease control and the availability of fungicides.

Specific potential disease problems which ADAS identified include mildew, yellow rust and the brown rusts of wheat and barley which overwinter on living cereal plants. The fungi survive the period between the harvesting of the previous crop and the emergence of the new, autumn-sown crop on late tiller and volunteer plants. In the case of mildew, they survive as black spore cases which eject their spores with the rains of early autumn.

Non-ploughing techniques can favour these diseases by not destroying the 'green bridge' and by leaving trash bearing mildew spore cases on the surface of the soil. There are several other factors which contribute, including earlier autumn sowing made possible by reduced cultivations, but softness, lushness, density and vigour all add to the problem. In organic systems, the lower availability of nitrogen has a significant impact on the susceptibility of the crop to disease, and this is aided by rotation design and the awareness of the need to delay autumn sowings to reduce disease carry-over. For similar reasons, take-all is also not a major problem; the effects of the rotation and the activity of antagonists and competitors in the biologically rich surface layers of the soil are usually sufficient to control the disease. In certain situations, such as where a couch grass problem develops, or there is a deterioration in soil structure which restricts rooting and aggravates the effects of the disease, then take-all may become a problem, but few organic farmers would identify it as one. Under anaerobic conditions, certain saprophytic fungi which are more numerous on undisturbed soils can affect seedling establishment. In these situations soil management needs to be aimed at preventing the development of anaerobic conditions in the soil.

There are certain pests which may also represent a greater problem with reduced cultivations. One is the increase in slug numbers because of trash on the surface, although this is likely to be a greater problem with direct drilling techniques where crop residues are not actually incorporated. There is also some evidence of increased numbers of wireworms with direct drilling. In both cases, cultivation practices and the rotation may need to be reconsidered if a serious problem arises.

Straw incorporation

The 'problem' of straw and other crop residues becomes significant under situations where burning is not acceptable. On most organic farms the available straw is usually required for livestock enterprises on the farm and is therefore baled and removed from the field. Increasingly, however, mainly arable or stockless organic systems are being developed. The need to return organic matter to the soil in these situations means that the farmer can ill afford to sell straw off the farm, even if it were economically viable to do so. The importance of ploughing-in or incorporating straw is a matter of dispute. Various trials in conventional systems have suggested that the long-term contribution to organic matter levels in the soil is minimal. There may, however, be some benefit from the annual incorporation of straw where soils have poor structural properties or low organic matter content. Farmers who regularly practise straw incorporation often claim this to be the case. Increasing organic matter levels is not the only reason for returning carbon-rich materials to the soil. Carbon represents the main form of energy for soil micro-organisms and is essential if a biologically active soil is to be maintained.

Work at Rothamsted has compared the effects of different methods of straw removal on the populations of soil invertebrates including earthworms. They found that straw removal, particularly by burning, affected the numbers of soil invertebrates adversely. This seemed to be due mostly to decreased availability of surface organic matter although there were also some direct effects of burning on surface-living invertebrates. Related work at Letcombe found that the response to different straw treatments depended on species; significantly larger numbers of deep-burrowing species were found on the chopped and spread area, whereas where straw was burnt, the number of surface living earthworms was significantly greater. This would again seem to relate to the availability of surface organic matter as a source of nutrients for the burrowing earthworms.

Where straw or other material with high carbon:nitrogen ratio is added to soil, the microorganisms decomposing the straw utilise available nitrogen; this is thus immobilised and unavailable to the subsequent crop until the C:N ratio is lowered. Although this can, under certain conditions, affect establishment, there may also be benefits in terms of reducing nitrogen leaching losses.

More importantly, problems can arise through the production of phytotoxins from crop residues decaying under wet conditions, either leached out of the straw or produced by microbial activity. Various organic acids such as acetic, butyric and proprionic acids may act on their own to deter seed germination and plant growth, but it is thought that they probably combine with microbial pathogens to produce harmful, synergistic effects. Dusting seeds with chalk to increase the pH can help reduce the effect of acetic acid, but careful cultivations should avoid the problem occurring in the first place.

Successful incorporation of straw depends on it being mixed into the soil so the bacteria can work on it effectively. There is a danger with ploughing that the straw can get buried at the bottom of the furrow to form an anaerobic layer of slowly rotting straw, creating a barrier to root development. Where a clean field is not an essential requirement, shallow cultivation is likely to achieve more thorough mixing. ADAS trials work suggests that ploughing in chopped straw is the most effective method of straw incorporation, although on heavy land little difference between ploughing and the use of tined cultivators was found. In recent years, ploughing has not always come out on top and the extra costs of ploughing have to be set against any yield benefits. A new implement designed specifically for straw incorporation by the Scottish Institute of Agricultural Engineering consists of a series of staggered pto (power take off) driven rotors which lift and mix the soil with the surface straw and then deposit it on one side. Burial was as good as with ploughing, but the new implement achieved a better distribution of straw through the topsoil. In non-ploughing situations where it is not possible to produce a clean seed bed, specially adapted drills are required. As straw incorporation is more widely adopted, the availability of the drills should improve.

Summarising the organic approach

Organic farming systems emphasise the use of cultivations which seek to maintain soil structure and allow the soil to have vegetative cover for as long as possible within the rotation. Shallow cultivations and mixing of only the surface layers of the soil are an important element of this approach, but there are also arguments in favour of traditional cultivations under certain circumstances.

Whether a ploughing or no-ploughing approach is used, deep cultivations to loosen and aerate the soil are essential on most soil types. This should only be done in dry conditions, for example after harvest. Deep cultivations when the soil is wet will create compact layers in the soil which can seriously limit plant growth. By not turning the lower layers, the bringing of subsoil and stones to the surface and the mixing of subsoil and topsoil can be avoided.

Figure 2.4 Deep loosening without turning aerates the soil
but keeps organic matter on top.

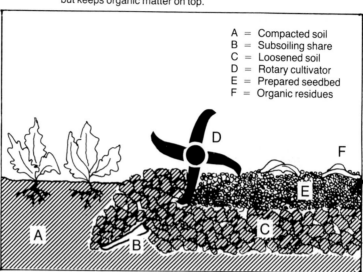

Source: Firma Weichel.

Mechanical deep loosening of the soil needs to be stabilised by biological means as soon as possible. The roots of green manures, catch crops, or even the main crop will achieve this. Any late autumn or spring cultivations should be carried out only in the top 10–12 cm of the soil, so maintaining the biologically stabilised soil structure in the lower layers.

Implements

There is a wide range of implements available to carry out the types of cultivations which are discussed here. Many of these are in common use and little would be gained by considering them specifically. They should, however, be used with due caution. Whatever the criticisms, deep ploughing is not the only problem area; deep cultivation with other implements such as spring-tined cultivators and rotovators can also lead to the undesirable mixing of soil layers and excessive loosening of the soil, or smearing of the base of the cultivated layer. There are, however, some more recent implements which may be able to make a significant contribution to improved soil management. Some of these have even been designed to combine the 'shallow turning and deep loosening' concept favoured by organic producers, although virtually all the work concentrating on the needs of organic farmers has taken place in continental Europe.

One example is the Weichel Terravator system, a version of which has been used at Elm Farm Research Centre. The Terravator system consists of several elements which can be used separately or in combination, making single-pass operations possible. The main elements are power-driven rotary cultivators which achieve the 'shallow turning' aim

Figure 2.5   The Weichel Terravator.

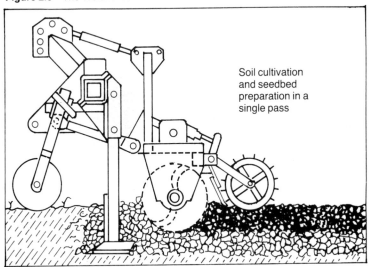

Soil cultivation and seedbed preparation in a single pass

Source: Firma Weichel.

Plate 2.6   Weichel Terravator (without sub-soiling bodies)

and sub-soiler type bodies fitted with goose-foot shares for 'deep loosening'. In the initial period of use, a worsening of the weed situation may be experienced, as weed seeds are not buried. Experience over time, however, suggests that weed seeds are not brought to the surface from lower layers, and rotational effects combined with surface cultivations can bring about effective weed control as the system becomes established.

Other examples include implements for conservation tillage such as the Paraplow which is designed to loosen soil below the surface and kill weeds, while keeping much of the stubble and straw on the surface. The Paraplow is essentially a cultivator fitted with wide sweeps, so they cut and loosen the soil 5–10 cm below the surface, but can also work at depths up to 35 cm. Rotary weeders which have rotating knives that cut and loosen the surface, or light reciprocating harrows, can achieve similar effects. None of these implements, however, allow for deep loosening activities and therefore structural problems may develop on certain soil types. In these situations, sub-soiling and similar activities need to be considered.

Many of the implements mentioned above, including the Weichel Terravator, the McConnel Shakaerator, the Paraplow and also the Wallace Soil Reconditioning Unit which has been popularised in Australia (Widdowson, 1987; Anon, 1988), can be used to aerate soil under grassland. This practice is potentially of greater significance in organic systems because of the need for plant roots to exploit the full soil profile for nutrients, rather than relying on surface fertiliser applications.

The power harrow is pto driven and consists of a number of contra-rotating rotors. Each is equipped with two vertical tines which rotate in the soil. This action ensures good crumbling even when working at high speeds. A free rolling crumbler is fitted behind to consolidate the soil. Unlike the rotovator, which chops and mixes, power harrows achieve a seed bed by stirring alone. They are therefore much less likely to damage the soil structure. Rotovators are, however, more suitable for breaking up hard ground without ploughing. Both the rotovator and the power harrow have the advantage over ploughing that a seed bed can be created in a single pass, but this has to be balanced with the mechanical impact on soil organisms such as earthworms. An alternative may be the slower working rotary digger or spading machine.

In horticulture, an approach developed by Hans Kemink is frequently advocated by organic growers. Essentially, the Kemink system is based upon the use of frequent cultivations to aerate and loosen the soil to a significant depth, while at the same time the frequency of the operations enables good weed control. The system has its critics for two main reasons: the aeration of the soil causes the breakdown of organic matter and mobilisation of nutrients at a rate which may not be sustainable

Plate 2.7
Howard Paraplow

Plate 2.8
McConnel Shakaerator

Plate 2.9
Kemink machine

47

over long periods, and the structure achieved is brought about more through mechanical means than through genuine biological structuring. Those who use the system have expressed satisfaction with the ease with which the soil can be cultivated once the system has become established.

Where ploughing is used, skim boards can help with the incorporation of crop residues and may avoid the problem of the burial of residues as a single 'mat'. Various other variations on the theme of the mouldboard plough may also be worth considering. Shallow or skim ploughs have low energy requirements and high work rates, cultivating to a depth of 10–15 cm, which is often sufficient to incorporate trash and provide acceptable tilth. Ploughs fitted with subsoiling tines are also potentially appropriate for a 'shallow turning, deep loosening' approach.

The choice of implement will also be influenced by the tractor power required and the number of passes required. The trend toward heavier machinery has been far from beneficial to the soil. Due consideration has to be given both to the ground pressure and the number of passes needed to prepare a field.

Unfortunately, there is no simple recipe for soil management. The wide variability in individual circumstances can mean that even the most favoured approaches are inappropriate in come cases. The approach must, however, be one of looking at soil management not just from the point of view of the crop, but also from that of the soil itself and the life in the soil. It is the perspective which is important, as it enables appropriate decisions to be made on the basis of the information, incomplete as it is, which is currently available to us.

REFERENCES AND FURTHER READING

General

Balfour, E. (1975) *The Living Soil and the Haughley Experiment*. Universe Books; New York
Howard, A. (1943) *An Agricultural Testament*. Oxford University Press
MAFF (1970) *Modern Farming and the Soil*. HMSO
Russell, E. W. (1973) *Soil Conditions and Plant Growth*. 10th Edition. Longman
Sykes, F. (1946) *Humus and the Farmer*. Faber and Faber
Turner, F. N. (1951) *Fertility Farming*. Faber and Faber
Wild, A. W. (ed.) (1988) *Russell's Soil Conditions and Plant Growth*. 11th Edition. Longman

Soil degradation and erosion

Hodges, D. (1983) *Soil degradation—a global problem*. Soil Association Quarterly Review, October: 2–5
Hodges, D. (1984) *Soil erosion in Britain*. Soil Association Quarterly Review, September: 10–12

Hodges, D. (1986) *The erosion of Britain's soils*. Soil Association Quarterly Review, September: 19–21

Hodges, D. & Arden-Clarke, C. (1986) *Soil Erosion in Britain*. Soil Association; Bristol

Long, E. (1989) *Soil Erosion*. Farmers Weekly, 10 March: 85–87

Morgan, R. P. C. (1986) *Soil Erosion and Conservation*. Longman

Reganold, J. P. , Elliott, L. F. & Unger, Y. L. (1987) *Long-term effects of organic and conventional farming on soil erosion*. Nature 330 (6146): 370–372

Reganold, J. (1989) *Farming's organic future*. New Scientist, 10 June: 49–52

Sampson, R. N. (1981) *Farmland or Wasteland*. Rodale Press; Emmaus, PA

Seymour, J. & Girardet, H. (1986) *Far from Paradise*. BBC

Worldwatch Institute (1984) *Soil Erosion: Quiet Crisis in the World Economy*. Worldwatch Paper, 60

Soil ecology

Edwards, C. A. *et al.* (1988) *Biological Interactions in Soil*. Elsevier, Amsterdam

Fitter, A. H. (ed.) (1985) *Ecological Interactions in Soil—plants, microbes and animals*. Blackwell Scientific Publications

Richards, B. N. (1974) *Introduction to the Soil Ecosystem*. Longman

Tate, R. L. (1986) *Soil Organic Matter—biological and ecological effects*. John Wiley & Sons

Tinsley, J. & Darbyshire, J. F. (eds.) (1984) *Biological Processes and Soil Fertility*. Developments in Plant and Soil Sciences, Volume II. Martinus Nijhoff/Dr W. Junk Publishers, The Hague

Soil structure

Stewart, V. I. & Salih, R. O. (1981) *Priorities for soil use in temperate climates*. In: Stonehouse, B. (ed.) *Biological Husbandry—a scientific approach to organic farming*. Butterworth

Stewart, V. I. & Adams, W. A. (1968) *The quantitative description of soil moisture states in natural habitats with special reference to moist soils*. In: Wadsworth, R. M. (ed.) *The Measurement of Environmental Factors in Terrestrial Ecology*. Blackwell

Stewart, V. I. (1984) *Functions of the soil*. Introductory paper for European Environmental Bureau conference on Soil Protection in the European Community, Rue Vautier 31, B-1040 Bruxelles. Also published in the Royal New Zealand Institute of Horticulture Annual Journal 13 (1985): 45–48

Earthworms

Appel, G. (1980) *Einfluss von alternativen und konventionellen Bewirtschaftungsmassnahmen auf das Leben der Regenwürmer in Ackerböden*. Diplomarbeit. University of Göttingen

Bauchhenss, J. & Herr, S. (1986) *Vergleichende Untersuchungen der Individuendichte, Biomasse, Artdichte und Diversität von Regenwurmpopulationen auf konventionell und alternativ bewirtschafteten Flächen*. In: *Vergleichende Bodenuntersuchungen von konventionell und alternativ bewirtschafteten Betriebsschlägen*. Bayerisches Landwirtschaftliches Jahrbuch 63: 1002–1012

Darwin, C. (1882) *The Formation of Vegetable Mould Through the Action of Worms*

Dorman, K. & Meier, T. (1986) *Die Regenwürmer des Versuchsbetriebes in Neu-Eichenberg*. Projektarbeit. Gesamthochschule Kassel, Witzenhausen

Edwards, C. & Lofty, J. R. (1982) *Nitrogenous fertilisers and earthworm populations in agricultural soils*. Soil Biol. Biochem. 14: 515–521

Ramshaw, G. (1985) *The effect of farming methods on the rehabilitation of opencast coalmining land in South Wales*. PhD Thesis, UCW, Aberystwyth

Satchell, J. E. (ed.) (1983) *Earthworm Ecology—From Darwin to Vermiculture.* Chapman and Hall

Scullion, J. (1985) *Rehabilitation of former opencast coal mining land—the Bryngwyn Project.* Soil Association Quarterly Review, June: 9–11

Scullion, J. (1984) *The assessment of techniques developed to assist the rehabilitation of restored opencast coal mining land.* PhD Thesis, UCW, Aberystwyth

Scullion, J. , Mohammed, A. R. A. & Ramshaw, G. A. (1988) *Changes in earthworm populations following cultivation of undisturbed and former opencast coal-mining land.* Agriculture, Ecosystems and Environment 20: 289–302

Scullion, J. & Ramshaw, G. (1987) *Effects of various manurial treatments on earthworm activity in grassland.* Biological Agriculture and Horticulture 4: 271–282

Stewart, V. I. , Salih, R. O. , Al-Bakri, K. H. & Strong, J. (1980) *Earthworms and soil structure.* Welsh Soils Discussion Group 21: 103–114

Stewart, V. I. , Scullion, J. , Salih, R. O. & Al-Bakri, K. H. (1988) *Earthworms and structure rehabilitation in subsoils and in topsoils affected by opencast mining for coal.* Biological Agriculture and Horticulture 5: 325–338

Wallwork, J. A. (1983) *Earthworm Biology.* Studies in Biology No. 161

Walthemathe, V. (1985) *Vergleichende Untersuchungen über die Auswirkungen konventioneller und alternativer Bewirtschaftung auf das Bodenleben von Grünlandstandorten, gemessen an der Regenwürmpopulation, der Bodenatmung und der Stickstoffmineralisation.* Diplomarbeit. University of Giessen

Soil microbial activity and farming practices

Bolton, H. (1983) *Soil microbial biomass and selected soil enzyme activities on an alternatively and a conventionally managed farm.* MSc Thesis, Washington State University

Bourguignon, C. (1989) *Biologische Aktivität im Boden—vergleichende Untersuchungen.* Zum Beispiel 10/89: 4–6

Dietz, Th. *et al.* (1986) Various papers in: *Vergleichende Bodenuntersuchungen von konventionell und alternativ bewirtschafteten Betriebsschlägen.* Bayerisches Landwirtschaftliches Jahrbuch 63: 979–1002

Elmholt, S. and Kjøller, A. (1989) *Comparison of the occurrence of the saprophytic soil fungi in two differently cultivated field soils.* Biological Agriculture and Horticulture 6: 229–239

Fraser, D. (1984) *Effects of conventional and organic soil management practices on soil microbial populations and activities.* MSc Thesis. University of Lincoln, Nebraska

Fraser, D. G. *et al.* (1988) *Soil microbial populations and activities under conventional and organic management.* Journal of Environmental Quality 17: 585–590

Gehlen, P. (1987) *Bodenchemische, bodenbiologische und bodenphysikalische Untersuchungen konventionell und biologisch bewirtschafteter Acker-, Gemüse-, Obst- und Weinbauflächen.* PhD Thesis. Rheinischen Friedrich-Wilhelms-Universität, Bonn

Helweg, A. (1988) *Microbial activity in soil from orchards regularly treated with pesticides compared to the activity in soils without pesticides (organically cultivated).* Pedobiologica 32: 273–281

Hodges, R. D. & Scofield, A. M. (1983) *Agricologenic Disease. A Review of the Negative Aspects of Agricultural Systems.* Biological Agriculture and Horticulture 1: 269–325

Lopez-Real, J. M. & Hodges, R. D. (1986) *The Role of Micro-organisms in a Sustainable Agriculture.* AB Academic Publishers, Berkhamsted

Schroeder, D. (1980) *Stroh und Celluloseabbau sowie Dehydrogenaseaktivität in 'biologisch' und 'konventionell' bewirtschafteten Böden.* Landwirtschaftliche Forschung, Sonderheft 37: 169–175

Trolldenier, G. (1987) *Influence of cultivation measures on soil life.* Plant Research and Development 26: 114–126

Soil management

Allen, H. (1981) *Direct Drilling and Reduced Cultivations*. Farming Press

Anon. (1988) *George Wallace—Farmer profile/The Wallace Soil Reconditioning Unit*. Acres Australia 1 (Spring): 6–9

Culpin, C. (1986) *Farm Machinery*. 11th Edition. Collins, London.

Davies, D. B. , Eagle, D. J. & Finney, J. B. (1982) *Soil Management*. 4th Edition. Farming Press

EFRC (1984) *The Soil: Assessment, Analysis and Utilisation in Organic Agriculture*. Elm Farm Research Centre

MAFF/ADAS (1974) *Cereals without ploughing*. Profitable Farm Enterprises No. 6

MAFF/ADAS (1983a) *Soil examination for autumn cultivations*. Leaflet 802

MAFF/ADAS (1983b) *To plough or not to plough*. HMSO

MAFF/ADAS (1985) *Straw Use and Disposal*. Booklet 2419

McRobie, G. (ed.) (1989) *Tools for organic farming*. ITDG Publications; London

Widdowson, R. W. (1987) *Towards Holistic Agriculture—a scientific approach*. Pergamon Press

Woodward, L. (1985) *Understanding the message of the soil*. New Farmer and Grower 7 (Summer): 23, 24, 33

Recent publications

Cook, H. and Lee, H. (in preparation) *Soil management in sustainable agriculture*. Proceedings of a conference at Wye College, September 1993. Wye College; Ashford.

Reganold, J. P.; Palmer, A. S.; Lockhart, J. C.; and A. N. McGregor. (1993) Soil quality and financial performance of bio-dynamic and conventional farms in New Zealand. *Science*. 260:344–349.

Chapter 3

Crop Nutrition

Nutrient Cycling

As part of the aim of creating a sustainable agricultural system, the principle of operating as far as possible within a closed system for crop and livestock nutrition is fundamental to organic systems. This entails minimising the use of nutrient inputs from outside as well as losses from within the system. The closed cycle principle need not necessarily apply at an individual farm level; the effective recycling of nutrients at regional level will also be essential in the long term.

Nutrients are constantly sold off the farm in produce. Unnecessary losses must therefore be minimised by making use of natural recycling processes and biological nitrogen fixation. A healthy, biologically active soil is the basis of this recycling process, requiring the replenishment of organic matter and nutrients removed by crop harvesting.

The successful operation of an organic system therefore requires an understanding of all aspects of nutrient cycling; this will involve knowing which crops remove most nutrients, how to minimise losses and maximise returns, how to make the best use of the soil's recycling abilities and understanding the long-term effect of cultivations and rotations. This also means an important change in perspective with regard to the use of fertilisers and manures. Organic farming is not about simply replacing artificial fertilisers with organic manures in what might be termed a 'neo-conventional' approach. It is about feeding the soil ecosystem and making full use of the natural resources which exist within the farm.

The availability of nutrients to plants is determined as much by the rate at which nutrients are cycled within the system as it is by the level of inputs to the system. The most extreme example of this is the tropical rain forests, which maintain very high biomass production from soils which frequently have very low reserves of nutrients. A nutrient deficiency problem should therefore be solved by assessing and amending the physical and biological condition of the soil, the balance of the rotation and the suitability of farming practices before recourse is made to imported fertilisers, whether mineral or organic.

The role of the soil's physical and chemical properties

The soil must provide nutrients for plants in a suitable form and at a rate at which they can be used. Plants take up nutrients primarily in the form of electrically charged ions (NO_3^-, $H_2PO_4^-$, K^+, Mg^{2+}, Ca^{2+}). The soil must be a bank for these ions and hold them safely, yet make them readily available on demand.

The positive ions (cations) are held in the soil by negatively charged sites, these absorptive sites are referred to as soil colloids. The ability of the soil to hold cations is measured by the cation exchange capacity (CEC). Two important contributors to the soil cation exchange capacity are clay particles and humus (decomposed organic matter). Clay minerals are impure silicate structures; the impurities in the clay carry a smaller positive charge than the silica they replace, resulting in the clay carrying a net negative charge. Soil organic matter contains carboxyl groups which dissociate to provide negatively charged sites, these typically have twice the charge of clay minerals so even low levels of organic matter can make a significant contribution to the cation exchange capacity.

The process of chelation, by which organic matter can hold and buffer trace elements, is also important. Organic matter can act as a 'scrubber' to avoid toxic levels of trace elements and, in some cases, can be used specifically for this purpose.

The role of soil biological activity

Emphasis has already been placed in the previous chapter on the role of soil microorganisms in the soil ecosystem and in maintaining soil fertility. All except the algae derive the energy they require indirectly via plant photosynthesis. The organic residues from these plants are broken down by soil microorganisms to release both energy and nutrients in the form of inorganic ions which can then be taken up by plants.

The full range of soil life, from the soil fauna (which includes the insects such as springtails, termites, beetles, flies, ants, bees and wasps, the mites, various invertebrates such as flatworms, nematodes, snails, slugs and earthworms, and the centipedes and millipedes) to the soil microorganisms (covering a wide range of different types of bacteria, fungi and actinomycetes), is important both for the turnover of nutrients and energy contained in organic matter and for many of the chemical weathering processes which take place in the soil.

Maintaining the correct soil pH is important both for the faunal population like earthworms and for chemical nutrient availability.

Acidity means the presence of many positive hydrogen ions (H^+), which displace nutrients held by attraction to negative sites. In acid (low pH) situations soil biological activity is reduced, slowing the turnover of organic matter and reducing the release of nutrients.

The role of organic matter

Soil organic matter is often treated as one single substance, but in reality there are many different types of organic matter or humus which perform different functions in the soil. They may be divided into three broad categories:

- fresh and incompletely decomposed plant and animal residues (active organic matter);
- products of advanced decomposition of organic residues and products resynthesised by microorganisms (protein-like substances, organic acids, carbohydrates, gums, waxes, fats, tannins, lignins, etc.);
- high molecular weight humic substances: fulvic acids, humic acids and humin, which are generally resistant to further biological decomposition.

The term humus is generally applied to the latter two categories. It is the presence of functional groups such as amino, hydroxyl and carboxyl humic compounds which gives rise to a high cation exchange capacity and the ability to complex with clays, thus contributing to soil structure, although the highest molecular weight compounds are most stable and therefore play a lesser role. These humic substances, because they act as ion-exchangers, will also regulate plant nutrition to a certain extent. Some of these compounds may be taken up directly by the plant, possibly as a result of mycorrhizal associations, and activate physiological and biochemical processes in the plant. The effect of this may be to stimulate or regulate growth, or to assist the plant's resistance to pathogenic attacks. The presence of humus is therefore of considerable importance to the crop in its own right, not only as a potential source of nutrients.

Active organic matter, on the other hand, has been defined by Allison (1973) as 'organic matter that contains a considerable amount of plant materials still undergoing active decomposition in the process of being converted into microbial material and humic substances'. New sources of organic matter need to be added at regular intervals to ensure that the soil remains active biologically.

The role of plant roots

The movement of nutrients through the soil profile is usually very slow, particularly in the case of phosphate. This means that plant roots generally have to grow to where the nutrients are. Plant roots can be very effective at this process, but they need to be stimulated by the presence of nutrients, notably nitrogen and phosphorus, to branch out and explore new areas. Surface fertilising, where the nutrients remain near the soil surface, encourages surface rooting and leaves the crop vulnerable to drought conditions. In an organic farming system, the nutrients have to be obtained from further down the soil profile, necessitating a highly branched, deep rooting system. Some crops, such as lucerne, are particularly suited to this and will often help to raise nutrients from lower down the soil profile.

Plant roots have another important role, in terms of the distinct ecosystem which they create around them known as the rhizosphere. The rhizosphere includes the living, outer tissues of the root as well as the mineral particles of the soil which are in contact with the root. The plant roots exude a mixture of substances known as mucigel, which contains sources of nutrients and energy for the microorganisms living in the soil. In addition, cells are shed from root surfaces, adding to the organic material in the root zone. The microorganisms in turn release chemicals which may affect plant growth and organic acids which are needed to break down soil minerals and make nutrients such as phosphate available to the plant. Some of these microorganisms, such as the *Rhizobia* bacteria associated with legumes and nitrogen fixation, are reasonably well understood, but others have proved much more difficult to study or have not been considered of sufficient importance.

The root exudates contain a wide range of organic compounds, including carbohydrates (e.g. glucose), growth factors, organic acids and enzymes. Some of these are directly involved in the release of nutrients from the soil. Others may serve to inhibit the growth of other plants or in some cases stimulate the hatching of nematodes. Some of these organic compounds help to stimulate the growth of microorganisms which are beneficial to the plant.

An important, but frequently ignored, aspect of plant roots is their ability to take up a range of organic molecules, including herbicides, antibiotics and humic substances which may have molecular weights ranging from 500 to several thousands. This fact has already been mentioned, but it is important because conventional approaches to crop nutrition have concentrated virtually exclusively on inorganic ions, the most extreme form of which is hydroponics. In terms of plant nutrients, these organic molecules are not very significant, accounting for less than 5% of nutrient uptake. Their role in other respects, such as disease resistance, can be much more significant and they should therefore be considered an important, if poorly understood, part of an ecological approach to soil and crop husbandry.

The role of legumes and grass leys

The maximisation of internal cycling of nutrients and minimisation of external input use in an organic system is most easily achieved in a mixed farming system which uses legumes for the major nitrogen input and largely relies on livestock to recycle other nutrients. The primary fertility building stage in most organic rotations is the grass/clover ley, involving a mixture of deep rooting and shallow rooting plants. The deep roots will bring up nutrients from soil unreachable by other species. The balance, concentration and nature of minerals in herbage varies with plant species, stage of growth and the availability of minerals in the soil.

The ability of different species to extract different minerals means the use of herbs and other deep rooting species in ley mixtures or as herbal strips can increase the extent to which the nutrient proportions in forage match microbial and animal requirements. Herbs often have a higher mineral content than grass or clover; legumes have higher contents of calcium, magnesium, copper, zinc and selenium than grasses.

THE NUTRIENTS REQUIRED BY PLANTS

Chemical analysis of plants can give an indication of the nutrients which are required for their growth, but it was the work of the German chemist Liebig in the mid nineteenth century which led to much of the knowledge on which modern crop nutrition is based. In particular, Liebig identified carbon dioxide in the air, and not carbon from humus, as the source of carbon in plants. He also formulated what came to be known as the Law of the Minimum, which states in effect that growth will depend on the nutrient which is most limiting. If two or more nutrients are limiting, then the addition of only one of them will not increase crop growth. Liebig also identified ammonium in the soil as the source of nitrogen for plants, but did not recognise at the time the role of nitrates. Liebig's work, however, and that of other scientists who subsequently developed and re-evaluated his work, sometimes finding it, as in the case of ammonium, to be erroneous, formed the basis of the mineral approach to crop nutrition and the modern fertiliser industry. In later life, Liebig was to be very critical of the developments which had taken place as a result of his work. In particular, he criticised the way in which crop nutrition had been abstracted from the ecological context of agriculture, and it can be argued that, were he alive today, he would be a proponent of the holistic, organic approach to agriculture which is the subject of this book.

The major nutrients which are required by crops are now well known and well documented. Attention is paid here primarily to the major nutrients nitrogen, potassium, and phosphorus, while recognising the importance of others including magnesium, calcium, sulphur, iron and a wide range of trace elements.

Nitrogen

Nitrogen is a component of amino acids, the building blocks of proteins. The nitrogen content of most proteins varies between 14 and 18%. Lack of nitrogen is one of the most common causes of nutritional stress. Our atmosphere is approximately 78% nitrogen, but this exists as an inert gas and most living organisms are unable to exploit it directly, requiring it to be 'fixed' before use. The conventional farmer relies on artificially fixed bagged nitrogen, whereas the organic farmer relies on the *Rhizobia* bacteria in symbiotic association with legumes to make nitrogen available for plant nutrition. In a mixed organic system, with care and sensible stocking rates, the exploitation of the fixation of atmospheric nitrogen by these bacteria allows purchased inputs of nitrogen, mineral or organic, to be minimised.

For both conventional and organic farmers the difficulty with nitrogen management is not so much the accessibility or cost of the input as the ease with which nitrogen in its various forms can be lost from the system. Nitrogen is the only nutrient which can be lost in appreciable amounts to the atmosphere; gaseous nitrogen is lost as ammonia, nitrogen oxides and molecular nitrogen. Nitrogen is also very easily lost in solution in the form of nitrate ions. Much of the blame for the increasing pollution of water courses from nitrates has been placed on the artificial sources of nitrogen used in conventional agriculture. The organic farmer has no reason for complacency, for without careful management the organic system can contribute considerable amounts of nitrate to drainage water, and the loss is much more difficult to replace.

Nitrogen cycling

A convenient way of visualising the movement of nitrogen is to identify the internal pathways (the physical movement of nitrogen in its various forms in the soil and around the farms as crops, excreta, etc.) and the many sources of addition to, and loss from, a farming system (Figure 3.1). The major nitrogen inputs to a mixed organic system are fixation of atmospheric nitrogen and purchased inputs in the form of feeds, livestock and manures. Losses can take place in solid, liquid or gaseous form by leaching, volatilisation, denitrification[*], sales of produce and livestock, and losses down the drain from stockyards, livestock housing and poor storage of excreta. Cycling on the farm involves nitrogen moving from the soil into herbage, from herbage to livestock and from livestock via excreta either directly back onto pastures or into storage for later spreading when it again becomes incorporated into the soil.

Figure 3.1   A diagrammatic illustration of the nitrogen inputs, transfers, losses and outputs on an organic dairy farm.

The wider the arrow, the greater the amount of N in kg/ha per year

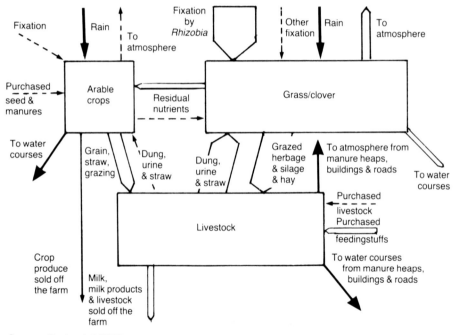

Source: Fowler *et al.* (1988).

There is a great deal of information on these individual processes and the factors which influence them, but it is often forgotten that the processes interact with each other and are greatly altered by changes in crop husbandry and soil management.

The nitrogen cycle in the soil (Figure 3.2) is an integral part of the

---

[*]'Leaching' is the physical removal of nutrients in solution by drainage from the soil, 'volatilisation' is the escape of ammonia gas produced by the hydrolysis of urea in animal excreta and 'denitrification' is the conversion by bacteria and subsequent loss of nitrate ions to either molecular nitrogen (N_2) or nitrous oxide (N_2O) gases when anaerobic conditions exist in the soil.

overall cycle of nitrogen on the farm and its working and efficiency is greatly affected by management practices. Nitrogen in the soil exists as organic compounds bound in the soil organic matter, which are almost completely unavailable to plants, and in inorganic form available to plants as ammonium (NH_4^+) and nitrate (NO_3^-) ions. The positively charged ammonium ion is relatively stationary, being adsorbed to organic matter or clay particles. The nitrate ion is negatively charged and is therefore not held to other negatively charged soil colloids and is completely mobile in the soil solution.

Soil bacteria, fungi and actinomycetes are responsible for the con-

Figure 3.2   The soil nitrogen cycle.

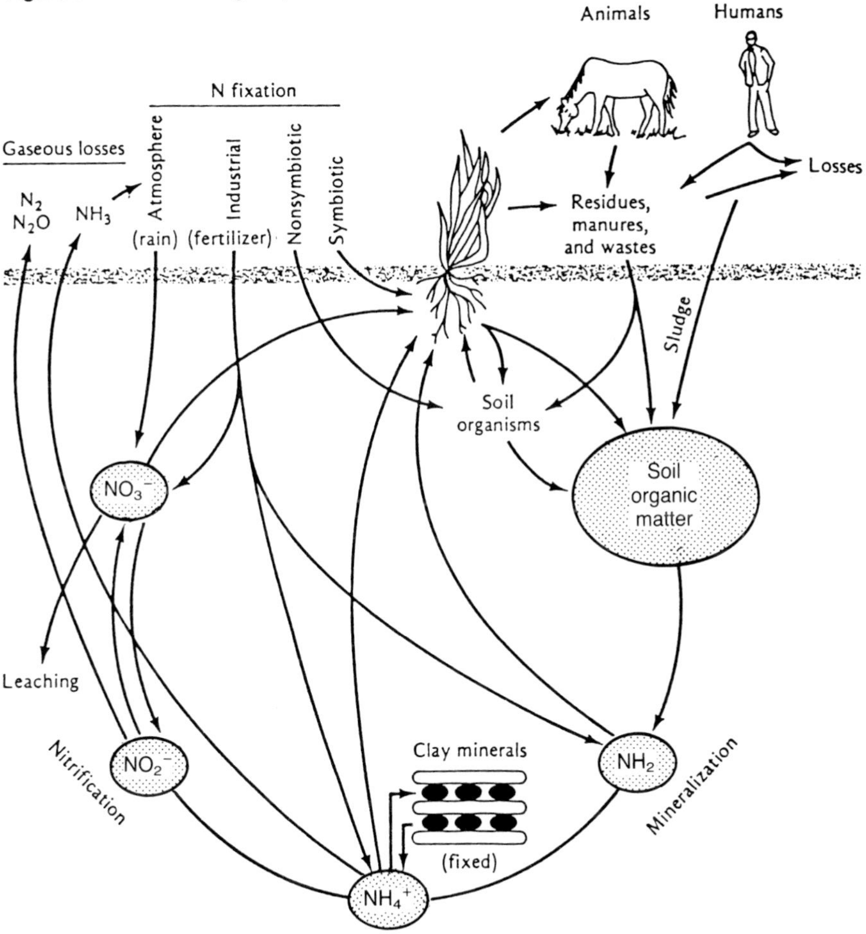

Source: Brady, N. C. (1984), *The Nature and Properties of Soils*. Macmillan; New York.

version of nitrogen from one form to another; the activities of these microorganisms, and therefore the form of nitrogen in the soil, is profoundly affected by cultivations, organic additions, crop sequence and root exudates.

The key feature of the soil nitrogen cycle is the turnover of nitrogen by mineralisation and immobilisation. Mineralisation is the conversion of bound organic nitrogen and other nutrients into the mineral (ionic or inorganic) form required for plant uptake. The inorganic nitrogen is then converted in a process known as nitrification from ammonium salts to nitrites and from nitrites to nitrates by nitrifying bacteria which obtain their energy from this oxidation process. Once free nitrate is formed the rapid recycling process offers many options: the nitrate may be immobilised, assimilated by plants, denitrified or leached.

Immobilisation occurs when inorganic ions are assimilated by soil microbes and once again bound organically. The processes of mineralisation and immobilisation occur simultaneously as a continual cycle. It is the net movement in one direction which determines if more or less inorganic nitrogen results and this is highly dependent on the availability of organic matter.

The carbohydrate content of the organic matter serves as an energy source for the microorganisms in the soil. If insufficient organic material is available, there will not be enough energy to maintain the process and a stage will be reached when inorganic ions accumulate, known as net mineralisation. These nutrients will then be readily available for plant uptake. Conversely, as may occur when straw residues are incorporated, the addition of large quantities of organic matter can result in too many nutrients being immobilised and insufficient being available for crop growth. This is particularly so for nitrogen and may occur with sulphur and phosphorus as well.

Mineralisation and subsequent nitrification or immobilisation by soil microorganisms are also affected by factors such as pH, nitrogen concentration and form, soil aeration, moisture and temperature. Mineralisation will occur under extremely acid conditions but the rate is considerably more rapid as neutrality is approached. Liming of acid soils will therefore increase the supply of available nitrogen. Acid conditions also cause the retention of ammonia in the soil and so slow nitrification. Mineralisation speeds up with increased aeration (e.g. by ploughing or rotovating) and with increased temperatures in the spring or in mild, wet conditions in the autumn.

Nitrate leaching

Leaching occurs when nitrate is present in the soil and rainfall exceeds evaporation resulting in net drainage, usually between October and

March/April. The extent of the losses is associated with the amounts of nitrogen in the form of nitrate in the soil profile. Once nitrate is beyond the root zone it is effectively lost from the system.

The causes of excessive nitrate leaching are complex and cannot be attributed to one single factor such as the use of artificial fertilisers (Addiscott, 1988; Harvey & Wilson, 1988; Royal Society, 1983; Department of Environment, 1988). Some leaching may even occur from undisturbed soils, often in February/March when freeze/thaw cycles of soil tend to increase nitrification and consequent loss to drainage waters. Nitrates can be held against leaching in well structured loams and clays because most nitrates are formed and held in crumbs; percolating water primarily moves down through cracks and coarse pores between the crumbs, so nitrates only get into the escaping water by diffusion, which is a slow process.

In organic systems, leaching losses will occur mainly from sudden, rapid nitrification of organic nitrogen, especially following cultivation when mineralisation will be enhanced. Autumn cultivations are likely to cause the largest losses especially if following a fertility-building ley. This is an important issue in organic systems; rotation design needs to consider how the large losses of nitrogen following the ploughing in of the grass/clover ley can be minimised.

Researchers at Elm Farm Research Centre (EFRC, 1988) found as much as 300 kg NO_3–N available per hectare during the winter period following the ploughing in of a lucerne-based ley (Figure 3.3). By the spring, this had fallen to less than 50 kg NO_3–N/ha, although it is important to realise that this decline is not necessarily due to leaching; the available nitrogen may be taken up by plants, immobilised or denitrified.

Similar results were found by Fowler et al. (1988). In one field, 72 kg N/ha was lost following the ploughing-in of a ley in autumn even though a winter cereal was sown. However, ploughing in mid-winter seemed almost to eliminate leaching loss, but obviously this means autumn sowing is not then possible.

In another study, Davies and Barraclough (1988) monitored actual nitrate leaching at different points in the rotation on an organic farm (Figure 3.4). They found that the magnitude of leaching was closely dependent on the position in the rotation. Grass/clover leys leached little nitrate; annual losses were always less than 2 kg N/ha. The largest losses corresponded to the ploughing in of the leys for arable cropping. In the year of ploughing, losses were 99 kg N/ha, but this had declined to 11 kg/ha by the second year.

Overall, the annual loss of nitrate from the farm averaged less than 20 kg N/ha. There was a suggestion in the results that nitrate leaching

Figure 3.3   Total NO₃–N (kg/ha) at three sites after cultivation of the ley for winter cereals.

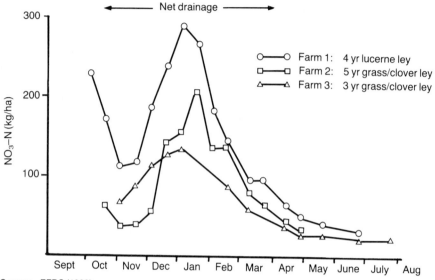

Source: EFRC (1988).

Figure 3.4   Nitrate leaching versus position in the rotation, Rushall Farm 1985–88.

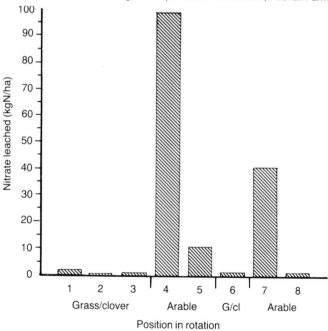

Source: Davies & Barraclough (1988).

declined the longer the field had been in the organic rotation, although no explanation was given as to why this might be the case.

It is interesting to compare these figures with data from conventional systems. According to Garwood and Ryden (1986), nitrate leaching under a grazed clover/grass ley may amount to 23 kg/ha. Under heavily fertilised grassland (420 kg N/ha) nitrate leaching may be as high as 162 kg/ha. In conserved leys the figures quoted are 2 and 29 kg/ha respectively. For arable farms, Davies and Barraclough suggest figures between 15 and 65 kg N/ha, although their figures for grassland are much lower.

It would therefore seem that, despite the problem of nitrate leaching following ploughing in of the ley, organic systems can contribute to the reduction of nitrate leaching overall. The difference between leaching rates under conventional and organic grazed clover/grass swards may be explained by higher biological activity and organic matter levels contributing to increased denitrification and immobilisation rates and hence lower leaching potential.

Despite these reassuring figures, organic farmers should not be complacent. Researchers are now trying to find ways of reducing leaching losses following the ploughing in of the ley. By far the simplest is early establishment of the following cereal crop, so that it has a chance to grow at the time in the autumn when nitrate is readily available. The use of catch crops has also been suggested. Researchers at Gleadthorpe Experimental Husbandry Farm compared a number of approaches and found considerable benefits from the use of a catch crop on sandy soils (Table 3.1).

Table 3.1   Leaching through sands at Gleadthorpe EHF in January 1989 after a cereal crop.

	mg/litre NO_3
Autumn catch crop after minimal cultivation	49
Straw chopped – left fallow	84
Winter rye sown	65

Source: Farmers Weekly 3/3/89.

Another possibility is the use of non-winter hardy green manures such as mustard or fodder radish sown with the cereal crop. Schoof (1988) found that fodder radish was better suited as a companion to wheat because of its growth habit (low-lying rosette), leading to less competition for light and a better microclimate. The results were, however, variable and oversowing also led to reduced yields. The best results, in terms of both yield and reduced soil nitrate levels, were achieved where crops were sown early, with or without green manures.

Denitrification

Losses through denitrification will occur under anaerobic conditions in waterlogged soils. The level of loss is related to the nitrate concentration, rate of biological activity and the time and extent of anaerobic conditions. Any factor increasing the rate of consumption of oxygen in the soil, such as increased microbial respiration from recent incorporation of crop residues, will increase the likelihood of denitrification; losses are highest during short periods of waterlogging during the growing season. In winter, biological activity is low so anaerobic conditions do not develop rapidly and losses through denitrification are unlikely. The process is difficult to quantify, some estimates give a maximum loss of 3 kg N/ha/year in unmanured soils.

Plant uptake

For most plants the nitrate ion is the primary source of nitrogen. Because of the extreme solubility of this ion the aim in an organic system must be to optimise, not maximise, the amount of free nitrate available to be taken up into plant material so as to minimise losses. The available nitrate content of soil is quickly reduced by a rapidly growing sward, so in practice the problem is to achieve sufficiently rapid mineralisation during the period of maximum demand. Due to the high mobility of nitrate ions the proximity of roots to ions is not important for the individual plant but roots are likely to affect each others' level of supply.

Biological processes are relied upon for nitrogen supply in organic systems, which means limitations to the choice of crop. For example, winter barley presents problems because it has a high nutrient requirement early in the season before biological activity has released sufficient nutrients. Nitrogen uptake is not constant throughout the year; evidence suggests cereals absorb 90% of the total nitrogen content of mature plants before the plant has achieved 25% of its final dry weight, although wet weather during the grain filling period may make soil nitrogen reserves more available to the plant and hence increase protein levels.

Care needs to be taken not to allow too much available nitrogen for nutritional as well as nitrogen economy reasons. If nitrogen is richly available in the soil while other elements such as calcium, magnesium and potassium are deficient, or if there is insufficient light and photosynthetic activity for the plant to make full use of the available nitrogen, nitrate-nitrogen will be accumulated in the plant cells instead of being used for protein building. In organic systems, excess of nitrates will only usually occur after heavy manuring and tillage of organically rich soils, but can also occur with winter production of leafy crops such as lettuce. High levels of nitrogen fertiliser use significantly increase the risk of nitrate accumulation in crops.

Rotation design should consider the nutrient budget on a field by field basis. Nitrogen fixing crops should alternate with high nitrogen demanding crops. Winter wheat is the most nutrient demanding cereal in organic systems and has relatively little root development, so it requires a preceding crop which will leave good levels of residual nutrients. The fertility-building ley is often used as the precursor to winter wheat, but as has already been mentioned the wheat crop does not appear to be able to take up enough of the newly mineralised nitrogen from an autumn cultivation to prevent serious leaching losses. If spring cultivation is possible it may be beneficial to plough in the ley in the spring and follow it with a crop like potatoes and then wheat; residual nutrients from manure applied to potatoes can then be used by the subsequent crop. Oats are suitable as a second cereal crop because they form deep rooting systems and can be highly productive in situations of low nutrient availability.

Sources of nitrogen

Atmospheric depositions
Nitrogen comes into the system from a variety of sources which include both atmospheric and biological processes. Rainfall provides a small, but significant, contribution estimated at 15 kg N/ha/yr up to as much as 30 kg N/ha/yr. Other atmospheric depositions have been calculated at 6–12 kg N/ha/yr; these amounts are obviously small compared with crop requirements.

Biological fixation
The biological fixation of atmospheric nitrogen is carried out by only a limited number of microorganisms; all other living organisms require combined nitrogen. Blue-green algae are free-living organisms which occur in almost every environmental situation where there is sufficient sunlight for photosynthesis; the amount of nitrogen these contribute on arable lands has been estimated at 10–15 kg/ha/yr. *Azospirilla*, free-living but non-photosynthetic nitrogen fixing bacteria, fix small amounts of nitrogen, but plants grow longer roots and more root hairs when the bacteria are present, which may contribute to improved exploitation of other nutrients in the soil profile by the plant.

Most important to the organic farmer, however, are the symbiotic *Rhizobia* bacteria associated with leguminous plants. These bacteria form nodules on the roots of leguminous plants and use carbon compounds produced by the plant as an energy source to fix atmospheric nitrogen. This is used by the plant and later released from decomposed material to become available to other plants. There are great problems involved in calculating the amount of nitrogen fixed by legumes, especially grass/clover leys in which the amount of clover is variable. Even with a known amount of legume, variable environmental conditions make quoted figures unreliable. The usual range of estimates of nitrogen fixed by grass/clover leys is between 60 and 200 kg N/ha/yr. Fixation by atmospheric nitrogen is not without 'cost' to the system; the use of energy supplied by the plant for the fixation process and the production and maintenance of root nodules can result in slower plant growth. The use of significant quantities of nitrogen fertiliser in conventional systems inhibits the natural bacterial processes, with the result that the presence of *Rhizobia* can even have a net detrimental effect.

Grain legumes such as peas and field beans can be useful break crops. Grain legumes take up more soil nitrogen and fix less atmospheric nitrogen than forage legumes. The nitrogen is stored in the grain as protein, so if the grain is sold off the farm little residual benefit will be obtained.

The fact that high levels of soil nitrogen inhibit nitrogen fixation means that the full benefit from fixation by legumes will not be realised if soil nitrogen levels are too high. Nitrogen fixation can be encouraged by growing legumes at points in the rotation with low nitrogen levels and also, in the case of grain legumes, by incorporating straw prior to sowing the crop so as to immobilise nitrogen.

In a mixed sward on a soil low in nitrogen and with no input of dung and urine, grass will be noticeably short of nitrogen and clover will dominate. In grazed swards the deposition of dung and urine exposes the clover to unevenly distributed high levels of soil nitrogen. Clover can match grass in its ability to take up soil nitrogen in high-nitrogen situations yet, in ley situations with high nitrogen, grass grows more vigorously and eventually eliminates the clover.

The problem for the farmer, as with all nitrogen additions, is to synchronise the nitrogen release from the legume residues to the requirements of the following crop.

Cultivations
Cultivation of the soil prior to drilling stimulates mineralisation and provides a ready source of nutrients for the new crop. The improved aeration, moisture distribution and exposure of new organic matter not previously accessible to microorganisms increases microbial activity and the release of organic compounds to soluble forms. A temporary increase

in respiration occurs each time an air-dried soil is rewetted; since considerable amounts of fresh soil are subjected to rewetting and drying through cultivation, losses of organic matter through this process are appreciable.

As has already been mentioned, the breaking up of a fertility-building ley creates the opportunity for dramatic mineralisation and release of nitrogen. As the demands of a newly sown crop are greatly exceeded by this supply, the negatively charged nitrate ion is liable to be washed from the soil when net drainage takes place, usually between November and the end of March. The earlier a winter cereal crop is planted, the more nitrogen is retained by the growing crop and the more the potential for leaching is reduced, although there are problems reconciling this requirement with weed, pest and disease control objectives. The use of fast growing, non-winter-hardy green manures such as mustard in combination with autumn cereal crops may help to trap greater quantities of nitrogen, releasing them in the early spring as the green manure residues decay. New techniques being developed at the Welsh Plant Breeding Station involve the direct sowing of cereals into a clover base without any cultivation at all.

Ley establishment is often achieved by undersowing the seed mixture with a cover crop, usually a cereal crop, which eliminates the necessity for autumn ploughing; this is important for nitrogen economy because turning the already depleted soil and exposing more organic matter for mineralisation just prior to the winter accelerates leaching losses. Undersowing cereal crops with clover can be useful to aid the smothering of weeds as well as to provide some extra nitrogen for the growing crop.

Livestock manures
The role of livestock in an organic system can be seen in terms of the profitable utilisation of legumes in the rotations, while facilitating the recycling of nutrients through manures. The potential for recycling of nutrients through the application of manure is high; both nutrients gathered from a large area during grazing and nutrients from conserved and purchased feeds are concentrated in dung and urine and so become available for redistribution. Grazing animals retain only 5–10% of the nitrogen they eat in herbage; of the remainder some 70% is voided in urine and 30% in dung. Dairy cows can excrete up to 250 g N per day.

Slurry and farmyard manure can effectively redistribute this concentration of nutrients but only with careful management. Losses occur from volatilisation, run-off, leaking slurry stores and leaching. During grazing, nutrients in excreta are deposited only on a small proportion of the field; the potential for loss from these areas is high, especially from volatilisation of urine patches; during warm windy weather up to 44% of the nitrogen can be lost, in cool and wet conditions losses have been measured at around 9%. Losses also occur from inorganic nitrogen accumulating in the soil beneath herbage killed by urine scorch. There are lower losses in the field from cut swards than grazed swards, although the loss is only deferred to the manure storage and spreading. Not all the nitrogen in excreta initially available for uptake is used by the herbage in producing harvestable dry matter; nitrogen may be temporarily retained in roots, immobilised into soil organic matter by soil microorganisms or lost by leaching or denitrification. The management of manures and crop residues to minimise nutrient losses is considered in greater detail in the next chapter.

Crop residues and green manures
Plant residues can be applied directly to the soil instead of via the compost or manure heap, the usual method being to plough in a growing crop as a green manure. This will have various effects on the soil depending on conditions. It may increase the organic matter of the soil, or increase available nitrogen, but not both at once. It may concentrate deficient nutrients in the surface soil and leave them more readily available for a following crop. Green manures can be grown in the autumn to reduce losses of soluble nutrients by taking them up before they are leached. Sowing green manures such as mustard or trefoil into a growing cereal crop benefits the crop by providing cover for the soil and raising the soil temperature.

The effect of a crop as a green manure depends on its maturity when it is ploughed under. Nitrogen availability will only increase if readily decomposable material high in nitrogen, such as young green plants, is ploughed in. Humus content of the soil will only increase noticeably if plant material fairly resistant to decomposition is ploughed in, but such material is typically low in nitrogen. Due to their greater degradability, green manures are not as effective as farmyard manure, per unit of carbon, in increasing the organic matter content of the soil. Research has indicated that the active decomposition of a green manure can actually reduce the soil organic matter by increasing the rate of production of carbon dioxide and mineral nitrogen from the humic matter in the soil. This effect, if it does occur, is generally thought to be small. Leguminous crops such as peas, clovers and some vetches are commonly used for green manuring because of their ability to increase soil nitrogen through symbiosis with nitrogen-fixing bacteria; these crops will normally only make adequate growth and fix enough nitrogen to make their cultivation worthwhile if the soil contains adequate supplies of carbon, potassium and phosphates.

The nitrogen liberated during the decomposition of the green manure crop can only benefit the subsequent crop if the latter is sufficiently developed to take up the nitrogen soon after it is released, although some of the nitrogen will form part of the soil organic matter pool as a constituent part of microbes. The protein of living green plants decomposes more rapidly than does that of dried or dead plant material; a long wet period between the ploughing in of a green manure crop and the establishment of the following crop can result in much of the nitrifiable nitrogen of the green crop being leached out of the soil. The period between release of nitrogen and leaching of nitrate can be short, especially in light soils under warm moist conditions. A further complication is the unfavourable conditions for germination and growth of very young plants brought about by the first flush of decomposition.

The return of crop residues is a logical step in the recycling of nutrients and serves to remedy the inevitable reduction in soil organic matter due to harvesting. Crop residues include more than just cereal straw; pea and bean haulms, roots and ground level residues are just as valuable. The incorporation of fresh crop residues as a method of building up soil organic matter or nitrogen must be done with caution. As with the use of green manures, the many benefits possible are balanced by problems when residues are badly applied. Large crop yield reductions may result from very large quantities of residues being incorporated into the soil if net immobilisation occurs. The carbon to nitrogen ratio of crop residues is often used as a guide to their composition and suitability for incorporation, but when planning their use caution is needed as the C : N ratio is not necessarily an indication of the actual availability of carbon or nitrogen to microorganisms.

The full value of any organic wastes as sources of plant nutrients is achieved by timing decomposition and mineralisation of the organic matter to coincide with the crop's nutrient requirements, although the rate at which organic materials decompose or mineralise in the soil is highly variable. Because of the slow-release nature of the nitrogen and phosphorus compounds, organic manures have a residual effect on soil fertility.

Overview
The many different components of both soil and whole-farm nitrogen cycles provide a challenge to the farmer to keep the system as tight as possible to minimise polluting, inefficient losses. In protein and other solid forms, nitrogen can be monitored easily; the difficulties arise when nitrogen is in solution or gaseous form. Some losses are inevitable due to the extreme solubility of nitrogen in solution, the escape of nitrogen as ammonia and the gases formed by denitrifying bacteria.

The best nitrogen management practices for organic farming systems are summarised in Table 3.2. In practice, a range of organic systems exist, differing in the extent to which they achieve the aim of nitrogen self-sufficiency and adopt these practices. Few, if any, organic farms can be said to be perfect in this respect. The reality is a spectrum of organic holdings which range from those which are 'balanced', i.e. come close to meeting the desired objective of nitrogen self-sufficiency, and those which are 'exploitive', i.e. are more heavily dependent on purchases of inputs such as manure from outside the system to make up for imbalances within the system itself. Mixed arable/livestock systems are characteristically on the 'balanced' end of the spectrum, whereas horticultural holdings, where land is often severely limited, tend to be at the 'exploitive' end.

Table 3.2 Summary of best nitrogen management practices for organic farming systems.

- Maximum reliance on legumes for biological nitrogen fixation.

- Minimum reliance on purchased manures.

- The use of on-farm manures as a means of recycling nutrients within the system, and for a range of other purposes including crop protection, not simply as a straight substitute for conventional fertilisers.

- The use of any manures limited to a quantity equivalent to that produced by livestock at a rate of 2.5–3 livestock units per hectare.

- The application of manures targeted at those points in the rotation where there is maximum nutrient offtake, especially potash (e.g. conservation leys, vegetables).

- Storage of manures under cover (fixed or temporary) and/or in situations where runoff can be collected and utilised.

- Ploughing and seeding either in *early* autumn or, preferably and when conditions allow, in late winter or spring.

- The use of green manures in combination with autumn sown cereals, or as a cover crop for the winter, so that nitrogen mineralised in the winter is taken up by the crop/green manure and not left liable to leaching.

- The soil should never be left bare over the winter.

- Judicious use of crop residues (e.g. straw) to lock up nitrogen in the autumn.

Potassium

Potassium is an essential plant nutrient which is necessary for the synthesis of amino acids and proteins from ammonium ions. Plants will usually contain as much potassium as nitrogen and it is the most abundant cation in plant cells. It is thought to be important in the photosynthetic process because potassium shortage in the leaf seems to lead to low rates of carbon dioxide assimilation. Potassium is essential for animals; their bodies contain more potassium than they do either sodium or chlorine. It is a major ionic constituent of intracellular fluid and other body fluids. Potassium contributes significantly to the osmotic properties of these fluids and is involved with muscle control and nerve activity. Potassium is not stored within the body; excess being excreted in urine. Deficiency in livestock is unlikely since most plants contain more than adequate potassium levels.

Deficiency symptoms
Identifying potassium deficiency is complicated because the effects are dependent on the relative concentrations of other elements, especially sodium and calcium, in plant tissue. Symptoms of potassium deficiency vary from crop to crop, but generally there is either a yellowing of leaf margins of older leaves followed by a brown scorch, or there is blue-green colouration followed by bronzing of the whole leaf. Potassium is highly mobile in plants, so it will be transported to young leaves. As a result, one indication of deficiency is premature death of older leaves. Clovers and lucerne develop whitish spots on the leaves as well as the marginal yellowing and browning. On light, sandy or chalky soils and on some peaty soils in situations of low nitrogen and low potassium, plants may become stunted, with leaves small and ashy-grey in colour and fruit and seed small in quantity, size and weight. An excess of nitrogen relative to potassium renders leaves susceptible to fungal and bacterial diseases and reduces resistance to drought.

Potassium availability
Problems with potassium generally relate more to the availability of the element to the plant rather than an inherent shortage. In the soil, potassium exists in four different forms: potassium as a component of soil minerals, fixed potassium, exchangeable potassium and water-soluble potassium.

The immediate source of potassium for crops is that of the soil solution, so uptake of potassium by the crop depends on the ability of the roots to exploit the soil, the concentration of potassium at the root surface and the rate at which potassium ions can diffuse from exchange sites to the root surface.

The exchangeable and water-soluble potassium between them form a pool of readily available potassium which is maintained in equilibrium at around 90% exchangeable and 10% in solution. The pool is strongly buffered, so local depletion of the soil solution is replenished by release of exchangeable potassium. It is these readily available fractions which are the usual forms of potassium analysed.

The fixed and mineral fractions are present in the soil in very much greater proportions and may provide a potentially inexhaustible supply to the water-soluble and exchangeable fractions. However, fixed potassium is normally released only in low exchangeable potassium conditions and mineral potassium is released only after weathering, which may be a very slow process.

In any soil the amount and availability of potassium is closely connected to the mineralogy of the parent rock and the origin of the soil. Potassium is one of the eight elements forming 98% of all rocks. It is common in acidic igneous rocks and is a constituent of feldspars and micas but most of this is relatively unavailable to plants; there is little reliable information on rates of weathering of rocks.

Clay soils will usually contain adequate potassium. In particular, the vermiculite, illite and smectite clays have very high natural levels of fixed and exchangeable potassium which is available to maintain water-soluble potassium. The sands and organic soils have much lower reserves of potassium.

It has been noticed in some organic systems that very low available potash levels according to soil analyses are not associated with potash deficiency symptoms in the growing plants or in lower subsequent yields, and that additions of mineral potash sources under these circumstances do not result in the expected yield increases. This may be due to the fact that the relationship between the different fractions is more finely balanced in an organic system, so that any available potassium is immediately taken up by the growing plant and is therefore not reflected in soil analysis figures.

Leaching
Potassium is a monovalent cation (K^+) held firmly by the soil so that when sufficient cation exchange capacity is present under normal conditions, very little loss through leaching occurs; however, the solubility of potassium means that in light or sandy soils leaching is a possibility. Light textured soils usually contain less potassium and are more quickly depleted of their reserves than heavier textured soils. If the ground is continuously covered by plant material leaching is unlikely to occur due to exchangeable potassium being held by soil organic matter. A deep rooting green manure will help to prevent losses by leaching by bringing potassium to the surface where it can be absorbed by the following crop.

Crop removals
The removal of crops from fields means the loss of considerable quantities of potassium, although if the crop is fed to farm livestock there is the potential for return through application of excreta. Considerable

potassium losses can occur, however, by leaching from uncovered manure heaps. Since there are no acceptable mineral sources of potash for routine use, losses through crop removal and poor manure management need to be avoided. In particular, the sale of large quantities of potassium off the farm in the form of straw or conserved forages should be avoided under all circumstances. In this sense, potassium is probably more of a problem in organic systems than nitrogen. A well-balanced system, however, should be able to minimise potash losses, and in some cases the losses which do occur may be compensated for by weathering of the soil and bedrock.

Crops vary in their effect on the potassium concentration of the soil. A crop removes potassium continuously during its period of active growth, thus reducing exchangeable potassium in the soil. The release of fixed potassium in the soil is slower than the rate of uptake so exchangeable potassium in the soil will fall during the growing season and rise again once growth ceases. Grass has very strong extraction abilities, a crop following a ley may suffer from this and may be limited by a shortage of available potassium. Through leaf fall and leaching, ripening cereals return to the soil more than half the potassium previously taken up, so that little potassium is lost by a cereal harvest if the straw is ploughed in but much more is lost if it is cut green for silage. One of the quickest ways of inducing potassium deficiency is to take repeated hay or silage cuts or any other leafy crops without returning potash in the form of manures or crop residues.

Crops also differ in their minimum potassium requirement for optimum growth and their requirements vary over the growing season. Potato tubers are reduced in size by inadequate supplies; tomatoes and soft fruit have high potash requirements. Leguminous pasture plants such as clovers and lucerne need adequate potassium to compete successfully with grasses; it may be that the grass exploits the limited potassium supply more successfully because of its more extensive root system. Potassium increases the winter hardiness of lucerne, possibly because it encourages the plant to store more carbohydrate and protein in its root system. Experiments have shown that the failure of lucerne to establish when undersown with barley is overcome with sufficiently high potassium status.

Sources of potassium
The return of excreta ensures the return of potassium removed in herbage and serves to transfer brought-in potassium (in the form of purchased feeds and bedding) to the land. Cattle return most of their daily intake of potassium in a readily available form. In grazed potassium-deficient communities the urine effect can be seen as dark

green foliage patches against a background of paler sward, in which the leaves show the characteristic tip burn of potassium deficiency. Return of excreta is therefore a vital component in potassium cycling.

Potential losses of excreta potassium can be as high as 60% under poor storage of manure, but if covered, losses can be minimised. Spread farmyard manure is variable in potassium content, depending both on the proportion of straw and the losses during storage. In farmyard manure 60% of the potassium is likely to be available for the crop in the year following application, with most of the nutrient content contributing to soil levels in the longer term. All the potassium in slurry is likely to contribute to maintenance of soil potassium levels, with 90% of potassium ions available for uptake by the first crop.

The effect of mulches on potassium supply can be considerable because the materials used for mulches, such as dried grass and straw, are high in potassium. Calcium, magnesium and sodium contents of typical mulching materials are low compared with potassium, resulting in preferential uptake of potassium by the roots of most crops. This may need to be considered carefully to avoid nutrient deficiency problems with other nutrients. The ability of grass to take up potassium strongly can be exploited by grassing land down and gang mowing it to keep it short in order to raise the level of available potassium in the surface soil.

There is no acceptable mineral source of potassium for routine use in organic systems available in Britain. Most commercially available potassium salts are readily soluble and are therefore not acceptable under organic standards. Sources of rock potash available in Britain tend to have extremely low availability and are therefore ineffective. Strict control must therefore be exercised to avoid unnecessary potassium losses from the system. Careful field by field monitoring of soil status and cropping is essential. During conversion, mineral sources may be employed to bring up the potassium status of the system. Where a clear deficiency exists, exceptions can be made at the discretion of the organic standards committee and potassium sulphate may be used. Chloride salts are never allowed, because the high concentration of the anion can harm growth of plants during dry weather and may have other adverse effects on the soil ecosystem. Wastes such as spent mushroom compost and seaweed, and by-products of industry such as wood ash and flue dust, are minor sources of both potassium and sodium. Full details of approved sources are contained in Appendix 1.

Phosphorus

In plants, phosphorus plays a fundamental role in the enzymatic reactions that depend on phosphorylation to make energy available for biochemical and other work. Deficiency symptoms can be difficult to diagnose. Crops can suffer without any obvious signs that phosphate starvation is the cause. Wheat and barley take up much of their phosphate in the early stages of their growth. Starvation during this period cannot be rectified by a good supply later. Excess phosphate over the amount required by the crop sometimes depresses yields. This is thought to be due to early maturation reducing vegetative growth, usually on light soils in dry years.

Phosphorus is also essential in livestock nutrition. Some 80% of the phosphorus in the animal body is in the skeleton. Phosphorus also occurs in many proteins and is necessary for the utilisation of carbohydrate. Phosphorus deficiency initially reduces animal productivity; serious deficiencies can result in bone disorders and infertility. Although phosphorus is a major plant nutrient, some plant species contain insufficient amounts to meet the high requirements of animals. Some pastures may be deficient and hay and straws are poor sources. Grains and their by-products, as well as materials such as meat and bone meals and fish meal, are rich in phosphorus.

In many respects the phosphorus cycle is analogous to the nitrogen cycle. Next to nitrogen, phosphorus is the most abundant nutrient contained in microbial tissue, making up as much as 2% of the dry weight. Partly for this reason phosphorus is the second most abundant nutrient in soil organic matter.

The phosphorus content of soils varies considerably depending on the nature of parent material, degree of weathering and the extent to which phosphorus has been lost through leaching. Phosphorus contents of common soil-forming rocks vary from as little as 0.01% in sandstones to over 0.2% in high phosphate limestones. On average the percentage of phosphorus in surface soils is about half that of nitrogen and one twentieth that of potassium. A soil containing 0.05% P will contain 1,120 kg P/ha to plough depth.

Phosphates exist principally as inorganic phosphates in the soil, either as definite phosphate compounds or films of phosphate held on the surface of organic particles. The soil solution contains small amounts of organic as well as inorganic phosphate although this is negligible in quantity compared with other forms. Plants take up phosphorus almost exclusively as inorganic phosphate ions, principally as the $H_2PO_4^-$ ion.

Maximum phosphorus availability occurs at intermediate pH values of 6 or 7 as this is above the pH range of maximum insolubility of iron and aluminium phosphates, but below the pH range of maximum insolubility of calcium phosphates. This can cause problems with phosphate availability on calcareous soils.

Phosphate compounds also exist in soil organic matter, although this is unlikely to be directly available to the plant. Liberation of this phosphate by decomposition is controlled by temperature—the warmer the soil the more rapid the rate of decomposition can be—and by the number of times the soil becomes very dry between rewettings. The release of soil organic matter phosphate is more important the lower the level of available inorganic phosphate, and hence may have greater significance in organic systems. Organic phosphate is important when attempting to build up soil humus, which requires a carbon to phosphorus ratio of between 100 and 200. If soils low in phosphate are put down to grass, inorganic or available phosphate may be converted into unavailable forms.

The phosphate concentration in the soil solution has been found to remain approximately constant when a crop is growing, which means that the phosphate is moving from solid to solution as quickly as the crop roots are extracting it from this dilute solution. Steep depletion contours exist near absorbing root surfaces due to the very low mobility of phosphate ions; during a season of plant growth a phosphate ion is unlikely to move a greater distance than the diameter of a root hair. Such extreme localisation of depletion zones minimises the chances that the rootlet of one plant will interfere with the availability of phosphate to another.

The ability to obtain the minimum phosphate concentration needed for reasonable growth depends in part on the microbiological environment of the root. Under some conditions the rhizosphere microorganisms will contain species that will solubilise phosphate from compounds of very low solubility and so increase the phosphate supply to the plant. If the phosphate concentration is low, however, some rhizosphere organisms will compete with the root for phosphate, thereby reducing availability to the plant.

The role of mycorrhizae

An important factor in phosphate uptake by the plant, particularly in organic systems, is the interaction between the plant roots and mycorrhizae. The term mycorrhiza applies to about 6,000 different fungi which behave in a similar manner. The fungal mycelia penetrate the plant roots enabling nutrient exchange to take place.

There are two groups of mycorrhizae which behave in different ways. The ectomycorrhizae surround the root surface and the mycelia penetrate only the outer cell layers (Figure 3.5). The endomycorrhizae, on the other hand, penetrate the cell walls and form fine branches in the host cells (Figure 3.6). The interactions between the plant's defence mechanisms and the fungi keep the association under control so that it does not become pathogenic.

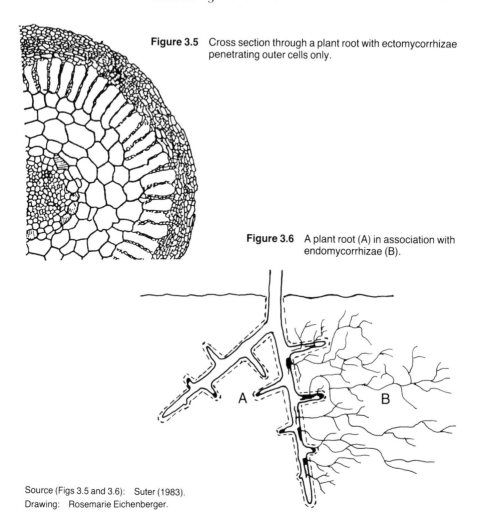

Figure 3.5   Cross section through a plant root with ectomycorrhizae penetrating outer cells only.

Figure 3.6   A plant root (A) in association with endomycorrhizae (B).

Source (Figs 3.5 and 3.6):   Suter (1983).
Drawing: Rosemarie Eichenberger.

The fine structure of mycorrhizal mycelia means that they are more effective than plant root hairs at absorbing plant nutrients including phosphorus, nitrogen, potassium and calcium as well as water. These are transported into the plant in exchange for carbohydrates. Roots infected with mycorrhizae have been shown to be more efficient scavengers than uninfected roots. In fact, although mycorrhizal associations are not generally considered to be important in conventional agriculture, because large quantities of soluble phosphates are generally applied anyway, their significance in organic systems could be much greater. By penetrating the plant roots, they provide the means by which organic molecules as well as phosphates can be transported from the soil into the plant and they may also assist the defence mechanisms of plants through the production of antibiotic substances.

New research is indicating an important link between mycorrhizae and the *Rhizobia* involved in nitrogen fixation. One experiment (Suter, 1983) with lucerne showed that when the plant was innoculated with both mycorrhizae and *Rhizobia*, yields were double those compared with rhizobial innoculation only. As with *Rhizobia*, however, large quantities of fertiliser tend to inhibit the activities of mycorrhizae, to the extent that if they are present, they may just end up being passive parasites on the plant.

Phosphorus removals and losses
Under cultivation some phosphorus is removed from the soil with the crop. Long-term cropping of soil without manurial inputs will therefore lead to a reduction in phosphorus content. Other losses occur through leaching and erosion. The strong tendency of phosphate to be adsorbed on colloidal surfaces and form insoluble complexes with cations means losses of phosphorus from agricultural soils in runoff are small, although over the years this can be appreciable. The main mechanism of loss from most agricultural soils is through erosion.

Phosphorus supply
Due to the slow rate of diffusion of phosphate, the phosphate supply to a plant depends much more on the size of the root system, the density of its root hairs and the intensity of its ramifications through the soil, than does the supply of most nutrients.

Mulching increases phosphate uptake by crops in two ways. Mulches may contain phosphate, but more importantly they encourage rooting in the parts of the soil where the decomposable organic phosphates are concentrated. Mulching also keeps the soil surface moist for longer, thereby increasing the time roots can take up phosphate.

Green manures can often exploit less available forms of phosphate and zinc than main crops, hence increasing the availability of these for the following crop. By rapid decomposition and liberation of large quantities of carbon dioxide, which dissolves in water to form carbonic acid, and other acids, green manures can help to lower the pH of alkaline calcareous soils and increase the availability of phosphates and trace elements to the succeeding crop.

Increases in soil phosphorus can be achieved through long-term applications of animal manures. The phosphorus content of soils has been shown to increase 14% over a 30 to 40 year period with annual applications of 5 to 11 t/ha of manure. The availability of total manure phosphorus is around 60% in the year following application, with the remainder contributing to maintain soil levels in the longer term. The long-term contribution of phosphorus in slurry is important in low phosphorus status soils, with all slurry phosphorus likely to become available over several decades. Current figures suggest 50% of the phosphorus content of slurry becomes available in the first year, but figures vary widely.

In situations of phosphorus deficiency in the whole farm system, mineral fertilisers (such as rock phosphate and basic slag where still available) and recycled organic materials can be used as a supplement to nutrient recycling on the farm. The solubility, and hence fertiliser value, of apatites in rock phosphates varies considerably depending on their composition. Calcined aluminium rock phosphate is appropriate for alkaline soils. Appendix 1 contains the list of Soil Association approved and restricted products.

Calcium

Of the other macronutrients, calcium is probably the most important. Calcium is needed both as a plant nutrient and for modifying pH levels by reducing the number of H^+ ions in circulation. Most plants will grow successfully only in a limited pH range, usually above 6, although some cereals such as rye and oats are suited to more acid conditions. The pH level is also of great importance with respect to the soil ecosystem, and in particular earthworm activity, which is reduced under acidic conditions (pH less than 5.5). Excess calcium may, however, restrict trace element availability and should be avoided.

Liming is standard practice in organic farming although certain sources of calcium such as slaked lime and quick lime are not permitted. Ground chalk, limestone and calcified seaweed are the major sources. The relative efficiency of each can be assessed from their quoted neutralising values or CaO equivalent (neutralising value of CaO = 100). Where possible, lime should be applied to green manures or during composting.

Other elements

Many of the other elements which are important to crop growth interact very closely with the major nutrients discussed above. The acidification caused by nitrogen fertilisers can have the effect of locking-up some nutrients, making them unavailable to plants. The same occurs with excess potassium availability, which displaces magnesium and results in reduced magnesium uptake by plants. The availability of trace elements can be affected in the same way, as well as by the presence of anaerobic conditions resulting from water-logging. In some cases, acid conditions may also result in the release of ions such as aluminium which are toxic to plants. In cases of deficiency, certain mineral sources such as dolomitic limestone or kieserite for magnesium may be acceptable, but attention should also be paid to correcting underlying faults in the farm system.

Rock dusts

In organic agriculture the term rock dust is applied to dusts from silicate rocks. These range from acidic (high in silicates) to basic forms (low in silicate) and include basalt, bentonite and serpentine. The use of rock dusts is still at an early stage, but initial results show promise; they may be useful as odour filters, to absorb ammonium-N from urine, as mineral feed supplements, in plant protection, or as soil amendments. It is suggested that they may be used to improve the texture of soils by adding coarse dusts or grits to fine, sticky soils of weak structure, or by adding very finely ground dusts to soils low in humus and clay minerals. Rocks, and hence rock dusts, are characterised by high contents of certain elements, often with a broad range of other minerals and minute quantities of trace elements. This range may be able to closely satisfy the nutrient requirements of most organisms in the soil.

The varying mineralogical composition of different rocks affects both their chemical and crystal structure. Crystal structure largely governs the susceptibility of a rock to weathering, that is the breakdown and release of nutrients. When rocks are ground to dust, nutrient release is also linked to the particle size; the larger the surface area which can be colonised by microorganisms or attacked by organic acids, the greater the rate of nutrient release.

If dusts are to be added directly to the soil it would be most appropriate to do so during a green manure stage. The incorporation of the residues promotes biological activity; this and the moist conditions will speed the release of nutrients from the dusts. Advantage may be gained by adding dusts directly to compost heaps where the high biological activity and temperatures accelerate their weathering.

OVERVIEW

In organic farming systems, the emphasis for crop nutrition must clearly be on cycling within the system and in particular on maintaining a biologically active soil which can release nutrients from the soil making them available for crop growth. If this can be achieved, then dependence on purchased fertilisers, whether mineral or organic, can be greatly reduced without necessarily entailing significant yield reductions. In any event, the existence of permitted fertilisers and manures under the Soil Association's organic standards should certainly not be taken to mean that they can be used as a direct replacement for other, prohibited fertilisers, without other husbandry changes first being implemented.

It can be argued, however, that the approach adopted by organic farmers results in 'mining' of the soil nutrients. This is supported to some extent by detailed research on 101 organic farms in Southern Germany (Weiss, 1988). The results were analysed on the basis of length of time since conversion to an organic system and compared with similar 'partner' conventional holdings (Table 3.3, page 82).

The study found that both available phosphorus and potassium were lower on the organic farms and that in the case of phosphorus, there was a clear reduction the longer the period since conversion. With potassium the decline was less marked. However, it was also found that higher stocking rates were associated with increased potassium levels and slightly increased organic matter and nitrogen levels. No influence was found on pH levels. Of particular interest is the potassium/magnesium ratio which is significantly different on the organic farms and provides supportive evidence for the argument that magnesium deficiency is not likely to be a problem in organic grassland management.

Another study (Kaffka & Koepf, 1989), however, found that on one dairy farm which had been managed biodynamically since 1929, soil tests for major nutrients showed only minor differences in phosphorus, potassium and magnesium between 1972 and 1982 and that after 50 years the potassium and phosphorus levels were still appropriate for good yields in spite of the fact that the only purchased inputs were feed and straw for bedding.

A farmgate nutrient balance for the 30-year period from 1952 to 1981 (Table 3.4) shows how well balanced the farm's nutrient management is. The negative balance for nitrogen indicates the contribution of bio-logical nitrogen fixation given stable organic matter levels. The positive balance for potassium is due primarily to purchases of straw for bedding. The land use was approximately 60% grassland and 40% arable.

Similar results were obtained for one of the three farms (Farm A) studied in detail by Fowler et al. (1988) (Table 3.5). The three organic holdings, however, varied widely in the amount of purchased nutrients. In cases B and C, the amounts purchased seemed to be unnecessarily high and above recommended organic practice and this is reflected in the nutrient balances. The rotation on Farm B has since been adapted and no manures are now imported onto the farm. Farm C is a horticultural holding and displays clearly the problems faced by growers with limited land forcing reliance on imported nutrients.

Table 3.3 Soil characteristics in the top 20 cm of the soil on farms classified by date of conversion to organic farming.

Characteristic		Conversion to organic farming				
		up to 1960	61–70	71–75	76–80	81–84
Number of farms		8	12	27	32	22
pH	org	6.8	6.7	6.6	6.6	6.7
	conv	6.6	6.4	6.7	6.5	6.7
Phosphorus	org	7.5	9.3	13.1	15.8	18.7
(mg/100 g)	conv	25.6	19.2	29.4	19.2	19.4
Potassium	org	23.8	16.0	18.6	22.1	24.4
(mg/100 g)	conv	38.6	28.0	29.4	29.0	30.1
Magnesium	org	18.4	14.1	14.7	12.3	14.4
(mg/100 g)	conv	15.1	13.4	11.8	13.1	15.0
K : Mg ratio	org	1 : 0.77	1 : 0.88	1 : 0.79	1 : 0.56	1 : 0.59
	conv	1 : 0.39	1 : 0.48	1 : 0.40	1 : 0.45	1 : 0.50
Organic	org	3.7	2.4	3.0	2.8	2.8
matter %	conv	3.4	2.5	2.7	2.8	2.8
Total N	org	0.22	0.17	0.20	0.17	0.18
	conv	0.22	0.16	0.17	0.17	0.15

org: organic farms.
conv: conventional farms.
Source: Weiss (1988).

Table 3.4 Farmgate nutrient balance on the Talhof – average over 30 years (1952–1981).

	N	P	K
		(kg/ha/year)	
Inputs purchased	5.9	1.0	5.3
Output sold in produce	20.0	3.2	4.3
Balance	−14.1	−2.2	+1.0

Source: Kaffka & Koepf (1989).

Table 3.5 Summary of purchases and sales of N, P and K on three organic farms (kg/ha/year).

Farm	Nitrogen			Phosphorus			Potassium		
	A	B	C	A	B	C	A	B	C
Purchases	19	137	358	7	31	94	11	65	313
Sales	24	51	25	5	10	5	5	12	33
Purchases minus sales	−5	+86	+333	+2	+21	+89	+6	+53	+280

Source: Fowler et al. (1988).

Table 3.6 Nutrient removal estimates for selected crops.

Crop	Nitrogen	Phosphate	Potash	CaO	MgO
	(kg/t fresh weight unless otherwise indicated)				
Rye	15	8	5	10	3
Wheat	15	8	5	6	3
Winter barley	20	8	5	10	4
Spring barley	18	8	5	10	3
Oats	17	8	5	6	4
Straw	5	3	13	5	1
Maize (+ straw)	30	15	35	8	8
Maize (− straw)	20	8	5	1	2
Oil seed rape	55	30	50	60	10
Peas	60	20	35	35	5
Field beans	65	20	45	35	8
Lupins	75	20	45	25	10
Potatoes (maincrop)	4.5	2	7	2	1
Potatoes (early)	3	2	5	0.5	0.5
Sugar beet (+ tops)	5	2	8	1.5	1.5
Sugar beet (− tops)	2	1	2.5	0.5	0.5
Lucerne*	30	8	20	25	4
Red clover*	30	7	25	22	5
White clover/grass*	10	3	5	20	5
Hay*	20	7	21	20	5
Silage*	25	7	25	20	5
Fodder rape	5.5	1.5	5.5	3.5	0.5
Forage rye	4.5	1.5	6	1.5	0.5
Fodder beet (− tops)	2	1	5	1	1
Fodder beet (+ tops)	4	1	6	1	1
Arable silage	5.5	2	5	3.5	0 5
Forage maize	3	2	4	1.5	1
Sunflowers	2.5	1	4	3.5	0.5
Mustard	4.5	0.8	4	4	0.5
Kale	6.5	2	7.5	4.5	1.5
Stubble turnips	2.5	1.5	5.5	2.5	1
Carrots	2	1	2.5	N/A	N/A
Carrot leaves	4	1	5	N/A	N/A
Beetroot	2	1	3	N/A	N/A
Beetroot leaves	4	1	5	N/A	N/A
Cabbages	3.5	1.5	5	2	0.8

* per tonne dry matter.
Sources: Various.

Nutrient Budgets

Published figures such as those in Tables 3.6 and 3.7 are available which detail the major nutrient removals by the harvest of common crops and potential rates of nitrogen fixation. These can be used to assess the need for replenishment of specific nutrients on a field by field basis and for rotation design (see Chapter 5).

Table 3.7   Nitrogen fixation by different crops (kg/ha).

White clover/grass	150–200
Red clover/grass	230–460
Lucerne	300–550
Field beans	150–390
Peas	105–245
Lupins	100–150

Source: Kahnt (1983) adapted.

There are difficulties involved; despite long-term research into legume nitrogen fixation there is still no reliable way of quantifying the amount of nitrogen fixed. Samples of other inputs such as cattle and chicken manure have been analysed and so estimates can be made regarding their nutrient content and availability, but these must be treated with caution as the materials are so variable and depend so much on storage conditions (see Chapter 4).

Nutrient budgets should be seen as only a rough guide indicating whether a rotation is balanced in nutrient terms or not. They are certainly not a replacement for monitoring the soil or the condition of the crop, which provides the best indication of the need for remedial action in practice.

References and Further Reading

Addiscott, T. (1988) *Farmers, fertilisers and the nitrate flood.* New Scientist 8/10/88: 50–54

Allison, F. E. (1973) *Soil Organic Matter and its Role in Crop Production.* Elsevier, Amsterdam

Davies, G. P. & Barraclough, D. (1988) *Nitrate leaching at Rushall Farm, Wiltshire, 1985–1988—a field monitoring study.* Unpublished mimeograph. Jealott's Hill Research Station; Bracknell, Berkshire. Reprinted in English language IFOAM Bulletin 7 (1989): 3–5

Department of the Environment (1988) *The Nitrate Issue.* HMSO

Dudley, N. (1990) *Nitrates—the Threat to Food and Water.* Merlin Press

EFRC (1988) *Nitrogen mineralisation in organic ley/arable farming systems.* Research Note No. 7. Elm Farm Research Centre, Newbury

Fowler, S. M., Watson, C. A. and Wilman, D. (1988) *The nitrogen cycle on organic farms*. Report commissioned by the Ministry of Agriculture, Fisheries and Food from the Department of Agriculture, University College of Wales, Aberystwyth

Garwood, E. A. and Ryden, J. C. (1986) *Nitrate losses through leaching and surface run-off from grassland: effects of water supply, soil type and management*. In: *Nitrogen flows in intensive grassland systems*. Eds H. van der Meer, J. C. Ryden and G. C. Ennick. Martinus Nijhoff, Doordrecht

Goldstein, A. H. (1986) *Bacterial solubilization of mineral phosphates*. American Journal of Alternative Agriculture 1: 51–57

Harvey, J. & Wilson, R. (1988) *Nitrates*. Farmers Weekly 30/9/88: 63–70

Kaffka, S. & Koepf, H. (1989) *A case study of the nutrient regime in sustainable farming*. Biological Agriculture and Horticulture 6: 89–106

Kahnt, G. (1983) *Gründüngung*. DLG Verlag, Frankfurt

Kononova, M. M. (1966) *Soil Organic Matter. Its nature, its role in soil formation and in soil fertility*. 2nd ed. Pergamon

Lampkin, N. & Woodward, L. (eds.) (1984) *The Soil: Assessment, Analysis and Utilisation in Organic Agriculture*. Elm Farm Research Centre, Newbury

Lopez-Real, J. M. & Hodges, R. D. (1986) *The Role of Micro-organisms in a Sustainable Agriculture*. AB Academic Publishers

Markus, P. (1988) *Untersuchung zur Wirkungsweise N_2 fixierender Bakterien auf das Wachstum und Ertrag von Sommerweizen in organischen Landbausystemen*. PhD thesis, University of Bonn

Markus, P. & Kramer, J. (1988) *Importance of nonsymbiotic nitrogen fixing bacteria in organic farming systems*. In: Klingmüller, W. (ed.) *Azospirillum IV. Genetics, Physiology, Ethology*. Springer Verlag, Heidelberg: 197–204

Richards, B. N. (1974) *Introduction to the Soil Ecosystem*. Longman

Royal Society (1983) *The nitrogen cycle of the United Kingdom*

Russell, E. W. (1973) *Soil Conditions and Plant Growth*. 10th edition. Longman

Schoof, W. (1988) *Winterweizen im ökologischen Landbau nach Luzernegrasumbruch — Versuche zur Reduktion von N-Verlusten*. Diplomarbeit, Gesamthochschule Kassel

Soil Association (1981) *The use of manures and mineral fertilisers in organic agriculture*. Technical Booklet No. 4

Suter, H. (1983) *Pilze und Pflanzen*. Zum Beispiel 7 (15/06/83): 4–6

Thoma, G. (1988) *Umweltrelevante Unterschiede im Stickstoffhaushalt 'biologisch' und 'konventionell' bewirtschafteter Böden*. PhD thesis, University of Hohenheim

Tinsley, J. & Darbyshire, J. F. (eds) (1984) *Biological Processes and Soil Fertility*. Developments in Plant and Soil Sciences Vol. II. Martins Nijhoff/Dr W Junk Publishers, The Hague

Weiss, K. (1988) *Vergleichende Bodenuntersuchungen in alternativ und konventionell bewirtschafteten Betrieben*. Lebendige Erde 3/88: 146–158

Wild, A. (ed.) (1988) *Russell's Soil Conditions and Plant Growth*. 11th edition. Longman

Recent publications

EFRC (1992a) *Nitrate reduction for protection zones: the role of alternative farming systems*. Report on research by Elm Farm Research Centre. R&D Note 108. National Rivers Authority; Bristol.

EFRC (1992b) *Assessment of nitrate leaching losses from organic farms*. Final report for MAFF, Contract CSA 2248. Elm Farm Research Centre; Newbury.

EFRC (1993) *The availability of water insoluble phosphorus and potassium sources in organic farming systems*. Final report for MAFF, Contract CSA 1486. Elm Farm Research Centre, Newbury.

Chapter 4

Management of Manures, Slurry and Organic Residues

Livestock Manures

The sound management of animal manures and crop residues plays a crucial role in organic systems. Manure is not simply a 'problem' which needs to be disposed of as cheaply as possible. It is a valuable resource which completes the nutrient cycle and allows much of the nitrogen fixed with legumes and harvested as forage to be returned to the soil where it can become available for subsequent crops. Organic manuring aims to improve the biological and physical/chemical properties of the soil and is important as a source of energy and nutrients for the soil ecosystem.

The main emphasis of this chapter is on the storage, handling and spreading of animal manures, which can occur in several different forms and need to be treated in different ways. There is no single approach which is the right one in all circumstances and there are many circumstances when composting, which is often regarded as *the* essential practice in organic farming, is inappropriate.

The main types of livestock 'wastes' which are found on farms in Britain are farmyard manures (FYM), either fresh or stockpiled, slurry and, occasionally, liquid manures such as separately collected urine. Before considering the management of individual types of manure and slurry in detail, it is useful to look at some data on rates of manure production and nutrient content/availability of different types of manure. Table 4.1 shows the quantities of excreta produced by different types of livestock under different circumstances, while Table 4.2 gives an indication of the nutrient content of different manures and slurries.

Although the data in Table 4.2 indicate the total nutrient content of the manures, not all of these nutrients are available to the growing crop in the year of application (Table 4.3). Nutrient losses during storage also occur because of leaching and volatilisation, but this depends to a large extent on how they are managed; this issue is considered in greater detail later in the chapter. Leaching losses of soluble nutrients may amount to 20% of the nitrogen, 7% of the phosphate and 35% of the

Table 4.1 Approximate amounts of excreta produced by livestock.

Type of livestock	Body weight (kg)		Amount of excreta – faeces and urine or droppings (litres/day)		Moisture content of excreta (%)
	Range	Approx. mean	Range	Best estimate	
1 Dairy cow	450–560	500	32–54	41	87
1 Beef bullock	200–450	400	19–28	27	88
1 Pig – dry meal or liquid fed (water : meal ratio 2.5 : 1)	20–90	50	2.0–5.5	4.0*	90
1 Pig – liquid fed (water : meal ratio 4 : 1)	20–90	50	4.0–9.0	7.0*	94
1 Pig – swill fed	20–90	50	–	14.0*	98
1 Pig – whey fed	20–90	50	9.0–15.0	14.0*	98
1 Dry sow	–	125	–	4.5	90
1 Sow + litter to 3 weeks	–	170	–	15.0	90
1,000 Laying hens	1,800–2,300	2,000	100–140	114	75
1,000 Broilers (+ litter)	100–2,000	–	56–63	68 kg	30

* Amounts of excreta produced over liveweight range 20–90 kg, i.e. production per pig place.
Source: MAFF/ADAS Booklet 2081, 1986 ed.

Table 4.2 Composition of farmyard manures and fresh, undiluted slurries (on fresh weight basis).

Type	Approximate dry matter (%)	Nitrogen (% N)	Phosphate (% P_2O_5)	Potash (% K_2O)
FYM				
Cattle	25	0.6	0.3	0.7
Pigs	25	0.6	0.6	0.4
Poultry				
deep litter	70	1.7	1.8	1.3
broiler litter	70	2.4	2.2	1.4
in-house air-dried droppings	70	4.2	2.8	1.9
SLURRY (fresh and undiluted*)				
Cattle	10	0.5	0.2	0.5
Pigs				
dry meal fed	10	0.6	0.4	0.3
pipeline fed	6–10	0.5	0.2	0.2
whey fed	2–4	0.3	0.2	0.2
Poultry	25	1.4	1.1	0.6

* Allow for dilution: for slurry diluted 1 : 1, divide by 2.
Source: MAFF/ADAS Booklet 2081, 1983 ed.

potash from FYM stored in the open, as is the case on most farms. The losses will be greater from shallow, flat heaps compared with a deep stack with steep sides. According to MAFF estimates, the benefits of covering heaps to reduce leaching losses would result in a saving in fertiliser use which can financially justify the costs of providing the polythene covering. Losses of nitrogen as ammonia or nitrogen gases from FYM may be of the order of 10%, ranging from hardly any when compressed in yards to 40% when stacked and turned. Gaseous losses from slurries may be of the order of 10–20%, more if agitated. On spreading, nitrogen will also be lost through leaching and volatilisation. Spring applications are more efficiently used, because nitrogen leaching losses are greater with autumn and winter applications (Table 4.4).

Table 4.3 Proportion of total nutrients available in the season of application (a).

Type	Nitrogen	Phosphate (% available)	Potash
FYM			
Cattle	25	60	60
Pig	25	60	60
Poultry (all types)	60	60	75
SLURRY			
Cattle	30(b)	50	90
Pig	65	50	90
Poultry	65	50	90

(a) These figures apply to manures and slurries spread in spring. When spread in autumn or winter less than these proportions of nitrogen are available.

(b) For slurry applied as a surface dressing to grassland. When incorporated into the soil soon after spreading, up to 50% of the nitrogen may be available.

Source: MAFF/ADAS Booklet 2081, 1986 ed.

Table 4.4 Relationship between time of application of manures and the amount of available nitrogen remaining for the spring growth.

Time of application	Available nitrogen effective for spring growth (%)
Autumn	0–20
Early winter	30–50
Late winter	60–90
Spring	90–100
Summer	(a)

(a) Crop response to summer application is very variable and is dependent upon the weather.

Source: MAFF/ADAS Booklet 2081, 1986 ed.

The figures in Table 4.5 provide a reasonable estimate of the quantity of nutrients available to the crop in the year of application for different types of manure and slurry. In terms of the rotation as a whole, the total nutrient contribution of these manures is likely to be higher. The nitrogen values are a guide to the amount of nitrogen available for crop uptake. The availability of nitrogen is variable and is affected by the rate and timing of application, the weather after spreading and the speed of incorporation into the soil. For phosphate and potash, the available nutrient figures in the table indicate the nutrient content available to the crop grown following application. The total nutrient content figures given in brackets should be used when considering overall effects over the rotation.

Table 4.5 Available nutrients in farm yard manures and slurries (spring application).

	Available nutrients in season of application		
	Nitrogen (N)	Phosphate (P_2O_5)	Potash (K_2O)
FYM		(kg/t)	
Cattle	1.5	2.0 (3)(a)	4.0 (6)
Pig	1.5	4.0 (7)	2.5 (4)
Poultry			
deep litter	10.0	11.0 (18)	10.0 (13)
broiler litter	14.5	13.0 (22)	10.5 (14)
in-house air-dried	25.0	17.0 (28)	14.0 (19)
UNDILUTED SLURRY (b)		kg/m^3 (kg/1000 litres)	
Cow (10% DM)	1.5 (c)	1.0 (2)	4.5 (5)
Pig (10% DM)	4.0	2.0 (4)	2.7 (3)
Poultry (25% DM)	9.1	5.5 (11)	5.4 (6)

(a) Values in brackets are for total nutrients available over rotation, derived from Table 4.2.
(b) Adjust values if diluted; if diluted 1:1, divide by 2.
(c) This value should be increased to 2.5 kg/m³ where slurry is incorporated.
Source: MAFF/ADAS Booklet 2081, 1986 ed.

It is important, however, to remember that organic manures contribute more than just nutrients to the soil, so these figures should not be interpreted too narrowly. Organic manures help to modify the physical condition of soils, by improving water holding capacity, aeration, drainage and friability, and the darker colour of organic matter means that soils warm up faster. Organic manures provide the energy needed for increasing microbiological activity and also help to protect crops from temporary gross excess of mineral salts and toxic substances and from rapid fluctuations in soil reaction by means of their high absorption capacity exerting a 'buffering' action. In this way, they can help to counter the effect of excess liming as far as the availability of micronutrients is concerned.

It should also be remembered that it is very easy to overmanure when using organic manures. For this reason, quantities applied over the rotation should be limited to the nutrient equivalent of the manure produced by a maximum of 2.5 to 3 grazing livestock units per hectare. In some countries, these limits are enforced by legislation, but they are also now a part of most organic production standards, although not yet part of the Soil Association and UKROFS standards. Using the MAFF tables above, this would be equivalent to 45,000 litres of fresh, undiluted cattle or pig slurry per hectare, containing approximately 225 kg total nitrogen, 90 kg phosphate and 225 kg potash, or alternatively 40–50 tonnes fresh farmyard manure (20–25 t composted), or 10–15 tonnes poultry manure per hectare. (The 1989 revision of the IFOAM standards specifies that the total amount of manure added, averaged over the rotation, must not exceed the quantity which could be produced on the farm if it were a self-sufficient livestock holding.)

FARMYARD MANURE

Farmyard manure (FYM) consists of animal excreta and bedding material, usually straw, in varying quantities and at varying stages of decomposition. In some housing systems, the manure is removed on a daily basis, requiring the handling of fresh material. In others, particularly straw yards, the FYM is allowed to build up gradually over the housing period. In poultry houses, wood shavings are the usual form of litter and decomposition may have barely started (e.g. broiler manure) or may have reached an advanced stage (e.g. deep litter). One form of solid poultry manure which is becoming more important is in-house, air-dried manure from deep pit houses. The droppings, without litter, collect on wooden slats or on the pit floor below stepped cages and are dried slowly in the air.

Nutrient content

Nitrogen availability
Because different types of bedding in varying quantities are used, and livestock are kept on sometimes widely differing diets, the nutrient composition of the manure can vary widely. In some cases, it may be useful to get the material analysed directly, but often the material is not homogeneous and generalisable results are difficult to obtain. The MAFF Bulletin No. 210 (Organic Manures) contains data on a wide range of different types of manure from different sources, as well as on other potential organic fertilisers, which can be used to supplement the tables above.

FYM contains less soluble nitrogen than slurry, which results in slower mineralisation, hence similar availability figures whether FYM is applied in autumn or spring. The carbon to nitrogen (C:N) ratio has been shown to be the main factor affecting the availability of nitrogen in manures. Fresh manure without straw can increase yields considerably, whereas the high carbon content of fresh FYM with high straw content can depress yields.

Knowing the C:N ratio does not, however, enable predictions of mineralisable nitrogen because organic manures with the same C:N ratio can be quite different. In aerobically composted manures, longer composting times increase the biological stability of nitrogen complexes and hence decrease nitrogen availability. This has implications for the use of manures in an organic system. If increased nitrogen availability is important, then fresher material should be used. If the contribution of the material to increasing soil organic matter levels is more important, then more mature, composted manure will be preferable.

Although higher rates of manure application do result in increased nitrogen yields, the plant's use of manure nitrogen decreases at higher levels. In field trials split applications have resulted in higher nitrogen uptake than one heavy application. One trial showed that repeated annual applications of farmyard manure increased the total uptake of nitrogen due to mineralisation of the accumulated residual manure, but single manure applications did not show any increase in the amount of mineralised soil-N in the following year. It should be remembered, especially when manuring horticultural plots, that luxury uptake of nitrate can occur, in which case the nitrate remains in that form in the plant material.

Nitrogen losses
Most losses from the storage of manures are gaseous. Ammonia is lost every time a heap is moved; additionally, the inside of well compacted heaps can become anaerobic so denitrification occurs. Losses increase with increased length of storage, and are lower in winter with a rapid increase in early spring. Initial C:N ratios affect the pattern of losses, a high initial C:N ratio (around 40) results in lower losses. Adding more straw to bedding materials does not tend to increase the C:N ratio in the final product because the extra straw serves to absorb more of the urine excreted. If a higher C:N ratio is desired straw must be added when the bedding is removed.

Leaching losses from unprotected heaps can be considerable. In a trial, sheltering manure from rain led to the heap drying, decomposition being retarded and ammonia volatilisation accelerating. Nitrogen losses from leakage were reduced; 4–6% under cover compared with 10–14%

unprotected. Leaching from manure heaps is mostly in organically bound forms, nitrate-N losses being insignificant in this case. The C : N ratio does not appear to affect leaching losses.

Losses on spreading are primarily by volatilisation, this is greater at increased temperatures so spreading should be avoided in hot weather. Trials have shown 60–90% of ammonium-N in cattle manure can be volatilised between 5 and 25 days after surface application to soil. Application losses can be reduced by incorporating the manure as soon as possible. Denitrification losses may result if manure decomposes rapidly in warm, moist soil causing the oxygen concentration to be depleted.

Toxic elements
It should be noted that the Soil Association's organic standards now prohibit the use of manures from ethically unacceptable livestock systems such as battery cages and intensive pig units. This also helps to avoid the risk of toxic elements in imported manures. Pig manure from fattening pigs can contain 300–2,000 mg/kg dry matter (DM) (mean 870 mg/kg DM) copper and 200–1,500 mg/kg DM (mean 600 mg/kg DM) zinc, both derived from feed additives. Antibiotics such as zinc bacitracin, where it is still used, are also found in poultry manures. The gradual build-up of these elements in the soil may lead to crop toxicity problems. Sheep are at particular risk from copper toxicity. MAFF recommends that at least one month should elapse before stock graze pasture which has received manures to avoid copper toxicity and disease risks, and the manure should be applied after and not before cutting silage. The Soil Association standards specify limits on the heavy metal content of brought-in manures.

The management of farmyard manures

There are essentially two approaches to the management of FYM used in practice. The first is to apply fresh FYM spread thinly at a rate of about 10 t/ha (sheet or surface composting). Alternatively, FYM can be stored, under a range of possible conditions, and then used when it has reached a more mature stage, but not usually older than six months. Fresh FYM contains higher levels of relatively easily available nutrients, but the organic matter contributes little to long-term organic matter levels in the soil and to soil structure. Because the nutrients are so readily available, attention needs to be paid to the weather and timing of application.

Sheet composting
Some organic farmers place great emphasis on the sheet composting approach using fresh FYM manure, because the microbiological activity

associated with decomposition takes place in the soil and not isolated in a heap. The increased microbiological activity also means that more nutrients may be released from the existing soil organic matter, supplementing those provided by the manures. Sheet composting can create a mulch effect and is also believed to benefit the antiphytopathogenic potential (inherent disease resistance) of the soil.

Stored manures
When stored, important chemical changes take place in the manure heap: urea is converted into ammonium compounds; carbohydrates contained in the litter are fermented, resulting in the production of heat, various gases (such as carbon dioxide, methane and hydrogen) and a decayed mass of organic matter which is richer in nitrogen and darker in colour than the original material; proteins in the litter and faeces are broken down into simpler compounds of nitrogen, such as ammonia; nitrogen is assimilated and fixed as protein in bacteria.

Manure can be stored under different conditions with respect to the availability of air and, in particular, oxygen. A common approach is to dump it all in a heap and hope for the best. This random dumping results in a variable product and can lead to high nutrient losses, especially through leaching. A mixture of aerobic and anaerobic conditions will exist, with the build-up of toxic compounds creating problems, not to mention the characteristic nasty smell of putrifying organic matter, but without necessarily obtaining the advantages of ammonia retention. Exposure to rainfall and surface run-off also mean that the pollution risks are quite severe.

Manure which is collected and stored under controlled conditions, as far as the siting and exposure to rainfall of the heap is concerned, can be referred to as stockpiled FYM. Most farmers should find that this type of storage is relatively easy to achieve, but some disadvantages such as localised anaerobic decomposition still remain.

Cold manure
A traditional approach to the storage of FYM in central Europe is the 'cold manure' technique, where the manure is carefully stacked and compacted so as to create completely anaerobic conditions where air is excluded from the heap. The temperature of the heap will remain at about 30°C. The advantage of this approach is that volatilisation losses of ammonia are minimal. There are, however, greater losses on spreading, because the material should be left on the soil surface if the toxic compounds produced in the process are not to impede root growth and soil microbiological processes. Under these completely anaerobic conditions, many weed seeds and pathogens are destroyed by the ammonia concentrations and lack of oxygen.

Warm manure

A compromise approach, which attempts to combine controlled aerobic and anaerobic processes, is represented by the 'warm manure' technique. This technique involves the gradual addition of layers of manure, which start to decompose aerobically with temperatures reaching 40–50°C, before further layers are added at 2–4 day intervals. As layers are added, the bottom of the heap becomes anaerobic, with the temperature falling back to about 30°C. This arises from natural rather than deliberate compaction. The daily production of fresh manure means several heaps have to be maintained simultaneously. Heaps can be used after three months. This approach maintains a greater carbon content than composting, but still achieves weed seed and pathogen kill. The process is, however, labour intensive and needs careful management and is therefore probably not suited to most British situations.

Compost

The alternative to controlled anaerobic decomposition is aerobic decomposition in the composting process. Composting attempts to recreate the conditions which would occur in an undisturbed ecosystem where organic matter builds up on the soil surface and is not regularly incorporated into the soil as in agricultural ecosystems. Reasonably mature compost quickly comes into equilibrium with the soil, whereas raw organic manures can cause a period of major disruption to soil processes.

Careful control of the conditions under which decomposition takes place allows the decay process to be optimised. Microbiological activity quickly raises the temperature to above 60°C. After a few weeks, the heaps are usually turned to allow a second heating-up to take place. The formation of humic substances during the composting process means that composted FYM generally provides a more stabilised form of organic matter than raw wastes, which is better suited to the long-term maintenance of soil organic matter, and the volume reduction of about 50% greatly facilitates spreading.

Nutrients are less readily available, but are maintained as a reservoir of slow release nutrients for mineralisation into the soil solution and plant uptake. Because decomposition is so far advanced, there is little in the way of readily available energy sources for soil microbes, so there is less tendency for soil organic matter to be mineralised when the manure is spread. If relatively insoluble nutrients such as rock phosphates or rock dusts are added to the compost heap, their availability, on the other hand, may be improved as a result of the intense microbiological activity during the composting process. Soil properties may also be improved, including water retention capacity, structure, drainage, aeration and workability.

The high temperatures involved with composting also help to kill weed seeds and disease pathogens. In addition, compost contains active ingredients such as antibiotics as well as antagonists to soil pests. For example, insects present in compost will eat cabbage root fly eggs, but the problem is distributing the compost at the right stage. There is a further advantage to composting which is that many pesticide residues are broken down before they are applied to the crop. This is one of the main reasons why the Soil Association's standards require that brought-in manures are composted before use, although the advantages in terms of weed and disease control are equally important.

Relative merits of the different methods of manure storage
Where manure is allowed to build up over winter, as in straw yards, a process similar to the 'warm manure' system described above takes place, with a slower breakdown at lower temperatures. As a result, there is little sense in recomposting this type of FYM and ideally it should be stored and spread directly from the shed.

There are, however, wide differences of opinion at to the best form of storage. Composting has many devotees, even though what passes for composting on some farms leaves a lot to be desired. In fact, if the issues are examined objectively, it becomes clear that there is no one system which is 'best' under all conditions.

Many of the issues have already been discussed above, but some deserve closer attention. One of these is the problem of nitrogen losses. Table 4.6 gives an indication from various studies of the levels of nitrogen loss which may occur under the different methods described above. This is largely associated with ammonia retention in the heap; 40% of the nitrogen in anaerobically stored 'cold' FYM is in the form of ammonia after 6–8 weeks' storage, compared to only 10% in aerobically composted manure. This does not, however, take into account losses on spreading, which can be significant (Table 4.7). The overall effect is that there may be little advantage between the different methods in this respect.

This does not, however, represent the full story. Ott (1980b), comparing compost and stockpiled feedlot manure in the Mid-West United States, found that using only a quarter of the quantity of compost relative to stockpiled FYM (4 tDM/ha against 16 tDM/ha), similar yields were obtained. The main soil characteristics (pH, organic matter, nitrogen, phosphorus, potassium and cation exchange capacity), as well as the plant tissue analyses (NO_3, P, K) performed on the leaf blades (wheat) at heading time were significantly improved by the application of composted manure.

These results call into question the conventional advice that the crop yield effect arising from the applications of organic manures can be

Table 4.6 Nitrogen losses during storage using different methods – data from various studies.

Material	N-loss (%)
Cold manure (1)	10–15
Warm manure (2)	22
Composted manure (3)	23
Stockpiled and packed (1)	25
Stockpiled not packed (2)	30–50
Manure added to top of layer of fermented manure (2)	17
Stockpiled and packed and slurry added (1)	16

(1) Anaerobic.
(2) Aerobic followed by anaerobic treatment.
(3) Aerobic treatment (composting).
Source: Vogtmann & Besson (1978).

Table 4.7 Average N (%) loss from FYM when applied to soil (Swiss data).

Time between application & ploughing	December sunny, frost	March 10.7 mm rain	April clear, windy 4.7 mm rain
6 hours	2	3	19
1 day	2	3	22
2 days	–	3	24
4 days	15	10	29

Source: Vogtmann & Besson (1978).

accounted for solely by their nutrient content. The differences in the nutrient contents of the two materials were not sufficient to account for similar yields being achieved with only one quarter the amount of compost compared with FYM. The beneficial effect of compost on soil organic matter and available nutrients was even more noticeable.

One possible explanation for the organic matter effect is that organic matter already in the soil was mineralised as a result of increased soil biological activity following the application of untreated manure. Another factor may be the plant disease control potential of composts discussed in greater detail in Chapter 7. Growth-promoting substances produced during the composting process may also play a role.

The results from this work confirm earlier findings by Sauerlandt (1956) (Table 4. 8), although in the Sauerlandt trials the same quantities of stockpiled and composted manures were used.

Table 4.8 Comparative trials with stockpiled and composted FYM.

	Sauerlandt trials			Ott trials		
	No FYM	Stockpiled	Composted	No FYM	Stockpiled	Composted
Crop – relative yield (%)	100	146	163	100	116	118
Soil						
Organic matter (%)	2.70	2.89	3.13	1.48	1.56	1.62
Total phosphate (mg/100 g)				70	76	85
Available phosphate (mg/100 g)	28	44	51	34	48	60
Available potash (mg/100 g)	37	70	91	260	326	356

Source: Ott (1980 a/b), Sauerlandt (1956).

Overview

The suitability of one approach or the other is determined by a wide range of factors, which include climate, soil type and cropping practices as well as the desired effects of manure applications. The issues involved can be summarised in the form of Table 4. 9. As Ott (1986) makes clear, the main issue is one of long-term versus short-term considerations, or ecology versus economics, in terms of the overall benefits. Farmers cannot, however, afford to be dogmatic and therefore individual circumstances need to be considered separately.

Table 4.9 Factors affecting the choice of composted or stockpiled farmyard manure.

	Choose composted FYM if	Choose stockpiled FYM if
(a) Objectives include	Long-term benefits Soil fertility	Short-term benefits Yield levels
(b) Specific factors include:		
Manure quantity	Surplus	Shortage
Manuring objective	Humus level	Potash requirements
Soil type	Sandy to loamy soil	Loamy to heavy soil
Crop rotation/legume proportion	N-surplus	N-shortage
Type of crop with respect to:		
vegetation length	Long (e.g. grassland)	Short (e.g. barley)
nutrient requirements	Light (e.g. cereals)	Heavy (e.g. potatoes)
nitrate risk	High (e.g. lettuce)	Low (e.g. cereals)

Source: Ott (1986) adapted.

In Britain, however, large quantities of manure are produced in straw yards, where they are allowed to accumulate over a period of time. Experience at Elm Farm Research Centre suggests that composting of stockpiled FYM from straw yards has little beneficial effect. The principal reason for this is that stockpiled FYM becomes compressed, and the building of a compost heap encourages secondary breakdown with anaerobic decomposition taking place in the large lumps. Adding straw at turning can help, but this involves more work. The case for composting has perhaps been made too strongly by some in the past, and it should only be used where appropriate for specific purposes.

Making good compost[*]

Composting, as has already been indicated, has a wide range of objectives (Table 4.10). If composted farmyard manure is appropriate to the farming system, and the full benefits of compost are to be obtained, then a knowledge of the composting process is essential. The following extract from a paper by Gray & Biddlestone (1981) describes the key biological changes which take place:

> (Figure 4.1) . . . shows the temperature and pH development during the composting process, which may be divided into four stages known as mesophilic, thermophilic, cooling-down and maturing. Initially, the mesophilic strains of microorganisms, which are present on the organic waste or in the atmosphere, start to decompose the materials; heat is given off and the temperature rises. The pH falls as organic acids are produced. Above 40°C, the thermophilic strains take over and the temperature rises to 60°C, where the fungi become deactivated. Above this temperature, the reaction is kept going by the actinomycetes and spore-forming bacteria. In this high-temperature phase, the more readily degradable substances such as sugars, starches, fats and proteins are rapidly consumed; the pH becomes alkaline as ammonia is liberated from the proteins. The reaction rate decreases as the more resistant materials are attacked; the heap then enters the cooling down phase. As the temperature falls, the thermophilic fungi reinvade the heap from the extremities and start to attack the cellulose. Later, the mesophilic strains of microorganisms reinvade. This process occurs fairly rapidly, over a few weeks. The final stage, maturing, requires several months; reactions occur in the

---

[*]The booklet *Composting*, by Pfirter *et al.* (1981), gives a more detailed, practical guide to composting and is available from Elm Farm Research Centre.

Table 4.10 Objectives of composting.

- Suppression of unpleasant odours
- Improvement of hygienic conditions
- Reduction of the germination capacity of weeds
- Maintenance and improvement of manurial value
- Increase of the biological activity of soils
- Positive influence on plant quality
- Minimum loss of nutrients during application
- Minimum additional investment expenditures
- Acceptable working conditions
- Minimum external energy requirements with regard to processing and use

Source: Vogtmann & Besson (1978).

Figure 4.1 Temperature and pH variations in a compost heap.

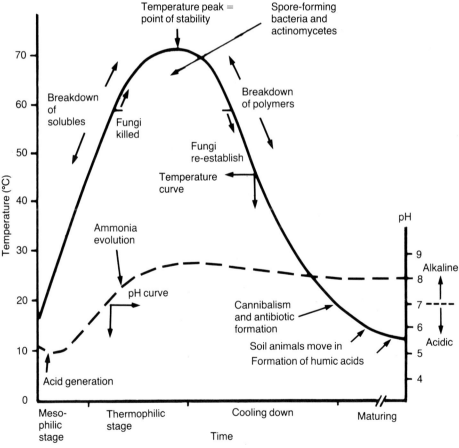

Source: Gray & Biddlestone (1981).

residual organic matter to produce the stable product of humus or humic acids. During this period, there is intense competition for food between the microorganisms: antagonism and antibiotic formation occur and the heap is invaded by macrofauna (mites, ants, worms, etc.) which contribute to the breakdown by the physical maceration of particles.

Raw materials, moisture and aeration
In order for the process as described to take place, a good compost heap needs to have the following elements:

- the right structure to allow sufficient aeration of the heap;
- the right moisture content;
- the right level of nutrients to allow the microorganisms to do their work.

The energy and nutrient balance of the material is indicated by the carbon to nitrogen (C:N) ratio. The ideal C:N ratio lies between 25 and 35:1. This is equivalent to about 7–8 kg straw per animal per day. In a cubicle system the use of 4–5 kg long or 3 kg short (chopped) straw per day provides a C:N ratio of 22:1. In a loose-housing system with economical use of straw, the C:N ratio will also be of this order. If the C:N ratio is too narrow, then there will not be sufficient carbon present for the microorganisms to make use of all the nitrogen, and losses from volatilisation will increase. Conversely, if the C:N ratio is too high, there will not be sufficient nutrients, especially nitrogen, available for the microorganisms to develop. Straw is not the only possible carbon source; the nitrogen contents and C:N ratios of a range of possible materials are listed in Table 4.11.

The moisture content is also very important. Moisture is essential for the composting process to proceed, but too much moisture will result in air being excluded and anaerobic decomposition will take over. The ideal moisture content is of the order of 55–70%. This can be tested by hand. When squeezed, if no water emerges, then the heap is too dry and needs to have water added. If water trickles out, then the heap is too wet. At optimum moisture content, water droplets should appear between the fingers, and the squeezed material should retain its form when the hand is opened. The less straw in the material, the higher the moisture content is likely to be. Moisture levels will also increase as a result of the break-down process. Under cover, moisture levels will decline over time.

The other main requirement of the compost heap is that air can get to it in sufficient quantities. A narrow, tall heap which allows air to get to the centre from the sides is preferable to a broader one which can

Table 4.11 Approximate composition of materials suitable for composting.

Material	Nitrogen (% dry-weight basis)	C : N ratio (x : 1)
Urine	15–19	0.8
Dried blood	10–14	3
Hoof and horn meal	12	–
Night soil, dung, sewage sludge	5.5–6.5	8
Bone meal	4	8
Grass	4	20
Brewers' wastes	3–5	15
Farmyard manures	2.2	14
Millet, pigeon pea stalks	0.7	70
Wheat, barley, rice straw	0.4–0.6	80–100
Coconut fibre waste	0.5	300
Fallen leaves	0.4	45
Sugar-cane trash	0.3	150
Rotted sawdust	0.2	200
Fresh sawdust	0.1	500
Paper	nil	infinity

Source: Gray & Biddlestone (1981).

become anaerobic in the centre. Holes in such heaps can, however, give a chimney effect to release moisture (bundles of branches have a similar effect), and air channels across the base of the heap may also help. The structural stability of the raw material is also important in this respect; material with low structural stability, such as vegetable residues, needs to be mixed with materials with greater stability, such as straw, woodchips and bark.

Constructing the heap or windrow
If manure becomes available at frequent intervals, such as with daily mucking-out, the heap can be built up in layers from above (using a side-unloading manure spreader). Layers of 20–40 cm are laid down every 14 days, so that each layer composts partially before the new layer is added, but unlike the 'warm manure' system, should still have access to air from the sides. If a rear unloading type spreader is used, the heap can still be built up in layers, but the height will be restricted. Another approach is first to build the heap upwards, and then to extend it sideways (Figure 4.2). This does entail the use of a muckspreader which can deposit manure to the side as a windrow. The worms and micro-organisms migrate upwards and sideways away from the heat and on to fresh manure as the older compost becomes mature.

Figure 4.2   Special windrows designed to save space and improve aeration.

Windrows in
stratified layers

Various layers 20–40 cm thick are spread
one upon another at 10–15 days interval

Windrows in
lateral layers

The principle is identical to the one at the
left, except that the layers are spread
laterally every 10–15 days

Source: Pfirter *et al.* (1981).

The composting process may be assisted by enforced aeration, where air is either blown through the heap or sucked into the heap. This speeds up the decomposition process, but is not usually practical on a farm scale. This system is only recommended for very large quantities and requires professional advice. Too much oxygen can lead to excessive nitrogen and organic matter losses.

Additives

Some practitioners advocate the use of compost 'starters' and other additives. The use of the composting process to assist with the chemical weathering of rock dusts and rock phosphate has already been mentioned and may be advantageous. The addition of clay and/or rock dusts may also contribute to the absorption of ammonia and thus the retention of nitrogen in the heap. The use of commercially available compost 'starters' on the other hand is likely to yield little benefit. There are more than adequate numbers of microbes around to get the process going so that no help should be necessary. However, if the heap is on an impermeable surface such as concrete and not directly in contact with the ground, then the use of soil or old compost as a starter may be beneficial.

Shelter

The heap needs to be sheltered from the elements. If it is not under permanent cover such as a fixed roof or a mobile polytunnel type arrangement, plastic sheeting or straw can be used to protect it. If plastic is used, the heap should only be covered after 10–14 days, once the first intensive heating-up process has taken place; the plastic should preferably be removed in dry weather to allow the heap to breathe. Straw needs to be effective at shedding rain; a good method is to use the compacted layers from unrolled round bales. A covering of soil or old compost will also help to protect the heap and has the advantage that all

Plate 4.1   Covering a compost windrow with straw helps to shed rainfall but is not fully effective

Plate 4.2   Windrows covered with plastic significantly reduce nutrient losses

Plate 4.3   Compost windrows under permanent cover

103

the material to be composted will heat up, not just the inner core, ensuring a better quality end product.

A one year trial comparing no covering, covering with straw and covering with straw and then plastic after six weeks, illustrates the importance of effective covering very clearly (Table 4.12). In spite of the low rainfall, straw on its own proved to be fairly ineffectual, but potash losses were reduced from 40% with no covering to only 8% when plastic was used.

Table 4.12 Water content (% fresh weight) and composting losses (dry matter, nitrogen and potash) as percentage of original levels with different types of covering.

	Water content	DM	N	K_2O
		reduction as % of original levels		
Without covering	82.4	46.7	21.9	40.5
Straw (round bale, 25 cm)	81.0	43.1	13.0	45.1
Straw, then plastic	75.9	39.4	7.1	8.2

Composting period: August–April (8.5 months).
Rainfall: 346 mm (low!).
Effluent: max. 180 litres/m² composting area.
Source: Gottschall & Vogtmann (1988).

Location
The best site for the composting area is within or close to the farmyard. The weight of material involved declines considerably over time, and it makes little sense transporting a high quantity of material to a field before composting, when a much smaller quantity could be transported afterwards. Composting in the farmyard also makes it easier for runoff to be trapped and utilised. A concrete base provides the best means for preventing nutrient losses into the ground, but it is expensive. If a permanent base is constructed, consideration needs to be given to the separate collection of rainwater so that the quantity of effluent needing to be stored can be minimised (Plates 4.4 and 4.5), but care should be taken to ensure that any arrangement does not interfere unduly with machinery manoeuvrability. The collection system can be linked to the silage clamps as well, so that silage effluent can also be trapped. Semi-porous concrete blocks have also proved to provide a useful base for compost windrows, reducing pollution risks while at the same time allowing aeration from underneath (Heilmann, 1989) (Figure 4.3).

Inverting the heap
It may be necessary to invert the heap after a couple of months. This helps the composting process to get going again. Machinery is available

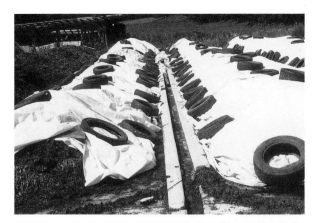

Plate 4.4 Permanent concrete base for compost windrows with separate effluent and rainwater collection and plastic covering. Note that the design of the rainwater collection channels is not ideal because they impede the movement of agricultural machinery

Plate 4.5 Tank for collection of silage and manure heap effluent

Figure 4.3 Semi-permeable base (Bioterra) for collecting effluent and permitting aeration.

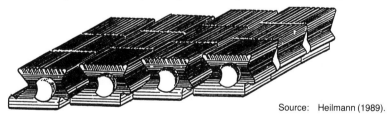

Source: Heilmann (1989).

to turn the compost (driving over the windrows, and depositing it back in the same place). Inverting the heap can also be done using a fore-end loader and a muckspreader if specialised equipment is not available.

Maturity
When the compost is ripe, usually after about 4–6 months, it can be spread on the crops which need it. A good compost should be tolerated

Plate 4.6 Specialist machinery for inverting compost windrows. The Trenkle inverter is drawn by a Unimog with a water tank mounted on the back to add water and recycle effluent in the early stages of composting when the windrows may dry out

readily by growing crops and should not interfere with root growth and development in the way which fresh manure can do. The optimum C : N ratio of finished compost is about 15 : 1. A sensitive test for compost maturity can be carried out using cress seed; if the compost is not yet mature, the cress will not germinate. It is important to ensure that compost has reached the right stage of maturity, especially for horti-cultural crops.

The pollution issue

Recently, the problems of pollution caused by agricultural wastes getting into water supplies have been attracting more attention. Other potential problems from ammonia and methane gases released from manures contributing to acid rain and the greenhouse effect have also been identified (Armstrong, 1988; Pearce, 1989). There is clearly a risk

106

that runoff from manure and compost heaps can exacerbate this problem (see Colour plate 5). It has been advocated here that where manure is stored outside, it should be kept covered, to protect it from rainfall, and it should preferably be on a water-impermeable surface with the runoff collected in a tank.

In the past few years, many organic farmers in Europe who have carefully laid out compost heaps along field edges have been threatened with prosecution, although there is no real evidence of pollution occurring. In response to this, researchers have tried to analyse what happens to the nutrients which are leached out of a heap into the soil.

One major concern is nitrate leaching, but the studies have shown that the soil underneath a heap rapidly becomes anaerobic, so that the nitrates leaching into the soil become part of a denitrification process resulting in nitrogen gas being released back to the atmosphere eventually. The penetration of nitrate into the lower soil profile has been found to be minimal, amounting to about 5 kg NO_3–N/ha on sandy soils and on a limited area of land. Also, the presence of large quantities of organic matter enhances the cation exchange capacity of the soil, allowing the other nutrients such as potash to be trapped in the soil by the clay—humus complex.

Research by the Swiss Research Institute for Biological Husbandry (Berner, 1988) found that up to 60% of the potash content of manure heaps could be lost from the heap, although this is retained in the vicinity of the heap. In order to reduce nutrient losses, the heap should contain absorbent material such as chopped straw to hold water. Relative to other sources of contamination, such as mineral fertilisers and the ploughing in of leys, the nitrate leaching from compost heaps was minimal. The risks can be further reduced by siting the heaps, if no purpose built composting area is available, on top of growing vegetation (e.g. grassland) which can take up nutrients, not on bare soil. The Swiss work shows that when this is done, the 'nutrient effect' is only present up to about one metre from the heap. Wistinghausen (1989) also concludes that the risk of nitrate leaching from compost heaps on field margins is low, so long as the site is not sloping and the effluent is not allowed to flow into surface water courses.

These results showing that nitrate leaching from compost heaps is not a significant problem are contradicted to some extent by Paffrath (1989). The nitrate leaching measured in this study was equivalent to what might be expected from land under intensive arable cropping, but with the difference that it was coming from a concentrated point source. In addition to site selection and covering of the heap, construction of the heap was also found to be important, with stacking in vertical layers to the side of a windrow resulting in greater nutrient losses than stacking in horizontal layers.

It is important to emphasise in this context the crucial difference between composting and dumping, where surface runoff from poorly stacked heaps represents a much more serious risk. In any event, the siting of manure heaps should be away from water courses and, particularly in high rainfall areas, the heaps should be covered in order to minimise the risk of surface runoff.

Spreading farmyard manures

Fresh or partially rotted FYM should be spread in the late winter or spring to avoid leaching losses and minimise the possibility of pollution. Although spreading during frosty weather can help to avoid structural damage, there is a significant risk of pollution and nutrient loss from surface runoff if semi-solid manures or slurries are spread on frozen ground. FYM should be applied at least six weeks before and immediately after cutting grassland for hay or silage, using partially rotted manure rather than fresh FYM, to allow time for it to be incorporated and avoid contamination. Mature stockpiled FYM and composts can be applied in the autumn, but should be incorporated into the soil as soon as possible.

Frequent spreading in thin layers is more useful than heavy dressings. Muckspreader design is important in achieving a thin, even spread with small-sized pieces. A manure spreader developed by the Weichel firm in West Germany with this in mind has a series of blades on rotating disks which cut the manure finely as it is passing out of the back of the spreader. The ECON side-discharge spreader available in Britain is also effective at achieving a fine, even distribution. Normal machinery produces a more uneven spread, but is better able to cope with a wider range of moisture contents.

Spreading should be done in such a way as to minimise compression by tractor wheels. The aim should also be to incorporate the manure as quickly as possible. An example of the way this can be achieved during ploughing is to have one tractor ploughing and another pulling a muckspreader on the unploughed land, but spreading the manure onto the ploughed area. The manure can then be harrowed in, thus avoiding burying it at the base of the plough layer.

SLURRY AND LIQUID MANURES

Slurry is a difficult material to manage; the provision of safe, long-term storage is expensive, so spreading is often carried out at inappropriate times of the year to make room. This has serious pollution implications

Plate 4.7   The Weichel rear-discharge muckspreader

Plate 4.8   The ECON side-discharge muckspreader

and is a waste of valuable nutrients. Slurry applications from autumn to January, when crop nitrogen uptake is small and rainfall exceeds evaporation, are generally less efficient due to denitrification or nitrate leaching. If at all possible, slurry should not be spread over the winter months.

Nitrogen availability and losses

Most of the nitrogen losses from slurry occur during spreading, although leakage from slurry stores can also result in significant wastage. About half the nitrogen in slurry is usually present in the form of ammonium nitrate or urea; the rest is in the form of undigested proteins and other organic molecules. The ammonium nitrate nitrogen is readily available for crop uptake, but the remainder is only slowly incorporated into the soil biomass, giving a slow release of nitrogen for crop growth over many months.

It is difficult to obtain a representative sample of slurry for nutrient analysis, and figures would be for total nutrients including the organically bound ones. Published figures suggest 0.5% N, 0.2% P_2O_5 and 0.5% K_2O in fresh undiluted cattle manure. The efficiency of slurry nitrogen depends on many factors, but low levels of application are likely to be better utilised than heavier spreads.

Published figures suggest that low rates of slurry application in spring may give efficiencies of 38% decreasing to 17% for high summer application rates. Summer applications are prone to losses by volatilisation; the only way to prevent this loss is by rapid incorporation into the soil, either by injection or when reseeding. Injection has been shown to increase losses by denitrification but losses are not as great as those incurred by spreading. The MAFF/ADAS Booklet 2126 covers injection techniques in more detail. It should be noted that injection in springtime can reduce yields, possibly because of the effect of toxic products of anaerobic decomposition in the root zone. There is little effect on yields from autumn applications, but soil organisms such as earthworms may still be harmed. Pastures can also be grazed sooner when slurry is injected because they are not contaminated and there is less risk of salmonella and other pathogens affecting livestock.

A cheaper alternative to slurry injection is bandspreading, where the slurry is placed in bands on the surface of the sward at approximately 400 mm intervals. Although blockage can be a problem when cattle slurry is used, there is evidence for improved slurry nitrogen utilisation and reduced smell during the spreading operation. A reduction in the smearing of grass with slurry compared with injection is also an important benefit.

Pollution risks
The pollution risks of slurry spreading, particularly at inappropriate times, are severe. The main problems arise from the flow of slurry into water courses, where the high biological oxygen demand (BOD) can be fatal to aquatic life, and where there is a risk of salmonella contamination of water supplies. Legislation exists to control this type of pollution, and the water authorities are keen to prevent it happening. The risks of pollution when spreading are greater on heavy soils, where slurry can run down fissures, than on sandy soils which can act as a filter. The legislation pertaining to agricultural pollution is summarised in the MAFF/ADAS Booklet 2077. Pollution may also result from nitrate leaching from autumn applications to stubble, and there is

increasing concern that the ammonia released on spreading could be a significant contributory factor to the acid rain problem. The other major pollution problem is the one of odour, euphemistically termed 'country air', which is a common, if avoidable, problem and perpetrators can be prosecuted under the Public Health Acts. In fact, for every 100 odour complaints which local health inspectors receive, 45 are a direct result of slurry spreading.

The pollution problem is tied in very closely with the way slurry is handled, including the way it is spread and the length of time and the manner in which it is stored. On farms where animal manures accumulate in the form of slurry, storage for periods of 4–6 months or more is desirable if pollution risks from spreading at the wrong time are to be avoided.

Slurry storage

The various types of straightforward storage of slurry in tanks, lagoons, etc., are more than adequately described in conventional sources such as MAFF/ADAS Booklets 2077, 2081, 2356 and 2273, and Grundey (1980).

Storage in above ground tanks is environmentally preferable to lagoon or compound storage, although serious problems can still occur. These include the formation of a thick crust, reducing tank capacity, allowing weed growth on the surface and creating difficulties in the pumping and spreading of slurry onto fields. In addition, the presence of the crust creates anaerobic conditions in the slurry resulting in high levels of ammonia and other compounds being released at spreading time. This in turn leads to scorching of plants and killing of earthworms, and contributes to the offensive odour. In addition, there is a risk to agricultural workers and their families from the potentially toxic gases such as ammonia, hydrogen sulphide, carbon dioxide and methane produced by uncontrolled anaerobic decomposition.

In the absence of aeration, the aim with slurry storage should be the reduction of nutrient losses through leaching, volatilisation or denitrification. The ideal situation is one where all the straw used is incorporated into the slurry, improving the $C:N$ ratio. A high $C:N$ ratio provides the energy needed by bacteria to incorporate the nitrogen as bacterial protein. Storage in a tank will also help to reduce N losses, largely because a carbon dioxide layer on top helps to hold the nitrogen in. This can be further helped by covering the slurry store; suitable covers are gradually becoming commercially available. A cover has the added advantage of keeping rainfall out.

Runoff from lagoons should be prevented under all circumstances because of the pollution danger. The storage space problem can be eased considerably if washings and dirty water from concrete yards used by animals are minimised, and water, especially rain water, from clean areas is collected and disposed of separately.

Plate 4.9   Poor slurry storage in an overfull lagoon contributes to slurry being one of the most significant sources of agricultural pollution

Slurry aeration

Research has clearly demonstrated the potential of slurry aeration for reducing odours, reducing disruption to soil life and crops, and reducing weed and pathogen survival in the slurry itself. At the same time, it increases the value of the slurry as a fertiliser.

Aeration reduces nitrogen losses at the time of spreading since much of the nitrogen is in the form of bacterial protein rather than ammonium compounds. The proportion of total nitrogen originally present which is lost during aeration of slurry has been estimated at less than 30%, so that assuming negligible loss at spreading, a fairly high efficiency of nitrogen cycling to crop plants can be achieved. Elm Farm Research Centre found significant increases in nitrogen availability through simply reducing the crust and keeping the material in a spreadable form. When slurry is not aerated, nitrate tends to be denitrified in the anaerobic conditions and considerable quantities of nitrogen are lost as odourless, gaseous nitrogen. Temperatures above 35°C increase nitrogen losses considerably. Temperatures between 18 and 20°C are more desirable: it is not really necessary to go for higher temperatures and excessive aeration should be avoided.

Without aeration, anaerobic decomposition takes place, forming noxious chemical compounds such as butyric acid which, along with ammonia, are responsible for the bad odours and damage to crops and soil fauna. Slurry aeration is therefore strongly recommended to organic farmers in central Europe and is widely employed. The Soil Association's standards now require that all brought-in slurries are aerated, and the treatment is recommended for slurry produced on the farm itself. However, it is carried out in smaller tanks and with less slurry in other parts of Europe compared with Britain.

Research into slurry aeration has taken place at various AFRC and Scottish Research Institutes, including the Institute for Engineering Research at Silsoe. Various techniques exist, all of which are variants of either batch or continuous flow processes. Generally, batch treatments are regarded as less efficient because the demand for oxygen is more variable. With continuous flow methods, oxygen and slurry are fed into the vessel or tank at rates designed to match supply and demand for oxygen. With this system, the AFRC suggests that one day's aeration should be adequate for up to two weeks' storage before spreading.

Most storage systems, however, are not set up for continuous flow treatment, and types of batch treatment are more commonly used. These are of two basic types; forced-air systems which effectively pump air into the tank, but do not agitate, and mechanical propellor type aerators which do. Forced aeration systems are usually better suited to very liquid materials and in most practical situations, therefore, are not as successful as mechanical propellors, but have the advantage of being less noisy. The latter come in two forms: 'surface' aerators and 'deep' aerators. Both throw liquid into the air at the same time forcing air into the liquid. The deep aerator has some form of tube so that the rotor draws material from the bottom of the tank up past the blades and out over the top of the aerator. Surface aerators work just underneath the surface of the liquid, but are not as efficient at drawing up liquid from the bottom. Deep aerators usually have to be built in and rigid, whereas surface aerators can be either rigid or floating, kept in position by chains (Figure 4.4).

Details of the main types of aeration system are contained in various books on livestock waste management such as Grundey (1980) and Hobson & Robertson (1977). The effectiveness of the aeration process is dependent on the rate at which oxygen can be added, and the ratio between the surface area and the volume of air being added. Small bubbles have a higher surface area per volume of air and are therefore more effective. They do not escape so easily, particularly if sucked in rather than blown through, and this reduces the rate at which ammonia is released. The power level is also crucial; the rating of aerators is measured in terms of weight of oxygen added and should be in the order of at least 1, preferably 2, kg oxygen per kWh in order to provide sufficient oxygen to aerate the whole store.

From the various studies which have been done, it would appear that

Figure 4.4 Modes of action of two types of surface aerator: floating surface (above) and sub-surface down draught (below).

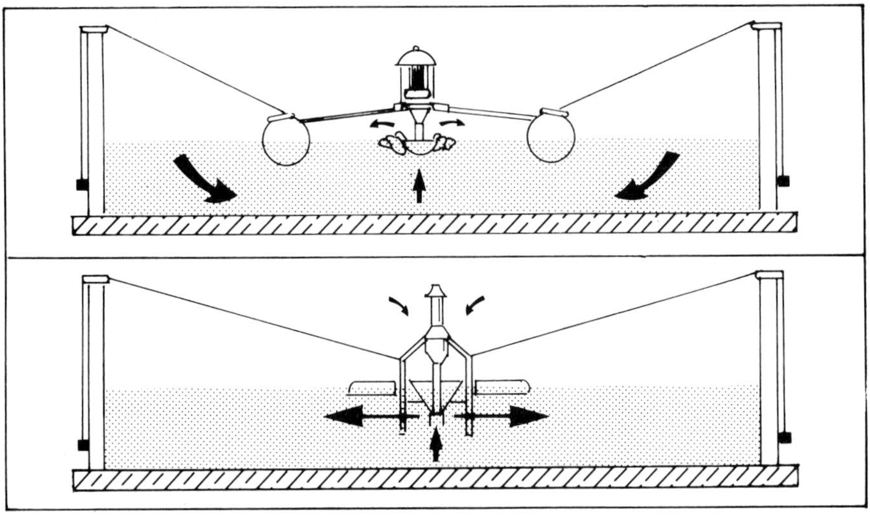

Source: AID paper 1201 (1988) Umweltgerechte Flüssigmistbehandlung. Auswertungs und Informationsdienst; Bonn.

aeration for periods of one hour four times daily is sufficient to achieve the benefits in terms of odour reduction, nitrogen conversion and weed and pathogen control. Brief aeration to start with allows the microflora to adapt to the new conditions. The frequency can then be increased for two or three weeks, after which it should be reduced again.

During this time, the carbon reserves are used up. If the slurry is aerated too much, there will be insufficient nutrients available and the nitrogen bound in bacteria will start to decay. Continual aeration over long periods can lead to a 30% reduction in organic matter and up to 90% loss of nitrogen. Nitrogen losses may also be exacerbated by the increase in pH which takes place, encouraging volatilisation. With intermittent aeration, the carbon levels can be maintained without the bacterial flora suffering unduly. Too long intervals, however, lead to anaerobic conditions again, the build up of toxic materials, and the denitrification to nitrogen gas of nitrate formed from the ammonia in the initial stages of aeration.

Carbon sources can be added, but to be effective, straw must be very finely chopped so as to increase the surface area. Sawdust is another potential carbon source which can be used. The addition of clay powders, such as bentonite, may help to reduce the amount of ammonia released. For aeration to be effective, the slurry also needs to be diluted,

Plate 4.10   Aeration of cattle slurry in an above ground tank   (Elm Farm Research Centre)

although as aeration progresses and organic matter is used up, the mixture will become more dilute anyway.

Weed and pathogen control (which may allow cattle to graze sooner where slurry is spread on grassland) is closely linked to the temperature build-up, caused, as in composting, by the increased microbial activity when oxygen is added. At 8°C, 85% of weed seeds will still germinate. This falls to 51% at 19°C and to zero at 28°C. Two weeks at these temperatures will kill most weed seeds, although wild oats and tomatoes appear to be more resistant.

There is also evidence of other benefits of aeration, including the possibility that phosphate availability may be increased by encouraging microbial activity. Aeration also helps to reduce the pollution risk by bringing about a substantial reduction in biological oxygen demand, although this will be greater in liquid manures and separated slurry liquids than in straight slurry.

In spite of the advantages, the cost of slurry aeration systems currently available is very high, and it is difficult to make a purely economic case for them. One recent German study, however, has shown that the costs of aeration may be covered both by fertiliser saved as well as by the additional yield benefits from using aerated as against unaerated slurry. A decision to install such equipment is likely to be taken mainly on the basis of environmental and husbandry considerations, rather than any immediate prospect of financial returns. One inexpensive, farmer-developed option is the use of a PVC pipe perforated with 0.05 mm holes coiled on the base of the slurry store. A compressor feeds air through the pipe, but it is not known whether the crust can be successfully controlled using this system.

Slurry separation and liquid manures

Separation techniques for the handling of slurry also have potential benefits. Mechanical separation of slurry prior to storage provides a stackable, fibrous solid and a free flowing liquid. Around 80% of the weight and nutrients are retained in the liquid, crust formation is minimal and aeration is made easier.

There are two main types of separation technique: mechanical separation using a fine mesh screen; or seepage through a coarse filter such as a slurry lagoon strainer or through the walls of a weeping-wall compound. The liquid and solid fractions from a mechanical separator both contain similar concentrations of plant nutrients to the slurry from which they were obtained, except that, in some cases, the phosphate content of the solids and the potash content of the liquids may be higher than in raw slurry. In the case of in-store separation, experience suggests that about 10% of the volume of slurry entering the store, plus rainfall, will drain off over an extended period. This liquid, because of rainwater dilution, tends to be lower in plant nutrients than the liquid produced by a mechanical separator. The liquid separated is still a serious pollutant with high biological oxygen demands, but is suitable for land spreading and contains approximately 1.5 kg N, 0.3 kg phosphate and 30.0 kg potash per 1,000 litres.

Urine from cattle can also be collected separately from the faeces, possibly also with quantities of rainwater or washings, and stored for use in diluted form as a readily available nitrogen source on crops like maize or barley. This is seldom done on organic farms in Britain, but the practice is more common in mainland Europe.

The principles of separate collection of liquid manure can also be applied to the collection of silage and/or manure effluents. Liquid manures contain about 5 kg nitrogen and 5.5 kg potash per 1,000 litres; the phosphate content, however, is negligible. The nitrogen is easily lost through volatilisation, especially if uncovered at high temperatures. A layer of chopped straw, possibly also fine compost, on top of the tank can act as a seal and provides additional organic matter which may contribute to binding ammonia. High losses can also occur on spreading. The best time for application of liquid manures is in early spring when their main role is as a direct plant feed. MAFF recommends that applications should be limited to 20,000 litres/ha on grassland, or 50,000 litres/ha if applied in increments during the season. Trials have shown that the use of liquid manures can have favourable effects on the botanical composition of grassland and will stimulate clover growth, even when used at high rates of nitrogen. This may be due to the potash content and the absence of large amounts of chloride. There is, however, a risk of hypomagnesaemia if used in large quantities on grassland.

Spreading slurry and liquid manures

A range of systems exists for spreading slurry, the main considerations being the need to incorporate the material as quickly as possible to avoid losses through volatilisation and the need to avoid compaction from heavy slurry tankers in unsuitable conditions. Injection systems which place the slurry into the soil are the most effective for reducing volatilisation losses and come in several forms (see MAFF/ADAS Booklet 2126).

New spreading techniques to avoid taking heavy slurry tankers on to the field are being developed. These include the use of 'umbilical' systems, where a long PVC hose links the slurry tanker with a spreader mounted on a separate tractor (Figure 4.5). This is designed to allow slurry application to row crops without damage to the growing plants. Ideas like this are often still experimental but are likely to have wider application in the future.

Figure 4.5   Umbilical system for slurry spreading.

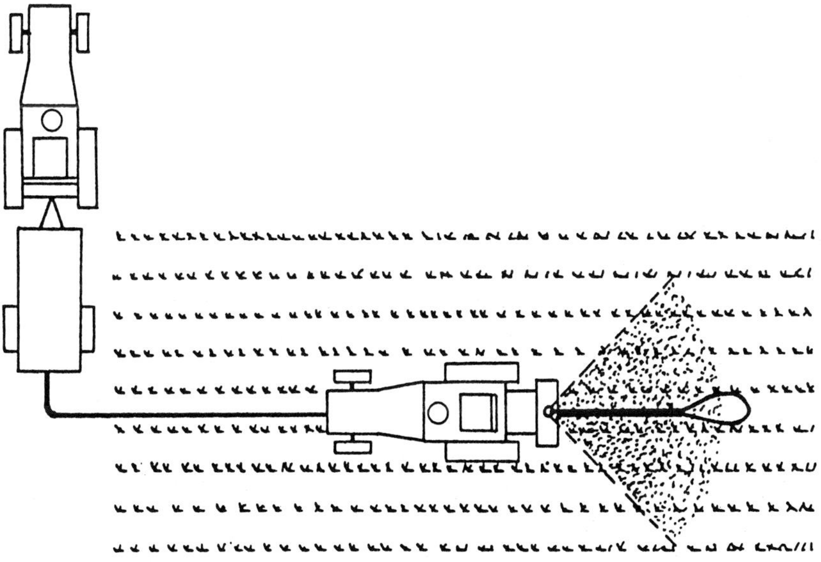

Source: Zum Beispiel, 1/89 p.16.

Straw/slurry composts

In arable areas where straw is plentiful, compost production using slurry and straw provides an alternative way of managing slurry. This is often a highly controlled, almost industrial process, involving innoculation with worms which are sometimes recovered and used as a protein feed for livestock. But it need not be so complicated and most farms, if they have a suitable composting area, could use this approach.

Typical quantities are 20 tonnes of straw to 100,000 litres of slurry, or in some cases sewage sludge. A 'starter' such as old compost or chicken manure may also be helpful. In a German trial, it was found that barley straw decomposed more easily, with a more rapid temperature increase, than wheat straw, although the latter gave better results in germination tests. The longer the maturation period, however, the less significant the differences became.

Anaerobic digestors (methane gas production)

An alternative approach to slurry treatment, referred to as anoxic, is the production of methane under conditions in which all oxygen is deliberately excluded and at temperatures around 35°C. This approach is also known as biogas production, and is carried out in methane digestors.

There are few working examples on farms, but research on these is being conducted by the Institute of Engineering Research and the Institute for Grassland and Environmental Research. The digested sludge is a stabilised organic fertiliser with most of the N, P and K of the original slurry. It has little odour and can be used in the same way as slurry. The organic matter level declines rapidly, with less than 60% left after 30 days. This means that the remaining dry matter has a greater concentration of nutrients than the original feedstock.

The increased nitrogen concentration is largely in the form of ammonia rather than organically bound nitrogen; in fact the levels of organically bound nitrogen are lower than before the process started. The low C:N ratio of the end product means that it makes little contribution to long-term soil organic matter and is therefore more akin to mineral fertilisers containing readily soluble nutrients. The digested sludge cannot usually be spread immediately and therefore requires further storage with the risk of high nitrogen losses through volatilisation. When it is spread, there may be further high losses of ammonia. The end product may not, therefore, be very suited for use in organic systems.

The economics of the process are dependent on maximising gas yields. Various gases are produced, which include methane (about 70%) and contaminants such as carbon dioxide and hydrogen sulphide. These need to be removed, and cleaning processes exist to purify the gases. A new process relevant to developing countries involves bubbling the gas through water, allowing the carbon dioxide to dissolve in it, and then using this water as a feed for a type of blue-green algae called *Spirulina*. Nitrogen is also supplied from the fermenter, with the result that the combined process provides both energy (methane) and protein (the algae) which can be fed to fish or even humans.

Brought-in Organic Manures

Sewage sludge

In an ideal world, it would be possible to recycle all human wastes and thus help to stem the loss of nutrients from the land. The use of sewage sludge would appear initially to be a step in the right direction. Sludges contain nitrogen and phosphate, but little potash; most of the potash and some of the nitrogen is lost in the liquid during separation (Table 4.13). The organic matter can make a useful contribution to soil improvement, although this may not compensate fully for the problems of soil compaction caused by heavy tankers.

Table 4.13 Fertiliser value of sewage sludges (approximate weight (kg) of available nutrients per tonne or fresh material).

	Nitrogen	Phosphate	Potash
Raw sludge – air dried	3.2	3.0	1.8
Digested sludge – air dried	3.6	5.0	0.9
Liquid digested sludge	2.0	0.8	0.1

Source: MAFF/ADAS Bulletin 210.

There is, however, still considerable concern about the levels of heavy metals of industrial origin which can be found in sewage sludge (Long, 1989; Giller & McGrath, 1989). These are not only potentially harmful to crops, livestock and humans, but they also have a serious effect on soil microbiology and a growing body of evidence suggests that the activity of nitrogen fixing bacteria may be severely restricted where sewage sludges have been applied in large quantities.

In particular, tests at Rothamsted (Brooks & McGrath, 1984) found that in plots which had received sewage sludge at high levels (albeit levels no longer currently acceptable) more than 20 years previously, the total biomass of soil organisms was generally about one half and sometimes only a third of the normal level. Since the biomass and protozoan populations were shown to be similar, the difference in biomass is presumed to show a big difference in fungal populations. The soils were also shown to be poor natural fixers of nitrogen, suggesting that blue-green algae were also severely affected. Heavy metal concentrations remain over long periods of time, possibly thousands of years, largely because the only way they are removed is through the harvesting of crops and hence into the food chain.

Because of the dangers from contamination with diseases such as salmonella, parasites such as beef tapeworm and the accumulation of toxic heavy metals in soils, strict limits on the use of sewage sludge are set by the Department of the Environment and the water authorities. These are based on recommended maximum soil concentrations and the total quantities of sewage sludge applied over a period of 30 years. The limits are summarised in Table 4.14, including a comparison with Soil Association and EC standards (see also Table 4.15). Full details of the MAFF recommendations and restrictions are contained in Booklet 2409, which should be consulted before any sewage sludge is applied.

If sewage sludge is used, then it must be used with extreme caution. For the future, it can only be hoped that sensible sewage systems will be developed so that human faeces and wastes can be kept free of industrial contaminants.

Other organic and mineral fertilisers

Purchased organic and mineral fertilisers should be regarded as supplements to nutrient recycling within the farm, not as a straight alternative to conventional fertilisers. In a balanced organic system, the purchase of organic manures onto the farm should be the exception, not the rule. Where they are purchased, generally only organic or mineral fertilisers that release nutrients through an intermediate process, such as weathering or the activity of soil organisms, are allowed. In certain very restricted cases, where serious deficiencies exist, soluble mineral fertilisers, e.g. potassium salts (but not chloride salts), are allowed as a remedy, but not for routine use.

Most organic fertilisers, such as dried blood, effectively act as straight fertilisers, feeding the plant directly and therefore their use should be restricted. Examples of the nutrient values of different organic fertilisers include: hoof and horn (7–16% N), meat and bone meal (5–10% N, 18% P_2O_5), fishmeal (6–10% N, 5–9% P_2O_5), dried blood (up to 13% N), shoddy (2–15% N), castor meal (5–6% P_2O_5). Also, by-products of industry such as brewery wastes contain low levels of N and possibly also phosphate. Calcified seaweed is sometimes used; it may be useful

Table 4.14 Soil concentrations and limits of application of elements to safeguard human or animal health.

Element	Soil Assoc.	Maximum soil concentration (mg/kg) United Kingdom	European Community	West Germany
Zinc (Zn)	150	300	300	300
Copper (Cu)	50	135	140	100
Nickel (Ni)	50	75	75	50
Cadmium (Cd)	2	3	3	3
Lead (Pb) (a)	100	300	300	100
Mercury (Hg)	1	1	1.5	2
Chromium (Cr)	150	N/A	N/A	100

Element	Soil Assoc. (mg/kg) (b)	Limits of application UK DoE (kg/ha) (c)	EC (mg/kg)
Zinc (Zn)	1,000	560	2,500–4,000
Copper (Cu)	400	280	1,000–1,750
Nickel (Ni)	100	70	300–400
Cadmium (Cd)	10	5	20–40
Lead (Pb) (a)	250	1,000	750–1,200
Mercury (Hg)	2	2	N/A
Chromium (Cr)	–	1,000	N/A

(a) Specific limits apply for lead where crop may be utilised by humans.
(b) Soil Association standards limit only the heavy metal concentration of the material applied, but prohibit applications more than one year in three and also prohibit use on all crops for human consumption.
(c) Department of Environment (1977). This is the total limit for applications over thirty years. Not more than one-fifth is to be applied in any period of six years.
Source: Soil Association; Giller & McGrath (1989).

source of potash and trace elements. The cytokinins contained in some seaweeds may also have a growth promoting effect and it has been claimed that foliar applications of seaweed extracts can result in better Hagbergs and higher protein contents in wheat.

For most of these products, the costs are not often justified by the benefits, except on horticultural holdings. Full details of acceptable products are contained in the Soil Association's standards in Appendix 1.

Composting household wastes

Another option for improving the return of nutrients to the land is the composting of household wastes. In several towns in Europe, organic household refuse is collected separately and composted by adding it to straw or other similar carbon containing materials. The finished compost has been shown to be of high quality and low in heavy metal contaminants. The possibility exists for these composts to be sold to gardeners or farmers, and the result may be financially attractive to hard-pressed local authorities. Such composts, where the organic waste is collected separately at source, are not the same as traditional municipal composts where all the wastes are composted together. The latter have much higher levels of heavy metal contamination, as can be seen from Table 4.15.

Table 4.15 Comparison of heavy metals content of Witzenhausen compost compared with traditional municipal composts (mg/kg DM).

	Witzenhausen compost	Municipal waste compost
Lead	86	513
Chromium	28	71.4
Copper	40	274
Cadmium	0.5	5.5
Mercury	0.17	2.4
Nickel	17	44.9
Zinc	255	1,570

Source: Fricke (1988).

In Witzenhausen in West Germany, all households are now involved with the separate collection of organic wastes, with the result that contamination is negligible (Fricke, 1988; Gottschall & Vogtmann, 1988). Separate collection also means that the remaining material can be mechanically separated, increasing the potential for recycling.

Trials have also been conducted on an organic farm in Kent (Oehlschlaegel & Lopez-Real, 1986) to see if a similar approach can be adopted in Britain. In the long run, all methods of returning nutrients to the soil need to be examined if agriculture is to be in any way sustainable.

References and Further Reading

General
Grundey, K. (1980) *Tackling Farm Waste*. Farming Press
Hobson, P. N. & Robertson, A. M. (1977) *Waste Treatment in Agriculture*. Applied Science Publishers
MAFF/ADAS *General information on farm waste management*. Booklet 2077
MAFF/ADAS *Profitable utilisation of livestock manures*. Booklet 2081, 1986. New edition with revised data due 1990/91.
MAFF/ADAS *Equipment for handling farmyard manures and slurry*. Booklet 2126
MAFF/ADAS *Storage of farm manures and slurries*. Booklet 2273

MAFF/ADAS *Slurry handling: useful facts and figures.* Booklet 2356
MAFF/ADAS *The use of sewage sludge on agricultural land.* Booklet 2409
MAFF/ADAS *Organic manures.* Bulletin 210. HMSO
Soil Association (1981) *Use of manures and mineral fertilisers in organic agriculture.* Technical booklet no. 4
Steele, J. (1989) *Muck to methane.* Farmers Weekly 20/10/89: 63–66

Composting
Bertoldi, M. *et al.* (1986) *Compost: Production, Quality and Use.* Elsevier
Gottschall, R. (1984) *Kompostierung.* Alternative Konzepte 45. C. F. Müller Verlag, Karlsruhe
Gottschall, R. & Vogtmann, H. (1988) *Bedeutung und Verwertungsmöglichkeiten von Kompost in den Grünen Bereichen.* IFOAM Sonderausgabe Nr. 24. Stiftung Ökologischer Landbau; Kaiserslautern, West Germany
Gray, K. R. & Biddlestone, A. J. (1981) *The composting of agricultural wastes.* In Stonehouse, B. (ed.) *Biological Husbandry—a scientific approach to organic farming.* Butterworths
Gray, K. *et al.* (1987) *Soil Management: Compost production and use in tropical and sub-tropical environments.* FAO Soils Bulletin 56. Food and Agriculture Organisation of the United Nations.
Heilmann, H. (1989) *Eine Rotteplatte kann die Mist-und Kompostflege unterstützen.* Lebendige Erde 1/89: 16–21
Kitto, D. (1988) *Composting: the organic, natural way.* Sterling, New York
Ott, P. (1990) *The composting of farmyard manure with mineral additives and under forced aeration and the utilisation of FYM and FYM compost in crop production.* PhD Thesis, University of Kassel, West Germany.
Pfirter, A., Hirschheydt, A. v., Ott, P. & Vogtmann, H. (1981) *Composting—an introduction to the rational use of organic waste.* Migros; Aargau, Switzerland
Poincelot, R. P. (1972) *The biochemistry and methodology of composting.* Connecticut Agricultural Experiment Station Bulletin 727
Sauerlandt, W. (1956) *Stallmistkompostierung.* Reihe Landwirtschaft—angew. Wissenschaft No. 57
Vogtmann, H. & Besson, J. M. (1978) *European Composting Methods: Treatment and Use of Farm Yard Manure and Slurry.* Compost Science/Land Utilisation. Reprinted in: EFRC (1980) *The Research Needs of Biological Agriculture in Great Britain.* Elm Farm Research Centre Report No. 1

Manuring strategies
Ott, P. (1980a) *A comparison of raw and composted manure from beef feedlots in the Mid-West States of the USA.* In: EFRC (1980) *The Research Needs of Biological Agriculture in Great Britain.* Elm Farm Research Centre Report No. 1
Ott, P. (1980b) *Utilisation of composted feedlot manures in the Mid-West (USA).* In: Hill, S. & Ott, P. (eds) *Basic Technics in Ecological Farming.* IFOAM conference proceedings, Birkhäuser Verlag
Ott, P. (1986) *Utilisation of farmyard manure and composted farmyard manure—a manuring strategy.* In: Vogtmann, H. *et al.* (eds) *The Importance of Biological Agriculture in a World of Diminishing Resources.* Verlagsgruppe Witzenhausen

Livestock manures and pollution
Armstrong, S. (1988) *Marooned in a mountain of manure.* New Scientist 26/11/88: 51–55
Berner, A. (1988) *Einfluss des Sickersaftes von Kompostmieten auf die Umwelt.* Forschungsinstitut für biologischen Landbau, Oberwil, Switzerland.

Brookes, P. C. & McGrath, S. P. (1984) *Effects of metal toxicity on the size of the soil biomass.* Journal of Soil Science 35: 341–346

Giller, K. & McGrath, S. (1989) *Muck, metals and microbes.* New Scientist 4/11/89: 31–32.

Long, E. (1989) *Metal that spoils all soils.* Farmers Weekly 26/5/9: 56–57

Paffrath, A. (1989) *Mistkompostmieten—eine Gefährdung für das Grundwasser?* Lebendige Erde 2/89: 86–91

Pearce, F. (1989) *Methane: the hidden greenhouse gas.* New Scientist 6/5/89: 37–41

Wistinghausen, E. von (1989) *Untersuchungen zur Belastungen von Grundwasser durch Nitrataustrag aus Stallmistkompost auf unbefestigtem Boden.* Lebendige Erde 5/89: 342–354.

Composting household wastes

Fricke, K. (1988) *Grundlagen zur Biobafallkompostierung.* PhD Thesis, University of Kassel, West Germany

Oehlschlaegel, U. & Lopez-Real, J. (1986) *Composting—a recycling role for organic farmers in society.* Soil Association Review, December: 16–19.

Vogtmann, H., Fricke, K., Schüler, C. (1989) *Nutrient recycling through off-farm organic wastes.* IFOAM English language Bulletin 9: 3–7.

Recent publications

MAFF (1991) *Code of good agricultural practice for the protection of water.* Ministry of Agriculture, Fisheries and Food, London.

Chapter 5

Rotation Design for Organic Systems

The use of rotations has fallen in popularity in recent years due to the development of techniques which have allowed strict rotations to be abandoned and simpler cropping systems to be adopted. The development of biocides to control weeds, pests and diseases, as well as the easy availability of readily soluble fertilisers, means that many of the constraints on cropping which rotations helped to alleviate no longer exist. Much of the technology which has enabled this shift in husbandry practices to take place, however, has undesirable long- and short-term consequences, both for agriculture and the environment. The development of specific new weed problems, the increasing resistance of many weeds, pests and diseases to chemical control, and the dangers of chemical pollution of the environment are all major reasons which lead organic farmers to reject the use of biocides as part of their farming system.

The decision not to use these aids cannot be taken in isolation. The cropping practices currently used in many parts of the country are heavily dependent on the use of agrochemicals if yields are to be maintained. Abandoning their use without also changing cropping practices can only result in a severe breakdown of the system and serious yield losses. For this reason, the issue of rotation design has to be foremost in the mind of the farmer when considering how to convert to an organic system, or when farming organically, if production is to be maintained and the viability of the farm is not to be endangered.

The Historical Background

Rotations formed the traditional base for agriculture in Britain and mainland Europe for many centuries, dating back to Roman times and beyond. Ever since population pressures brought about a settled agriculture and an end to systems of shifting cultivation, where an area was cropped until exhausted and then abandoned, some form of rotation has been practised.

In Britain, the three course rotation of autumn corn, spring corn and fallow held sway for more than 1,500 years. The growth of the use of roots and clover as farm crops in the eighteenth century was the first major impetus towards change. Fallowing could be omitted altogether on lighter soils and the extent of the fallow could be reduced on the heavier soils. The best known example of the change which took place in the eighteenth century is the Norfolk four course rotation, which originally took the form of:

Roots—Barley—Seed—Wheat

where the seed part of the rotation was generally some form of legume (mainly red clover, sometimes with ryegrass; grain legumes and crops such as sainfoin and trefoil were also used). Sheep folded on the arable crop and beef over-wintered in yards were very important elements of the system.

Economic pressures and increasing demand for other agricultural products brought an end to the Norfolk four course rotation at the beginning of this century. In many cases, it was abandoned completely; in others it was modified by the introduction of additional courses (with an emphasis on high-value crops such as sugar beet, potatoes and cereals) or existing crops such as turnips and swedes were replaced by more profitable alternatives. In some cases, six course rotations were adopted with the form:

Roots—Barley—Seeds—Potatoes—Wheat—Oats

Modifications of this approach were made depending on the area of the country in which the system was being operated and, particularly, the soil and climatic conditions.

The Norfolk four course rotation and its derivatives were most appropriate for the drier, eastern parts of the country. Further west, the one-year seeds ley was extended into a short-term two-year ley, or into the medium- to long-term leys which form the basis of alternate husbandry. In fact, many of the rotations currently used by organic farmers in Britain are of this type.

The presence of a ley in the rotation allows soil fertility to be restored, particularly in terms of organic matter and nitrogen. In addition, the activities of earthworms in combination with grass roots and other biological activity in the soil has been shown to have beneficial effects on soil structure. Much of the damage caused by mechanical cultivation and arable cropping can be repaired in this way. Ley-farming was popularised largely as a result of the efforts of agriculturalists such as Stapledon and Elliot and it could be argued that this particular approach became a religion for many devotees.

It is illuminating to compare the development of rotations in Britain with the corresponding developments in central Europe. The influence of Roman culture created the same three course rotation (the 'old three-field cropping' system) which became widespread in Britain: winter cereals, spring cereals and fallow. Again, it was only towards the end of the eighteenth century that new crops were introduced enabling the development of the 'improved three-field cropping' system, involving a much more intensive use of the land and livestock.

The basis of this improved three course rotation is the principle of two thirds straw and one third leaf crops together with the abandonment of the fallow. An example of this is the following nine course rotation which still retains the basic three course idea:

1. Red clover
2. Oats
3. Winter wheat

4. Potatoes
5. Winter wheat
6. Winter rye

7. Fodder beet
8. Spring barley
9. Winter rye (with
    clover undersown)

Even in this context, concern was expressed about the effects on soil fertility and weeds of having similar crops (i.e. cereals) succeeding each other. The concept of alternating the type of crop each year, for example straw crop/leaf crop, and where possible including catch cropping, attracted interest towards the end of the nineteenth century, but never became widespread because of high labour and capital costs. The following example illustrates the concept (50% leaf/50% straw):

1. Red clover
2. Winter cereal
3. Pulses, rape or roots

4. Winter cereal
5. Roots
6. Spring cereal (undersown)

More recently, rotations consisting of two years leaf crops and two years straw crops developed, allowing certain benefits which were not possible when crop types were alternated annually. The different direction which the development of rotations took in central Europe, compared with Britain, has resulted in cropping systems which are still far more complex than we are used to, with a greater emphasis on catch cropping, green manuring and short-term leys. The ley-farming, alternate husbandry approach is found much less frequently, although similar systems do exist and have existed over long periods of time in some of the wetter areas less favourable to arable cropping.

As in Britain, these developments allowed intensification and in-creases in productivity. The importation of fertilisers into the farm system and later the development of improved machinery allowed further advances to be made, but were also accompanied by increasing economic pressures, particularly in terms of labour and capital costs. These economic factors were the major causes of specialisation; i.e. concentration on only a few crop or livestock enterprises. In this way, economies of scale could be achieved, allowing incomes to be maintained.

The 'freedom' which modern technological developments have given to farmers has resulted in a move away from the concept of a strict rotation to the idea of cropping systems and monocultures with perhaps an occasional break crop. This progression, particularly in the arable areas of eastern Britain, has brought with it further increases in yields and profitability of arable systems, but not without agricultural and environmental consequences some of which are no longer socially acceptable and may even pose a threat to the long-term future of agriculture.

THE IMPORTANCE OF ROTATIONS

Soil sickness and soil fertility

The development of rotations was spurred on by the realisation that crops grown in a rotation normally yielded more than if the same crop was grown continuously for a period of years. The increase in yield would often compensate for a reduction in the frequency of occurrence of the crop in the rotation, and the effect would be one of increased overall cropping efficiency. This 'rotation effect' has long been recognised by practising farmers, but an understanding of the causes has taken longer to evolve and some aspects are still poorly understood, if at all.

Seen from a slightly different perspective, the decline in yields resulting from monocultures and poorly designed rotations has been attributed to a general condition sometimes referred to as 'soil sickness'. This is a situation where the growth and development of the crop is hindered, with increased incidence of plant diseases and pest damage as a result of reduced crop resistance, resulting in lower yields. Soil sickness was initially thought to be caused by some toxic substance which reduced the self-tolerance of certain crops; i.e. caused a loss of yield as a result of growing a certain crop more than one year in succession. The discovery of the importance of individual plant nutrients by the German scientist Liebig and others, and the development of the concept of limiting factors, particularly nitrogen and water, provided a partial explanation for the phenomenon, but not the complete answer.

Soil sickness is now regarded as a general concept which has several components, the most obvious of which are nutrient deficiencies and imbalanced nutrition. From the discussion of nutrient cycling in Chapter 3, it is clear that without the addition of fertilisers from an external source certain crops will remove large quantities of nutrients from the soil. Continuous growing of these crops will result in a reduction of nutrient availability in the soil and a fall in the yield of subsequent crops. Similar considerations apply to organic matter and the carbon cycle. At the same time, other crops are able to add to the levels of nitrogen and organic matter in the soil and, to some extent at least, can restore the balance. A distinction can thus be made between crops which contribute to soil nutrient and organic matter status, those which are neutral and those which cause a decline in the level of these soil constituents.

Less obvious are the effects of crop residues on the subsequent crop. Plants produce a range of toxic substances, which are released in significant quanitities when crop residues are incorporated. Many of these are antibiotic in behaviour and therefore block the activities of soil microbes. This reduces the biological 'buffer capacity' of the soil and can, in certain situations, allow soilborne pests and diseases to gain a foothold where they might otherwise have been kept under control by other soil organisms. Toxic substances may also be produced, as a result of microbial action on crop residues, which inhibit the germination and growth of the new crop.

The other main aspect of soil sickness is the toxic root exudates which are specific to the crop species and which create an allelopathic effect, i.e. have an inhibiting effect on the growth and development of other crops. The question of allelopathic effects is of considerable interest in terms of weed control, but the negative effects on subsequent crops need also to be recognised. Even a small weakening effect on the following crop can increase its susceptibility to pathogens, so that yield losses still occur, albeit indirectly.

One aim of the rotation must therefore be to build up high natural resistance to pests and diseases in the soil. This can best be achieved by promoting a high level of biological activity, in particular the microflora, so that harmful pathogens can be neutralised. This natural resistance and high biological activity is often reduced by the use of biocides and readily soluble fertilisers, creating a situation where problems are actually caused by the farming practices adopted.

Weeds, pests and diseases

Rotations are the principal means for controlling weeds, pests and diseases in an organic system. They allow for different types of cultivation to take place at different times of the year, so that no one weed species can become dominant. The same is true for the interactions between crop plants and weeds: certain crops have a weed suppressant effect (either by direct competition or by allelopathic interactions) while others are less able to compete successfully. The rotation provides the opportunity to alternate these types of crop so that the overall effect is one of minimising weed problems.

Pest and disease carry-over from one crop to the next can be reduced, as can the survival of soilborne pests and diseases, if the host crops are alternated with non-host crops. Potato cyst nematodes, wireworms, club-root, cereal cyst nematodes and take-all and blackleg in cereals are all examples of pest and disease problems which can be controlled effectively by good rotation design.

Diversity, rotations and polycultures

The relationships which have been discussed here can be understood in an ecological framework, and in particular using the principle of bio-logical diversity. Biological diversity, where as many different plant and animal species are present as possible, provides the most favourable conditions for an equilibrium to become established in an ecosystem, whether natural or agricultural. This means that the natural control of weeds, pests and plant pathogens is more likely to be effective and that the occurrence of specific population explosions of a pest or disease is less likely.

Any cropping activity inevitably involves a simplification of the eco-system, and the inter-relationships which usually keep potential pest or disease outbreaks under control are therefore adversely affected. Rotations are one way of attempting to restore biological diversity to a cropping system, while enabling annual cropping of single species to take place.

Rotations are not, however, the only way of achieving this. Mixed or inter-cropping of one or more plant species (polycultures) also provides a significant element of diversity. Crop mixtures tend to have greater cropping efficiency (greater total yield) than monocultures. This is associated with more efficient utilisation of resources (light, water, nutrients), the benefits of nitrogen fixation where legumes are involved, the reduction in spread of pests and diseases as well as increased opportunities for natural predators and improved weed control through increased competition.

The presence of certain 'weed' species within the crop or on the headlands may provide a niche for beneficial insects which can keep pests under control, or may act as decoys for other crop pests. Poly-cultures also have the advantages of increased protection of the soil from erosion, the reduction of financial risk and uncertainty, improved yield stability through diversity and, in subsistence farming situations, the provision of a varied diet; see, for example, Harwood (1985), Altieri & Letourneau (1982), Altieri (1987) and Voss & Shrader (1984).

Although closer to the natural ideal, the disadvantages of polycultures in modern agricultural systems are largely related to mechanisation and the other benefits of specialisation, including the 'optimal' management of individual crops. Combinations of crops therefore need to be selected which are adapted to current constraints, for example cereal/grain legume mixtures. Variety mixtures in cereals is another example of applying the advantages of polycropping under commercial constraints.

The primary application of polyculture techniques lies in permanent cropping such as viticulture or in orchards, where a diverse understory is crucial to pest management, and in horticulture where companion planting is now widely recognised to have considerable potential once the interactions are better understood. Examples include dwarf beans and brassicas, where the beans divert and confuse brassica pests, onions and carrots reducing the attraction of carrots to carrot fly, and strip cropping of wild varieties of brassica which have a stronger attraction for flea beetles than cultivated forms. If farming without the use of agrochemical inputs is to be successful and productive, then a much better understanding of soil–plant, crop–weed, crop–pest and other interactions is essential.

ROTATION DESIGN

With current technology and economic constraints, well designed rotations rather than polycultures are the most essential component of commercial organic farming systems. In short, the rotation has to maintain soil fertility, soil organic matter levels and soil structure, while ensuring that sufficient nutrients, especially nitrogen, are available and that nutrient losses are minimised. Rotations are the main means of minimising weed, disease and pest problems, by achieving crop diversity, both in space and time. At the same time, the rotation has to produce sufficient feed for livestock and maintain the output of livestock and cash crops so that the farmer can obtain a satisfactory income.

The key to a successful organic farming system lies in ensuring that the rotation is carefully designed so that it can meet all these objectives as far as possible. Inevitably, an element of compromise is involved to balance the ideal agricultural and ecological elements with economic considerations such as income and labour and capital distribution and requirements.

The starting point for the design of a rotation should always be the capabilities of the farm and the land in terms of soil type, soil texture, climatic conditions and the effect which these considerations have on the type of crops and livestock which can be produced on the farm. The suitability of certain crops will also be determined by the requirements of livestock for fodder or the availability of markets for cash crops.

Within the cropping limitations imposed by the environmental constraints on the farm system, the following basic guidelines should be observed (see also Table 5.1):

- Deep rooting crops should follow shallow rooting crops, helping to keep the soil structure open and assisting drainage.
- Alternate between crops with high and low root biomass—high root biomass provides soil organisms, particularly earthworms, with material to live on. The grass/clover ley can be valuable in this respect.
- Nitrogen fixing crops should alternate with nitrogen demanding crops—ideally it should be possible to meet all the farm's nitrogen requirements from within the system.
- Wherever possible, catch crops, green manures and undersowing techniques should be used to keep the soil covered as much as possible, protecting it from erosion risks and reducing nutrient leaching, particularly in winter.
- Crops which develop slowly and are therefore susceptible to weeds should follow weed suppressing crops.
- Alternate between leaf and straw crops (important for weed suppression).
- Where a risk of disease or soilborne pest problems exists, potential host crops should only occur in the rotation at appropriate time intervals; to some extent this is indicated by the concept of self-tolerance of crops.
- Use variety and crop mixtures where possible (suitable for on-farm use, but marketing may be more problematic).
- Alternate between autumn sown and spring sown crops (distribution of work load, different weed species germination).

The following points should also be considered:

- suitability of individual crops with respect to climate and soil (Table 5.2);
- balance between cash and forage crops (and market situation);
- seasonal labour requirements and availability;
- cultivations and tillage operations.

The points listed above can be brought together and summarised in schematic form as in Table 5.3, which gives an indication of the suitability of different crops in relation to each other. This does not

Table 5.1 Characteristics of some crops in the context of rotation design.

Crop	Rooting depth	Residual biomass (tDM/ha)	Soil structure	Contribution to Organic matter	Nitrogen balance	Weed control	Pests & diseases Self-tolerance	Break (years)	Winter soil cover
Wheat	o/+	} 0.9–1.7	–/+	–/o	–	–/o	–	2–4	–/+
Barley	o/+		–/+	–/o	–	–/o	–/o	2–4	–/+
Oats	o/+		–/+	–/o	–	–/+	– –	5	–/+
Rye	o/+		–/+	–/o	–	o/+	+	–	–/+
Field beans	o	} 0.5–2.3	o	o	+	–/o	–/+	4–5	–/+
Field peas	o		o	o	+	– –	– –	6–7	–
Potatoes	–	} 0.6–1.0	–/o	– –	– –	–/+	– –	4–5	– –
Beet	–		–/o	– –	–	–/+	– –	4–5	– –
Carrots	–		–/o	– –	–	–/+	– –	3–4	– –
Maize	o	1.8–2.2	– –/o	o	– –	–/+	+	–	– –/+
Rape	o	} 1.3–1.5	o	o	–	+	– –	3–4	–/+
Turnips	o		o	o	–	–/+	– –	3–4	–/+
Green manures									
– non-leguminous	–/+	} 0.9–3.0	o/+	o/+	–/o	+	+/–	–	+
– leguminous	–/+		o/+	o/+	o/+	+	+	–	++
Red-clover based short-term ley	+	4.5–5.5	++	++	+++	++	– –	6	++
White-clover based longer-term ley	o	} 6.0–8.0	++	+++	++	++	o/+	–	++
Lucerne	++		++	+++	+++	+/–	– –	5	++

+++   Excellent.   o   Neutral/average/medium.
++   Very good/very deep/very large.   –   Bad/shallow/small.
+   Good/deep/large.   – –   Very bad.
Sources: Various

Table 5.2 Suitability of crops for different soils and climates.

Soil/climate type	Suitable crops
Lightest sands	Rye, lupins, carrots, kidney vetch.
Light soils	Barley, roots for folding, sugar beet, potatoes, peas, horticultural crops, short leys.
Light chalks	As above except potatoes; sainfoin, lucerne and trefoil very suitable.
Medium loam	Practically all crops.
Heavy clay	Wheat, oats, beans, mangolds, leys, permanent grass.
Fens and silts	Wheat, potatoes, sugar beet, horticultural crops, root crops for seed, buckwheat.
Acid soils	Oats, rye, potatoes.
Wetter areas	Oats, turnips, longer leys.

For any crop, the wetter the district the lighter the soil needed to achieve top yields.

Note: varieties may be suited to different soils, so the above generalisations need to be interpreted with caution.

Table 5.3 The suitability of different crop combinations in the rotation.

Following crop		wh	wb	sb	r	o	m	pe	fb	lr	ley	mc	ep	be	br
								Preceding crop							
Winter wheat	(wh)	--	--	--	○	○	○	++	++	○	○	++	++	○	○
Spring wheat	(wh)	--	--	--	○	○	++	+	++	++	++	++	+	++	++
Winter barley	(wb)	○	--	--	--	○	○	--	++	-	○	○	--	++	--
Spring barley	(sb)	○	--	--	○	○	○	++	-	-	--	○	++	+	++
Winter rye	(r)	○	○	○	○	○	○	○	++	++	○	○	○	++	-
Spring rye	(r)	○	○	○	○	○	++	+	++	++	++	++	+	++	++
Oats	(o)	○	○	○	○	-	++	++	++	++	++	++	++	+	++
Maize	(m)	++	++	++	++	++	-	++	++	++	++	++	+	++	++
Peas	(pe)	++	+	++	++	++	++	--	--	--	++	++	+	++	++
Field beans	(fb)	++	+	++	++	++	++	--	--	--	++	++	+	++	++
Lucerne/red cl	(lr)	+	○	++	++	○	○	--	--	--	--	++	++	++	++
Ley	(ley)	○	○	++	++	++	○	++	++	○	○	++	++	++	++
Maincrop pots	(mc)	++	+	++	++	++	++	++	++	++	++	-	-	++	++
Early potatoes	(ep)	++	+	++	++	++	++	++	++	++	++	-	-	++	++
Beets	(be)	++	++	++	++	++	++	++	++	++	++	++	+	--	--
Brassicas	(br)	++	++	++	++	++	++	++	++	++	++	++	+	--	--

++ Good.

  + Good, but unnecessary. Other crops make better use of the preceding one. Could be used in combination with catch crop or green manure.

  ○ Possible.

  − Limited applications – not advisable if preceding crop harvested late, in dry areas, if pest risk exists (mainly nematodes), or if danger of lodging (e.g. spring barley after legumes).

−− Inadvisable.

Source: Faustzahlen für Landwirtschaft und Gartenbau.

allow, however, for the role and position of individual crops in the rotation to be considered in any detail, which is the aim of the following section. Specific details, such as drilling depths or implements for weed control, have been omitted, either because they are readily obtainable elsewhere, or because they are considered in greater detail in later chapters.

Cash Crops

Cereals

Wheat

Although probably the most important cereal cash crop in organic systems, wheat is also the most demanding in terms of nutrients. In addition, its relative lack of root development and late spring growth give weeds a good opportunity to flourish. The most suitable preceding crops are those which leave good levels of residual nutrients and have a weed suppressing/controlling effect. In many respects, the grass/clover ley and lucerne are ideal, but mineralisation of nitrogen when the ley is ploughed in can be very rapid. There is some doubt that the winter wheat crop can take up sufficient nitrogen in the autumn to prevent serious leaching losses, especially on light, sandy soils.

If spring cultivations are feasible, it may be preferable to plough in the ley in the spring, following it with a crop like potatoes and then wheat. Wheat is also well suited to follow crops such as grain legumes and root crops which have received manures and leave high residual fertility. Early harvested crops allow a clean seedbed to be prepared and crop residues to decompose before the wheat is drilled. In the case of grass leys, considerable time is needed for this; the opportunity might also be used for a bastard fallow for weed control.

Wheat can be grown two years in succession, but ideally should not occur more frequently than once every three or four years in the rotation, despite the economic temptations, because of disease risks like take-all and eyespot and pest problems such as cereal cyst nematode. Most cereals, except maize and to some extent oats (which are not susceptible to take-all), are poor preceding crops for wheat, because they also act as hosts to the same diseases and pests. Spring wheat is somewhat more self-tolerant than winter wheat, because of the greater gap between successive cereal crops. If the spring crop is following an autumn harvested crop, however, then a non-winter hardy green manure should be used to protect the soil and reduce nitrate leaching.

There is an argument that older varieties bred under conditions closer to those of organic farming perform better in organic systems. Various trials (Marriage, 1985; Stopes, 1987; Stöppler *et al.*, 1988; Stöppler, 1988), however, have shown that modern (post 1960) varieties do perform better, although the yield ranking may be different from the ranking in NIAB recommended varieties lists. Maris Widgeon, for example, is a relatively modern variety which has a good reputation in organic systems because it often yields well and is of high breadmaking quality. The long straw which contributed to its phasing-out in conventional systems is not a problem where there is low nitrogen availability. Stöppler (1988) found that high yields in organic systems were closely related to high above-ground biomass and high straw yields, and that these factors were associated with greater root penetration and development, which is important for the nutrition of the plant and for anchorage where mechanical weed control is used. In organic systems, the short-strawed wheat varieties are also disadvantageous because they leave the crop open, allowing weeds to compete more effectively, and because on most organic farms straw is an important raw material.

Broad spectrum disease resistance is a very important characteristic and varietal mixtures can be effective in reducing disease and pest incidence. This aspect of variety assessment is receiving increasing attention from NIAB, and details of suitability of different varieties can be found in their recommended lists.

Barley
Barley is generally less popular in organic systems because there has not been an identifiable premium market for the produce, at least until recently, and because in many cases oats are preferred to barley as a livestock feed. In addition, winter barley can present problems in organic systems because of high nutrient requirements early in the spring, ahead of availability from biological activity in the soil. In view of this, the less demanding two-row varieties may be better suited. Barley is sensitive to poor drainage and acid conditions, but the lodging problem associated with high fertiliser use is not significant in organic systems.

Traditionally, barley was taken at two points in the rotation: as a spring cereal crop following a folded root crop, and as a second straw crop, usually after wheat. Weed control was a major reason for taking it after turnips or swedes which, because of the cultivations and folding which take place, can generally be regarded as cleaning crops. There may be a volunteer wheat problem with barley following wheat.

Similar disease considerations apply as for wheat, with eyespot, take-all and cereal cyst nematodes all representing potential problems. These can be controlled by rotation with a short break so long as couch and other similar hosts are not present to form bridges to the next crop. In some respects, because they are resistant to take-all, oats may be a more appropriate preceding crop for barley, but there is a risk of greater incidence of cereal cyst nematodes.

In any case, it is important that the preceding cereal crop is harvested in good time to allow for the removal of crop residues and for effective weed control before drilling, as winter barley does not compete well against weeds. Where a couch problem exists, the early harvesting of winter barley provides an opportunity for controlling weed rhizomes by desiccation immediately after harvest. If the field is very dirty, these operations may be followed by the use of drag and chain harrows to pull the rhizomes to the surface and collect them for burning.

Oats

The demand for milling quality organically produced oats is currently very high in Britain, making the crop as financially attractive as wheat. Oats are eminently suitable as a second cereal crop, although their position in the rotation is not as fixed as wheat or barley. They form deep rooting systems and can be highly productive in situations of low nutrient availability. Oats after oats, however, is not a suitable combination because of soilborne pest risks, in particular cereal cyst nematodes and stem eelworm, although they do help to control root-knot nematode in barley in cereal/grass systems.

Oats are attracting increased interest from researchers, with strong emphasis being given to the (re)development of naked oat varieties, in which the grain threshes free from the husk and which have a higher nutritional quality, although yields have tended to be variable.

Rye

Although rye is not widely grown in Britain, there is considerable demand for organically produced rye in other parts of Europe and the crop therefore deserves a closer look. It is the most self-tolerant of the cereals, with few disease problems and only low requirements for nutrients. For this reason, it tends to be restricted to poorer, lighter land. In organic systems, rye is most often found as a second cereal crop after wheat, where manuring is seldom required. If the preceding wheat follows another cash crop, then a small amount of manure (10 t/ha) to help establishment in autumn may be necessary.

Its extensive rooting system, autumn tillering, rapid spring development and long straw have a marked suppressant effect on weeds, with fewer viable weed seeds being produced as well. Because of tillering, mechanical control of weeds in the autumn should not be attempted. The deep rooting system means that rye is also drought resistant. Unlike the other cereals, rye is not self-pollinating, with the result that varieties are less pure.

Rye is well suited to mixed cropping with grain legumes such as field beans. The field beans, or a field bean/lupin mixture, are sown immediately after the harvest of the previous crop (mid August) and rye is over-seeded into this mixture mid–end September. The legumes are killed by frost in the winter, leaving the rye to grow on alone. It is necessary to ensure good weed control in the field beans, otherwise weeds can get ahead of the rye. If the field beans are sown too early, or in a mild autumn, they can get ahead and become too dense, so that the rye tillers poorly. In this situation, the beans need to be topped and left lying on the surface as a light mulch. The advantages include reduced need for cultivations, some nitrogen fixation, green manuring effects and reduced nitrogen leaching, but at the cost of greatly reduced opportunities for weed control.

Triticale

Triticale, a cross between wheat (*Triticum*) and rye (*Secale*), combines the yield and quality of wheat with the less demanding and winter hardiness characteristics of rye. Although the yields may not always be as high, it has higher crude protein and essential amino acid (lysine, methionine, cystine) levels than many cereals and is therefore well suited as a concentrate replacement. Its place in the rotation is very similar to that of rye and, like rye, it can also be used as a forage crop or green manure.

Maize

Maize, either for grain or as a forage crop, has only limited applications in organic systems, because of its high nutrient requirements, weed control problems and the risk of soil erosion. It does, however, form a useful break between cereals, as it is not susceptible to the same range of pest and disease problems. Maize can also help with the control of beet nematodes. In an organic system, however, maize should follow a ley or an intensive green manure (ploughed-in in mid April) to make use of and avoid leaching nutrients. The high nutrient requirements of maize mean that some form of manuring is necessary for the growing crop. This may involve spreading 20–30 t/ha of well rotted FYM at ploughing and top dressing with slurry subsequently. The crop is sown mid May, which although late allows for the crop to get away rapidly. About 5–6 days after drilling, the crop can be blind harrowed, when the seedlings are still at least 2 cm under the surface. Later cultivations aerate the soil, mineralising nitrogen, and help to keep weeds between the rows under control. Weeds within the row are still a significant problem, although some of the problems can be overcome with advanced inter-row cultivation and flame-weeding techniques currently being developed and described in more detail in Chapter 6.

The undersowing of a legume once the crop has become established is a widespread and important practice in organic systems, protecting the soil from erosion and from compaction and churning when the crop is harvested. A growing crop, which can be used for forage or green manuring, remains over the winter until the ground is prepared for the next crop. If the grain is harvested, the chopped straw should not be spread evenly over the surface as this can adversely affect the undersown crop. Uneven distribution allows the legumes to grow through.

There is a wide range of suitable crops for undersowing. Subterranean clover has a good weed suppressant effect, competes well with the maize but fixes less nitrogen than white clover, which grows more slowly but is winter hardy. Trefoil is good for neutral to alkaline soils. Vetches can be used, but need to be drilled in and are sensitive to wheel pressure during harvesting. Grasses can also be used, but a mixture of clovers and grasses is preferable. In general, fast growing plants keep up with maize better and are more tolerant to shading. Strip cropping with grain legumes serves a similar purpose and has been shown in American research to increase overall yield compared with cropping maize and beans in subsequent years, although maize yields are increased at the expense of the beans.

Oil seeds

Oil seed rape
Oil seed rape is hardly grown in organic systems because there is no specialist market for the crop and because it is very demanding with respect to nutrients. In practice, however, it is possible to grow oil seed rape organically and there is increasing experience with this crop in central Europe. If oil seed rape were to be included in an organic rotation, it would ideally follow a clover ley or other similar crop which leaves high levels of residual nitrogen. It is a good crop before cereals, because its extensive rooting and early harvesting mean that the soil can be prepared well for winter cereals.

The highest nitrogen demand occurs later in the spring and summer when the soil is more biologically active and nutrients are being mineralised, but top dressing with slurry or liquid manure will be required. Manures have to be carefully distributed, as uneven manuring will lead to uneven maturity at harvesting.

The main problem with weeds is black grass and wild oats, which would normally be fairly well controlled by the organic rotation. Broad leaved weeds are less of a problem as they do not compete very successfully, especially with spring sown varieties.

As a crop, oil seed rape is fairly self-tolerant, but it is a host plant for nematodes which attack cruciferous crops generally. A break is therefore needed, not only between successive oil seed rape crops, but also between oil seed rape and other crucifers, of at least four years and ideally six years in the longer term. The residue after harvesting should be chopped and ploughed in to minimise the spread of stem canker and the new crop should not be planted in fields adjacent to old stubble. There are special problems where sugar beet is also grown because certain nematodes are common to both crops.

The development of double-low varieties has helped to improve the quality of the crop for oil production, but has also been linked to wildlife fatalities, notably deer. This should be an issue of serious concern to organic farmers considering producing oil seed rape.

Linseed
Linseed is relatively much less important in the UK than oil seed rape, but it has greater potential in organic systems because of much lower nitrogen requirements. It used to be grown widely, but it is only now that the market for the crop is supported by the Common Agricultural Policy that it is receiving serious attention again. It suffers from few disease problems, except the fungus *Alternaria*, but its relatively high potash requirement means that good soil reserves are necessary. The main technical problem is the long straw, which becomes entangled in harvesting equipment. This can be resolved, allowing the straw to be baled and used elsewhere on the farm.

Grain legumes

The most important grain legumes currently grown in Britain are peas and field beans. These crops are useful break crops with the important advantage of significant levels of nitrogen fixation. If, however, the protein rich grain is sold off the farm, then the benefit from nitrogen fixation will be limited.

Field beans
The field bean is characterised by a strongly developed tap root which is capable of penetrating to considerable depths and of adding large amounts of organic matter to the soil. The high moisture requirement of the bean means that it is better suited to loamy or clay soils, where the crop is sometimes taken as a cleaning crop in place of roots. In some rotations, beans replace a clover break and they are usually followed by wheat. Fertiliser is not usually needed, but manure may help establishment and contribute to the potash requirements of the crop.

Field beans come in both winter and spring forms. There are three types of spring bean: tick beans, which are small and are usually grown as a cash crop for pigeon feed, and the larger horse beans and large-seeded beans which are usually used as livestock feed. The choice between winter and spring varieties is largely determined by the prevalence of the chocolate spot fungus and the black aphid pest. Winter varieties are not likely to be damaged significantly by the black aphid, because they are usually well developed when the pest attacks in June, but they are more susceptible to the chocolate spot fungus than the spring varieties.

Winter beans are harvested early August to late September, giving a late but satisfactory entry to winter cereals, whereas the spring sown crop is not usually ready for harvesting before mid-September. Winter beans are, however, at risk from frost damage, water-logging and birds over the winter months. The tolerance of beans for other grain legumes and lucerne is limited and they should not be grown more than one year in five because of the soil-borne fungus disease and nematode risk.

Beans do not suppress couch successfully and can therefore exacerbate problems of the carryover of diseases such as take-all between cereal crops. The technique of blind harrowing and inter-row cultivations can help to keep other weeds under control, which is important in the early stages of growth.

Higher yields will be obtained if pollination is assisted by the presence of bees, particularly bumble bees, which should be encouraged.

Other beans

Broad beans are suited to situations where vining peas are already produced as the same harvesting equipment can be used, but they are usually limited to horticultural holdings. They are of interest to farmers because, if they are not sold fresh, they can be combined when mature and used for stock feed, especially the newer tannin-free varieties. Chocolate spot is still a problem, as is the risk of *Fusarium* on wet or compacted ground.

Soya beans are widespread in many parts of the world, but not yet in Europe. New varieties may be suitable for the UK, but their nitrogen fixing ability is lower than that of field beans and they are not an ideal alternative.

Field peas

Peas are traditionally associated with loams or loamy sands with high fertility. They are harvested as a green crop in areas where there is a market or where facilities exist for canning and freezing. In organic systems, they tend either to be grown on horticultural holdings for the fresh market, or grown to maturity for livestock feed. Threshing peas are often used as a cash crop break between two cereals and may also precede potatoes instead of a clover ley. Fodder peas are generally only grown on lighter soils, often as part of an arable silage or green manure mixture.

Although they usually require no manuring, good potash availability is important, and peas are intolerant to soil acidity. Weed control can be achieved by harrowing the crop just as the shoot begins to show above the ground, and again when about 75 mm high until the rows meet. The timing for these operations is crucial and therefore a clean seedbed is necessary. A certain level of weeds in the crop may actually be helpful, because they provide support and help to reduce the problems of lodging. Pest and disease problems can be significant; the overall content of legumes in the rotation, and of grain legumes in green manures, should be limited. Six to seven year intervals between successive pea crops are recommended to avoid the build-up of nematodes and insect pests such as pea root nematode and pea and bean weevils. Planting spring barley on the headlands has been suggested as one way of helping to prevent the inward migration of black aphids from field boundaries.

Cereal and grain legume mixtures
Mixtures of cereals and grain legumes such as spring barley and peas, oats and peas, and winter wheat and field beans are receiving increasing attention. The benefits of crop mixtures generally, including greater cropping efficiency and improved pest and disease control, apply to these mixtures as well. In addition, oats or barley and pea mixtures provide support to the growing pea crop which is otherwise lacking. there is a small benefit in terms of nitrogen fixed and this may lead to higher protein contents in the cereals.

Similar maturation dates mean that the crop can be combined and, if required, separated after harvest. Early spring barley ripens just before or with the peas; the presence of the barley stimulates podding and aids drying by equalisation of moisture levels if the barley moisture content is lower. Best results are obtained if barley is broadcast and harrowed in and then the peas are drilled. The mixtures are not as effective as a straight legume break, because of the risk of perpetuating take-all. Further research on grain legume/cereal mixtures is currently underway at Elm Farm Research Centre. Initial results (for one season only) showed a mixture of winter wheat and winter field beans, each sown at 75% of their recommended sowing densities as individual crops, giving higher financial and physical output as well as enhanced weed suppression (Bulson, pers. comm.).

Lupins
Modern varieties of sweet lupin do not contain the alkaloids which traditionally made this grain legume unpalatable for stock feed. The grain contains high levels of protein and is potentially able to replace imported soya in livestock feed. It can also be ensiled at 30% moisture, and fed whole. The large-seeded albus (white flowering) types are best for grain and are suited to heavier soils, although traditionally lupins were grown on poorer, sandy soils and used as a green manure. They are susceptible to water-logging, have a long growing season (March— October) and have a non-aggressive response to weed infestation. In spite of these disadvantages, lupins are potentially a useful crop in an organic system. Further information can be found in Belteky & Kovacs (1984).

Root crops and field vegetables

Potatoes
Potatoes can be very valuable as a preceding crop, particularly before winter cereals. Residual nutrients from manure applied to the potato crop can be effectively utilised by a subsequent crop, and inter-row cultivations and ridging during the growing season can be effective in reducing weed incidence later. A four to five year break between subsequent potato crops is necessary to avoid the build-up of potato cyst nematodes, although this can be reduced to once every other year in the case of early potatoes, which are harvested before the potato cyst nematode has time to complete its cycle. Potatoes usually follow a grass ley, over-wintering green manure, grain legumes or other crops which leave high levels of residual nutrients and organic matter, although there can be a problem with wireworms in some instances. Some varieties are more resistant to wireworm and other pests and diseases than others. To avoid disease carry over problems, however, potatoes should not follow maize or beet crops (even if there were adequate residual nutrient levels to do this).

Sugar beet
This crop is hardly grown in organic systems, largely because the absence of a developed market and high weed control costs make it financially unattractive. The crop is not self-tolerant; beet cyst nematode is a major problem which requires a four to five year break between crops, although currently limited by contract to not more than one year in three. For the same reason, the crop should not be grown in the same rotation as oil seed rape. A significant break between subsequent crops is also needed to avoid the build-up of weed beet, the seeds of which can

survive several years when buried. Minimum tillage techniques tend to be more successful in controlling this problem. High residual nitrogen levels and fertiliser use can reduce the sugar content; for this reason, sugar beet is usually grown after a cereal crop rather than a ley.

Other field vegetables
The high prices which can be achieved for field vegetables such as carrots, cabbages, cauliflowers, table swedes and turnips mean that these are an attractive option to organic farmers, although frequently insufficient consideration is given to their high labour requirements. They usually take the place of a root crop in the rotation, although, like sugar beet, some of the crops are not well suited to very high residual nutrient levels and are better placed after a cereal crop, preferably preceded by an over-wintering green manure. Field vegetables have been grown successfully following leys, but the wireworm risk needs to be considered. In some cases, field vegetables can be grown for two successive years, but care has to be taken to avoid the build up of the usual pests and diseases of brassicas, which also include green manures and crops such as oil seed rape. Crops such as fodder beet may benefit from the weed control achieved during the production of field vegetables, many of which are important as cleaning crops.

Alternative crops

There is little experience with organic production of 'alternative crops' such as borage and evening primrose, but there may be a role for exploring the possibilities if the market develops.

FORAGE CROPS

On a mixed farm with a ley/arable rotation, 30–50% of the tillable land will be devoted to the production of forage at any time. A mixed ley of three to four years' duration is the pivotal point of most rotations, which will also include cereals or other arable crops, although the length of the ley usually depends on the location of the farm and the importance of livestock to the system. The ley contributes fertility to subsequent crops and acts as the main break to allow weed, disease and pest control. An additional short ley of one or two years is often an important part of a rotation which maximises the proportion of cash crops. Finally, other forage and fodder crops may be used to provide bulky feeds early and late in the season, times when there may often be a shortage of feed available. Most farms will also have a proportion of permanent or long-term grassland which is never (or not normally) part of the arable rotation. From all of this land, sufficient grass must be available for grazing and enough must be conserved as hay or silage for winter feeding along with any other fodder crops.

Legumes in leys

Legumes form an essential component of leys on organic farms, and their value is increasingly being recognised in conventional agriculture where productive grass/clover swards offer a low cost alternative to high nitrogen demanding grass swards. The nitrogen fixation ability of the legumes is the greatest benefit obtained from their inclusion in leys, since this confers increased productivity on the grass species included, as well as a high level of residual fertility to benefit the subsequent crops. White clover based swards may fix 150–200 kg N per hectare per year and provide the basis for good production of milk or meat, whether grazed or conserved. The food quality and digestibility of most legumes for grazing and conservation is very high, with red clover having a crude protein content of 18–19%.

Red clover
Of the red clovers, both the broad and late flowering varieties are well suited to short, one or two year leys on organic farms. They do not persist well beyond this. Broad red clover is best suited to conservation leys, while the late flowering varieties perform best under grazing. Newer tetraploid hybrid varieties have been bred which are highly productive and well suited to silage making. Red clover is more susceptible to clover rot disease (*Sclerotinia*) than is white clover, although the most resistant varieties can considerably reduce the risk from this disease. Some varieties contain high levels of oestrogens which can affect fertility in breeding stock, so care needs to be taken in this respect.

Lucerne
Lucerne can perform well under dry conditions, producing a large and deep root system. With good establishment, the long growing season of lucerne makes up to four crops possible in the season with good management. Problems can occur as a result of slow establishment in the first year and there is a risk of *Verticillium* wilt which may limit the life of the crop.

Lucerne cannot perform adequately in wet or acidic conditions and is primarily a crop for conservation, being better suited to ensiling than hay making, due to its fragility when dry. If it is grazed, then sufficient time for recovery should be given after each grazing and great care should be taken to avoid the occurrence of bloat. Lucerne can be combined with both timothy and meadow fescue, which are less competitive varieties, and is well suited to undersowing in a nurse crop such as oats.

Other legumes

Other legumes can be important components of leys under different situations, for example Alsike clover can be a substitute for red clover if there are problems of clover rot. It can also perform well in acid and wet conditions. Sainfoin can be useful on lower fertility and alkaline chalk or brash soils, with giant sainfoin establishing rapidly but persisting for only two to three years. In contrast, common sainfoin reaches peak production in the third year. Sainfoin has the advantage over lucerne that it is bloat free. It has high protein levels and very high intake rates, but is not very vigorous and should not be overgrazed.

Trefoil is an annual legume which can be undersown into cereals, particularly on alkaline chalk soils. Winter grazing and residual fertility are two advantages of this use of a legume. Vetches or tares can be used as keep for sheep and green fodder for cattle and horses in spring, and are useful in arable silage mixtures. As a main crop, they generally follow a white straw crop. In light land districts, they may also be grown as a catch crop. Cultivated winter vetch is the most common, although Goar or summer vetch is stronger growing, as hardy, but earlier maturing. They are relatively self-tolerant as fodder crops, but as a seed crop they should be grown at three to four year intervals. Hairy vetch is also sometimes used as a catch crop. Other annual clovers such as crimson clover and subterranean clover can be useful under drier, warmer conditions.

Other fodder crops

Fodder crops have an important role to play in the rotation, as an energy source to balance the high protein, legume-based forage, as a cleaning crop for weed control and as a pioneer crop after grassland, providing an opportunity for thorough soil cultivations and a potential replacement for the bastard fallow.

Fodder crops can be distinguished as between annual or multiennial main crops and catch crops, and between forage legumes, grasses and cereals, root crops and leaf crops. Forage legumes such as lucerne, red clover, sainfoin, annual clovers, forage peas and annual grasses and cereals such as Westerwold ryegrass, triticale and maize have been touched on in previous sections.

Arable silages containing cereals and grain legumes, clovers or vetches are a good alternative to single crops like maize. They are less frost sensitive and can be sown and harvested earlier, in some cases allowing for another catch crop to follow. Arable silages and catch crops usually follow cereals or a first silage cut and generally aim to make use of the growing season left after harvest, although sometimes they are used to replace other main crops which have failed.

Many fodder crops can be grown as catch crops or as main crops after late harvested preceding crops, and are therefore suited to filling gaps in cropping and providing additional fodder at times when grass is not available. As main crops, they usually occupy the root break in the rotation. It should be noted, however, that the late harvesting of maize, fodder beet or kale can place the following winter cereal crop in jeopardy. These crops could therefore be followed by another forage crop, as illustrated by the following examples:

a) first cut silage, kale, stubble turnips following spring, autumn reseed;
b) winter barley, stubble turnips, rape or rye, pea/cereal mix under-sown with grass in spring;
c) kale followed by pea silage then into winter wheat;
d) winter barley, rye for spring grazing, followed by maize or fodder beet.

In recent years the management problems, particularly labour requirements for weed control and harvesting, have tipped the balance in favour of conserved grass, usually silage. As a balance to protein-rich conserved clover/grass, home-produced, energy-rich fodder crops still have an important role to play on organic farms. More detailed information, albeit from a conventional perspective, is contained in Jarvis (1985), as well as the various MAFF and NIAB leaflets on fodder crops and suitable varieties.

GREEN MANURES

Green manures have a very important role in the design of rotations for organic systems. Not only do they help to retain and in some cases accumulate nitrogen and other nutrients, thus reducing leaching losses, they also maintain ground cover, protecting the soil from erosion, and can make a contribution to pest and weed control.

The main benefits are summarised in Table 5.4. Although this list represents the possible advantages from green manures, it is not usually possible to achieve all of them simultaneously. Fast growing green manures, such as mustard or rape, conserve nitrogen which might

Table 5.4 Potential benefits of green manures.

- Nitrogen accumulation/maintenance
- Carbon accumulation/maintenance
- Reduction of nutrient leaching (N, Ca, K)
- Reduction of soil erosion
- Improved utilisation of rainfall (water)
- Shading of soil
- Aeration of soil
- Weed control
- Pest control
- Providing cost savings in a crop rotation as a result of:
  lower fertiliser use
  improved nutrient utilisation
  easier cultivations
  reduced plant protection requirements

Source: Kahnt, 1983.

otherwise be leached, but will decompose very quickly when ploughed in, with little effect on soil humus. Fresh material may even lead to a decline in soil organic matter by stimulating soil biological activity. Green manures which make a greater contribution to soil humus may not release as much nitrogen and the decision as to which green manure to use depends on what is the main priority. Generally, green manures will tend to contribute more to the nutritive and less to the stable fraction of organic matter in the soil.

Green manures can also contribute to the biological stabilisation of soil structure following mechanical cultivation, to the extent that plant roots are involved in this process. Depending on soil type, the roots of red clover or lupins can extend down to 1.5–2 m, whilst the roots of white clover, trefoil, common vetch, fodder radish and stubble turnips extend to 0.8–1.5 m and hairy vetch, mustard and rape up to 0.8 m.

Weed control can be achieved through fast, competitive green manures which smother weeds by competing for space, light, water and nutrients, or frequent cutting and mulching. Nematode control is achieved through the use of either antagonist plants over long periods, or of host plants over short periods, to stimulate hatching but remove the food source before the nematode's life cycle is completed. For example, some control of fodder beet nematodes can be obtained using cruciferous green manures as a preceding crop, which encourages the larvae to emerge from the cysts and they can then be ploughed in. Field beans have a similar effect prior to sugar beet and a mustard green manure can help with wireworm control.

There is a range of potential green manures which may be undersown, as has been described in the section on maize above, or stubble sown and they may be either winter hardy or non-winter hardy. Winter hardy types include winter rye, hairy vetch and winter rape. They are usually stubble sown after cereals, potatoes or other crops which are not harvested too late in the year. In the spring, they can be grazed, but are usually topped and mulched, left to wilt for a few days, then rotovated or ploughed in, with a three to four week delay before the next crop.

The non-winter hardy types include mustard, annual clovers, summer vetch, rape, lupins, *Phacelia* (bee's friend) and fodder radish. These types die off over the winter and are easy to cultivate the following spring. It may be possible to use them in combination with winter cereals, for example field beans and winter wheat or rye, so that nutrients mineralised in the autumn can be trapped and released in the early spring as the crop begins to grow. There is, however, a risk in mild winters, such as the one experienced in 1988/89, that there will be insufficient frost to ensure complete destruction of the green manure and this can lead to problems later in the season; mechanical topping or other techniques of foliage destruction will be required.

Green manuring has a particularly important role in stockless arable systems, which, apart from crop roots and straw incorporation have no other regular supplies of organic matter. An example of an appropriate green manure is a mixture of field beans, red clover and other legumes. The mixture is cut and mulched several times during the year, and the whole lot is ploughed in before the next cereal crop.

Green manures need not occur every year in the rotation, and usually would only be grown every three—five years. They are more likely to be used in simplified rotations such as stockless systems based on cereals and grain legumes, where a diverse range of green manure types can help to compensate for the lack of diversity in the rotation. Generally, a cereal rich rotation will have limited dependence on grass-type green manures, but there is a need to watch the extent of cruciferous species and clovers to avoid the build up of pests and diseases. In dry seasons, green manures may extract too much water and have a damaging effect on the subsequent crop. There is also a risk, as with straw incorporation, that residues may become buried and decompose anaerobically, or, if the material is too mature and the carbon/nitrogen ratio is too high, nitrogen will be locked up while decomposition takes place.

The benefits of green manuring are hard to quantify financially, as they cover a wide range of aspects and are often long-term. They should not be seen solely in terms of nutrient supply to the following crop. The detailed principles of green manuring and suitable crops to use are beyond the scope of this book, but some additional information can be found in Woodward & Burge (1982), Soil Association (1980) and HDRA (1989).

Examples of Suitable Rotations for Organic Farms

It should be recognised that no one rotation is suitable for all farming situations, neither is it likely that one rotation will be suitable for the whole of an individual farm. Flexibility is required when implementing a rotation. Rigid adherence to a pre-ordained timetable can mean that leys are ploughed-in before their useful life is ended, other cropping opportunities are missed, or even that weeds are allowed to get out of hand because changes are not implemented when the situation dictates. Most farmers spend many years developing a rotation which is suited to their own individual circumstances. The examples given in the following sections are done so with these points in mind and are intended to be illustrative, but helpful.

Arable rotations

These very much depend on soil type, but will usually consist of:

a) 2–3 year short-term ley (red clover or lucerne on calcareous soils)
  Wheat (or potatoes)
    Green manure
  Potatoes/roots
  Wheat
  Rye or oats

or b) 2–3 year short-term ley
  Wheat
  Rye or oats
  Grain legume
    Green manure
  Wheat
  Barley or oats undersown with ley

These rotations can be extended by the use of an annual clover or grain legume break, before returning to the short-term ley, but one would not usually expect to find more than two years cereals in succession (for weed control reasons) and less than 35% legumes of various types (to ensure sufficient nitrogen availability). Green manures would be used wherever there is a gap between autumn harvested and spring sown crops, and leys would be undersown if at all possible. Grain legume mixtures could be used to help maintain nitrogen levels while keeping the emphasis on arable cash crops. Other arable rotation examples include:

c) Wheat
   Green manure (annual clover/grass/arable silage)
   Fodder maize (silage) or potatoes/roots
   Wheat
   Barley, undersown with
   Annual clover/perennial ryegrass mixture
   Wheat
   Green manure (annual clover/grass/arable silage)
   Fodder maize, may be undersown with
   2 years red clover/grass ley

d) Wheat
   Rye
   Green manure/stubble turnips/*Phacelia*/ bitter lupins
   Potatoes/beet/turnips
   Wheat or rye or winter barley
   Green manure
   Maize or field beans/oats or
   2 years clover or
   Leguminous green manure followed by carrots
   Wheat
   Rye, undersown with
   3 years red clover/grass ley

Stockless arable rotations

The trend towards specialisation in farming has meant that many arable farmers are not in a position to consider converting to organic systems if the reintroduction of livestock is involved. Various efforts have been made to develop stockless systems. A very simple example is a three year rotation: two years cereals and one year of legumes (e.g. a mixture of field beans, red and white or annual clovers). The legumes are cut at regular intervals during the growing season and left lying on the ground. Experience shows that the plants will grow back through the mulch quite readily. This rotation will provide sufficient nitrogen for the cereal crops, and also enables a good level of weed control. The straw is incorporated to avoid excessive organic matter and potash loss.

The problem with this rotation is whether the financial viability of the operation can be maintained with one third of the farm out of action at any one time, but if the price achieved for the cash crops in the rest of the rotation is high enough, then it may be worthwhile. The rotation could be extended by the use of a grain legume course, or by using cereal/grain legume mixtures.

The main aim in designing a stockless rotation is to attempt to include as high a proportion of legumes as possible, while balancing the advantages with the disease and pest risks and financial constraints. The following example indicates how this may be achieved:

Field beans or peas or oats/vetches
Winter wheat
   Green manure (undersown)
Spring cereal
   Green manure (catch crop)
Field beans or row crop (potatoes, beetroots, carrots)
Winter wheat or cereal/grain legume mixture
Winter rye or oats
   Green manure

The course which 'carries' the rotation is the grain legume or a mulched leguminous 'green fallow' which replaces the missing forage crop course. The choice of legume depends on the soil type and other constraints. Red clover or lucerne for seed production could also be used as the first course. The choice of green manure will depend to a large extent on the harvesting date of the preceding crop. Where a crop is harvested late, and there is insufficient time for a catch crop, consideration should be given to undersowing. A roots crop could follow the second grain legume course, but the residual fertility may prove limiting and it is unlikely that there will be sufficient residual fertility for a further two cereal crops. To minimise the risk of legume pest and disease problems, the type of legume used should be varied and diverse mixtures of species should be used as green manures. It will probably also be necessary to include strictly non-leguminous green manures such as mustard or forage rye.

It should also be remembered that although legumes can potentially fix sufficient nitrogen to maintain a stockless system, great care has to be taken over the way a crop is utilised. A catch crop will not fix nearly as much nitrogen as a crop which is down over a longer period, as the nodules do not have sufficient time to develop. Grain legumes will fix a reasonable amount of nitrogen, but a very large proportion of this is removed in the grain. Normally, this would be recycled through livestock, but this is obviously not possible with a stockless system. If the grain is sold off the farm, then there will be little benefit to other crops in the rotation. Clovers grown for herbage seed do not present the same problem, and it may well be worth investigating this option further. Legumes such as lucerne and red clover can fix large amounts of nitrogen, but there is no real use for these crops in a stockless system, other than conservation as hay which can then be sold. In this situation, very large amounts of both nitrogen and potash are sold off the farm, and it is questionable whether this kind of deficit can be sustained for very long. Finally, straw contains significant quantities of potash, and the aim should always be to incorporate it back into the soil, rather than be tempted to sell it off the farm.

There may be significant problems with weed control in stockless rotations, as only the competitiveness of the field beans, the green manures and to some extent the rye will serve to keep weeds under control. Beneficial management practices such as topping grass leys or cultivating row crops are not a normal part of this type of rotation, so direct mechanical intervention may be necessary.

There are few working examples of stockless systems in Britain, other than horticultural holdings which depend to some extent on imported manure. The Ballybrado farm run by Richard Auler and Josef Finke in Ireland (Chapter 11) does come close and trials are also taking place on a farm in Kent and at Elm Farm Research Centre. Because there is very little practical experience with them, anyone attempting to set up such a system is definitely in a pioneering situation and will have to depend very much on trial and error, with all the attendant risks.

Alternate husbandry systems

Alternate husbandry systems are characterised by a four to five year medium-term white clover/grass ley followed by a period of arable cropping, the extent of which is largely determined by climatic conditions and the relative importance of livestock and cash crops. This type of rotation is the one most commonly found on British organic farms and is closely related to traditional ley farming. A typical example would be:

4–5 year ley or lucerne
Up to 3 years cereals, possibly undersown with the new ley

In areas with low rainfall and/or calcareous soils, lucerne may be a better bet than the usual white clover/grass ley. The length of the ley will depend to a large extent on its productivity; if it is still highly productive, then it makes little sense ploughing it in, especially considering the lack of production in the establishment year.

The advantage of this type of rotation lies in the considerable build-up of fertility under the ley, as well as the opportunities which it presents for weed control. It does, however, mean that less than 50% of the land area is available for cash cropping. There are also likely to be problems with insufficient nutrient availability and weed control in the third cereal course. These problems have led to various attempts to modify and extend the cash cropping part of the rotation by means of a short-term ley or grain legume break, as well as the increasing use of cereal/grain legume mixtures and catch cropping. A typical 'extended' rotation might be of the form:

4–5 year ley
Winter wheat
Winter oats (or barley)
2 year red clover/Italian ryegrass ley (or grain legume)
Winter (or spring) wheat
Cereal/grain legume mixture (or cereal undersown with ley)

The break allows for new nitrogen to be fixed and provides an opportunity for weed control. It could be followed by a root crop such as potatoes. The overall effect of extending the rotation is to increase the proportion of cash crops closer to 50%. In some areas, there may be a problem with clover diseases, as the time between the grass/clover ley and the short-term ley is insufficient to deal with the problem. If this is the case, then other options should be considered.

Predominantly grassland rotations

In the wetter western and northern parts of the country, grassland systems predominate. If land is a major constraint, as on small dairy farms, then grassland only rotations may be encountered. In some cases, this will involve alternating between longer-term grazing leys relying on white clover and short-term conservation leys based on red clover.

Where sufficient land is available, reseeding provides an opportunity to cash in on built up fertility. Traditionally, feed cereals, turnips, swedes and frequently field vegetables would be grown for a period of up to two years before the land was put down to a long-term (seven to ten year) ley again. In these types of mainly grass rotations, green manures have a very limited role.

Permanent grassland should not be ploughed up simply for the sake of establishing a rotation, as permanent grassland containing clover generally provides a stable system of nutrient supply as well as weed, disease and pest control and may be of ecological significance. Where clover is not present, techniques are available to assist with the introduction of clover without the need for a complete reseed.

Horticultural rotations

On many farms, the land which is suitable for vegetable production is limited and it may be necessary to develop a special rotation. This is frequently limited to disease considerations only, particularly with brassicas. Some holdings do manage to include a short-term ley, but if no livestock are present then an alternative solution is required. The following is an example of a possible horticultural rotation on a farm with livestock:

Arable silage mixture (oats and vetches)
    followed by rape and vetches
Carrots
Onions
Arable silage mixture
Leeks
Brassica, beets and sweet corn

The two arable silage breaks make weed control possible and allow the land some possibility to rest with only limited cultivation. With this rotation, the nitrogen available from the arable silage will be very limited and the vegetable crops will be dependent on manure or compost brought in from elsewhere. Another rotation, again from a farm with livestock, but with no forage crops, is the following:

Onions
Potatoes
Carrots
Brassica
Fallow (with legume for fertility building phase?)

This rotation is simple, but also effective, and the crops would appear to be in the right sequence in terms of weed control priority. This system is still dependent on manure and nutrients transferred from other parts of the farm. If the fallow were to include high yielding legumes (a mixture of field beans and red clover for example), which were then cut and mulched, the nitrogen budget would be helped considerably and there might also be some benefit in terms of organic matter replacement, while still allowing for weed control. Some consideration should be given to green manures to protect the soil and retain nutrients over the winter months.

ALLOCATION OF MANURES

Previous chapters have emphasised the role of livestock in the cycling of nutrients in organic systems, and the importance of high standards of manure management. The allocation of manures to the most appropriate points in the rotation is an important part of this.

There is seldom sufficient manure available for an application to every field every year, so selectivity becomes important. In general, manure and slurry is applied to crops which remove greatest quantities of nutrients, such as conserved forage, and to those which are most nutrient demanding, such as root crops and maize. Excessive applications of manure or slurry to the leys, however, will reduce the nitrogen fixation effectiveness of clover. Although crops like wheat will respond to manure applications, they are usually able to make good use of the residual fertility from the previous crop in the rotation. Application of manures to wheat may not, therefore, make most effective use of the available nutrients in terms of the farm system as a whole, even though the short-term financial gains may seem attractive. In certain cases, such as barley, top-dressing with slurry at tillering may be appropriate.

Various studies have looked at the issue of rates of manure application and their effect on yield in organic systems. Besson *et al.* (1988) in a long-term comparison of conventional, organic and biodynamic systems with three levels of manure use equivalent to average stocking rates of zero (control), 0.6 and 1.2 livestock units/ha (including both forage and cash crops) found variable responses to manure applications to winter wheat (Table 5.5). Wegrzyn (1983) examined the impact of supplemental nitrogen applications on maize yields on a longstanding organic farm with lucerne in the rotation and concluded that 'in general, supplemental nitrogen applications (mineral or organic) resulted in luxury consumption rather than increased yield. Increasing plant populations increased yields as much or more than supplemental nitrogen.'

Table 5.5 Relative yields for winter wheat with different levels of manure application (conventional, 1.2 LU intensity equivalent = 100).

	Control (0)	Intensity (LU/ha equivalent)	
		0.6	1.2
Organic	N/A	86	112
Biodynamic	90	95	86
Conventional	101	93	100

LU: Livestock units.
Source: Besson *et al.* (1988).

Manuring plans can provide a systematic approach to determining the appropriate allocation of manures in the rotation. It is necessary first to estimate the total quantities of manure and slurry available, based on total stocking, length of time that stock is housed and data such as that contained in Chapter 4. An example of a manuring plan is given in Table 5.6.

Table 5.6 Manuring plan for 10 course rotation[a].

Course number	Crop	FYM (t/ha)	Nutrient value (kg/ha)[b]		
			Nitrogen	Phosphate	Potash
1	White clover/ryegrass Silage	15	22.5	49.5	100.5
2	White clover/ryegrass Silage	15	22.5	49.5	100.5
3	White clover/ryegrass Silage	15	22.5	49.5	100.5
4	White clover/ryegrass Silage	15	22.5	49.5	100.5
5	Wheat				
6	Oats				
7	Red clover (cut)	20	30	66	134
8	Red clover (cut)	20	30	66	134
9	Potatoes	20	30	66	134
10	Wheat				
	TOTAL	120	180	396	804

(a) Average farm stocking rate = 1.2 LU/ha = 120 t FYM/ha over 10 courses.
(b) Nutrients available over rotation, immediate availability will be lower.

NUTRIENT BUDGETS

As mentioned in Chapter 3, nutrient budgets can give a crude indication of the nutrient gains and losses of the farm. They can be used when planning a conversion to help assess the suitability of a rotation, but inevitably, the figures will not be very accurate and the results will need to be treated with caution. An example is given in Table 5.7 using the same rotation as was used for the manuring plan in Table 5.6 and the crop nutrient removal data in Table 3.6.

In this example, there is a positive nitrogen balance, and the phosphate and potash deficits are within reasonable limits, given the limited accuracy of the approach. Many factors will influence the amount of nitrogen fixed and the quantity of nitrogen and potash lost from the system. It is assumed here that nitrogen losses resulting from volatilisation and denitrification are equal to gains from rainfall and free-living nitrogen fixing bacteria. Potash losses through leaching will depend on the clay content of the soil and will be most severe in light sandy soils. Potash may be replenished from soil reserves at rates greater than the deficit indicated here. In most cases, therefore, soil analysis will provide a more reliable indication of whether action is required to correct any nutrient imbalance.

Table 5.7 Nutrient budget for 10 course rotation.

Course number	Crop	Yield (t/ha)	Nitrogen fixation (kg/ha)	Nitrogen	Phosphate (kg/ha)	Potash
				Nutrient removals		
1	Clover/grass ley grazed	4*	120	40	12	20
	Silage	3*	90	75	21	75
2	Clover/grass ley grazed	4*	120	40	12	20
	Silage	3*	90	75	21	75
3	Clover/grass ley grazed	4*	120	40	12	20
	Silage	3*	90	75	21	75
4	Clover/grass ley grazed	4*	120	40	12	20
	Silage	3*	90	75	21	75
5	Wheat	4.5	0	68	36	23
	Straw	3.5	0	18	11	46
6	Oats	4	0	68	32	20
	Straw	3.5	0	18	11	46
	Leg. ley u/sown		100	0	0	0
7	Red clover (cut)	7*	245	210	49	175
	Red clover (grazed)	3*	105	30	9	15
8	Red clover (cut)	7*	245	210	49	175
	Red clover (grazed)	3*	105	30	9	15
9	Potatoes	25	0	75	50	125
10	Wheat	4.5	0	68	36	23
	Straw	3.5	0	18	11	46
	Leg. ley u/sown		100	0	0	0
	Total for rotation		1,740	1,271	434	1,087
	Allowance for leaching			300	30	100
	Nutrients supplied by organic manures			180	396	804
	Nutrient balance					
	rotation			350	−68	−383
	per ha and year			35	−7	−38

*Dry matter.

References and Further Reading

General
Robinson, D. H. (ed.) (1972) *Fream's Elements of Agriculture*. 15th Edition. John Murray
Spedding, C. (1983) *Fream's Elements of Agriculture*. 16th Edition. John Murray. (Note: The 15th and earlier editions are much more useful for indicating possible husbandry options in organic systems.)
Many pre-1950 books on cropping and crop rotations
Various MAFF/ADAS booklets on individual crops
Various NIAB lists of recommended varieties

Rotations and mixed cropping (polycultures)
Altieri, M. A. & Letourneau, D. K. (1982) *Vegetation management and biological control in agroecosystems*. Crop Protection 1 (4): 405–430

Altieri, M. A. (1987) *Agroecology: The scientific basis of alternative agriculture.* Intermediate Technology Publications; London

Harwood, R. R. (1985) *The integration efficiencies of cropping systems.* In: Edens, T. *et al.* (eds) *Sustainable Agriculture and Integrated Cropping Systems.* Michigan State University Press

Voss, R. D. & Shrader, W. D. (1984) *Rotation effects and legumes sources of nitrogen for corn.* In: Bezdicek, D. *et al.* (eds) *Organic Farming: Current Technology and its Role in a Sustainable Agriculture.* American Society of Agronomy Special Publication 46

University of Saskatchewan (1988) *Crop Diversification in Sustainable Agricultural Systems.* Proceedings of conference held on February 27, 1988. Extension Division; University of Saskatchewan

Cereals

Marriage, M. (1985) *New wheat varieties crop well under organic conditions.* New Farmer and Grower 7 (Summer): 10–12

Dreyer, W., Reuter, W. & Thimm, C. (1988) *Roggen im ökologischen Landbau.* Arbeitspapier 124. KTBL; Darmstadt

Stopes, C. (1987) *Myth of old varieties must be finally buried.* New Farmer and Grower 15 (Summer): 15–16

Stöppler, H. (1988) *Zur Eignung von Winterweizensorten hinsichtlich des Anbaues und der Qualität der Produkte in einem System mit geringer Betriebsmittelzufuhr von aussen.* PhD Thesis, University of Kassel, West Germany.

Stöppler, H. , Kölsch, E. & Vogtmann, H. (1988) *Investigations into the suitability of varieties of winter wheat in low external input systems in West Germany.* English language IFOAM Bulletin 6: 12–18. Also in: Proc. 3rd Int. Symp. on Genetic Aspects of Plant Mineral Nutrition to be published in *Developments in Plant and Soil Sciences,* Martinus Nijhoff Publishers.

Legumes

Belteky, B. & Kovacs, I. (1984) *Lupin—The New Break.* Panagri Ltd. , Wiltshire BA15 1NB

Morrison, J. (1982) *The potential of legumes for forage production.* Soil Association Quarterly Review, June: 9–13

Sheldrick, R., Thompson, D. & Newman, G. (1987) *Legumes for Milk and Meat.* Chalcombe Publications

Various MAFF/ADAS Booklets and NIAB variety leaflets

Vegetables

Geier, B. & Vogtmann, H. (1988) *Der biologische Möhrenanbau und Möglichkeiten der Beikrautregulierung.* EDEN-WAREN GmbH, Bad Soden. (Only available from B. Geier, Ökozentrum Imsbach, D-6695 Tholey-Theley, West Germany)

Srässle, B. & Eichenberger, M. (1989) *Ertragsbildung und Nitratgehalt verschiedener Gemüsesorten und arten in unterschiedlichen Düngungssystemen.* Schweizerische Landwirtschaftliche Forschung, 28: 97–130.

Forage crops

Jarvis, P. (1985) *Alternative Forage Crops.* MMB FMS Information Unit Report 44

Pearce, S. (1983) *Fodder Beet.* MMB FMS Report No. 35

NIAB Recommended varieties leaflets nos 1–7

MAFF/ADAS Booklets on individual forage crops

Green manures

HDRA (1989) *News from the research grounds.* HDRA Newsletter 117: 14–17. Also: *Gardening with Green Manures.* Step by Step Organic Gardening pamphlet series. Henry Doubleday Research Association; Coventry.

Soil Association (1980) *Green Manuring.* Technical Booklet No. 5

Woodward, L. & Burge, P. (eds) (1982) *Green Manures.* Elm Farm Research Centre Practical Handbook No. 1

Pieters, A. J. (1927) *Green Manuring.* Wiley: New York

Barney, P. A. (1987) *The use of Trifolium repens, Trifolium subterraneum and Medicago lupulina as overwintering leguminous green manures.* Biological Agriculture and Horticulture 4 (3): 225–234

MacRae, R. J. & Mehuys, G. R. (1987) *Effects of green manuring in rotation with corn on the physical properties of two Quebec soils.* Biological Agriculture and Horticulture 4 (4): 257–270

Kahnt, G. (1983) *Gründüngung.* DLG Verlag; Frankfurt (Main)

Renius, W. & Lütke-Entrup, E. (1985) *Zwischenfruchtbau zur Futtergewinnung und Gründüngung.* DLG Verlag; Frankfurt (Main)

Manure applications

Besson, J. M. *et al.* (1988) *Vergleich biologisch-dynamische, organisch-biologische und konventionelle Wirtschaftsweise anhand der DOK Versuch.* Unpublished mimeograph, Research Institute for Biological Husbandry, Oberwil, Switzerland

Ott, P. (1986) *Utilisation of farmyard manure and composted farmyard manure—a manuring strategy.* In: Vogtmann, H. *et al. The Importance of Biological Agriculture in a World of Diminishing Resources.* Proc. 5th IFOAM Conference. Verlagsgruppe, Witzenhausen

Wegrzyn, V. A. (1983) *Nitrogen Fertility Management in Corn—a Case Study on a Mixed Crop-Livestock Farm in Pennsylvania.* PhD Thesis, Pennsylvania State University

Recent publications

EFRC (1993) *The use of legumes for nitrogen accumulation by green manuring in organic systems.* Final report for MAFF, Contract CSA 14660. Elm Farm Research Centre; Newbury.

Stopes, C. and Millington, S. (1992) Organic stockless rotations. *Elm Farm Research Centre Bulletin* 3:6–8.

Chapter 6

Weed Management

A New Approach to 'Weeds'

Weeds have traditionally been seen as the enemy, in many cases not without good reason. They interfere with the production of crops, often causing yield losses if not total crop failure. They have to be combatted at all costs or they will win. The attitude is one of war; articles and books with titles such as *The Pest War* and *Recognise and Control the Enemy* are commonplace. Herbicides are sold under trade names such as *Commando*, *Avenger*, *Crusader* and *Harrier* (at least until recently, when the agro-chemical industry recognised the fact that such militaristic imagery might not be helping farming's already tarnished public image). This may seem a somewhat trivial issue, but the imagery certainly does reflect a long-standing attitude towards weeds which has been counter-productive.

Increasing resistance of some weeds to herbicides and the changing spectrum of problem weeds is a clear indication that the 'dominance' strategy towards weed control is failing. A fundamental change in attitude is required if weeds are to be controlled successfully in the future. Such a change involves accepting that weeds cannot be 'beaten', the war cannot be 'won', even if success is achieved in some of the battles along the way. The organic farmer has to learn to live alongside weeds, to understand why and how they grow and how farming practices affect them, even to appreciate the benefits which weeds can bring. Weeds are, after all, nature's way of responding to human interference with the soil.

The nature of weeds

Weeds have been defined in a number of different ways, the most common being of the form 'a weed is any non-crop (unsown) species' or 'a weed is any plant growing where it isn't wanted'. A plant only becomes a weed, however, in relation to human activities, and in particular when it interferes with agricultural or horticultural processes. Thus a better definition of a weed is 'any plant which is adapted to man-made habitats and interferes with human activities'.

161

This human-centred definition of weeds is important because it relates to our attitudes to weed control. At other times and in other places, certain plants are seen in a completely different light. They may have valuable medical properties; foxgloves, camomile and comfrey are good examples. They may have nutritional advantages such as dandelions and wild white clover in grassland. Non-crop plants also have aesthetic and ecological qualities which are appreciated by a wide range of people and often form the centre of many conservation battles.

Weeds as a problem

Weed plants, those plants which do interfere with agricultural activities, can be detrimental in several respects. Some may be directly parasitic on crop plants, such as the witchweeds which affect sorghum and maize in Africa, India and parts of the United States, although this problem is not significant in Europe. Others are poisonous, such as ragwort in grassland which although usually avoided by grazing livestock may become incorporated into hay or silage. Problems may also be caused as a result of weeds being unpalatable, nutritionally poor or tainting animal products, even if they are not actually poisonous.

Weeds compete with crops for space, light, water and plant nutrients. The extent to which individual weeds are directly competitive depends on the actual crop being grown; in many cases cultivated crops are more aggressive towards each other than are weeds. In other cases, such as wild oats, the growth habit of the weed is very similar to that of the crop and competition can be very effective.

During the growth of the crop and after harvest, weeds can act as hosts for pests and diseases which affect crop plants (Table 6.1). At harvesting, the presence of weeds such as knotgrass can cause problems with farm machinery and the value of a crop may be seriously affected if weed seeds are present in any quantity. Cultivations can also be made more difficult by the presence of weeds such as couch.

The value of weeds

These negative aspects of weeds are well known and widely recognised. But what about the positive, beneficial aspects which have already been alluded to?

Non-crop plants provide ground cover, protecting the soil when it might otherwise be bare and liable to erosion, particularly after harvest and under permanent crops. A balanced weed population can provide a favourable microclimate, and the activities of the plant roots help to improve soil biological activity and structure. Weeds can thus be useful as green manures.

Certain chemicals produced by weeds have been shown to have

Table 6.1 Examples of weeds acting as hosts for diseases and pests of plants.

Type of pathogen or pest	Weed	Crop plant
Fungi		
Ergot	Blackgrass	Rye
Take-all	Couch	Cereals
Clubroot	Cruciferae	Brassicas
Viruses		
Cucumber mosaic	Chickweed	Many crop plants
Raspberry ringspot	Chickweed, creeping thistle	Raspberry, strawberry, redcurrant
Nematodes		
Stem and bulb		
cyst nematode	Many weeds	Many crops
Insects		
Black bean aphid	Fat hen, many legumes	Broad and field beans

Source: Hill (1977).

beneficial effects on crop plants. For example, agrostemmin, a chemical produced by the corncockle, can increase the yield and the gluten content of wheat (Gajic and Nikocevic, 1973). These same chemicals are also of interest to the pharmaceutical industry and form the basis of some traditional herbal remedies.

Wild plant species can help alleviate the monoculture character of certain crops. Many insects depend on non-crop plants as sources of food. Although some of these insects are pests, others include natural predators and parasites which contribute to biological pest control, helping to keep pest populations within acceptable limits (Altieri & Letourneau, 1982). In fact, the complete removal of weeds from a crop may mean that insects have no alternative but to attack the crop itself. In certain cases, the use of weed strips between crop rows is being actively recommended as a means of controlling pest problems. For example, undersown ryegrass and other weeds provide conditions for beneficial insects such as ladybirds to breed and control aphids in cereals, as research at Rothamsted has shown.

Birds and butterflies depend on these insects for their food (Table 6.2). Evidence from the Game Conservancy Council suggests that partridge chick mortality has increased because of the serious decline in insect food under intensive herbicide regimes for weed control.

Changes in agricultural practices have resulted in weeds which were once common, such as darnel, the cornflower and the corncockle, becoming extremely rare and in some cases threatened with extinction. At the same time, other mainly grass weeds, such as blackgrass, have

Table 6.2 Biological significance of individual weed species as food for birds and insects.

	Birds	Hymenoptera		Beetles		Butterflies	
		Adult	Larvae	Adult	Larvae	Adult	Larvae
Creeping thistle	×	×	×	×	×	×	×
Weld		×		×	×	×	×
Small mallow	×	×		×	×	×	×
Mugwort			×	×	×		×
Common toadflax	×	×	×	×	×	×	×
Ground elder		×		×	×	×	×
Stinging nettle				×	×		×
Shepherd's purse	×	×		×	×		
Cleavers	×			×	×		×
Prickly lettuce	×	×	×	×	×	×	×
Creeping buttercup		×		×	×	×	×
Viper's bugloss		×		×	×	×	×
Deadly nightshade		×		×	×	×	
Scentless mayweed					×		×
Knotgrass	×			×	×	×	×
Chickweed		×		×	×	×	×
Hedge mustard				×	×	×	×
Dead nettle		×		×	×	×	×
Wild carrot	×	×		×	×	×	×
Hedge bindweed	×	×		×	×	×	×

Source: Zum Beispiel 28/3/84: p. 5.

become major problems. This has had a significant impact on the extent and variety of wild flowers and animal life in Britain.

The presence of individual weed species and communities may also indicate problems with soil structure or nutrient status, reflecting the habitats to which individual weeds are adapted. The ability of weeds to act as indicators in this way is, however, limited by the extent to which their presence reflects cropping practices rather than soil conditions. The use of fertilisers and lime, for example, tends to eliminate weeds which are adapted to infertile or extreme acid or alkaline conditions; unfortunately, these are also often the rarest and most valued wild plant species. Table 6.3 lists various weeds and the soil characteristics with which they are associated. Reference books such as Hanf's *The Arable Weeds of Europe* provide more detailed descriptions of the natural habitats and indicator potential of individual species.

A new aim for weed control

It is clear that credit needs to be given to the beneficial role which weeds can play, yet in many cases weeds still represent a significant problem. Weed control, though essential, should not be aimed at the complete

Table 6.3 Weeds as indicators of soil structure and nutrient problems.

Structure/moisture	Soil friability	Nutrient status	pH
Severe compaction	*Very low*	*Moderate*	*Calcareous soils*
Silverweed	Vernal grass	Field foxtail	Sainfoin
Wild tansy	Horsetail	Scentless mayweed	Field pansy
Greater plantain	Toad rush	Field lady's smock	Dwarf spurge
Procumbent pearl-	Venus's looking-	Parsley piert	Cornflower
wort	glass	Lesser stitchwort	Ground-pine
		Corn spurrey	Small bur-parsley
Poor drainage	*Low*	Jagged chickweed	Field mouse-ear
Creeping buttercup	Charlock	Forget-me-not	Hare's ear cabbage
Field mint	Wild chamomile	Shepherd's purse	Venus's looking-
Horsetail	Field lady's smock	Field pansy	glass
Silverweed	Silky bent		Thorow-wax
Coltsfoot	Scented mayweed	*High nitrogen*	Blue pimpernel
Bog pimpernel	Parsley-piert	Amaranth	Common poppy
	Wild radish	Groundsel	Charlock
Well aerated, but moist	Annual knawel	Annual mercury	Field madder
Speedwells	Creeping bell-	Stinging nettles	Bladder campion
Fumitory	flower	Fat hen	Night-flowering
Red dead nettle	Viper's bugloss	Annual nettles	catchfly
Forget-me-not		Ground elder	
Chickweed	*Moderate*	Gallant soldier	*Neutral soils*
Corn spurrey	Stinking mayweed	Docks	Venus's looking-
	Wild oats	Sowthistles	glass
Open, loose soils	Consound	Black nightshade	Corn gromwell
Consound	Corn gromwell		Chickweed
Red morocco		*Nutrient rich*	
Campions	*Friable*	White campion	*Lime deficient*
Night-flowering	Black bindweed	Cleavers	Sand spurrey
catchfly	Fat hen	Dead nettle, red	Bugloss
Ruptureworts	Dead nettles	Cockspur	Jagged chickweed
Creeping bell-	Knotgrass	Chickweed	Procumbent pearl-
flower	White campion	Borage	wort
Sand spurrey	Pale persicaria	Small-flowered	Hare's foot clover
Common amaranth	Water peppers	catchfly	Parsley piert
	Common amaranth	Common mouse-ear	Shepherd's cress
Dry, stony soils		Loose silky-bent	
Venus's looking	*Very friable*	Procumbent pearl-	*Acid soils*
glass	*(high humus)*	wort	Day-nettle
Common storksbill	Spurges	Prostrate pigweed	Field woundwort
Basil thyme	Garden mercury	Creeping thistle	Holly
Small alison	Chickweed	Sun spurge	Hemp nettle
Prostrate pigweed	Gallant soldier	Fumitory	Corn spurrey
Sheep's bit	Veronica	Charlock	Sorrel
Viper's bugloss	Annual nettle	Common orache	Plantain
	Groundsel, common	Field pennycress	Pansy
	Fat hen	Redshank	Horsetail
	Stinging nettle	Speedwell	Field lady's smock
	Smooth sow thistle		Lesser stitchwort
	Dandelion		Common mouse-ear

(continued)

Structure/moisture	Soil friability	Nutrient status	pH
		Nutrient deficient	Acid soils
		Legumes	Wild radish
		Bladder campion	Annual knawel
		Ruptureworts	Speedwell
		Spurreys	Corn chamomile
		Ground pine	
		Small alison	
		Cudweeds	
		Sheep's sorrel	
		Annual knawel	
		Shepherd's cress	
		Wild pansy	
		Wild radish	
		Spring speedwell	
		Sheep's bit	

Sources: various.

elimination of weeds and achieving a perfectly clean field, especially given the implications this has for the loss of biological diversity and the benefits which weeds can provide. Instead, weed control should be aimed at creating a balance between weed and crop species. Intensive, selective weed control can result in other, less controllable weeds surviving and becoming more widely estalished. The more varied the weed population is, the more competition there is likely to be between weeds, making mechanical control easier and reducing their effects on the crop.

Low levels of weeds in a crop do not necessarily represent an economic threat to the crop and the cost of their removal may exceed the advantages of any yield increase. This concept of an 'economic threshold', below which further control of weeds is not worthwhile, makes sense even in situations where herbicides are considered acceptable. But an ecological approach to weed management involves more than just consideration of the economic threshold: the full range of benefits are derived from all the interactions within a diverse crop and non-crop flora and fauna.

THE ECOLOGY OF WEEDS

Agricultural activities, and arable cropping in particular, represent the complete opposite of a stable, natural ecosystem. Agriculture takes place in the context of a disturbed ecosystem. Each time a field is cultivated, the process of ecosystem regeneration begins again.

Certain plants occur more frequently as weeds because their physiology and behaviour is better suited to the re-colonisation of bare soil and, generally speaking, they are better suited to this than crop plants. The inherent variability of weeds enables quick adaptation to hostile environments, both natural and man-made. In undisturbed soils, weed species actively influence their own environment, bringing about changes in soil conditions over time.

The most successful weeds are those which have adapted closely to the life-cycle of crop plants and to farming practices. In addition, they have several characteristics which give them an advantage:

- weeds live in communities rather than monocultures, enabling them to exploit available resources more effectively;
- competition causes changes within the community (between species) as well as within the species, which means that control aimed at one particular weed may enable other species to become dominant;
- they have high genetic variability which gives them the ability to adapt rapidly to hostile conditions;
- many grow rapidly and form large quantities of seed or perennating parts which give them high regenerative capacity;
- the seeds often have organs which help distribution, and many remain viable for long periods of time, even when buried;
- in general, they are more resistant to diseases and therefore better able to compete with highly bred crop plants.

Weed identification

To be able to manage weeds effectively, these characteristics of weeds need to be understood in some detail. An important aspect of this is a knowledge of the biology of individual weeds and their correct identification. Certain weeds, such as field bindweed and black bindweed, annual knawel and corn spurrey, fat hen and common orache, creeping thistle and perennial sow thistle, may look similar but have significantly different growth patterns.

Individual weeds may be classified as either grass or broadleaved types, which may be annual, biennial or perennial, and which may have distinct periods of germination in the spring or autumn. Some of the main weeds are considered in detail below, but readers are also referred to other sources of information on individual weeds such as Hanf's *The Arable Weeds of Europe* and various publications from the Weed Research Organisation (now part of Long Ashton Research Station) including the Farmers Weekly Weed Supplement of 6 February 1987.

Weed competition

Weeds interfere with crops primarily by competition for growth factors such as light, space, water, air and plant nutrients. The total biomass produced by a mixture of crop and non-crop plants tends to be relatively constant, so that the presence of weeds usually means direct competition with the crop and a reduction in yield. This is, however, closely related to the lifecycle of the crop, with the most severe impact being felt in the early stages of the crop. Weeds which emerge after about one third of the crop's lifecycle do not usually reduce yield.

Access to light is the most important factor in weed competitiveness. This is closely related to leaf area, with the result that broadleaved weeds tend to cause greater losses than grasses. The relative demand for soil factors (moisture and nutrients) is largely determined by the soil volume occupied by roots and the rate at which water is extracted. Because nitrogen is relatively mobile in the soil solution, its depletion zone is similar to that for water and competition for this nutrient is much more important than for potash or phosphate. In fact, where fertiliser is added, weeds tend to take nitrogen up in larger quantities and at a greater rate than crop plants. However, competition is seldom restricted to a single growth factor and it is the interactions between several factors which determine the prevalence and impact of weeds.

The composition of weed communities

The composition of weed communities (mixtures of different species) is influenced primarily by climate and by interactions between land use, soil characteristics and modifying influences such as the frequency of tillage, mowing or grazing. Other factors which influence weed population and development include the influence of soil cover on soil temperature, humidity and light, differences in growth form or habit of both roots and aerial parts, and microclimate effects associated with different practices such as ridging compared with flat-bed systems. The adaptability of certain weeds like field bindweed and creeping thistle to a wide range of conditions accounts to a large extent for their notable success as weeds.

The spectrum of weeds on a given farm or field may also change as a result of the introduction of new species from outside, as a result of physiological alteration within a species, and as a result of production practices. Of these, production practices are the most important influence on changes in the weed population. Arable cropping tends to favour annual weeds and some perennials, especially those which have a similar lifecycle to the crop being grown, while intensive cultivation as in horticulture favours short lifecycle annuals such as groundsel and small nettle. In grassland, perennial species which propagate vegetatively are favoured, and the weed species present can be significantly influenced by sward height.

Each new farming practice will eventually have its own complement of weeds. The effects of herbicide use often overshadow those of other practices, although there has been a significant increase in perennial weeds in response to reduced tillage practices as well as the shift to monocultures and decreased reliance on traditional cleaning crops and grass leys. Many annual weeds have been encouraged by improved crop nutrition, resistance to herbicides (e.g. blackgrass and triazine) and changes in mechanisation such as the widespread use of combine harvesters.

The marked increase in the importance of volunteer crops as weeds, including wild oats, weed beet and more recently weed potatoes, is a further reflection of changing agricultural practices and in particular simplified rotations. Weed potatoes, for example, survive as tubers after harvest and as true seed, which can remain viable for up to ten years, much longer than current conventional rotations.

Another example of the adaptation of weeds to agricultural practices is the consequences of widespread use of atrazine and simazine in maize, which has resulted in various wild grasses, usually millet species, but also amaranths, becoming problem weeds. The specific control of one wild grass species, cockspur, which was previously outcompeted by broadleaved weeds, resulted in another, finger grass, taking over, itself to be superceded by bristle grass.

The reason for the succession of weed problems as control over one is achieved is that, generally, weed communities are characterised by the dominance of a few species. The removal of a dominant species tends to be more disruptive to the weed—crop system than the removal of a non-dominant species, with a high likelihood that the dominant species will be quickly replaced by another. Even so, there is usually a fixed upper limit on the density of any given weed, determined by competition with the crop and with other weeds, as well as by their susceptibility to toxins and other potential causes of mortality, including husbandry practices.

The development of resistance to a particular control method can be avoided if a range of practices is used. Selective weed control measures generally allow the strongest plants to survive, develop and reproduce. The more diverse a weed population is, the more weeds will compete with (and help to control) each other.

Allelopathy

The term allelopathy is used to refer to 'the direct or indirect influence of one plant upon another through the production of chemical compounds that escape into the environment' (Rice, 1974). Chemicals produced by plants may be liberated by rain wash from leaves, root exudates and volatilisation from leaves. These chemicals can then have an influence on other plants which may promote or inhibit growth, and which can be either direct or indirect, with microorganisms playing an intermediary role.

Various temperate crops, including wheat, barley and oats, have been found to produce growth inhibitors. Horseradish, carrot and radish contain inhibitors in the root sap. Allelochemicals produced by the *Labiatae*, characteristically aromatic plants (including mints), have been shown to be biocides of wide activity.

Alkaloids, which occur frequently in plants and are widely used in medicine and recognised as a means of plant defence against insects, microorganisms and animals, have seldom been implicated in allelopathy. Recent investigations, however, suggest that plant-produced alkaloids contribute to interference with other plants. Gramine and hordein from barley have been shown to inhibit the growth of various plants, including chickweed and shepherd's purse under greenhouse conditions.

There is, however, some difficulty ascertaining the actual importance of allelopathy relative to other influences, such as competition for resources, on crop growth. Part of the difficulty lies in the separation of cause and effect by intermediate stages. For example, work at Rothamsted has shown that the length of time grass has been down changes the ratio of ammonium to nitrate nitrogen, and that ryegrass exudates inhibit the rate of nitrification by as much as 84%. Wheat roots have also been shown to do this, but to a lesser extent. The reduced nitrification may then affect the growth of other plants.

Russian researchers have reported that root exudates of white lupin and maize inhibited the growth of two associated weed species, fat-hen and pigweed. At the same time, an exudate obtained from the roots of the weed species stimulated the two crop plants. Giant foxtail, on the other hand, has been shown to reduce maize yield partly through allelopathic effects. The analysis of allelopathic effects in grassland has proved much more difficult because of the diverse plant communities which exist and the problem of isolating any of the myriad interactions.

Allelopathic effects can also be observed as a result of the decomposition of crop residues. Toxic substances such as phenols which inhibit plant growth can be formed in anaerobic conditions. In many cases there has been little attempt to distinguish between toxic substances from living plants and products of degradation. In some cases, such as the use of mulches and green manures, the effects of toxins from decomposing crop residues can be used effectively. Mulches act to control weeds mechanically, by impeding growth, and through allelochemicals which are washed into the soil, inhibiting the germination of weed seeds and limiting growth.

The mechanisms of allelopathy are not well understood. The practical applications of allelopathy in agriculture are therefore limited at present, although future research, if pointed in this direction, could yield some useful results.

Weed reproduction and dispersal

The ability of weeds to colonise ground rapidly is directly related to their mechanisms for reproduction and dispersal, which range from self-pollination and high levels of seed production with wide dispersal typical of most annual weeds, to vegetative reproduction by individual plants (including most perennials) which tend to be highly exploitive and competitive. Most weeds, however, produce seed at some stage in their life (except certain species such as bracken and horsetails). The distinction between annual, biennial and perennial plants, although a convenient one, is not therefore necessarily rigid.

Annual weeds usually depend on the production of large numbers of seeds, even under very adverse conditions, and are able to adjust the number of seeds per plant to compensate for losses in plant numbers. The number of seeds produced per plant can vary greatly, from 100 in the case of ivy-leaved speedwell to as many as 215,000 with black nightshade, although actual numbers of viable seeds will depend on competition and the level of pest damage. Even so, the control of seed production and dispersal is a crucial part of controlling annual weeds.

Many weeds, especially annuals, have a relatively short lifecycle, especially from emergence to flowering, and they quickly reach the point of having viable seeds. Some, such as groundsel and shepherd's purse, flower within six weeks of germination whereas others take much longer and may overwinter before flowering. In such situations, day length and vernalisation are important factors. Some weeds can, however, flower all year round.

Similarly, some will produce and set seed continuously, others only at certain times of year. The production of seeds is most influenced, however, by the availability of light; shading by crop plants can have considerably more impact on seed production than it does on vegetative growth.

Dissemination and germination of weed seeds

Weed seeds are disseminated in a variety of ways and these give an important indication of potential control strategies. Long distance transport of seeds is usually associated with human activity, such as contamination of crop seed through inadequate cleaning or the use of hay and straw for feed and bedding. The spreading of manures, slurry or sewage sludge which have not been properly composted or aerated to achieve high levels of weed seed kill, and the use of farm machinery such as combine harvesters are also important sources of weed contamination.

Natural agencies will also play an important role; many seeds have special adaptations which allow them to cling to animals or be carried by the wind, and in some cases they may be eaten by birds or animals and pass intact through the gut.

Once the seeds have been shed, their continued survival through to germination depends on the characteristics of the individual species and a range of environmental factors. The seeds of some species become viable before they have fully developed and usually germinate more readily than those which are mature. Others suffer significant losses in viability if they are left on the soil surface for extended periods of time. If buried, however, they can form part of the seed bank in the soil, which can contain several million seeds per hectare in the plough layer.

Although some of the seeds in the seed bank will remain viable for many years, the year following production usually accounts for the majority of total germination. Chickweed, for example, germinates rapidly in the surface soil (up to 95% in the first year) but can remain viable for up to 60 years when buried under grass. The longevity of seeds in the soil is affected by frequency of tillage and by the depth of seeds in the soil, but if no further seeds are added to the seed bank, the reserves in the soil will decline. The invasion of weeds from outside is of only minor importance as far as the addition of seeds to the seedbank in a given area is concerned.

Germination requires the presence of oxygen, moisture and suitable temperatures, usually in conjunction with light and nutrient sources such as nitrate ions. The mechanism which allows seeds to survive for long periods of time until conditions are suitable for germination is known as dormancy. Normally some factors such as further ripening, vernalisation, daylight, suitable temperatures or the correct degree of alternation between daytime and night time temperatures (diurnal fluctuations) is needed to trigger germination. Some seeds, such as coltsfoot, are viable immediately and have no dormancy, while others, for example barren brome and other annual grasses, can enter enforced dormancy if conditions, such as soil moisture or enforced shading by burial, are not favourable to plant growth.

Although many weed seeds require light for germination, infra-red light, which passes through leaves, often inhibits germination, preventing seeds from germinating in conditions where the seedlings are likely to be shaded-out by other plants. Seeds which require light for germination are likely to germinate only in the surface layers of soil, although the depth of germination is also related to the size of the seed and its ability to supply nutrients to the seedling until it becomes properly established (Figure 6.1).

Figure 6.1   Relationship between seed size and the maximum depth of germination from which seedling emergence can occur.

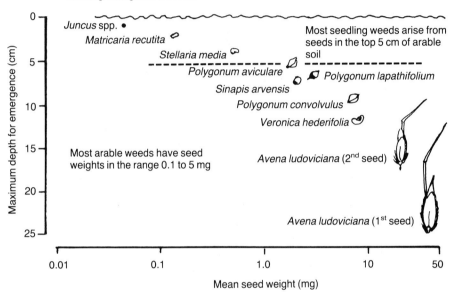

Source: Roberts (1982).

Seasonal germination

Many seeds germinate in particular seasons, some over very short periods. The time of germination is usually determined by day length, detected either by the penetration of light or by diurnal fluctuations in temperature. For this mechanism to be effective, the seeds need to be relatively near the soil surface. For example, the common hemp nettle has its main germination period in the calendar spring, even with low temperatures, and will germinate from a depth of 1 to 4 cm, followed by rapid, strong root development. Other spring-germinating weeds include knotgrass and fool's parsley, as well as fat hen and small nettle, which peak in the spring but continue germinating into the summer.

Weeds which germinate mainly in the autumn include cleavers, winter wild oats, and ivy-leaved speedwell. Cleavers, however, are capable of germinating throughout the year, even in winter, but not in the hot summer months. Germination is controlled primarily by temperature. The minimum temperature required is 1°C, although the optimum is 4–6°C. The optimal germination depth is 1–5 cm, maximum 20 cm, but the seed will not germinate on the surface. The seeds retain viability when dry for long periods, but are very susceptible to loss of viability in manures or silage. Other species which will germinate all year round include annual meadow-grass, field-speedwell, chickweed,

Figure 6.2   The main germination periods for some

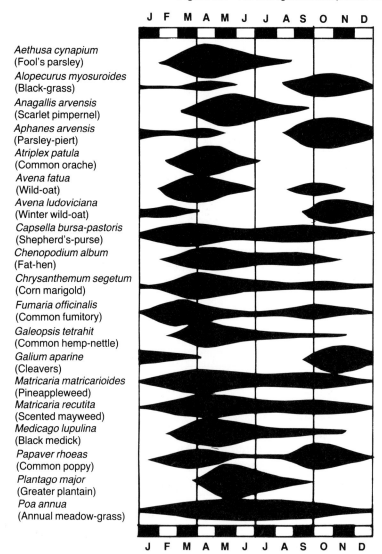

field pansy and common poppy, although there is a tendency for some of them to peak in the spring and early autumn.

Soil cultivation enhances, but does not alter, the tendency of species with specific periodicity to germinate at particular times of year (Figure 6.2). In addition, although cultivations will encourage germination, only a small proportion of the seed bank will be affected. Repeat cultivations

common annual weeds of arable land species.

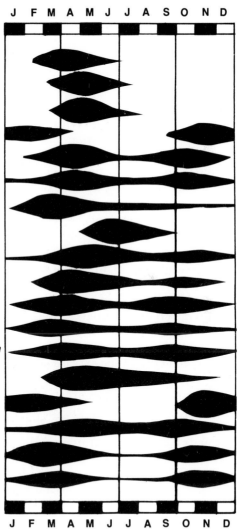

Source: Roberts (1982).

will bring new seeds to the surface. The stale seed bed technique for controlling weeds by cultivation relies on this principle; once the weeds have germinated, the soil should be disturbed as little as possible subsequently.

Vegetative regeneration

Perennial weeds such as thistles, couch and docks present a more intractable problem in organic systems, and the reason for this lies primarily in their ability to reproduce or regenerate vegetatively, making them more resistant to some of the common cultural control practices. Their characteristics therefore deserve particular attention and some of the most important species are considered individually later in the chapter.

Although most, except species like bracken and horsetail, flower and produce seeds as well, the importance of this method of reproduction varies from species to species. With docks, for example, seeds are a very important method of distribution, but others such as creeping thistle can survive quite happily with only occasional opportunities for flowering and setting of seed. This aspect of perennial weeds is important for avoiding further spread of a weed problem. Table 6.4 summarises the characteristics of the most important perennial weeds, while Figure 6.3 shows their main growth and flowering periods.

The vegetative or perennating parts which regenerate most readily vary from species to species and include creeping stems (including both rhizomes and stolons as in couch and creeping buttercup), creeping roots (e.g. creeping thistles and field bindweed), tap roots (e.g. docks) or bulbs (e.g. wild onion). The strong competitiveness of these weeds arises because their perennating parts are confined near to or are physically attached to the parent plant and therefore have greater access to and are better able to utilise resources in a given area of growth. This advantage is, however, achieved at the expense of increased susceptibility to control by cultivation and natural environmental factors because the plants are contained within a restricted area.

The production of perennating parts by the parent plant is strongly influenced by environmental pressures, such as competition for light and other growth factors. Favourable conditions, such as elevated nitrogen levels, will tend to reduce the production of perennating parts, but other factors such as low temperatures may also have this effect. Perennating parts do not generally survive as long in the soil as seeds, although the length of time a perennating part will survive varies directly with burial depth, due in part to protection against desiccation and low temperatures.

With creeping perennials, sprouting from the perennating parts is

Figure 6.3 The periods of growth, when green foliage is present (thin lines), and of flowering (thick lines) for some common perennial weeds.

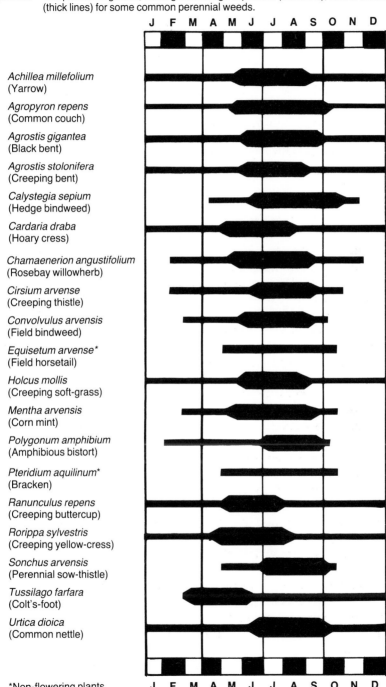

Achillea millefolium
(Yarrow)

Agropyron repens
(Common couch)

Agrostis gigantea
(Black bent)

Agrostis stolonifera
(Creeping bent)

Calystegia sepium
(Hedge bindweed)

Cardaria draba
(Hoary cress)

Chamaenerion angustifolium
(Rosebay willowherb)

Cirsium arvense
(Creeping thistle)

Convolvulus arvensis
(Field bindweed)

Equisetum arvense*
(Field horsetail)

Holcus mollis
(Creeping soft-grass)

Mentha arvensis
(Corn mint)

Polygonum amphibium
(Amphibious bistort)

Pteridium aquilinum*
(Bracken)

Ranunculus repens
(Creeping buttercup)

Rorippa sylvestris
(Creeping yellow-cress)

Sonchus arvensis
(Perennial sow-thistle)

Tussilago farfara
(Colt's-foot)

Urtica dioica
(Common nettle)

*Non-flowering plants

Source: Roberts (1982).

Table 6.4 Characteristics of some important perennial weeds.

Species	Reproductive parts and overwintering state	Depth of vegetatively reproductive parts*	Reproduction by seed
Achillea millefolium (Yarrow)	Stolons; terminal rosettes of leaves overwinter	Very shallow	Very important
Aegopodium podagraria (Ground-elder)	Rhizomes with dormant underground buds; aerial parts die	Shallow	Unimportant
Agropyron repens (Common couch)	Rhizomes with dormant underground buds; aerial shoots overwinter	Shallow	Moderately important
Agrostis gigantea (Black bent)	Rhizomes with dormant underground buds; aerial shoots overwinter	Shallow	Very important
Agrostis stolonifera (Creeping bent)	Aerial creeping stems that overwinter	Above ground	Importance unknown
Allium vineale (Wild onion)	Offset bulbs and bulbils that overwinter	Aerial or very shallow	Rarely produced
Armoracia rusticana (Horseradish)	Fleshy tap root that over-winters; leaves die	Deep	None produced
Arrhenatherum elatius (Onion couch)	Bulbous shoot bases that overwinter (some forms only)	Very shallow	Very important
Calystegia sepium (Hedge bindweed)	Rhizomes that overwinter; aerial shoots die	Deep	Rarely produced
Cardaria draba (Hoary cress)	Creeping roots; small rosettes of leaves overwinter	Deep	Important
Cirsium arvense (Creeping thistle)	Creeping roots that over-winter; shoots die	Deep	Occasionally produced
Convolvulus arvensis (Field bindweed)	Creeping roots that over-winter; shoots die	Deep	Important
Equisetum arvense (Field horsetail)	Rhizomes with tubers that overwinter; aerial shoots die	Deep	Non-seeding plant
Mentha arvensis (Corn mint)	Rhizomes; aerial shoots die	Shallow	Very important
Oxalis spp.	Bulbils, tap roots and rhizomes; leaves die	Shallow	Important in some
Poa trivialis (Rough meadow-grass)	Short stolons; a few leaves overwinter	Above ground	Very important
Polygonum cuspidatum (Japanese knotweed)	Rhizomes, dormant underground buds; aerial shoots die	Shallow	None produced
Polygonum amphibium (Amphibious bistort)	Rhizomes, dormant underground buds; aerial shoots die	Shallow	None in arable

Species	Reproductive parts and overwintering state	Depth of vegetatively reproductive parts*	Reproduction by seed
Pteridium aquilinum (Bracken)	Rhizomes; leaves die	Deep	Non-seeding plant
Ranunculus repens (Creeping buttercup)	Procumbent stems; a few leaves overwinter	Above ground	Very important
Rumex crispus R. obtusifolius (Docks)	Tap roots; rosette of leaves overwinters	Very shallow 7 to 10 cm	Very important
Sonchus arvensis (Perennial sow-thistle)	Creeping roots; aerial shoots die	Very deep	Important
Taraxacum officinale (Dandelion)	Fleshy tap roots; few leaves overwinter	Shallow	Important
Tussilago farfara (Colt's-foot)	Rhizomes; leaves die	Very deep	Important
Urtica dioica (Common nettle)	Rhizomes; short green shoots overwinter	Very shallow	Very important
Veronica filiformis (Slender speedwell)	Stems creeping on the surface	Above ground	None produced

* Depth may vary considerably; the categories are intended as an approximate guide only. 'Very shallow' indicates depths down to 15 or 25 cm, 'shallow' down to 30 or 45 cm, 'deep' down to a metre and 'very deep' down to 3 m or more.
Source: Roberts (1982).

usually controlled by what is known as apical dominance. One bud, normally the apical bud, has an inhibiting effect on the regrowth of others along the length of the creeping root or stem. The apical dominance which keeps these lateral buds dormant is broken by fragmentation, with the result that fragments with shoot buds along the whole length grow readily; there tends also to be an inverse relationship between the length of a rhizome and the number of buds that sprout. Creeping roots may also have buds ready formed, but are able to produce new buds rapidly should the need arise.

As with seeds, the emergence of perennating parts is inversely related to depth, although they often emerge from a greater depth because of increased nutrient reserves. The ability of perennating parts to support sprouting and growth is related directly to levels of nutrient reserves, which are reduced rapidly in the early stages. This means that there is commonly a period of days immediately following the flush of sprouting in the spring when a perennial is most easily destroyed by tillage.

Similarly, cutting as a control method needs to be carefully timed if it is to be successful. A plant derived from a perennating part which has separated from the parent plant may be controlled by the removal of top growth so long as it has not started to produce perennating parts of its own. During this period a perennial weed is no different from an annual, but again the time available for control at this stage is relatively short.

Creeping stems

Control of perennial plants with creeping stems (rhizomes and stolons), examples of which include couch, creeping cinquefoil, bracken, coltsfoot and yarrow, by cultivations or cutting is likely, in the early stages, to exacerbate the problem. The cutting of aerial shoots results in more but smaller shoots and ploughing or digging increases the number of plants through fragmentation and the removal of apical dominance. The weak point is the limited number of buds, so that once exhausted, no further growth may be made.

The aim of control by cultivation should therefore be to stimulate the growth of lateral buds by cutting up the rhizomes and then the shoots, but unless the process is seen through to completion when nutrient reserves and/or viable buds are exhausted, the problem will only be made worse.

Creeping roots

Creeping roots differ from creeping stems in that they are subterranean, grow more or less horizontally and usually the apex eventually turns down to become a vertical root. They can be distinguished from underground creeping stems by the absence of small scale leaves. They never turn upwards, but can, unlike stems, produce new adventitious buds whenever they want.

The time of year when new aerial shoots are produced depends on the species. For example, shoots that over-winter are produced in the autumn by hoary cress and creeping yellow-cress, while creeping thistle, perennial sow-thistle, and field bindweed produce new shoots in the spring. In undisturbed ground, rapid extension of the rhizomes is possible, although grassland generally suppresses growth. Perennial weeds of this type will succumb to intensive cultivation.

Control of perennials

Control strategies for perennial weeds therefore require a knowledge of the characteristics of the individual species, such as whether the perennating parts are on creeping roots or stems, at what depth the rhizomes are usually found, and at what time of year they regenerate and subsequently have least reserves left. For example, creeping thistles form root buds on horizontally extending rhizomes at 2.5–3 cm intervals at a depth just below or in the plough layer. Each 1.2 cm length of rhizome with a cross section of 3–6 mm in the growing season has a 95%

probability of forming a new plant. Only towards the end of flowering is the plant so weakened that only 5–10% of the pieces can grow into new plants.

Soil conditions and depth of burial also play a role. Field bindweed, for example, can send up shoots from 20–25 cm, although the optimal depth is 2–5 cm. Growth occurs from 2°C, the optimal temperatures lie between 1° and 35°C. The minimum reserves (most suitable stage for mechanical control) occur 15–20 days after the beginning of rhizome development (in spring or after removal of green vegetation), at about the three to four leaf stage.

Plants with tap roots
Regeneration from tap roots is a feature of species such as hemlock, cow parsley, docks and dandelions. Such plants are able to regenerate in disturbed ground when the tap roots are fragmented or buried.

The ability of tap roots to regenerate when fragmented varies from species to species. Docks, for example, can produce new shoots only from the uppermost 7 to 10 cm, but dandelions can produce shoots from all portions. Cutting the aerial shoots of biennials before flowering causes them to perennate and in perennials stimulates production of more shoots unless the root is damaged or removed to a depth greater than that from which regrowth can occur.

CONTROLLING WEEDS BY MANAGING THEIR ENVIRONMENT

Husbandry practices

Soil conditions
The requirements of individual weeds for specific soil conditions for growth can provide an indication of possible control methods. Creeping thistles, for example, require sufficient water for growth. Compaction, accessible ground water and/or deep, loamy, well aerated, healthy arable soils can meet their requirement for moisture. Field bindweed thrives in nutrient rich, heavy soils, and requires a lot of light. Perennial sow-thistle requires heavy, dense, moist, cold soils or light soils with a high water-table; their presence may indicate damage to soil structure. Horsetail is a coloniser which likes open, loose, well aerated soils with a ready supply of water at depth (where the rhizomes develop). Chick-weed thrives in good soil conditions, with good water, nutrient and air supply, at least in the upper layers. Cleavers are not so demanding in terms of climate, but prefer good soil conditions, well supplied with moisture, and thrive in well aerated, nutrient rich, loamy soils.

Removal of the optimal conditions for growth therefore provides an important opportunity for controlling weeds, although many of the conditions which favour weed growth are also necessary for successful crop growth. For example, the typical acid indicators such as annual knawel, corn spurrey, or corn chamomile can be controlled by liming. Sub-soiling can remove the compaction and moisture conditions which allow certain weeds to grow in the plough layer. With perennial weeds, however, this must take place at a time which does not cause further vegetative reproduction. 'Biological' soil aeration, by encouraging deep penetration of roots using crops such as lucerne, is a good alternative. Drainage can help to control water-demanding weeds such as rushes, horsetail and coltsfoot.

The weed environment can also be modified by encouraging the biological breakdown of the seed bank through promoting increased biological activity in the soil. Loose silky-bent and runch (wild radish) can both be controlled to an extent in this way.

Rotations
Once the option of improving soil conditions has been exploited, the next most important aspect of weed control in an organic system is the rotation. A diverse rotation which allows for:

- alternating between autumn and spring germinating crops (and their respective weed complements);
- alternating between annual and perennial crops (e.g. cereals and leys);
- alternating between closed, dense crops which shade out weeds (e.g. field beans or rye), and open crops such as maize which encourage weeds;
- a variety of cultivations and cutting or topping operations (in particular the traditional cleaning crops, leys and green manures);

will help to prevent highly adapted weeds such as wild oats, blackgrass and volunteer crops from becoming dominant.

A good illustration of the effect on weeds of different crops in the rotation can be seen in Colour plates 14 and 15 taken on the same farm on the same day. The prevalence of common poppy in a field of winter wheat (Colour plate 14) is much greater than in the neighbouring field of spring oats (Colour plate 15). The common poppy, although capable of germinating in the spring, is mainly autumn germinating and the effect of this can be seen clearly. It also requires sufficient light for germination, which an over-wintering cereal crop provides. There may also be an allelopathic effect with the spring oats, because the poppy seeds were seen to germinate, but failed to develop to maturity.

The perennial crops such as long-term leys are important for the control of perennial weeds. If well managed, long-term leys can be used to suppress perennials like thistles through a combination of cutting and shading. One-year leys (e.g. clover and Westerwolds ryegrass), cut frequently, may have the same effect. Topping is particularly important for controlling weeds during the establishment of grassland, especially spring reseeds which can suffer dramatic infestations of annual weeds such as charlock, redshank and fat hen (Colour plate 16).

Cultivations and sowing practices
Timely and appropriate cultivations prior to the establishment of a crop form the next stage of preventive weed control in an organic system, either through burying weed seeds such as sterile brome which then die, or by bringing other seeds to the surface and encouraging them to germinate. The most important aspect of this is to allow sufficient time between successive cultivations for weeds to germinate, so that reserves in the seed bank are diminished. Various methods of preventive weed control are based on these properties.

Traditionally, a full fallow was a possible method of weed control, but very few farms are in a financial state where they can afford to take land out of production for a full growing season, and the effects on the soil and the environment are not always desirable.

Fallowing the land for part of the growing season, as a bastard fallow, can be advantageous and less costly. Barry Wookey, who farms on chalk and heavy loam in Wiltshire and whose methods are described in his book *Rushall—The Story of an Organic Farm*, uses a bastard fallow to great effect following grass before winter wheat. After a hay cut, the field is chisel-ploughed twice, then chain-harrowed twice. These operations are done over a period of time which allows the sod to rot down and weeds to germinate. The field is then ploughed and a seedbed worked down. Ideally, it is then left for ten days before drilling at the end of the first week in October. This allows further weed germination. The whole process is designed to germinate and destroy as many weeds as possible.

There is not enough time between harvest and sowing another winter cereal to have a bastard fallow. The creation of a false seedbed on which weeds germinate can be an effective alternative. Great importance should be placed on allowing time between the creation of the seedbed and drilling the crop to enable an effective weed strike to take place through the drilling and subsequent harrowing operation. This approach is suitable in the summer when there is no downward movement of water—later in the year the nutrients released as a result of the cultivations may be leached from the soil. Spring fallowing or cultivations, although releasing nutrients at a time advantageous to the crop, are usually more difficult because of soil conditions after the winter.

Pre-drilling cultivations can, however, also cause problems. The right approach will depend on knowing which weeds are a problem in a particular field and paying attention to timing. The fact that the germination time of many plants is related to annual temperature cycles has already been mentioned above, but a specific 'trigger' is required before they will appear in any number. Cultivation probably provides this trigger by exposing the seeds to light. Using the information in Figure 6.2 above, cultivations can then either be timed to stimulate germination of specific weeds, or to avoid their main germination period.

Consideration should also be given to the type of implement used for cultivation. Implements for reduced tillage, such as spring-tined culti-vators, tend to leave many seeds on the surface rather than burying them where they cannot germinate or are not eaten by microorganisms. They offer little scope for the control of perennials like couch grass. Some weeds, notably the weed grasses, benefit from this, whereas other weeds such as the poppy, whose seeds survive for long periods in the seed bank but need light to germinate, can be encouraged to germinate and therefore can be controlled by a later cultivation.

Ploughing and deep cultivations have the opposite effects of shallow or reduced cultivations. Seeds with a short life in the seed bank can be buried and should not re-emerge at a later date. Winter cereals following row crops which have been cultivated for weed control during the growing season, however, are best cultivated rather than ploughed so that no new seeds are brought to the surface.

It is difficult, however, to come to a firm conclusion about the best methods of seedbed preparation. Weichel, the West German machinery manufacturer and organic farmer, takes an extreme position when he argues that any deep cultivations or ploughing will result in a weed increase, because seeds do not germinate immediately, but come up later. Therefore, he argues, shallow cultivations are important to encourage germination without burial. Ploughing or deep cultivation should take place after germination.

An understanding of the ecology of weed plants should lead one to the conclusion, however, that a range of different approaches, including both burial and shallow cultivation, is necessary to ensure that individual weeds are not encouraged selectively by any one approach.

In addition to the creation of stale seedbeds and cultivations aimed at controlling weeds prior to sowing, specific drilling practices can also be helpful. Late sowing of slow growing crops such as maize (to encourage them to grow faster when soil temperatures are warmer), timing to avoid the flush of weeds, cross drilling of crops, higher sowing rates and the spacing of rows, not only to enhance the shading of weeds, but also to allow the use of implements such as hoes at a later stage, are important aspects of this. The issues of closer rows and allowing sufficient space for implements to be used need to be considered together, with sufficient space left for tractor wheelings and the possible use of double rows to reduce the area needing to be cultivated.

Increasing the competitiveness of the crop
The competitiveness of the crop can be increased by using a combination of the husbandry factors described here, such as rotations and cultivations, and other practices. Balanced nutrition using organic manures can help to stimulate the competitive ability of the crop (but may also stimulate weed growth under certain circumstances). An early start for the crop is important; this may be achieved by pre-germination, propagation and transplanting or by the use of pre-drilling and pre-emergence techniques such as stale seedbeds, blind harrowing and flame weeding. Crop varieties, in particular their growth habits, are also significant. For example, short-strawed cereal varieties are less effective at shading out weeds than long-strawed ones. Similarly, plants with an erect growth habit will allow more light to reach weeds than plants with a prostrate growth habit. The mixing of species with different growth habits, for example melons and plantains (Obiefuna, 1989) may also be an effective form of weed control.

Other techniques such as increased sowing rates designed to increase crop densities (e.g. 10% extra in cereals), the use of crop mixtures such as cereals and grain legumes, and undersowing open crops such as maize with clover or other suitable species (Werner, 1988), all help to enhance the ability of the individual crop to suppress weeds. Green manures in the rotation increase the competitive pressure on weeds by encouraging their germination prior to a main crop and through cutting and mulching. Sometimes similar species (e.g. mustard) can be used, but it is usually preferable to use other species. Further possible, but less predictable, effects may be derived from the allelopathic properties of certain crops which actively discourage the growth of competitive plants. These effects may also be achieved by the use of mulches, which work through light exclusion and, to a lesser extent, the chemical by-products of their decomposition (Weibel & Niggli, 1989; Lennartsson *et al.*, 1989).

Controlling the spread of weed seeds
The spread of weed seeds around and between farms is largely associated with the transport and use of crop seeds, manures, and bedding and feedstuffs for livestock, as well as distribution via machinery and the

redeposition of seeds in the wake of combine harvesters. Dealing with this problem is a very important preventive measure, particularly if other practices have resulted in a relatively clean field. Contaminated crop seed can very quickly result in a new weed problem such as docks becoming established. Mechanical cleaning of home-grown seed and quality control of purchased seed are therefore essential.

Little can be done to prevent the feeding of seeds to livestock in hay or the use of contaminated straw for bedding, but proper handling of manures and slurries should considerably reduce the numbers of viable weed seeds. The process of composting, described in detail in Chapter 4, encourages both germination in the surface layers and the destruction of seeds inside the heap by heat and the action of microorganisms. Slurry aeration also involves an increase in temperature which reduces the viability of weed seeds.

Even so, the prevention of crop and bedding contamination in the first place is obviously preferable, and can be achieved by cutting or topping to prevent seeding, by cultivations, and by avoiding invasion of weeds from headlands. The latter involves cutting or cultivating headlands to maintain a clean boundary, but the danger of invasion by headland species is often over-estimated as many of these species are not typical weeds. In fact, as has already been mentioned, perennials such as creeping thistles and docks are a much more serious problem, and hand-rogueing, though labour-demanding, can pay dividends in future years.

Biological control of weeds

The biological control of weeds using parasites and disease pathogens is an area which is being actively explored by scientists around the world. Biological weed control methods can be classified as follows (Wapshere et al., 1989):

- the classical or inoculative method, based on the introduction of host-specific, exotic natural enemies adapted to exotic weeds;
- the inundative or augmentative method, based on the mass production and release of native natural enemies usually against native weeds;
- the conservative method, based on reducing the numbers of native parasites, predators and diseases of native natural enemies; and
- the broad-spectrum method, based on the artificial manipulation of the natural enemy population so that the level of attack on the weed is restricted to achieve the desired level of control.

In some cases, these techniques may be of practical use to farmers, the most notable example in Britain being the possibility of using the larvae of a South African moth to control bracken, which is not easily controlled by other means. In Australia, a major programme is currently underway to control the weed *Echium plantagineum*, which is widespread in New South Wales and Victoria, using a leaf mining moth imported from France. In the United States, scientists are close to finding a successful biological control for the musk thistle (*Carduus nutans*) using species of fly and weevil imported from Europe. In New Zealand, the use of goats to control gorse is a more prosaic example. There is also considerable research effort aimed at genetically engineering fungi ('mycoherbicides') and bacteria so that they are more effective at controlling specific weeds.

Two major issues arise from this. One is the dangers inherent when alien organisms are introduced into an ecosystem where there is no effective control of the organism which is introduced. This applies whether it is a South African moth (which is not found in Britain), or a genetically engineered microorganism. The experience of the introduction of rabbits into Australia and New Zealand clearly illustrates the inherent risks involved. Stringent safeguards exist to try to avoid this type of problem occurring, but, as with the use of alien chemicals in the environment, it is not always possible to foresee all the possible environmental interactions which may occur.

The second is that the use of biological control methods to cure the symptoms of ecological imbalance, without attempting to change the system and the factors which cause the problem in the first place, is likely to be no more successful in the long term than the use of herbicides. Bracken has become a major problem partly as a result of the change from a low intensity, balanced mixture of sheep and cows, with large cloven feet to tread in the bracken, to the intensive, sheep-only systems which are now prevalent in the hills and uplands. Resistance to biological control can develop in the same way as resistance to herbicides has developed. Again, rabbits provide a good illustration of this; the introduction of myxomatosis was only effective temporarily and soon led to a rabbit population which is effectively resistant to the disease. Although the example refers to animals, similar principles apply to plants.

Biological control can therefore only be seen as a useful alternative to herbicides as part of an overall change in approach to ecosystem management and needs to be implemented with extreme caution where alien organisms are involved.

Direct mechanical and thermal intervention

Non-chemical weed control has to be approached from an holistic perspective. Rotations, seedbed preparation, manure management, green manures, fertilising, choice of varieties and seed rates all play a very important part in the overall weed control strategy. Direct intervention in the growing crop should therefore be seen as the last step in the process. The different mechanical and thermal approaches to direct intervention for weed control are considered here with respect to the main groups of crops and specific problem weeds.

There are, however, some general aspects which are worth noting. Firstly, although many of the practices such as blind harrowing are based on traditional experience, there have been considerable recent advances in the design of implements for weed control, some of which are described below. These new implements, although frequently expensive to purchase (ranging from £2,000 to £5,000), do not necessarily work out more expensive than the use of herbicides because of their increased effectiveness, and are usually better than traditional implements such as chain harrows.

Secondly, there are other effects associated with weed control cultivations which can help to stimulate crop growth. Cultivations help to aerate soil and make root penetration easier, increasing the rates at which nutrients are released and the ability of the plant to exploit them. Spring cultivations also increase the rate at which the surface soil dries out, allowing it to warm up faster.

There are also disadvantages to this approach. Cultivations for weed control are highly weather dependent and timing can be crucial. The additional labour and tractor use required may lead to work bottlenecks at certain times of year, and the machinery is not easy to use on slopes, stony soils, or under other such disadvantageous conditions. Excessive numbers of cultivations use significant quantities of fossil energy and can result in soil compaction and possibly soil erosion problems. There is, therefore, always a need to assess carefully whether direct intervention is really required and to balance the advantages against the disadvantages.

Cereals
The use of higher seed rates to increase the competitiveness of a cereal crop against certain weed problems also helps to compensate for any damage to the crop which may occur during the growing season as a result of direct weed control measures. This may be combined with attention to row spacings to allow for post-emergence cultivations (hoeing and harrowing) at a later date.

Blind harrowing is the practice of harrowing just prior to the emergence of the crop. The ideal time is about 24 hours before the crop can be seen. Spring-tined, chain or drag harrows can be used. This method is commonly used in Germany. The University farm at Kassel, for example, blind harrows all its arable crops provided the weather and

Plate 6.1 (Left)   *Detail of the Tearaway weeder with individually sprung tines*

Plate 6.2 (Above)   *The Einböck spring-tined harrow, available in working widths up to 12 metres*

Plate 6.3 (Below) *The Hatzenbichler spring-tined harrow, available in working widths up to 12 metres*

soil conditions are satisfactory. They stress, however, that blind harrowing at the wrong time can produce more weeds. It is necessary to look at the weed seeds and see whether the first tiny white roots have emerged. If they have, then the operation can proceed. If not, then harrowing may well cause them to germinate at a time when the emergence of the immature cultivated crop prevents further passes.

Harrowing established crops sounds a little daunting, but under the right conditions can be effective with minimal damage. Although the traditional chain harrow can be effective, the ideal implement is a harrow with long, thin, spring-loaded tines which are kinder to the growing crop. A German firm, Rabewerk, designed the Hackstriegel especially for this purpose (Figure 6.4). A similar implement is marketed in Britain as the Tearaway weeder and is also suitable for use in grassland. Two similar implements manufactured by the Austrian firms Einböck and Hatzenbichler are based not on rigid tines, individually spring-loaded, but on springy tines, with or without a loop in the tine to give added springiness.

Figure 6.4   The Rabewerk Hackstriegel.

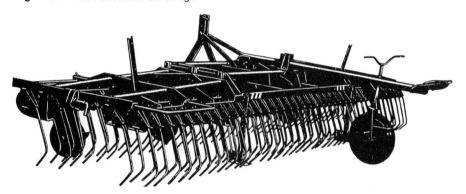

Source: Firma Rabewerk.

Once the crop is established and is past the three-leaf stage, harrowing can be effective. Before this stage, the crop may be seriously damaged. Passes can be made across and along the cereal rows. It is a quick operation, and provided the ground is not too wet, little damage will be done, although the crop does look sorry for itself immediately afterwards. Wet conditions will lead to damage from tractor tyres and many plants will be torn out if the operation is attempted then.

This technique is effective against immature weeds that can be ripped out by the tines. If, however, the weeds themselves are well established, then harrowing has little effect on them. This approach is most successful on lighter soils and not so suitable on heavy soils, although new versions of the spring-tined harrow are being developed which are more robust (Plate 6.4).

Although inter-row hoeing of root crops is common practice in Britain, very few farmers use the technique in cereal crops. In other parts of Europe, however, there is a good deal of experience in hoeing cereals. Hoeing is less effective under wet conditions, but is able to deal

Plate 6.4 (Above left)   *The Reinert harrow*

Plate 6.5 (Above right)   *Hoe for cereal crops with 'goose-foot' shares mounted on individual parallelogram linkages*

with certain weeds such as vetch, mayweed and hemp-nettle which are resistant to harrowing. The type of hoe most commonly used in cereals consists of 'goose-foot' shares mounted on individual parallelogram linkages which follow uneven soil surfaces.

The establishment of a good seed bed and accurate drilling are prerequisites for successful hoeing. The British tradition of harrowing in seed can move the seed enough to alter the width between rows, making hoeing extremely difficult and causing high plant damage. Unless drilling is accurate, the hoe width should not be greater than drill width because the crop rows are likely to be out of alignment.

Depending on the type of hoe used, it is likely to be necessary to use unconventional row spacings. One possible arrangement is 32–36 cm spacing to allow for tractor wheels, 8 cm between individual rows, and 22 cm at intervals to allow for the hoe (Figure 6.5). Seed rates should be about 10% higher to compensate for mechanical damage.

A mid-mounted tractor hoe is used at Kassel University and elsewhere in Germany. This approach has also been used at Elm Farm Research Centre with a Fendt Tool-carrier. This has a conventional

Figure 6.5 Spacing of cereal rows to enhance weed suppression and allow for post-emergence cultivations.

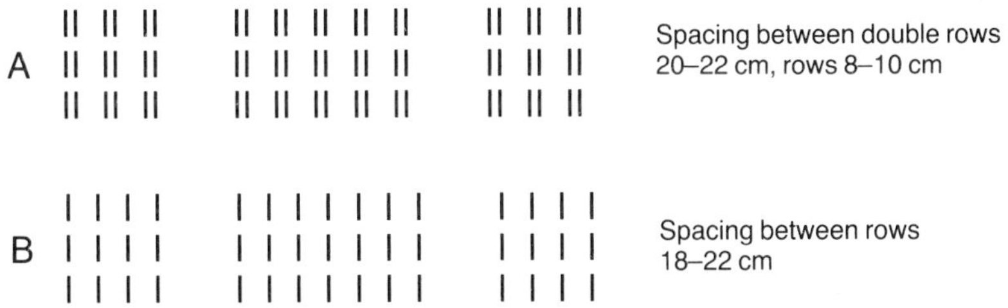

```
     II II II    II II II II II    II II II      Spacing between double rows
A    II II II    II II II II II    II II II      20–22 cm, rows 8–10 cm
     II II II    II II II II II    II II II


     I I I I    I I I I I I I    I I I I        Spacing between rows
B    I I I I    I I I I I I I    I I I I        18–22 cm
     I I I I    I I I I I I I    I I I I
```

three-point linkage at the rear of the machine. However, the engine is placed below the cab, allowing various implements to be placed in the middle of the tractor body. This is especially good for hoeing, as the tractor driver can steer the implement accurately. However, as long as the rows are straight and evenly spaced, rear mounted hoes will be effective. Steerage hoes require an extra operator.

A new implement has been developed by the Baertschi company (developers of the brush-weeder described below) which is currently being trialled in Switzerland. It is called the Rollhacke and is a rotating combination of tined harrow and shared hoe. This implement is suitable for use in cereals and looks very promising for heavy soils, but further details were not available at time of writing.

Research work on organic farms in Switzerland (summarised by Schmid & Steiner in Hoffmann & Geier, 1987) has come up with some interesting results on the interactions between rotations and post-drilling cultivations and their effects on weed populations in winter cereals (Figure 6.6). In particular, it was found that harrowing or hoeing an established cereal crop could reduce weed incidence by 40–50%. The influence of the grass ley on the original level of weeds also comes through clearly. The farms studied fell into three categories:

- Those where the weed cover in the control plots (no mechanical intervention) was around 40%. Mechanical intervention reduced this to about 20%, which was still over the tolerance level.
- Those where the weed cover in the control plots was around 20%. In this case, where control was effective, the weed cover could be reduced to below the tolerance level.
- Those where the weed cover was only 10%. In these cases, additional mechanical weed control was not really necessary. Even so, cultivations took place to prevent individual weeds seeding and to stimulate soil activity and nutrient mineralisation.

The effectiveness of the individual implements has also been assessed by the Swiss researchers; the results are presented in Table 6.5. The most effective weed control was achieved by a combination of spring-tined harrow and hoeing, although the yield advantages were insignificant.

It is well to remember in this mechanical age that hand-rogueing is still an effective and acceptable method of weed control. The odd dock or thistle, or patch of weeds in a crop of wheat, can be the cause of tremendous problems in the future. Localised weedy patches spread into and across the field quickly. The damage caused by walking into a standing cereal crop and rogueing a weedy area is minimal compared to

Figure 6.6   Weed populations on nine sites on organic farms in Switzerland.

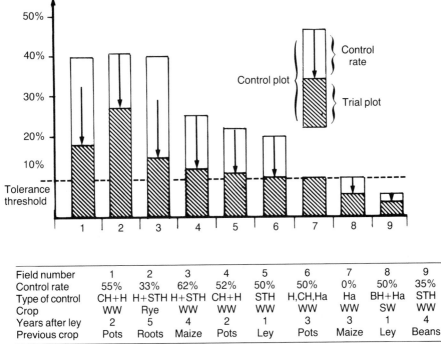

Field number	1	2	3	4	5	6	7	8	9
Control rate	55%	33%	62%	52%	50%	50%	0%	50%	35%
Type of control	CH+H	H+STH	H+STH	CH+H	STH	H,CH,Ha	Ha	BH+Ha	STH
Crop	WW	Rye	WW	WW	WW	WW	WW	SW	WW
Years after ley	2	5	4	2	1	3	3	1	4
Previous crop	Pots	Roots	Maize	Pots	Ley	Pots	Maize	Ley	Beans

Key	CH	= Chain harrow	Ha	= (Zigzag) harrow	SW	= Spring wheat
	H	= Mechanical hoe	BH	= Blind harrowing	Pots	= Potatoes
	STH	= Spring-tined harrow			WW	= Winter wheat

Definitions
Ground cover	=	percentage cover of ground not occupied by crop plants
Tolerance threshold	=	threshold where the financial losses due to no control are equivalent to the control costs
Control rate	=	difference between the ground cover in the treated plots compared with the control plots as percentage of the control ground cover

Source: Zum Beispiel 28/3/84 pp. 10–11.

the problems of allowing the weed to seed. In some cases, cutting a particularly bad patch may be a wise preventive move.

It is useful to consider leaving a strip along the field sides that can be cultivated during the season. This will provide a barrier against colonisation from the sides. Furthermore, it would enable any hedgerow bottoms to be cut before weeds seed into the field.

The various approaches discussed above may not be suitable for all farming situations. There are three points to remember, however, that do have general application:

Table 6.5 The effect of different implements on weed levels in winter wheat.

Weeds	Spring-tined harrow	Hoeing & s-t harrow	Hoeing	Brush weeder	Flame weeding
All weeds:					
Light soils	1–2	1–2	2–3	1	1
Heavy soils	4	3	3–4	3	n
Hemp-nettle, common	1	1–2	2	1	1
Cleavers	2–4	1–2	3	2–3	1
Red dead-nettle	5	2	3	3	1
Scented mayweed	4	2	2–3	2	n
Forget-me-not	3	2	3	2–3	2–3
Black bindweed	2–3	1–2	2	2	1
Chickweed	3–4	2–3	4	2–3	1–2
Ivy-leaved speedwell	2–3	2–3	3	2	n
Hairy vetch	4	2	3	2	n
Black grass	3	1	1	1	n
Loose silky-bent	3	2	4	1	1

Key:
1 = Very good   (more than 80% reduction)   2 = Good   (60–80% reduction)
3 = Moderate   (40–60% reduction)   4 = Limited   (20–40% reduction)
5 = Bad   (less than 20% reduction)   n = No results.
Sources: various cited in Hoffmann and Geier (1987).

- Any mechanical operation within a cereal crop inevitably results in some plant losses. Increasing the seed rate is a wise precaution.
- Growing cereals in a field that is so dirty that weeds are uncontrollable should not be attempted. A crop that is overwhelmed by weeds is not only uneconomic and unsightly, but multiplies the problem and makes it more difficult to solve in the future.
- There is no single organic alternative to herbicides. The farming system itself—rotation, manure management, tillage methods, etc. —is the most important factor in controlling weeds on an organic farm.

Row crops
The use of inter-row cultivators and hoes for keeping weeds down between rows is well established practice for crops such as potatoes and maize. It may also present a viable option for conventionally managed crops like sugar beet as herbicide prices increase. More effort is now going into the development of larger cultivators suitable for large-scale root crops and field vegetables. A Danish-built 12-row cultivator was on display at the British Royal Agricultural Show in 1985 specifically for this purpose. Its main features were heavy spring tines, disc shares and

leaf protector plates so that the maximum area between rows could be cultivated, as well as the option of hydraulic stabiliser equipment to allow its use across slopes.

Various other types of inter-row cultivator are commonly used in horticultural systems, most of which require some form of protection for the crop in its early growth stages. This can be achieved using tunnels, rotating discs or simple plates. Many of the implements also incorporate ridging bodies to bury weeds within rows with soil. Crops such as carrots are grown on ridges for this reason, although some of the newer implements make mechanical weed control in bed systems considerably easier. The advantages of ridges for weed control may still not be sufficient to outweigh the yield reductions caused by lower crop densities compared with bed systems.

Plate 6.6 (Above left)   *A four row inter-row cultivator from Hatzenbichler*

Plate 6.7 (Above right)   *A four row pto-driven inter-row cultivator*

The use of some of these implements is described elsewhere (for example Parish, 1987 and Mattsson *et al.*, 1989). Emphasis is placed here on the new developments which are likely to make mechanical weed control more effective.

In maize, the Howard Roll-Culi (Plate 6.8) and Haruwy rolling cultivators are important developments. They consist of a series of ground driven 'stars' or 'spider tines' which are attached in groups of four to individual parallelograms. The direction in which the soil is moved can be altered, so that in the early growth stages the soil is moved away from the crop rows allowing the young crop to grow undisturbed; later the soil can be directed into the rows, smothering late-germinating weeds which may compete with the crop. The ridging up of the crop also encourages increased root development. The weeds are not cut off from their roots, but are turned out onto the soil surface. Damage to the shallow side roots of the crop is reduced compared with other techniques. The soil itself is crumbled, breaking up any crusts which may have developed and allowing increased air and water penetration,

Plate 6.8   The Howard Roll-Culi rolling cultivator

as well as increasing nitrogen mineralisation rates. An important advantage of the rolling cultivator is its work-rate, which at 10–12 km/hour is twice that of a spring-tined harrow and nearly four times that of the brush-weeder described below.

Alternative implements for use in crops like maize include inter-row cultivators with protective plates and pto driven rotary band hoes which cultivate only the strips between the rows. The latter are expensive to purchase and operate, and harder on the soil, so are used only in exceptional cases to deal with specific problems.

The brush weeder
Many of the implements discussed so far are based on old designs and concepts, with few if any technical innovations or improvements. The brush-weeder, or multiple brush hoe, is a completely new concept in mechanical weed control, developed in Switzerland by the Baertschi company in conjunction with Swiss growers. Although widely used in continental Europe, it is only now receiving attention in Britain. The machine was awarded the HORTEC medal at the 1984 Karlsruhe Horticultural Show, an indication of the interest that the implement is also attracting from conventional growers.

The crop is covered by a protective tunnel over a length of 60–80 cm, depending on the crop. Between the rows, flexible brush rolls rotate on a horizontal axis, working the soil to a depth of 5 cm. The cultivating depth is adjustable and can be maintained at a constant level. Depending on the tractor speed and rotation rate, ground speed is approximately 3 km/hour. This will of course be determined by the crop and the desired working accuracy.

The brush rolls consist of strong, 2–3 mm diameter, polypropylene bristles which are rigid and very resistant to wear. They brush the weeds out of the soil, complete with roots, and throw them against a flexible buffer-curtain. The soil is knocked from the roots and the bare plants are left lying on the soil surface. In warm weather, it only takes an hour or so before the roots have dried out and the plant is dead. This is the decisive advantage of this system compared with conventional hoes, where frequently the uprooted plants are able to grow again.

Silting-up or capping with fine earth thrown up by the brushes does not occur, because the brushes do not leave a smooth layer below the soil surface. The individual bristles leave an uneven surface and the soil capillaries remain open. They are not blocked off as often happens with normal hoeing. The fine earth which is thrown up acts as a sponge and stimulates the aeration of the soil. The danger of erosion and silting-up is therefore greatly reduced.

The implement comes in working widths between 1.5 and 3.5 m suitable for both field vegetables and bed systems. Row distances from 16 cm upwards are easily adjustable. Both machines can be rear or mid-mounted (on suitable tractors). With the rear mounted version, a second person is required to steer the implement. Given a suitable pto, front mounting is also possible.

The brush weeder is mainly used for vegetables such as carrots, beetroot, onions, garlic, celery and leeks, but in trials it has also been used for soya and dwarf beans as well as maize, cereals and grain legumes. Because of its high price and high labour requirement, it is most appropriate for high-value crops, but if it is already available on the farm, it can be used for most if not all crops. However, it cannot deal with biennial or perennial weeds such as thistles and docks, and in cereals there is a possibility of damage to the leaves of the crop which come in contact with the brushes.

The implement is most effective during early growth stages when inter-row cultivations would not be possible because of sideways pressure on the crop rows. The brushes can get very close to the row without damaging the crop roots. The brushes also remove weeds such as cleavers and chickweed which are growing within the rows. This is because the height of the protective tunnels is variable and they only touch the soil in the area of the brushes. Under normal conditions, early use with narrow tunnels is preferable, as this saves on the number of passes which are required.

Because the brushes clean themselves, the implement can be used after rain as soon as the field can be driven upon. On stony soils, the

Plate 6.9 Three-point-linkage mounted Baertschi horizontal-axis brush weeder with steerage attachment

Plate 6.10 Mid-mounted Baertschi horizontal-axis brush weeder

Plate 6.11 A prototype vertical-axis brush weeder with the added advantage that weeds can be brushed out of crop rows or soil brushed into the row to smother them

flexible brushes have proved very advantageous, but care has to be taken to avoid stones jamming in the tunnels. Where the soil surface is already very hard, cultivator attachments can be added to break the surface. Similarly, cultivator shares can be used to break up wheelings.

An aspect which is important in terms of general weed management is the ability to concentrate on only those weeds which directly interfere with the crop. Other weeds are left as bands between rows which can then be mulched as a green manure, or provide some of the beneficial weed–crop interactions which have been discussed earlier in this chapter.

Trials with the brush weeder at the University of Kassel compared the use of this implement in carrots with that of a conventional hoe (Geier & Vogtmann, 1988a). Under good weather conditions (dry and warm), little difference was found between the two implements. Under wetter and cooler conditions, however, the brush weeder was significantly better. The researchers raise the question whether the investment in a brush weeder (at more than £3,000) could be justified when its only advantage was under poor conditions. For many growers, however, the ability to rescue a crop from disaster in a bad season could provide more than adequate return on the investment.

The flame weeder

Although weed control between rows is relatively easy even with the implements which are currently available, a more significant problem is weed control within the row. For many vegetable and root crops, hand hoeing or pulling of weeds may be the only solution, but this approach involves high labour costs. New techniques are, however, being developed.

The most important of these is the field scale flame weeder (Plate 6.12). Flame weeding is often essential for slow emerging row crops but should ideally be seen as a last resort when the rotational and mechanical control of weeds has not proved sufficient. The technique involves exposing the plant tissue to temperatures in excess of 90–100°C for one tenth of a second. The weeds are not actually burnt. Death is caused in two ways: by dehydration due to the expansion of the cell contents and subsequent bursting of the cell membranes, and the coagulation of protein at cell sap temperatures above 50–60°C. There is a simple test to see whether or not the weeds are receiving the right degree of heat, which involves pressing a leaf gently between finger and thumb. A dark green pressure mark on the leaf will indicate that there has been sufficient cell degradation.

There are four ways in which flame-cultivation can be used:

Plate 6.12   Reinert four row, gas-phase flame weeder being used post-emergence for weed control in maize

Plate 6.13   A simple, hand-operated, gas-phase flame weeder for use in horticultural systems

1. Pre-emergence. This technique is widely used by organic producers of carrots, beetroot and other high-value crops. The crop row is treated just before the crop emerges (Figure 6.7), which can be detected by leaving a glass plate on part of a row—the crop under the glass will come up sooner. The soil under the glass should be kept moist. Weeds in inter-row spaces are mechanically treated later. As the propane fuel is the largest part of the cost of this treatment, this can result in a considerable saving of both money and a non-renewable resource. Although the cost is high, flame weeding enables a much greater saving in hand weeding costs. Timing is crucial for its effectiveness. Research is continuing to develop more energy efficient designs of flame weeder.

Figure 6.7   Principle of pre-emergence flame weeding.

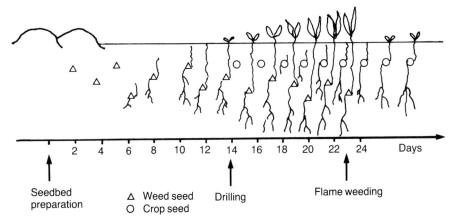

Seedbed preparation

△ Weed seed   Drilling

O Crop seed

Flame weeding

2. Post-emergence. Inter-row treatment is possible using shields to protect the crop from the heat. As long as the actual weed leaves are dry, it is possible to flame weed when the soil would be considered too wet for mechanical control. With sugar beet, the shields can be so designed to lift up the leaves of the crop at the six leaf stage in order to treat the inter-row spaces.

Selective treatment relies on the fact that some crops are less sensitive to thermal shock, especially monocotyledons. For example; maize can be treated at two stages—from above at the 'match' stage (2–3 cm), when the leaves are still folded, and at 25 cm crop height with the burners aimed under the leaves at the base of the plant. On the basis of trials (Geier & Vogtmann, 1988b), it is likely that flame weeding at the 'match' stage is most effective, because this is when competition from weeds is most severe. Cereals in their early growth stages and dry bulb onions, from sets, can be treated in much the same way. Set onions can be flame weeded at three stages—when the onions are 5, 20 and 40 cm tall, considerably reducing the amount of hand labour required, but with a risk of slightly reduced yields (Ascard, 1989).

3. Pre-harvest. The equipment can be used as a defoliant to aid the harvesting of potatoes and onions.

4. Prophylactically. The incidence of *Botrytis cinerea* in strawberries has been reduced by destroying the inoculum with heat before the buds emerge in the spring and after harvest.

The best time for treatment, using the least amount of fuel, is when the leaves are dry, for example on a fine, sunny afternoon. Most weeds are easily killed up to their two true-leaf stage. Some weeds like couch and thistles have a certain amount of heat resistance. These require earlier and further treatments.

Flame weeding techniques are being used increasingly by organic growers in Britain, particularly for crops such as carrots and parsnips. Most of the research, however, is taking place in central Europe and Scandinavia, where several different flame weeding systems are now available. The main difference between the systems lies in the way the gas (usually propane) is supplied to the burners; this can be either liquid- or gas-phase.

The liquid-phase system (Plate 6.14), where the bottles are often inverted and the liquid evaporates only when it reaches the burner, avoids the problem of icing-up at the neck of the bottles which can affect the gas-phase system at high rates of gas use. However, because of the rate at which gas is supplied, and the potential danger of leakage of large quantities of very inflammable fuel, liquid-phase systems are not recommended on safety grounds, particularly those with inverted bottles.

To avoid the problem of icing-up, gas-phase systems (Plate 6.12) have been developed which involve the gas bottles being placed in a warm-

Plate 6.14   Four row, liquid-phase flame weeder being used pre-emergence for weed control in carrots

water bath (also heated by gas). A further development is the use of infra-red burners, which use much less gas, but have slower work rates.

A comparison between the different systems at the University of Kassel in West Germany (Geier in Hoffmann & Geier, 1987; Geier & Vogtmann, 1988b) found no significant differences in the yield of carrots using the different systems. The older, liquid-phase, and from a safety perspective questionable, version caused significant heating of the soil surface, with potentially harmful effects on macro and microorganisms on and in the soil. There was, however, a difference in the number of weeds per unit area between the different systems (Table 6.6).

Table 6.6   Weed populations before and after a period following flame weeding using different flame weeding systems.

	Weichel (liquid-phase)	*Infra-red*	*Reinert (gas-phase)*	*Control*
	Number of weeds per m^2			
Before	100	150	100	160
After	150	320	170	480

Source: Geier in Hoffmann and Geier (1987).

The reason for the higher weed incidence with the infra-red system is explained by the fact that the burner covers a much narrower and more closely defined area, which did not completely fill the counting frame. The weeds outside this area were dealt with mechanically anyway.

Related trials at the University of Kassel suggest that the requirements for hand weeding in crops such as carrots can be reduced by at least 50% if flame weeding is applied effectively at the right time. Some growers in Britain have managed without any hand weeding at all where conditions have been ideal.

Further information on flame weeding is contained in CRABE (1985); Desvaux & Ott (1988); Hoffmann & Geier (1987) and Hoffmann (1989).

Specific weed problems

While annual weeds are fairly easily controlled by a combination of the rotation and mechanical control, certain biennial and perennial weeds present more intractable problems, both in arable systems and grassland under organic management. Such weeds really need to be considered in terms of their own biological characteristics and the specific circumstances of the infestation. The main problem weeds and potential control methods are considered here, but in practice further information may be required before action is taken.

Table 6.7 Overview of different mechanical weed control implements.

System	Application	Working width (m)	Workrate (ha/hour)	Purchase price (£)	Operating cost (£/ha & pass)
Chain harrow	Cereals & potatoes Good on slopes	3.0–5.0	2.0	500–1,000	10
Spring-tined harrow	Cereals & row crops Limited use on heavy/dry/stony soils Useable on slopes	4.5–6.0	2.0	2,000–2,500	20
Inter-row cultivator	Cereals & row crops Heavy & light soils OK on stony soils Limited use on slopes	3.0	0.7	500–2,000	40
Rolling cultivator	Row crops Heavy & light soils Good on stony soils Also on slight slopes	3.1	1.0	3,000	35
Brush weeder	Vegetables & row crops Heavy & light soils Good on stony soils Also on slight slopes	3.0	0.7	4,000	60
Hoe with rotating blades	Row crops & vegetables Heavy & light soils Bad on stony soils	3.0	0.7	3,500	60
Gas-phase flame weeder	Vegetables & maize Any soil type	3.0	1.0	4,000	100

Sources: Zum Beispiel 25/4/89 p. 8; Parish (1987).

Common crouch and other creeping grasses
Couch spreads mainly by means of rhizomes in the soil and stolons on the soil surface. These stem structures act as storage organs for food reserves. The majority of the buds which are distributed along their length at frequent intervals remain dormant unless the couch is disturbed, when fresh aerial shoots will arise from many of them. The seeds are not always fertile, but seed spread should be avoided.

Couch spread is strongly encouraged by spring cultivations with implements like disc harrows, power harrows and rotavators. Minimal autumn cultivations leave established plants relatively undisturbed. The presence of rhizomes hinders soil cultivations and the harvesting of some root crops, while the dense aerial parts encourage lodging of cereals and can make harvesting difficult.

The spread of couch is also encouraged by open crops including poorly established leys and cereal crops. Specific crops such as winter rye and winter barley, however, provide a good degree of shade and can be useful in inhibiting the spread of this weed. Couch will catch up quickly after harvest, though, so where it is a problem, a programme of cultivations should follow soon after.

The best time to tackle couch problems is in July-August after an early cereal crop such as winter barley. At this stage, the plants have utilised their reserves in seed formation and have not yet begun to rebuild them for the winter. Repeated cultivations are required, and they need to be carried out when the soil is dry and the weather warm. The ground should be ploughed or rotavated at 10 cm, then harrowed and the couch rhizomes allowed to dry out in the sun. The process should then be repeated at 20 cm and several times more as necessary. This will usually exhaust the weed.

It is best followed up with a fast growing, strongly shading forage crop or green manure which can compete effectively with any surviving rhizomes, and also take up any nutrients released as a result of the cultivation process. Finally, the green manure should be ploughed under to a depth of at least 15–20 cm. This approach should not be used when the harvest is late since there will be insufficient time for a full programme of cultivations before inclement weather or dormancy sets in. Failure to exhaust the rhizomes completely results in the weed being spread more widely throughout the field.

Deep ploughing on its own may also be effective, but the couch must be completely buried under 15 to 20 cm of soil, involving ploughing to a depth of at least 30 cm.

Apart from cultural controls, there is some evidence from research at Rothamsted that undersowing of cereals and grain legumes with a mixture of Italian ryegrass and broad red clover can appreciably retard the spread of couch grass and prevent its rapid spread after harvest if cultivations are delayed. These findings may also apply to other undersown crops.

Implements designed specifically to remove couch are currently under development in Germany and Denmark (Plate 6.15).

Creeping thistle
Another perennial weed which spreads mainly by its roots is creeping thistle. Undisturbed, the thistle will grow as a single plant and die off after flowering in the second year. Seeds play a relatively minor role in its spread, but viable seeds can travel for quite long distances and for this reason creeping thistles and spear thistles are listed as injurious weeds under the Weeds Act, 1959.

Plate 6.15 Prototype implement for couch grass control developed at Tholey-Theley, West Germany

The process of flowering and seed formation exhausts the plant to such an extent that no further shoots are produced. If the plant is cultivated or cut in the spring before flowering, when it still has significant reserves available, it will spread underground and send up new shoots which can grow into fully fledged plants (Figure 6. 8), which only makes the problem worse. The horizontal roots lie just below the plough layer—they like damp, compacted conditions and thrive in places where perhaps cultivation has taken place when the soil is too wet. Soil compaction should therefore be avoided. They also like high levels of available nutrients and will thrive in areas of high fertility.

The principal control method is the rotation—deep-rooting crops like lucerne break up the plough layer and disrupt the conditions which encourage thistle growth. Thorough winter cultivation followed by an inter-row cultivated root crop or a heavy smothering crop such as arable silage or a green manure can greatly reduce infestation. Cultivations in July-August, when the plants' reserves are low, may be effective. The reserves are also low when the young shoots are about 5–10 cm high, and this provides a good opportunity for extensive cultivation. In some cases, it may be appropriate to cut the plant just before flowering, but timing is crucial as the seeds set within 10 days of flowering. Grazing can also help, but only if it is concentrated in late spring when grass and thistles are growing together and stock are unable to be selective. Grazing too

Figure 6.8   Effect of cutting creeping thistles in spring.

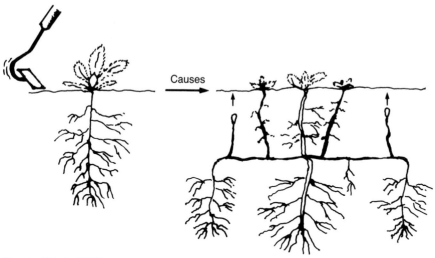

Causes

Source: Plakolm (1984).

early will retard the grass and allow thistles to take a strong hold.

The spear thistle only spreads by seed, so the essential control method is to prevent seeding. Plants should be spudded out below ground (in the young or rosette stage) or cut by mower just before flowering.

Blackgrass

Where cereals are prominent in rotations, weeds such as black grass may become a problem, particularly in winter wheat and barley. The species requires a fine seed bed and light for germination. Seeds buried deeper in the soil stay dormant for many years. Blackgrass grows at the same rate as grain and cannot be harrowed out. It sheds its seeds well before harvest. A degree of control is possible by means of a long and varied rotation or alternation of arable crops with long leys. Spring varieties of susceptible cereals are preferable.

Cultivation before drilling is an important element of blackgrass control. But if cultivation is carried out under damp conditions and the blackgrass seedling is simply transplanted, the number of seed heads it produces in the final crop may be considerably greater.

Docks

Docks can be a very serious problem, particularly in grassland, and are also scheduled as injurious weeds under the 1959 Weeds Act. While the problem is not limited to organic farming, their control in an organic system can be very difficult. The difficulty lies in the ability of docks to produce large numbers of seeds which can retain their viability in the soil for long periods, remaining dormant for many years, even decades, combined with vigorous tap roots which can resprout easily when the stem or the root is cut.

The major cause of dock infestations is seed spread, especially where the seeds have been ingested by livestock and excreted again as dung. Slurry-based systems, which provide nutrients in the quantities that allow docks to thrive, contribute more to the spread of seeds than straw-based manure or compost. Composting can, in fact, help to reduce the numbers of viable seeds which are returned to the field. The viability of the seeds is greatly reduced; 72 hours at 60°C is sufficient to achieve 100% control. The difficulty is maintaining this level of temperature long and consistently enough throughout the heap.

Seed spread is also caused by contaminated purchased and home-grown seed such as red clover and lucerne, and at harvesting, when the seeds are deposited behind the combine. Where a problem exists, or is likely to exist, these factors should be given very careful attention.

A dense, well managed sward can minimise infestation since seedling docks are poor competitors. However, if for any reason the sward is opened up and light reaches the soil surface, docks readily germinate and establish. Poaching of grassland and high soil fertility contribute to this problem.

At the seedling stage, control can be achieved by mechanical cultivation before the tap root develops. If this is done regularly, the seed bank may eventually diminish.

The bigger problem is dealing with the roots, which can quickly re-establish following cultivations. One option is hand-pulling or spudding—collecting the roots and removing them physically from the field. Although this process is very labour intensive, it may be the only way of preventing a few docks causing a major problem later. Even when they are removed from the field in this manner, the dock roots should be burnt rather than left lying in heaps, as they can very quickly re-establish.

A second option which is occasionally mentioned is to carry out a rough cultivation in early-mid summer (a type of fallow) which leaves the roots on the surface, but still embedded in clods which can be dried out by the sun thus (hopefully) killing the plant. A more promising line which has been suggested is a series of passes with a rotavator at progressively deeper settings, starting initially at only 1 cm depth. Until the leaves are about 5 cm long, the plant will be living on reserves from the tap root. Subsequent cultivations, 2 cm deeper each time, should exhaust the root reserves. However, care needs to be taken not to damage soil structure excessively.

Grassland management measures include adequate drainage, the use of high-tillering and persistent grasses, minimal grazing during winter and the avoidance of treading and poaching, using a 'sacrifice' area if necessary. Silage making will prevent the main flush of docks from seeding in the early summer, but some plants can produce seed heads later in the year. The ensilage process will kill dock seeds. Late hay gives docks an ideal opportunity to seed. More frequent defoliation of conservation crops and tight-stocking on grazing land will help to restrict dock growth, but can result in seed being set at lower levels (a few cm) than would otherwise be the case. Docks can adapt to almost any circumstances!

In reality, control of docks will require different measures to be undertaken at different points in the rotation, but this should be part of a continuous strategy and not one which is haphazardly implemented from time to time. Also, in a situation where a farm is being converted from conventional to organic management, one should avoid converting fields which are already heavily infested with docks, at least until some form of control has been achieved.

Some further information on docks can be found in Foster (1989).

Bracken

Bracken has now become a very severe weed of upland areas, colonising large areas of land with few control techniques having proved successful. It is also highly poisonous to livestock. Bracken is not, however, a significant problem for lowland, mixed arable/livestock farmers and has not therefore featured as a significant weed problem in organic systems. With increasing interest in organic systems now being shown by upland livestock farmers, it is still worth discussing the implications.

Bracken does not produce seeds, but spores. Its main form of spread is achieved by a widespread network of underground rhizomes which are capable of sending up new growth to replace fronds destroyed during growth, either mechanically or by frost. The new fronds sprout in May and die with the onset of autumn frosts. The maximum altitude at which bracken is found is about 500 m, due to the adverse effects of frost and wind damage.

Where ploughing is feasible, this can be used as a method to control bracken. To result in a good kill, ploughing should be carried out between late June and early August, with a deep furrow completely inverted. This should then be worked down immediately for sowing. Cultivations should be repeated in the following summer, which should ensure complete eradication.

Mechanical cutting can be used, but this needs to be repeated frequently over several years to exhaust the plants effectively. The treading of infested areas by stock, particularly cattle, can also help to restrict growth. The damaging effect is limited to the buds and developing fronds which lie close to the surface. Out-wintering cattle can be encouraged onto bracken infested areas by the strategic placing of fodder, but sufficient should be available to ensure that the stock does not eat the dead fronds. The stock should be removed as soon as any new fronds appear to avoid poisoning. Pigs can also be used to root for the rhizomes, which are palatable and nutritious, but should be supplied with supplementary feed to avoid the risk of poisoning.

Biological control may prove in the future to be an appropriate way of dealing with bracken, but the risks inherent in this approach need to be considered carefully.

Overview

Weeds are often cited as the most significant problem in organic farming systems and they are certainly the problem which most concerns farmers considering converting. In fact, if all the preventive husbandry techniques including sound rotations, manure management, cultivations, etc. are implemented, then they need not be significant at all. Weeds are a prime example of the need for a change in attitude: to learn to live with nature rather than to dominate it and to appreciate the benefits which a certain level of 'weeds' can bring.

Various studies have shown the serious decline in the range of plant species which are to be found in intensively managed farmland. Many former 'weed' species are now threatened with extinction, along with the animal life which depended on them for its existence. More recent studies comparing the non-crop species on organically and conventionally managed farms (e.g. Hampl & Hermann in Hoffmann & Geier, 1987; Plakolm, 1989; Elsen, 1989; Hermann, 1989), have produced clear evidence that a wider range of species exist in organic systems—in the studies cited, nearly 50 different species were found compared with only 20 in conventionally managed fields. Several species threatened with extinction were found only on the organically managed land. In spite of the wide range of species in the organic systems, only a few were dominant and these tended to be the same as in the conventional systems. The use of herbicides encouraged these dominant species and aided their selection as problem weeds, while destroying those of little economic but significant ecological importance.

REFERENCES AND FURTHER READING

General

Andres, L. A. (1984) *Opportunities for reducing chemical inputs for weed control*. In: Bezdicek, D. *et al.* (eds) *Organic Farming: Current Technology and its Role in a Sustainable Agriculture*. American Society of Agronomy, Madison, USA. pp. 129–140

Hance, R. J. & Holly, K. (eds) (1990) *Weed Control Handbook*. 8th Edition. Blackwell Scientific Publications

McLeod, E. J. & Swezey, S. L. (1979) *Survey of weed problems and management technologies in organic agriculture*. University of California, Berkeley, Appropriate Technology Program. Mimeo. 119pp

Roberts, H. A. (ed.) (1982) *Weed Control Handbook*. 7th edition. Blackwell Scientific Publications

Weed ecology

Aldrich, R. J. (1984) *Weed—Crop Ecology—Principles in Weed Management*. Breton Publishers

Altieri, M. A. (1987) *Agroecology—the scientific basis of alternative agriculture*. Intermediate Technology Publications, London

Altieri, M. A. & Letourneau, D. K. (1982) *Vegetation management and biological control in agroecosystems*. Crop Protection 1: 405–430

Altieri, M. A. & Liebman (eds) (1988) *Weed Management in Agroecosystems—Ecological Approaches*. CRC Press; Boca Raton, Florida

Elsen, T. van (1989) *Ackerwildkrautbestände im Randbereich und im Bestandesinnern unterschiedlich bewirtschafteter Halm- und Hackfruchtäcker*. Proc. 3rd IFOAM International Conference on Non-Chemical Weed Control, October 1989. Bundesanstalt für Agrarbiologie; Linz

Gwynne, D. & Murray, R. (1985) *Weed Biology and Control*. Batsford

Herrmann, G. (1989) *Zustand und Entwicklung von Ackerunkrautvegetation und Samenpotential in ökologisch bewirtschafteten Umstellungs- und Altbetrieben*. Proc. 3rd IFOAM International Conference on Non-Chemical Weed Control. October 1989. Bundesanstalt für Agrarbiologie; Linz

Hill, T. A. (1977) *The Biology of Weeds*. Studies in Biology No. 79. Edward Arnold

New Zealand Soil and Health Association (1987/1988) *Working with weeds*. Soil and Health, Summer issue, pp. 20–25; *Just dandy (edible wild plants)*. Soil and Health, Autumn issue, pp. 30–35

Plakolm, G. (1989) *Unkrauterhebungen in biologisch und konventionell bewirtschafteten Getreideäckern Oberösterreichs*. Proc. 3rd IFOAM International Conference on Non-Chemical Weed Control, October 1989. Bundesanstalt für Agrarbiologie; Linz

Sagar, G. R. (1974) *On the ecology of weed control*. In: Price Jones, D. & Solomon, M. E. (eds) *Biology in Pest and Disease Control*. John Wiley & Sons

Soil Association (1982) *The value of weeds*

Allelopathy

Gajic, D. & Nikocevic, G. (1973) *Chemical allelopathic effect of* Agrostemma githago *upon wheat*. Fragm. Herb. Jugoslav. XXIII.

Gliessman, S. R. (1983) *Allelopathic interactions in crop—weed mixtures*. Journal of Chemical Ecology 9: 991–999

Levitt, J. & Lovett, J. V. (1985) *Alkaloids, antagonisms and allelopathy*. Biological Agriculture and Horticulture 2: 289–302

Lovett, J. V. (1981) *Allelochemicals in a future agriculture*. In: Stonehouse, B. (ed.) *Biological Husbandry*. Butterworths

Lovett, J. V. (1982) *Agro-chemical alternatives in a future agriculture*. Biological Agriculture and Horticulture 1 (1): 15–28

Lovett, J. V. & Weerakoon, W. L. (1983) *Weed characteristics of the Labiatae, with special reference to allelopathy*. Biological Agriculture and Horticulture 1: 145–158

Putnam, R. A. & Duke, W. B. (1978) *Allelopathy in agroecosystems*. Annual Review of Phytopathology 16: 431–51

Rice, E. L. (1974) *Allelopathy*. Academic Press

Swain, A. (1977) *Secondary compounds as protective agents*. Annual Review of Plant Physiology 28: 479–501

Biological control

Bannon, J. S. (1988) *CASST^{TM} herbicide (Alternaria cassiae): a case history of a mycoherbicide*. American Journal of Alternative Agriculture 3: 73–76

Huffaker, C. B. (1957) *Fundamentals of biological control of weeds*. Hilgardia 27: 101–157

Obiefuna, J. C. (1989) *Biological weed control in plantains with Egusi melons*. Biological Agriculture and Horticulture 6: 221–228

Templeton, G. E. (1988) *Biological control of weeds*. American Journal of Alternative Agriculture 3: 69–72

Wapshere, A. J., Delfosse, E. S. and Cullen, J. M. (1989) *Recent developments in the biological control of weeds*. Crop Protection 8: 227–250

Werner, A. (1988) *Biological control of weeds in maize*. In: *Global Perspectives on Agroecology and Sustainable Agricultural Systems*. Allen, P. & van Dusen, D. (eds), University of California, Santa Cruz

Direct control

Appel, J. (1982) *Unkrautregulierung ohne Herbizide. Erfahrungen auf Betrieben der biologisch-dynamischen und organisch-biologischen Wirtschaftsweisen*. PhD Thesis, University of Hohenheim, West Germany

Ascard, J. (1989) *Thermal weed control with flaming in onions*. Proc. 3rd IFOAM International Conference on Non-Chemical Weed Control, October 1989. Bundesanstalt für Agrarbiologie; Linz

CRABE (1985) *Flame Cultivation for Weed Control*. Proceedings of 1st IFOAM Conference on non-chemical weed control. Available from CRABE, Rue des Wastines 7, 5974 Opprebais, Belgium

Desvaux, R. & Ott, P. (1988) *Introduction of thermic weed control in southeastern France*. In: *Global Perspectives on Agroecology and Sustainable Agricultural Systems*. Allen, P. and van Dusen, D. (eds), University of California, Santa Cruz

Foster, L. (1989) *The biology and non-chemical control of dock species* Rumex obtusifolius and R. crispus. Biological Agriculture and Horticulture 6: 11–25

Foster, L. & Woodward, L. (1989) *The ecology and biological control of docks*. New Farmer and Grower 23: 30–32

Geier, B. & Vogtmann, H. (1988a) *Der biologische Möhrenanbau und Möglichkeiten der Beikrautregulierung*. EDEN-WAREN GmbH, Bad Soden. (Only available from B. Geier, Ökozentrum Imsbach, D-6695 Tholey-Theley, West Germany)

Geier, B. & Vogtmann, H. (1988b) *Weed control without herbicides in corn crops*. In: *Global Perspectives on Agroecology and Sustainable Agricultural Systems*. Allen, P. & van Dusen, D. (eds), University of California, Santa Cruz

Hoffmann, M. (1989) *Abflammtechnik*. KTBL; Darmstadt

Hoffmann, M. & Geier, B. (eds) (1987) *Beikrautregulierung statt Unkrautbekämpfung. Methoden der mechanischen und thermischen Regulierung.* (Methods of mechanical and thermal weed control.) Proceedings of 2nd IFOAM Conference on non-chemical weed control. Alternative Konzepte 58. C. F. Müller, Karlsruhe

Lennartsson, E. K. M. *et al.* (1989) *The use of light exclusion techniques for clearing grass pasture in organic horticultural systems.* Proc. 3rd IFOAM International Conference on Non-Chemical Weed Control, October 1989. Bundesanstalt für Agrarbiologie; Linz

MAFF/ADAS and Scottish Agricultural College Advisory Leaflets on specific weed problems—the older versions concentrate more on cultural control methods, e.g. MAFF Leaflet 51 (1985): Thistles and MAFF Leaflet 50 (1937): Thistles in Grassland

Mattsson, B., Nylander, C. & Ascard, J. (1989) *Test of seven inter-row weeders.* Proc. 3rd IFOAM International Conference on Non-Chemical Weed Control, October 1989. Bundesanstalt für Agrarbiologie; Linz

Parish, S. (1987) *Weed control ideas from Europe visit.* New Farmer and Grower 16: 8–12

Plakolm, G. (1984) *Die Acker-Kratzdistel und ihre ökologischen Bekämpfungsmöglichkeiten.* German-language IFOAM Bulletin 50: 17–18

Weibel, F. P. & Niggli, U. (1989) *Unkrautkontrolle mit organischem Bodenbedeckungen in Apfelanlagen: Auswirkungen auf Unkrautbewuchs, Dynamik des Stickstoffs in der Bodenlösung und mikrobielle Bodenatmung.* Proc. 3rd IFOAM International Conference on Non-Chemical Weed Control, October 1989. Bundesanstalt für Agrarbiologie; Linz

Chapter 7

Pest and Disease Control

PESTICIDES: MORE OF A CURSE THAN A BLESSING?

The one aspect of modern agricultural technology which attracts most attention from environmentalists and other critics is the use of pesticides. Rachel Carson's book *Silent Spring* served to highlight, at an early stage, the environmental consequences of increasing pesticide use. Concerns at that time were largely centred around the use of chlorinated hydrocarbons such as DDT and their effect through the food chain on bird populations in particular. Although some of the worst offenders have been banned subsequently, others have continued in use until very recently and many of those which have been banned in developed countries are still actively promoted in developing countries.

Much has been written elsewhere about the problems associated with the use of pesticides (e.g. Dudley, 1987; Pesticides Trust, 1989), but it is still worth summarising some of the main problems.

- Pesticides are often chemicals which are alien to the environment and have the potential to disrupt a wide range of ecosystems, from the soil and its microbes right through to the higher animals. This disruption may be caused by direct toxicity, but usually the effects are more subtle: disruption of food chains, weakening of immune systems or confusion of the chemical signals by which many organisms communicate. Changes in behaviour can be as important as toxic effects.

  For example, Kickuth (1982) describes the effect of the herbicide active ingredient Phosphonylsarcosine on the water flea. The water flea relies on a chemical, sarcosine, produced during the decay of organic matter, to lead it to food sources. Phosphonylsarcosine is a closely related chemical which has the effect of confusing the water flea so that it no longer responds to sarcosine and hence cannot detect potential sources of food, which eventually leads to disruption of the food chain.

- Pesticide residues accumulate in foodstuffs for human nutrition, a factor being linked increasingly to allergies and other diseases (e.g.

Monro, 1984); these problems are now causing concern at the highest levels.

● Some pesticides have been linked to cancers, leukaemia in children and birth defects. Only a few have been banned; some, such as 2,4,5-T are banned in other European countries, but not in Britain.

● Many pesticides are directly toxic to humans: poisonings by pesticides are commonplace and affect both workers and people eating severely contaminated food. In 1983, the United Nations Economic and Social Committee for Asia and the Pacific estimated that two million people were poisoned by pesticides every year, 40,000 of them fatally.

● The process of manufacturing pesticides also pollutes the environment, whether by accident (for example Seveso in Italy and Bhopal in India) or through routine emissions; the severe contamination of the Rhine by the accident at the Sandoz plant was followed by releases at several other chemical plants along the Rhine, including the BASF plant at Ludwigshafen.

Governments are gradually beginning to take action to reduce the pesticide problem by introducing more rigorous environmental standards and in some cases, notably developing countries, banning considerable numbers of existing pesticides. Indonesia, for example, has recently banned 57 different pesticides after the realisation that they may be causing more problems than they were solving in the control of a rice pest, the brown plant hopper. The wolf spider and other natural predators were found to be much more effective, but were also destroyed by the pesticides. Greater emphasis is now being placed on integrated pest management.

In Europe, action by the European Commission has forced the British government to introduce maximum residue limits for pesticides (but only in certain foods), and to ban chemicals such as aldrin and dieldrin which have continued to be available many years after they should have been banned. The new constraints have also meant considerable difficulties for the agrochemical companies in introducing new pesticides; from February 1986 until the end of 1987, no new agrochemical ingredient was approved or provisionally approved in the UK.

New problems for farmers

The killing of predators of the Indonesian rice pest is just one example of the other important side to the use of pesticides, which is that they actually create additional problems for the farmer. The problem of increasing resistance to chemicals is now well known. Much publicity has been given to the MBC fungicides which were used for eyespot and *Fusarium* foot rot disease control in cereals, but are now virtually ineffective. This is thought to be due to resistant strains of the fungi becoming predominant through selective pressure. One controversial explanation which has been put forward is that these chemicals are also nematicidal in action, killing worms in the soil and reducing the rate of crop residue decomposition, thus increasing the probability of infection. Elements of fungicide resistance are also appearing with chemical controls for potato blight.

There are, however, many less obvious ways in which the use of pesticides creates problems for farmers.

- The use of herbicides eliminates the host plants for pest predators.
- Some pesticides disrupt the soil ecosystem, killing the microorganisms which would otherwise keep soilborne pests and pathogens under control.
- Some herbicides have been linked directly to increased pest problems, for example the active ingredient diallate has been linked with increased incidence of beet cyst nematode in sugar beet.
- Some fungicides have been shown to affect adversely the quality of bread making wheat by increasing alpha-amylase activity.
- Pesticide spray drift can adversely affect non-target crops.

The organic alternative

For these reasons, organic farmers reject the use of pesticides and favour instead a holistic approach to pest and disease control which integrates a wide range of cultural practices including:

- crop diversity through rotations and mixed cropping;
- organic manuring to stimulate soil biological activity; and
- the careful use of selected biological control techniques, naturally occurring plant extracts and minerals.

It has to be said, however, that pests and diseases are not generally a significant problem in well established organic systems, although certain specific problems do require remedial action. This stems from the fact that a healthy plant, given optimal soil conditions and balanced nutrition, will be better able to resist pests and pathogens. Disease (literally, dis-ease) arises when organisms are stressed or their environment is unbalanced. A healthy plant is one which not only shows no disease symptoms but can also actively resist the onset of an infection.

INFLUENCES ON PEST AND DISEASE PROBLEMS

The first stage of minimising pest and disease incidence in an organic system is to recognise that many of the problems which arise are actually caused directly by human intervention. The need for technological solutions to control pests and diseases is often a result of technological problems caused by simplified rotations, the lack of varietal diversity, increased pest resistance to chemicals, and the direct and indirect impacts of pesticides on beneficial insects and plant health.

In natural ecosystems, pests and diseases have a specific role: attacking weak points in the system and making space for better adapted species. Stability is achieved through balance, with pests and disease pathogens themselves being controlled by other organisms. Usually, therefore, their role is beneficial. It is only when they get out of hand that they are a problem, and that is much more likely to be a symptom than a cause of the disturbance. The existence of a pest or disease problem may therefore be seen as an indicator of inappropriate husbandry.

Agricultural chemicals

Griffiths (1981) used the term 'iatrogenic' plant diseases to refer to those diseases which result specifically from the use of agricultural chemicals. Herbicides and pesticides are compounds with high biological activity which, despite careful selection, tend to have some consequences on non-target organisms or physiological processes. Undesirable side-effects are common and these include the induction of 'new' diseases or, more usually, the exacerbation of diseases already present.

One effect which they can have is to change sugar concentrations, making the plant more or less prone to attack by pathogens such as fungi, depending on their preference for high or low sugar levels.2,4-D, for example, decreases sugar content while maleic hydrazide increases it and simazine increases nitrogen content. Also some biocides can increase leakage of metabolites from roots, making the root vicinity more attractive for pathogenic infestation.

There is some indication of biocide effects on natural defence mechanisms. Agrochemicals may stimulate the growth of pathogens by selectively inhibiting competitive or antagonistic organisms. There may also be more complex effects on the ecosystem which counteract the desired impact of the treatment, such as the development of 'secondary' pests which were initially unimportant, but have been able to develop because the primary pest or pathogen has been removed by the use of a pesticide.

This 'boomerang' effect, where the disappearance of a pest or pathogen is followed by its re-appearance in even greater numbers, is the main subject of a review by Coaker (1977) giving a wide range of examples of the ways pest problems may be exacerbated by the use of pesticides. The aphid problem is a good example of this, as has been demonstrated by the Boxworth Experimental Husbandry Farm comparisons of cereal production under high, medium and low intensity pesticide use. Clear adverse effects on predatory insects such as money spiders and the small ground beetle from the full pesticides regime have been shown, as well as on the springtails which support populations of aphid predators before the aphids arrive. Some of these effects became apparent after only one year, others after three years.

The late French plant physiologist Chaboussou, however, placed greater emphasis on the impact of pesticides on plant metabolism (Chaboussou, 1977 & 1985). In his last book, published shortly before his death, he brings together a considerable body of evidence to support his thesis, which is, summarised simply, that:

- healthy plants are able actively to resist pest and disease pathogens;
- this ability to resist attack is related directly to the synthesis of protein by the plant;
- protein synthesis can be disturbed either by the direct effect of pesticides or by unbalanced crop nutrition resulting in excess uptake of certain nutrients and in some cases the suppression of others (this may also happen when environmental factors such as temperature and humidity are not optimal);
- when protein synthesis is disrupted, there is a build-up of water-soluble sugars and nitrogen compounds as well as free amino acids in the plant tissue;
- these soluble compounds provide an ideal nutrient source for pathogens;
- given these ideal nutrient conditions, pathogens are able to reproduce at a faster rate, even to the extent of out-growing the natural predators which would normally keep them in check.

To illustrate his arguments, he not only examines evidence of the effect of pesticides (such as the organo-phosphorus compounds, carbamates and dithiocarbamates) on the nutrient content of plant tissue and their susceptibility to pathogens, he also points to the working of two commonly used pesticides, the copper-containing Bordeaux mixture and sulphur. Both of these help to reduce the incidence of fungal and bacterial diseases and are widely used, but the mechanism of their action is not well understood. These compounds are not directly toxic to the pathogen, so they must work in some other way. Chaboussou argues that it is the role of copper and sulphur in protein synthesis which is the key factor, so that the use of these substances, which are able physically to enter the cell tissue, actively stimulates the resistance of the plant to infection, rather than working directly on the pathogens themselves.

Much of Chaboussou's work concentrates on the physiology of the plant itself, rather than the complex interactions with other environmental and cultural factors, although he does recognise their role. For example, soils which have high organic matter levels and biological activity are known to suppress plant pathogens. This is usually believed to be due to the increased number and complexity of pest—predator relationships maintaining stability in the soil ecosystem (see below). Chaboussou argues, however, that the action of suppressive soils can also be explained by the fact that high microbial activity is essential for balanced nutrition, and therefore the protein synthesis theory still applies. He goes further to argue that the negative impact of pesticides on soil microbial activity also affects the plant by limiting nutrient availability.

Fertiliser type

Pesticides are by no means the only cause of pest and disease problems. Many other ecological factors and agricultural practices play a role. Hodges & Scofield, in their review of this topic (1983) coined the term agricologenic disease to refer to those diseases which were caused by agricultural practices in general, rather than the use of agricultural chemicals in particular (iatrogenic diseases).

The type and quantities of fertiliser used, both organic and inorganic, are an important aspect of this. Nutrient imbalances can occur if readily available mineral fertilisers are used in significant quantities, either because of luxury uptake of nutrients such as nitrates which are subsequently stored in the plant cells until use can be made of them, or because the presence of high quantities of nutrient ions in the soil solution blocks the release and uptake of other nutrients. In particular, the nitrate, ammonium and chloride ions have been shown to be harmful in this respect, with ammonium and chloride ions being directly toxic to young plants.

In one of the studies conducted at Haughley, the Soil Association's former experimental farm, the incidence of cereal thrips was linked to potassium and phosphorus nutrition. The effects of nutrient imbalances can also be seen in livestock, for example hypomagnesaemia, which is due to excess potassium and nitrogen uptake by grass at the expense of magnesium and is rarely, if ever, found on organic farms. Other minor or trace elements, such as sulphur, copper and boron, may be lacking and lead to deficiency symptoms or diseases.

Nitrogen has been studied in detail and it is widely recognised that the high content of soluble nitrogen compounds in plant cells leads to increased incidence of aphids and fungal diseases. This may be due to the increased attractiveness of the crop (increased cell size and thinner cell walls enabling easier penetration to extract the nutrients), but the primary effect is on reproduction rates given the increased availability of nutrients.

In a review of this subject, Huber & Watson (1974) concluded that: 'although a wide range of interactions of pathogens and their hosts are involved, it is generally the form of nitrogen available to the host or pathogen that affects disease severity or resistance, rather than the amount of nitrogen'. Nitrogen is assimilated by plants in both ammonium and nitrate forms (Chapter 3), together with very small amounts of organically bound nitrogen. Ammonium-nitrogen is rapidly converted to amino acids whereas nitrate-nitrogen can be stored. It is the nitrate-nitrogen which forms the main source of nutrients for pathogens.

The relative quantities of ammonium-N and nitrate-N which are taken up by plants is determined by a wide range of ecological and cultural factors, such as water availability, pH and temperature. The use of organic manures, and the emphasis on nutrient release by microbial activity for plant nutrition, means that a greater quantity of the nitrogen taken up by plants will be in the form of ammonium-N in organic farming systems. Corroborative evidence for this can be seen from studies of nitrate levels in organically produced crops (Chapter 15).

In a recent study, Eigenbrode & Pimentel (1988) set out to test again the assertion that plants supplied with nutrients from organic sources are more resistant to insects than those grown using chemical fertilisers. They compared two plots with fresh manure at two different levels, one with sheet-composted manure, one with chemical fertiliser and one with no added nutrients. They found that, during population peaks, flea beetle densities were significantly higher on plants receiving the chemical fertiliser compared with plants receiving similar amounts of macro-nutrients from manure, although the flea beetle population was higher on plants grown with sheet-composted manure than on those which received fresh manure.

In another study (Patriquin et al., 1988) aphid infestation of field beans on an organic farm was studied to test the effects of nitrogen fertilising and weed control on the problem. The researchers found an increase in aphid population where nitrogen fertiliser was used, and where plots were weeded. The results led them to suggest that in addition to enhancing the control of pests through their effects on natural enemies, weeds or intercrops can reduce the susceptibility of the host to pests by consuming soil nitrogen and restricting luxury uptake of nitrogen by the crop.

Organic manuring and a biologically active soil

The term organic manuring is intended here to cover all forms of organic soil amendments, including both livestock wastes and crop residues. Its importance lies not only in the form of the nutrients which plants receive, but also in that organic manures are a source of nutrients and energy for the soil ecosystem, with microorganisms subsequently making the nutrients available to plants in balanced proportions and distributed throughout the growing season. Another important feature of organic soil amendments is their ability to stimulate the complex of predatory microbes which help to keep potential pests and pathogens under control.

The role of organic matter in protecting crops from disease first became apparent because of the realisation that improvements in yield as a result of applying organic manures and composts were greater than could be explained in terms of nutrient content alone. Further investigations have shown that this 'humus effect' is associated with increased microbial activity, reduced aggressiveness and infestation of pathogens, increased viral resistance and a reduction in soil 'tiredness' or toxicity. The use of organic manures also allows the direct uptake by plants of specific chemicals, such as phenols, which are needed for the development of the plant's immune system.

The application of organic manures therefore makes a direct contribution to the anti-phytopathogenic potential of soils (i.e. their ability to keep soilborne pathogens in check). This is particularly important in the case of the fungal damping-off diseases such as *Rhizoctonia*, *Fusarium* and *Pythium*.

In some soils it has been found that high levels of organic matter and biological activity are directly associated with low levels of disease incidence, even when a pathogen is introduced in the presence of a susceptible host; such soils are known as 'suppressive soils'. It has also been found that crops in sterile conditions are sometimes subject to fungus diseases which in other circumstances would not be a problem.

One widely quoted example of suppressive soils (Broadbent and Baker, 1974) relates to a root rot (*Phytophtora cinnamoni*) of avocado which was found to be successfully controlled in recently reclaimed Australian rainforest soils by maintaining the suppressiveness of the soil through regular organic manuring, while avocado producers using mineral fertilisers were suffering serious losses from the disease. The researchers found that the addition of *P. cinnamoni* in amounts sufficient to cause severe root rot of plants in other soils produced little or no damage in the suppressive soil, which had higher populations of bacteria and actinomycetes than soils conducive to root rot. The suppression of sporangium formation was found to be due to microbial activity and was not related to the nutrient level of the soil leachate.

The mechanisms involved in soil suppressiveness and the contribution of organic manuring to the soil's anti-phytopathogenic potential have been the subject of increasing interest in recent years, many examples of which can be found in the bibliography at the end of this chapter.

The most important mechanism is the antagonism of soil microorganisms towards each other, which may take the form of production of toxins and antibiotics, competition for nutrients and energy, and/or parasitism. A soil with an active microbial population is likely to have individual microorganisms kept in balance by the action of their antagonists, which suppress the emerging parasite or hasten the digestion of propagules and pathogen-infested remains. Organic matter is essential for supplying the nutrients and energy (carbon) to maintain this biological activity.

When a fungus is short of food in a competitive environment, it will either be destroyed by its own enzymes or it can form spores which survive until new nutrients are available in a process known as fungistasis. The spores will germinate in the presence of plant root exudates (containing carbohydrates, amino acids, organic acids, other plant growth substances and vitamins) or other readily available nutrient sources.

The addition of fresh organic matter acts as a source of nutrients causing the spores to germinate, but it is important that the material has a wide carbon to nitrogen ratio so that there is no surplus nitrogen to feed the fungi, and that time is allowed so that the microorganisms can suppress the recently germinated fungi. The addition of organic matter also reduces the tendency of 'facultative' parasites, which are saprophytic as well as parasitic, to attack living material by increasing the availability of decaying organic matter. In one study, a crop of lettuce sown immediately following the turning in of a green manure was immediately wiped out by *Pythium*, whereas lettuce planted two weeks later survived. The *Pythium* was still present in similar numbers, but antagonists to the pathogen had more time to respond.

Very fresh manures or crop residues may also make problems worse because of the toxins which are produced during the early stages of the decay process.

One group of fungi which has a particularly important role in the plant's defence mechanism is the mycorrhizae. In addition to their role in plant nutrition, the ecto-mycorrhizae assist in disease control by forming a protective fungal jacket around the plant root. Plants with roots infected by mycorrhizae are known to be less disease susceptible than those without (Baker, 1968; Lynch, 1983; Cook, 1984) and are also more resistant to nematodes.

Various chemicals in the soil are also known to contribute to the anti-phytopathogenic potential of the soil. The breakdown of organic manures results in the release of carbon dioxide, high concentrations of which are harmful to some pathogens. At the same time, toxins produced by the plants themselves against other plants (allelochemicals) and by microbes which lead to the phenomenon of soil 'sickness' or 'tiredness' can be broken down by an active microflora or absorbed by humus colloids and thus rendered inactive.

The final mechanism which is thought to be involved is the improved 'vitality' of the plant itself as a result of physical/chemical improvement of the soil (structure and nutrient availability) and the fact that the resistance of the individual plant is aided by the uptake of phenols, phenolic and other compounds such as salicylic acid which have an antibiotic effect and work directly on pathogenic germs. Plants can take up organic molecules of high molecular weight such as aromatic polymers from lignin which have a catalytic influence on metabolic processes. In addition, specific antibiotics such as penicillin, strepto-mycin and bacitracin (molecular weight 1,500) taken up from the soil have been found in upper plant foliage.

Compost and compost extracts for disease control

The value of organic manuring for fungal disease control, and in particular the use of composts, has been confirmed in recent work in Germany (Gottschall *et al.*, 1987; Schüler *et al.*, 1989), where the incidence of *Pythium* infection of beetroot, peas and beans was reduced from 80% to 20% through the use of compost (Plate 7.1). Research in the United States using composted tree bark (Nelson & Hoitink, 1983) showed not only the benefits of using composted bark as an alternative to peat, but also the detrimental impact of compost sterilisation on the control of *Rhizoctonia solani* (Figure 7.1). Bruns and Gottschall (1988), however, also emphasise the need to carefully control the composting process to ensure that disease problems are not made worse.

These results have led to further work investigating the possible use of filtrates of soil bacteria and even compost extracts as a natural fungicide. The extracts can be made by mixing compost with water in a ratio of 1 : 4. The mixture is then shaken mechanically for two hours, allowed to settle and filtered.

At the University of Bonn in West Germany, researchers have used compost extracts to control mildew in sugar beet (Samerski & Weltzien,

Plate 7.1   Comparison of the influence of compost (right) and mineral fertiliser (left) on the germination of beetroot seedlings infected with the damping-off disease Pythium ultimum (S Ahlers)

1988), *Botrytis cinerea* in strawberries (Stindt & Weltzien, 1988) and blight in potatoes (Weltzien *et al.*, 1989). In the potato blight trials in 1987, both compost extract and a compost extract enriched with selected microorganisms were used. The unenriched extracts were not particularly effective although they did delay infection, but the enriched extracts proved to be more effective than even conventional fungicide treatment. In 1988 blight incidence was much less severe generally, so the differences observed were not so marked.

Figure 7.1   Comparison of *Rhizoctonia solani* infection of radishes grown in soils amended with compost and with peat.

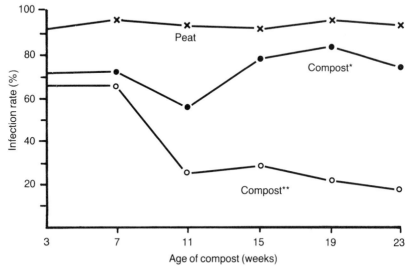

* Compost sterilised at 60°C after decomposition.
** Compost without additional sterilisation.

Source: Nelson & Hoitink, 1983.

In trials at the University of Kassel, the compost extracts reduced the incidence of powdery mildew (Thom & Möller, 1988) and soilborne root rot *Pythium ultimum* in peas and beetroot. However, results in many of the trials have been variable and highly dependent on the nature of the compost used and the method of extraction. The researchers at Kassel have not been able to duplicate some of the results found at Bonn, particularly with respect to potato blight. Research is continuing to resolve these difficulties and to see if the compost extract approach can be developed for practical application.

Plate 7.2   Comparison of compost extract (left) and water (right) on cucumber leaves infected with the mildew fungus Erisyphe cichoracearum   (R Gottschall)

Further developments
Much scientific effort is now going into identifying the individual fungi, bacteria and other organisms which help to keep plant pathogens in check. In particular, various strains of the parasitic *Trichoderma* fungus and other fungi such as *Gliocladium virens* have been identified which will control *Sclerotinia* and *Rhizoctonia* as well as the damping off diseases *Fusarium* and *Pythium* and black scurf in potatoes. In addition to physical penetration and strangulation of the organism it is attacking, *Trichoderma* produces anti-fungal substances which poison it.

The fungus *Verticillium chlamydosporium* is being used to control a variety of nematodes including cereal, potato and beetroot cyst nematodes. The fungus attacks the nematodes, either as the eggs before they hatch or the adult females before they can form root cysts. Another fungus, *Nematophtora gynophila*, is probably the main organism which keeps cereal cyst nematodes under control in Britain. An isolate of the actinomycetes, *Streptomyces ochraceiscleroticus*, has been shown to be effective against 19 different soilborne fungal diseases.

Various beneficial bacteria exist in the vicinity of plant roots, such as the plant growth promoting bacteria which produce compounds that bind iron, immobilizing it in the rhizosphere and so depriving harmful rhizobacteria and pathogens of a vital trace element, although some of these bacteria also restrict iron uptake by the roots. The *Agrobacterium* (which produces a chemical called agrobactin), is closely linked to the defence mechanism of plants and has been shown to promote iron uptake, while the *Pseudonomads* (which produce pseudobactin) restrict it. Pseudobactin is antibiotic, which may explain the ability of this bacterium to control other organisms. Work at Washington State University indicates the possible role of *Pseudonomads* in the control of take-all. Researchers at Bristol University have been looking at the *Bacilli* bacteria in this respect, but the response has so far been inconsistent. The literature (e.g. Cook, 1984 & 1986; Lynch, 1983 & 1988; Schneider, 1982; Chet, 1987) gives a wide range of other examples.

The next step for researchers is to look at how they can be introduced artificially into the soil, possibly as 'biological' seed dressings or fungicides. In Britain, work on this approach is being conducted by researchers at the Glasshouse Crops Research Institute, Littlehampton, on the use of *Trichoderma*, and at the University of Bristol on take-all control. This work is looking very promising, with *Trichoderma* already available commercially as Binab T.

These developments do, however, raise the question whether the solution really lies in this direction, even without taking into account the expense of producing bacterial or fungal preparations for widescale commercial applications. In organic farming the emphasis should surely be on creating the biological conditions in the soil so that the organisms are present in the first place.

Cook (1986, 1988) makes it clear that various cultural practices promote the destruction of plant pathogens partially or entirely by naturally occurring, antagonistic microorganisms, including, in addition to organic amendments, crop rotations, tillage, soil flooding and solarisation (a process of heating the soil to destroy pathogens by covering with clear plastic). There are also physical influences, such as well-structured soils which allow deep rooting and help avoid moisture-stress, and chemical factors such as sources of exchangeable calcium and the availability of nitrogen. All these influences need to be considered together, integrating the various biological, physical and chemical factors, in order to achieve optimal conditions for plant growth.

Rovira (in Schneider, 1982) argues that ' ... we have been naive in believing that the introduction into soils of suppressive microorganisms will lead to suppression; we have forgotten that such introductions will not succeed unless suitable niches exist or are created for the immigrant organism. '

Crop rotation in particular allows time for the destruction of pathogens and the alternation between host- and non-host crops in the case of parasites. Under certain conditions, however, soil suppressiveness may actually be induced through monocultures. The phenomenon of take-all decline, where the incidence of the disease diminishes after several years of cropping with the same crop, is now believed to apply to cereal cyst nematodes and common scab of potatoes as well, when these crops are grown in monocultures over long periods of time.

Work at Rothamsted and Woburn has shown that cereal cyst nematode decline is linked to the cereal root microorganism complex, with fungal parasites of the nematodes responsible for their control. The same fungi, which take some time to build up in the soil, will attack a range of crop cyst nematodes, including sugar beet, cereals, carrots, clovers and peas, but not potatoes. The phenomenon of pathogen decline is a strong indicator of the resilience and adaptability of natural systems to changed circumstances.

Soil suppressiveness can also be induced by tillage practices, by the innoculation of conducive soils with suppressive soils (which works in a similar way to the addition of organic matter), and by manipulation of other environmental factors. Of these, tillage is probably the most important. Aeration of soil increases microbial activity, thus controlling pathogens, but there is a clear conflict between intensive tillage and soil conservation.

Monocultures, diversity, complexity and stability

It is a widely held view among organic farmers that they suffer less from pest and disease problems than might be expected, a view which has been confirmed by comparative research between conventional and organic systems. For example, Motyka and Edens (1984) conducted a study of onion fly incidence where it was found that populations of the pest were higher and fluctuated widely in plots where insecticides had recently stopped being used, but in plots under organic management for a longer period of time (several seasons), they were much less significant. The stable organic systems also had lower pest incidence than conventional (high input) systems.

Part of the explanation why pest and disease problems may not be so widespread in organic systems lies in the ecological argument that species diversity (including both plants and animals) leads to greater stability in the agroecosystem, and therefore to reduced risk from sudden outbreaks of specific pests or diseases. The monocultures which characterise conventional farming systems (simple rotations, lack of genetic diversity within a species) leave such systems much more exposed to attack by pests and disease pathogens.

Some scientists, however, have questioned the theory that greater diversity of itself leads to greater stability. While agreeing that it should be possible to design agroecosystems which reduce the severity of insect pest problems by using ecological theory on pest population dynamics, the creation of diversity as such may not be very useful. They argue that the marked instability of agricultural ecosystems, compared with natural communities, results from the frequent disruption of crops by humans and from the lack in crop systems of species which have evolved with the crop itself. Physical complexity, such as patchiness in fields, may play more of a role in creating stability than species diversity. Pests may even occur because of too much diversity in the form of alternative food sources and refuges that are essential at some stage of the lifecycle of the pest.

Studies by scientists such as Pimentel (1977) and Altieri (1985) support the view that species diversity enhances stability, but that lack of genetic integration also plays a role. Many species have evolved together (co-evolved) and thus have an element of dependence on each other which is disrupted in simplified agricultural ecosystems.

A good example of the importance of co-evolution is provided by Patriquin and Baines (1989) in a study of the incidence of aphids in field beans. The presence of weeds was associated with lower aphid incidence and 55% higher yields in the absence of added nitrogen, because competition for nitrogen resulted in reduced luxury uptake by the beans and/or increased nitrogen fixation. In addition, the presence of aphids for short periods of about three weeks (as is common before natural control by predators becomes effective) actually increased yields by contributing to the reduction of apical dominance in a manner similar to pruning. It was only when they were present for longer periods that they caused significant yield losses, a situation which may occur when insecticides are used because of the destruction of beneficial insects. These beneficial crop–weed–pest interactions are a clear indication of the importance to crop plants of the plant and animal species with which they have evolved.

Genetic diversity is also an important factor because of the ability of pathogens to overcome host resistance, and can be seen in the proven benefits of the use of varietal mixtures described in more detail below. Diversity is known to help in the control of insect pests by providing camouflage for the crop or acting directly as a barrier or hazard, by providing alternative, more attractive hosts than an at-risk crop and by directly benefitting the activities of natural enemies. Considerable evidence in support of diversity providing an element of protection against pests in agricultural systems comes from the work on rotations and the benefits of the 'rotation effect' (diversity over time) (Sumner, 1982) and the work on mixed-cropping (diversity in space) which has been carried out over many years.

Ecological Pest and Disease Control

Any ecological approach to pest and disease control which is not dependent on the use of chemicals requires a recognition that there is no single factor which is responsible for a pest or disease problem and will rely on a range of husbandry practices which promote stability and balance between crops and their pests.

In particular, an ecological approach should seek to enhance the activities of natural enemies of crop pests including other insects, animals such as hedgehogs and birds, as well as fungal, bacterial and viral pathogens. A good example of the role of natural enemies or 'beneficials' is in the control of aphids in cereals. In addition to the ladybirds and the hoverfly larvae, there are more than 300 potential predators of aphids, including lacewings, beetles, spiders, mites and various other surface insects, as well as parasitic fungi and virus diseases.

The diversity and stability of insect pests and their predators will be influenced by:

- the diversity of plant species and structure within the field, including temporal and spatial arrangements;
- the composition, management and permanence of surrounding plant communities;
- the soil type and surrounding environment;
- the type and intensity of management;
- the distance of the crop from sources of colonists;
- the permanence of crops and the time available for colonisation;
- the complexity of trophic relationships between crop and non-crop plants, herbivores and natural enemies.

Various authors, including Pimentel (1977), Speight (1983), Brown (1984/5), Altieri & Letourneau (1982) and Altieri (1985, 1986, 1987) have considered the husbandry practices which can positively influence the crop ecosystem by altering the availability of the crop for pest exploitation in space and time, thus contributing towards the control of pests and diseases. These are summarised in Table 7.1 and considered in greater detail below.

Table 7.1 Non-chemical crop husbandry practices which influence the crop ecosystem with the aim of controlling pest and diseases.

Optimal site conditions (soil, climate, environment)
Cultivations.
Organic manuring and crop nutrition.
Stalk and residue destruction.
Soil moisture and irrigation.

Diversity over time
Discontinuity of monocultures.
Crop rotations.
Use of short-maturing varieties.
Use of crop-free or preferred host-free periods.
Manipulation of sowing and harvesting dates.

Diversity in space
Varietal mixtures.
Resistant cultivars.
Crop mixtures.
Strip/inter cropping (companion planting).
Mixed cropping.
Undersowing.
Soil cover.
Management of wild plants (weeds) in and around crops.

Altering pest behaviour
Use of trap crops.
Use of green manures, e.g. to stimulate hatching of nematodes at the wrong time.
Size, planting density and shape of crop, e.g. use of small scattered fields to create
     mosaic; spatial arrangement of plants.
Pheromones.

Biological control
Augmentative releases of beneficial insects and pathogens.

Optimal site conditions

The way in which the health and vigour of the crop can be influenced by soil conditions and by the sources of plant nutrients has already been discussed in some detail. Account also needs to be taken of conditions at the time of planting (e.g. soil moisture levels) and whether or not the crop being grown is suited to the environment in which it is being grown.

Crops introduced into a new region are likely to suffer from new pest or disease problems. The potato is a good example of this. Its introduction into North America allowed the potato beetle to become a pest, where previously the potato beetle had only fed on the wild sand bur, a relative of the potato. In Europe, the potato's susceptibility to blight is a further indication of a general tendency to produce crops in sub-optimal conditions. Conditions at specific locations within a region are also important: for example, potatoes should not be grown in soils which are too heavy; damp valley sites should be avoided for wheat and barley because of the risk of fungal disease; and carrots should not be grown on sites where there is little wind movement because of the risk of carrot fly.

Site conditions can be improved through soil cultivations and other husbandry practices to provide optimal conditions for germination and healthy growth. Examples include consolidation of the soil to prevent the development of nematodes, and cultivations to bury pupae or destroy the preferred habitats of snails and slugs. Stalk and residue destruction can play a similar role by removing sources of food and shelter. Irrigation to keep the soil moisture deficit above 18 mm for six weeks after tuber initiation has been shown to greatly reduce the incidence of potato scab.

Diversity over time

Rotations
Rotations have already been discussed at length in Chapter 5. They will remain the prime means of achieving diversity on a commercial organic holding because of the mechanical, labour and financial implications of more complex approaches such as polycultures. Their main advantage lies in the control of soilborne pests and diseases and other restricted feeding insects by creating a break between susceptible crops. The length of break may range from a few weeks, to avoid a green bridge for foliar diseases from one cereal crop to the next, to several years in the case of certain nematodes (see Table 5.1).

Where rotations are likely to be less successful is in the control of the more mobile insects and of pathogens which survive saprophytically in the soil or of foliage pathogens which are spread by airborne spores, although the initial primary inoculum in crop residues or in the soil may greatly influence the progress of a disease. In the case of bacterial plant pathogens, primary infections normally come from sources other than the soil, and thus with certain exceptions crop rotations have only a limited control over such bacterial diseases.

The manipulation of sowing, planting and harvesting dates
The other techniques which can be used to achieve diversity over time, such as the use of short-maturing varieties, the use of crop free or preferred host free periods and the manipulation of sowing, planting and harvesting dates, generally rely on the principle of depriving the pest of a suitable host at important stages of its lifecycle and reducing the time available for the pest to exploit the crop. These approaches can also have

the effect of altering the availability of the crop in relation to other types of vegetation.

In particular, changes to sowing and planting dates can help with the avoidance of the egg laying period of certain pests, the establishment of tolerant plants before an attack occurs, the maturation of the crop before a pest becomes abundant, the synchronisation of insect pests with their natural enemies, the synchronisation of crop production with climatic conditions adverse to pests and the production and destruction of the crop before the pest can enter its diapause.

Diversity in space

Varietal mixtures

One of the simplest ways in which diversity in space can be achieved in commercial organic farming systems is through the use of varietal mixtures. Modern plant breeding techniques have resulted in the abandonment of the genetically diverse ancient land races and the widespread use of a few, high-performing varieties which are very susceptible to the breakdown of pathogen resistance and provide ideal conditions for the spread of newly introduced exotic parasites (Marshall, 1977). A recent example of the narrowing of the genetic base is oil seed rape which has become the third largest arable crop in Britain, but is based primarily on the use of only two highly productive cultivars; now light leaf spot, alternaria and septoria have become significant problems.

Professor Martin Wolfe, formerly of the Plant Breeding Institute in Cambridge, and scientists at the National Institute for Agricultural Botany have put considerable effort into the development of varietal mixtures for cereals, and more recently into the development of cereal/grain legume mixtures such as winter wheat and field beans, barley and peas and oats and peas.

Wolfe (1985) found that mixtures usually yield at least as well as the mean of their components on their own, often more so, and sometimes exceed the highest yielding component. It is uncommon for a mixture to yield less than the mean of its components and rare for it to be as low or lower than the worst component. This fact means that recommended lists, based on results from single-variety trials, may not in fact predict the highest yielding varieties and that farmers generally might better be advised to grow mixtures of several of the best available varieties, rather than going for pure stands of a single variety.

The reasons why mixtures can be successful in reducing disease incidence are several. Different varieties generally have different disease resistance patterns, with some being more susceptible to specific diseases than others. Certain diseases, such as powdery mildew on cereals, rely on close contact with neighbouring susceptible plants for dispersal and this affects the rate at which spores multiply and the disease spreads. The ideal spatial arrangement, therefore, is one in which plants susceptible to the same pathogen race do not occur as neighbours. In addition, resistant plants can create a barrier effect by filling the space between susceptible plants. Resistance induced by non-pathogenic spores can also mean that when normally pathogenic spores land in the same area, they are prevented from infecting the plant or are limited in their productivity.

However, trials with winter wheat mixtures in organic systems in Germany (Stöppler *et al.* 1989) have had mixed results, possibly due to lower disease infection pressures. Further research is needed in the context of organic systems to fully assess the potential of varietal mixtures with respect to disease control and to the milling/baking quality of the end product.

The information published in the NIAB lists of recommended cereal varieties can be used to select suitable variety mixtures. Table 7.2 illustrates the concept with respect to mildew in spring barley. To use the table, first find the diversification group number for the preferred variety from the NIAB lists and then check that any subsequent varieties chosen are in compatible diversification groups. Care should be taken to ensure that the end product is acceptable to the purchaser, especially where milling wheat is concerned. Even where varietal mixtures are not used, the principles of variety diversification can be applied to the choice of varieties in neighbouring fields.

Table 7.2 Varietal diversification scheme to reduce the spread of mildew in spring barley.

Diversification Group	1	3	4	5	6	7	9	10	11
1	+	+	+	+	+	+	+	+	+
3	+	m	m	+	+	+	m	m	m
4	+	m	m	+	+	+	m	+	+
5	+	+	+	m	+	+	+	m	+
6	+	+	+	+	m	+	m	+	+
7	+	+	+	+	+	m	+	+	+
9	+	m	m	+	m	+	m	+	+
10	+	m	+	m	+	+	+	m	+
11	+	m	+	+	+	+	+	+	m

+ = good combination, low risk.
m = risk of mildew spread.
Source: NIAB.

Mixing crop species

The principles which apply to variety mixtures can be applied to different crop species as well. A variety of approaches exists including border-cropping, strip-cropping, inter-cropping, mixed-cropping or polycultures and undersowing. In the case of border-cropping, plants of a different species are planted around the edge of the crop. Strip-cropping refers to the use of strips of a different species at intervals within the crop, while inter-cropping is used to refer to alternate rows of the different species. Mixed-cropping or poly-cropping refers to the use of two or more species combined at random.

In terms of pest control, these techniques rely on the fact that pests recognise a suitable crop either by sight or by smell and that mixtures can be structured so as to confuse the pest. For example, thrips are white flies which are attracted to green plants with a brown (soil) background, but they will ignore areas of complete vegetation cover including mulches and weeds. Other pests may be attracted under the opposite circumstances. Even so, the size, planting density and shape of the crop can be significant in controlling pests.

Pests such as the carrot fly and the flea beetle do not rely, or rely to a lesser extent, on visual recognition. They depend instead on 'smell' or chemical signals from the plant. Trap crops such as wild mustard can be used as borders or in strips to attract flea beetles away from cultivated brassicas, because the wild species has significantly higher concentrations of the chemical allylisothiocyanote, a powerful attractant of flea beetle adults. This technique may not always be reliable or safe and could under certain circumstances exacerbate the problem and should therefore be used with caution. Flea beetles also rely to some extent on the recognition of bare soil and research has shown that their incidence can be reduced by increasing planting densities.

Mixing species can also assist in pest control in a number of other ways, including influencing the availability of light which can affect insect behaviour; creating barriers, for example using tall non-host plants; increasing the distance between host plants; influencing the microclimate; reducing the likelihood of the selection of resistance-breaking genes; and acting as alternative hosts for natural enemies. Polycultures generally contain a greater abundance of natural enemies.

Many crop combinations are possible and each can have different effects on insect populations. The choice of tall or short, early or late maturing, flowering or non-flowering companion crops can magnify these effects. Beneficial insects are more likely to be attracted to flowering plants, for example.

The mixed-cropping of different cereal species, such as wheat and rye, and the mixed-cropping of cereals and grain legumes has proved to be potentially successful, although trials on organic farms by Martin Wolfe and Elm Farm Research Centre indicate a need for further research to determine the optimal seed rates. In the case of cereals and grain legume mixtures, the harvested crop can be separated as part of the grain cleaning process. In some cases, markets may exist for cereal grain mixtures, but these need to be identified in advance.

The mixtures need to be selected so that they mature at the same time and can be harvested together. Combinations of winter wheat and winter field beans as well as spring oats or barley and peas will mature together (see Colour plate 9). At Elm Farm, the winter wheat and winter field beans performed best when sown at 75% of normal seed rates for both components. With the barley/peas mixture, the barley seed rate needs to be reduced to only 3% (i.e. about 5 kg/ha) of the normal seed rate in eastern counties of the United Kingdom, producing just sufficient barley to act as a support crop for the peas, but resulting in a 10% reduction in the area needed to produce the same total quantity of peas and barley separately. In the west and north of the United Kingdom, the quantity of barley may have to be increased.

The crops do not necessarily have to be mixed to gain these benefits. Martens (1983) found that aphid populations were significantly reduced when oats were cropped in strips with peas as compared with oats on their own. The lower populations were also associated with increased predator numbers. Grass strips dividing large fields of cereals can also be beneficial, allowing pest predators to migrate further into the field.

The use of species mixtures is more widely practised in horticultural systems and in certain 'traditional' agricultural systems, particularly in developing countries. Altieri & Letourneau (1982) give several examples of crop mixtures, the pests controlled and the factors involved in their control (Table 7.3).

Work at Cambridge University's Department of Applied Biology has concentrated on brassicas such as cauliflowers and cabbage grown with both broad and dwarf beans. Aphids are attracted by the silhouette of the plant and this type of companion cropping provides a form of camouflage.

Another common 'companion crop' is onions with carrots to mask the smell of carrots from carrot fly. In the early stages, this can lead to a reduction of up to 70% in fly numbers, but once the onions start to form bulbs, they are less effective in repelling the pests and numbers are only cut by 30%. The onions need to be sown at the same time as the carrots. Although they are only fully effective in the first generation of the fly, the second generation will be correspondingly smaller. Marigolds represent another possible option.

Undersowing of brassica crops with clover has also shown promising

Table 7.3 Selected examples of multiple cropping systems that effectively prevent insect pest outbreaks.

Multiple cropping system	Pest(s) regulated	Factor(s) involved
Cabbage intercropped with white and red clover	*Erioischia brassicae*, cabbage aphids, and imported cabbage butterfly (*Pieris rapae*)	Interference with colonization and increase in ground beetles
Cotton intercropped with forage cowpea	Boll weevil (*Anthonomus grandis*)	Population increase of parasitic wasps (*Eurytoma* sp.)
Intercropping cotton with sorghum or maize	Corn earworm (*Heliothis zea*)	Increased abundance of predators
Cotton intercropped with okra	*Podagrica* sp.	Trap cropping
Strip cropping of cotton and alfalfa	Plant bugs (*Lygus hesperus* and *L. elisus*)	Prevention of emigration and synchrony in the relationship between pests and natural enemies
Strip cropping of cotton and alfalfa on one side and maize and soybean on the other	Corn earworm (*Heliothis zea*) and cabbage looper (*Trichoplusia ni*)	Increased abundance of predators
Cucumbers intercropped with maize and broccoli	*Acalymma vittatta*	Interference with movement and tenure time on host plants
Corn intercropped with sweet potatoes	Leaf beetles (*Diabrotica* spp.) and leafhoppers (*Agallia lingula*)	Increase in parasitic wasps
Corn intercropped with beans	Leafhoppers (*Empoasca kraemeri*), leaf beetle (*Diabrotica balteata*) and fall armyworm (*Spodoptera frugiperda*)	Increase in beneficial insects and interference with colonization
Intercropping cowpea and sorghum	Leaf beetle (*Oetheca bennigseni*)	Interference of air currents
Maize intercropped with canavalia	*Prorachia daria* and fall armyworm (*Spodoptera frugiperda*)	Not reported
Peaches intercropped with strawberries	Strawberry leafroller (*Ancylis comptana*) and oriental fruit moth (*Grapholita molesta*)	Population increase of parasites (*Macrocentrus ancylivora, Microbracon gelechise* and *Lixophaga variabilis*)
Peanut intercropped with maize	Corn borer (*Ostrinia furnacalis*)	Abundance of spiders (*Lycosa* sp.)
Sesame intercropped with corn or sorghum	Webworms (*Antigostra* sp.)	Shading by the taller companion crop
Sesame intercropped with cotton	*Heliothis* spp.	Increase of beneficial insects and trap cropping

Multiple cropping system	Pest(s) regulated	Factor(s) involved
Squash intercropped with maize	Acalymma thiemei, Diabrotica balteata	Increased dispersion due to avoidance of host plants shaded by maize and interference with flight movements by maize stalks
Tomato and tobacco intercropped with cabbage	Flea beetles (Phyllotetra cruciferae)	Feeding inhibition by odours from non-host plants
Tomato intercropped with cabbage	Diamondback moth (Plutella xylostella)	Chemical repellency or masking

Source: Altieri & Letourneau (1982)

results. Dempster & Coaker (1974) report some success in the control of cabbage root fly in Brussels sprouts and cabbages with increased numbers of predators present, although there were significant difficulties still to be overcome with the competition between the crop and the clover. White clover is better suited than red and other species such as annual or even subterranean clovers may be worth considering. Much of the benefit of undersowing clover is the attractant effect of the flowers on beneficial insects. This principle can thus be extended to many other crops including maize and even permanent crops such as vines.

Although all these techniques show significant potential, they are often difficult to implement in practice, particularly in commercial situations, and they may considerably complicate the process of keeping crops relatively weed free. The future development of ecologically appropriate species mixtures will therefore depend on full consideration being given to their mechanical, labour and financial implications.

Habitat enhancement for pest control
Habitat enhancement refers to the management of non-crop vegetation (weeds, wildflowers, hedgerows, windbreaks or understories in orchards) to influence the populations of agricultural pests and beneficial insects. For example, in the sheltered area on the leeward side of hedgerows and windbreaks, flying insects can be present in high numbers. Aphids and thrips can cause serious damage in such circumstances. The micro-climate of sheltered fields can also be altered by hedgerows and windbreaks, which may favour either the pest or its natural enemies. The level of infestation is also related to the height of the hedge, with the distance of insect penetration of the crop up to ten times the height of the hedge to leeward and only twice the height to windward.

Non-crop vegetation such as weeds can benefit pests by acting as secondary host plants, reservoirs of insect-borne diseases and preferred overwintering sites for insect pests. Pests which survive outside crops on unploughed headlands and in hedges include carrot fly, which commonly overwinters on nettles, the cabbage root fly which lives and breeds on cow parsley, the oat aphid which lives on the bird cherry and the black bean aphid which uses the spindle tree as an alternative host.

In a review of the role of non-crop vegetation in pest control, Altieri and Letourneau (1982) found that the risk of pest infestation from borders was greater the more closely related the border species are to the crop. Woody borders such as hedges are likely to reduce the risks of infestation where the predominant crops are grain, vegetables and forage plants. However, weeds and ornamentals can also act as trap crops, attracting pests and limiting damage to crop plants, or they may emit chemical repellants, and they play an important role in increasing the population of beneficial insects by providing nectar and pollen, nesting sites, alternate prey and/or overwintering sites. It is important in this respect to realise that natural vegetation cannot be categorised as either 'good' or 'bad'—the fact that a weed may act as a secondary host for a pest also means that there is a chance for beneficial insect populations to develop, possibly in time to be effective within the crop.

An illustration of the contribution of edges of natural vegetation to the dynamics of arthropod communities of adjacent cultivated areas is given in a study of Californian apple orchards by Altieri (1985). Altieri found that edges of the organic orchards supported considerably more natural enemies than the edges of the sprayed orchards. In the early season, considerably more aphids invaded the sprayed orchards than the organic ones.

In general, plant diversification based on natural vegetation ('weeds') can create significant environmental opportunities for natural enemies and thus improve biological pest control. Weed borders can have an enriching effect which may extend up to 40 rows into the field, indicating the possible benefits and use of strategic cultivations to enhance populations of certain weeds. Trials at Rothamsted found that avoidance of herbicides and undersowing of cereals significantly increased the number of aphid predators and helped with aphid control.

Most beneficial insects present on weeds tend to disperse to crops, but sometimes the prey found on weeds can prevent or delay dispersal. Allowing weeds to grow to assure concentrations of insects and then cutting them to force movement could be an effective strategy.

The use of habitat enhancement probably has most relevance in permanent crops such as orchards and vineyards, where rotations and other forms of species diversification are impractical. Robert Crowder of the Biological Husbandry Research Unit at Lincoln College, New Zealand, has achieved considerable success in the control of codling moth in apples through the use of flowering plants as an orchard under-story (Crowder, 1987), especially the *Umbelliferae* such as carrots, parsnips, etc. which support hymenopterous parasites of the codling moth larvae (see Colour plate 17). This is combined with greater use of shelter belts and headlands so that beneficial insects are encouraged. If suitable flowers are selected, they may provide subsidiary cash income. Crowder's work supports the findings of Altieri's study in Californian apple orchards reported above. Similar effects have also been reported from a comparison of extensive and intensive orchards in Europe (Mader, 1984) which also found a correspondingly greater use of the extensive orchards by birds which were influenced both by the physical shape and height of the trees and by the enhanced biological diversity.

Altering the behaviour of pests: trap crops and green manures

The use of trap crops as an alternative host for flea beetles has already been described above, but they can also be used as part of a green manuring programme for the control of nematodes. The basis of this approach is that when a nematode susceptible crop of mustard or rape is grown, the nematode cysts are encouraged to hatch and to invade the roots. If the crop is then killed, the invading juveniles will die as well, but if the catch crop is not destroyed at the right time, it could do more harm than good. An alternative is the use of nematode resistant catch crops which may encourage the eelworms to hatch but chemicals released by the plant prevent them from developing into mature adults.

Kahnt (1983) describes in some detail the use of green manures for this purpose. Examples include the use of oats ploughed in three to six weeks after sowing for cereal cyst nematode control, but this is risky because of host weeds which may survive in a non-host main crop. Rape as a green manure can help with the control of the wheat bulb fly and the cabbage gall weevil, but all these approaches could make the situation worse if the crop is left too long before being ploughed in, for instance as a result of a prolonged period of bad weather.

The choice of a suitable green manure will depend on the character-istics of individual species. For example, suitable host plants for a green manure (maximum four—six weeks) to control beet cyst nematode include sugar beet, fodder beet, beetroot, mangel, winter spinach, swede, rape, mustard, brassicas, turnip, fodder radish and weeds such as wild radish, charlock, shepherd's purse, chickweed, dandelion, and orache species. Antagonist plants which deter the beet cyst nematode include lucerne, clovers, winter rye, maize, field beans, peas, vetches, chicory, onions, serradella and linseed. Neutral, non-host species in-clude potatoes, barley, oats, wheat, lupins, dwarf beans, carrots, hemp, poppies and *Phacelia*. One potentially very good green manure species is the corncockle, which in trials has achieved 60% control of beet cyst nematodes, which is better than other resistant varieties.

Similarly, oats, barley, wheat, and to a lesser extent rye and some grasses, act as hosts for cereal cyst nematodes; potatoes and tomatoes act as hosts for potato cyst nematodes; and all cereals, maize, carrots, peas, lettuce act as hosts for grass cyst nematodes.

Infestations may be reduced by legumes and row crops in the case of cereal cyst nematodes, resistant varieties, clovers, buckwheat, oats, beet and sheep's fescue in the case of potato cyst nematode and beets, crucifers, potatoes and chicory in the case of grass cyst nematodes.

An alternative approach to the control of beet cyst nematodes is the use of resistant varieties of trap crops such as Emergo white mustard and Pegletta oil radish which allow only the male nematode to penetrate the roots. The female nematodes, which need 40 times as much food as males, are encouraged to hatch, but are denied a food source. Control is unlikely to be 100% as not every single plant is resistant, but the green manure does not need to be ploughed in quickly. There can, however, be a subsequent problem with the radish beet interfering with sugar beet production.

Plant resistance to diseases and pests

Like humans and animals, plants have their own sophisticated defence mechanisms to protect them from attack by pathogens and pests. These range from physical deterrents such as thorns to biochemical substances which either act as chemical signals in the ecosystem, sending messages via the senses such as taste and smell to discourage herbivorous activity (antifeedants), or are directly toxic or may cause sterility or failure to reach sexual maturity (Swain, 1977). In some cases, they may also encourage predators of pests and pathogens.

The substances which are known to influence pest activity are wide ranging and include amino acids, sugars, germination and growth inhibitors such as scopoleptin and trans-cinnamic acid, enzymes, phenols, alkaloids, saponines, glucosinolates and cyanogenic glycocides. Their activity is usually as a complex of substances, with no one compound solely responsible for a particular effect. This has the advantage that it is much more difficult for a pest or pathogen to overcome the plant's resistance. Some of these substances are also linked to the allelopathic interactions between plants which have been discussed in the previous chapter.

The alkaloids and saponines are of particular importance. Alkaloids occur widely in plants and in many cases probably act as repellents and toxins towards parasites, predators and competitors, a fact which is becoming more widely accepted as research progresses. They may also be toxic to alkaloid-containing plants themselves in certain situations.

The role of alkaloids in medicine is well known. Digitalis from foxgloves has been used for many years as a heart stimulant, similarly atropine derived from deadly nightshade which is used as a muscle relaxant. More recent examples include ricin from the castor oil plant, which is being used for treating tumors. Some of these are deadly when absorbed in too large quantities. Many foods, drinks and other substances which humans consume regularly also contain alkaloids, for example caffeine and nicotine.

In agricultural situations, examples include solanin (from potatoes) which repels potato beetles and the use of nicotine as a naturally occurring, but wide spectrum, pesticide for aphid control. Some of these substances disappear as the crop ripens. One example is tomatin (found in green tomatoes) which protects the tomato against storage pests. More recently, scientists have found a range of alkaloids that are analogous in shape to various sugars, yet act differently. One known as DMDP mimics fructose and low levels kill off the larvae of the bruchid beetles.

Disease pathogens may be kept under control by a similar range of mechanisms. An important factor in resistance to fungus development is the strength of the waxy cuticular layer and the morphology of the stomata. The trichomes of various types which cover the plant surface may act as a physical barrier or be a source of defensive compounds such as tannins (which can be quite potent antibiotics) and phytoalexins, a catch-all term for a variety of compounds with antimicrobial properties. For example, the potato is known to respond to an attack of blight with the production of phytoalexins and may also adopt a barrier defence, depositing lignin-like material in the tuber and leaves or encasing the fungal filaments.

Phytoalexins inhibit the development of fungal pathogens in hypersensitive tissue and are formed or activated only under stress conditions, when the host cells come into contact with the pathogen. The defensive reaction occurs only in living cells and the phytoalexin is non-specific in its toxicity towards fungi, although fungi may exhibit different sensitivity to it. The resistant state is not inherited, unlike the immune system in animals. Phytoalexins need to be produced in sufficient quantities and it is thought that disease susceptibility may be due to the inability of the infecting fungus to stimulate the formation of a phytoalexin or to its capacity to be tolerant of the level of phytoalexin produced. Much more work is required to understand the mechanisms which induce phytoalexin production, but this may provide a means to achieving control of problematic diseases such as potato blight in the future.

In addition to substances produced by the plant itself, certain bacteria in the soil produce their own 'natural' pesticides when stimulated by a plant at risk of attack. Opportunistic pathogens, mostly fungi, take advantage of pre-existing damage to invade the plant. The rhizosphere bacteria, such as *Agrobacterium*, are attracted to wounded plants by miniscule amounts of fluid leaking from the wound. Bacteria respond to these chemical signals by swimming towards the site of the wound where they produce toxins poisonous to the fungi.

Resistant varieties
One way in which all these various defence mechanisms are expressed is through the characteristics of individual varieties and in particular their disease resistance. The use of resistant varieties therefore forms a very important part of pest and disease control in an organic system. Variety selection also needs to be based on the physiology and growth pattern of different varieties, the timing of growth and maturity, locational constraints and, ideally, the availability of certified seed. However, care needs to be taken to ensure that resistant varieties are used in such a way so as not to encourage the development of pathogen strains or pests which are able to overcome the resistance. In particular, the cultivation of large areas of hybrid varieties represents such a risk, and concepts of varietal mixtures discussed above can be usefully applied.

Modern plant breeding objectives take account of this problem by distinguishing between vertical and horizontal resistance. Vertical resistance is where specific resistance against an individual race of a pathogen is contained in a single cultivar. This can be overcome relatively easily by a pathogen. Horizontal resistance is more general, aiming not for total resistance to a specific race, but to cover all races of a pathogen by concentrating on genetic material which confers partial resistance. It is believed that in this way it should be possible to produce new, resistant varieties whose resistance is longer-lived than in the past.

Plant extracts
The extraction and use of some of the substances involved in plant defence mechanisms may help to enhance the resistance of crops, particularly in the case of the more intractable disease and pest problems such as potato blight. In fact, the use of plant extracts has long been espoused by organic horticulturalists. Preparations of horsetail, onions, garlic and horseradish are used against fungal diseases, the latter against *Monilia* in fruit trees. Extracts of stinging nettles, comfrey, tansy, bracken, wormwood (absinthe) and chamomile have been used against aphids and other pests. Table 7.4 contains some typical examples.

Table 7.4 The use of plant extracts in organic horticulture.

Plant	Quantity/Concentration	Production	Application
A. For seed treatment			
Valerian	10 drops/litre water	From chemist	Put seeds in perforated container
Chamomile	Normal tea concentration	Hot water on flowers, allow to cool	and dip into extract several times, air dry
Horsetail and waterglass	50 g/litre water 50 g/litre water	Soak in water, boil, cool and add waterglass	
Horseradish	100 g/litre water	Leaves or roots (horseradish) or cloves (garlic)	Moisten seeds and allow
Garlic	100 g/litre water	chopped finely and mixed with cold water	to dry
B. For the control of foliar fungal diseases			
Horsetail	15 g/litre water (dried) 100 g/litre (fresh)	As for seeds (keeps up to 3 days)	Use on gooseberry,
Garlic	10 g/litre water	As for seeds (always use fresh)	roses, fruit, cucumber, tomatoes,
Onion	15 g/litre water	As horsetail (use fresh)	strawberries
Horseradish	30 g/litre water (leaves or roots)	As for seeds (use fresh)	Use on fruit at flowering against *Monilia*
C. For the control of insects			
Tansy	30 g/litre water (fresh)	Pour boiling water over material	
Wormwood	3 g/1 litre (dried)	and allow to cool	Use against sucking insects,
Stinging nettle	100 g/litre (fresh) 20 g/litre (dried)	Soak for 12 hours in water,	esp. aphids, partial effect
Rhubarb	200 g/litre water	strain	against mites and thrips
Quassia	75 g/litre water	Soak for 24 hours, boil, (poss. add 50 g/litre soft soap), cool	

Source: zum Beispiel 14/5/86 pp. 6−7.

Much of the effect of plant extracts on disease is believed to be through structural strengthening of the plant, increasing its resistance to the penetration of fungal mycelia and sucking insects such as aphids, or through encouraging vigorous growth to overcome an attack, rather than by any direct toxicity. For example, seaweed extracts contain nutrients, trace elements, growth substances and vitamins and are thought to increase the plant's resistance via a growth response. On the other hand, horsetail contains the fungitoxic saponin equisetonin, as well as silicic acid which confers structural benefits. There is not usually one single substance responsible, but a complex interaction between a range of agents.

With insect pests, the extracts may be repellant as in the case of tansy and wormwood, or they may be directly toxic as in the case of pyrethrum, derris and quassia. Repellant substances can be used prophylactically, possibly supplemented by rock dusts or rhubarb and stinging nettle extracts which are believed to act as irritants.

Considerable work has been done on the testing of plant extracts, but the results have been variable. For example, Matthews and Schearer (1984) found that tansy extracts deterred feeding by adult Colorado potato beetles, but these results were not confirmed in field experiments. Aqueous extracts of bracken have an inhibiting effect on both weeds and fungal development, but again their use in practice has been problematic. Onion extracts have been shown to be effective against potato blight, but only in the laboratory.

Researchers at the University of Kassel in West Germany tried plant extracts and rock dusts in trials to control potato blight and chocolate spot disease in beans without success. In subsequent trials, which also included the use of soft soap and soya lecithin, some plant extracts did have a limited effect on stem rust in blackcurrants. However, the soft soap preparation was more effective than soya lecithin on its own and soya lecithin plus seaweed extract at controlling this disease. Various extracts were also tried against black bean aphids, including nettle, garlic, field horsetail, wormwood and quassia. Only the quassia treatment proved to be effective and this may have been by dissuading aphids from approaching the treated area, rather than direct insecticidal action.

Tests in Switzerland on alternative seed dressings to control *Septoria* and *Fusarium* incidence in wheat found that horsetail extract compared well with mercurial seed dressings, partly because of a growth promoting effect, whereas garlic and a garlic/horseradish mixture had only a limited effect and reduced germination rates. Horsetail eliminated *Septoria* completely, which was not achieved by any other dressing. However, these tests were done under glasshouse conditions and further field research is needed.

There are several reasons for this variability. Many of the compounds are aromatic or water soluble or are rapidly degraded by sunlight and cannot be successfully applied in the field because they disappear too quickly. The quality and concentration of the active substances may vary by as much as 500% from season to season and with the location, age and maturity of the plant material from which the extract is taken. Not every plant of the same type contains an active ingredient and, if it is present, the substance is often concentrated is a specific part of the plant, although often for commercial reasons the whole plant is used. Further difficulties include the fact that the active substance is usually chemically bound to a sugar and needs to be separated by an enzyme before it becomes active, and that rapid drying prevents or reduces the splitting and therefore its effectiveness. Soaking may also destroy the substance or allow it to escape. As a result, accurate dosing is virtually impossible when using homemade preparations.

In addition to the problem of seasonal variability and short-term activity, side effects do exist. The extracts may work against beneficial insects and non-pathogenic microorganisms including those which are antagonists. Preparations which have an effect on leaf surface micro-organisms may therefore be beneficial, in that they delay the ageing process of the leaf and hence lead to higher yields, or harmful, in that the antagonistic microorganisms are also adversely affected allowing the pathogen to recover and develop unchecked. In addition, many fungi depend on the existence of plant wounds in order to penetrate tissue. It is probable, therefore, that the major effect of plant extracts will only be seen in already weakened or sick plants.

There are certain plants, however, whose extracts have proved to be reliably effective, even when prepared on the farm without sophisticated equipment or knowledge. One example is docks, whose intractability as weeds owes much to their resistance to virtually all pathogens and insect pests, which seems to indicate that they have a particularly well developed defence mechanism.

This led to recent trials at the Swiss Research Institute for Biological Husbandry to see if extracts would be effective against fungal diseases. The dock extract* proved to be the most effective of 40 different sub-stances tested in the laboratory against powdery mildew on cucumbers and apples, and was comparable with sulphur and the fungicide Bayleton (Figure 7.2). In field trials the effects were confirmed, but were not quite as good as in the laboratory experiments, because the active substance is quickly broken down by warmth and light and is washed off in heavy rainfall. In high rainfall periods, treatment needs to take place every seven to eight days. The extracts were not effective, however, against *Septoria* leaf disease and *Botrytis* in strawberries. Anthracinones

*Dock extract can be prepared by taking fresh, mature dock roots (at least two years old) and cutting them into small pieces. These can be stored deep frozen, or immediately pureed.15 g of puree is needed for 1 litre of end-product. Mix with some water and allow to stand for an hour. Filter and dilute the filtrate to the desired concentration. The extract can be used directly or deep frozen for later use. Under warm storage, it loses its effectiveness.

Figure 7.2   The effectiveness of dock extract for the control of powdery mildew on apples compared with more conventional treatments.

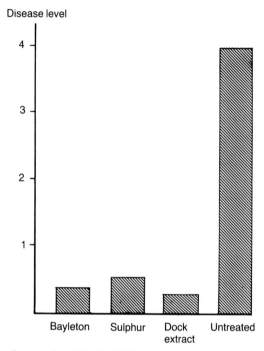

Source: Zum Beispiel, 15/3/88, p.19.

are the likely active agents. The extracts need to be used prophylactically because they work by restricting the formation of organs which allow the fungus to penetrate the leaf.

Another plant extract which has clearly demonstrated its value is derived from the seeds of the Neem tree which is widely distributed in dry regions of the tropics and the sub-tropics (Schmutterer and Ascher, 1987; Hellpap, 1989). The leaves, fruits and seeds all contain a range of active substances, the most important of which is azadirachtin. The substances act against a wide range of herbivorous insect pests such as beetle larvae, caterpillars, leaf miners and grasshoppers in a number of ways: as a repellant and anti-feedant, especially with grasshoppers and locusts; reducing reproductive rates; killing the larvae by disrupting growth and moulting; and affecting the insect's hormonal system.

The importance of Neem extracts compared with other potent plant-derived pesticides such as pyrethrum is that there is only a slight contact effect; the substance has to be ingested so that the effect on beneficial insects is limited, but this also means that sucking insect pests like aphids are not controlled. In addition, because a wide range of substances is involved, there is little chance of pest resistance developing, although this risk could occur if commercial production were to concentrate on one single active agent. Tests have also shown no toxic effects on humans and other mammals.

The major advantage of Neem extracts for developing countries is that they can be prepared without advanced technical equipment by the farmers themselves, although there is still a problem of variability in the content of active agents depending on the season, harvesting, drying and storage, which may mean that medium-scale industrial preparation is preferable.

Although there are many problems to be overcome, not least in terms of understanding the exact mechanisms involved and the conditions which are necessary for optimal utilisation of plant extracts, there is a strong case for further research and development in this area.

In Britain, the use of home made substances for pest and disease control is technically illegal under a strict interpretation of the Food and Environment Protection Act's Pesticides Regulations (1986), even if the substance is simply a water extract of plant or compost material, and the use of these substances may be subject to a fine of up to £2,000. This also covers substances which are currently commercially available, such as some soft soaps, but which have never been officially approved. However, if the substances are used as a foliar feed and not for pest control, then they are no longer covered by the regulations and may be used without restriction.

Some plant extracts are now available as commercial preparations, such as Bio-S which contains various herbs and also sulphur, which is directly toxic. Efforts are being made to find an alternative to sulphur. One such is Bio-Blatt which contains lecithin from soya beans and can be used prophylactically against powdery mildew. It should be realised, however, that plant extracts are prophylactic rather than curative and are not, therefore, a general solution. They should only be used when absolutely necessary and in the context of other cultural control measures.

Certain other plant-derived 'natural' products such as pyrethrum, rotenone (derris) and quassia are allowed only on a restricted, non-routine basis under the Soil Association's standards for the control of pests such as flea beetles, leather jackets and carrot fly. This is because they are either broad spectrum, killing beneficial insects as well as pests, or they have harmful side effects on fish or other organisms. The Soil Association has recently published a report detailing the risks associated with some of these substances (Dudley, 1988).

Pyrethrum is derived from the leaves of chrysanthemum species and can be used against sucking insects (and biting ones as well to some extent). Rotenone comes from the roots of certain legumes and can be used against various insects. Both pyrethrum and rotenone are poisonous to fish. Quassia is derived from the tropical 'bitter' wood and can be used to control aphids, but is toxic to all insects. Priority should therefore be given to alternative control methods, in particular under-standing the lifecycle of the pest and working around it where possible.

Permitted mineral/chemical sprays

Certain minerals and chemicals are allowed in organic systems for the control of problem pests and diseases. Some of these, such as waterglass (sodium tetrasilicate) and other silica containing preparations are used to physically strengthen the plant's resistance to attacks by pests and diseases. Silica is built into the walls of plant cells where it increases mechanical resistance to penetration by sucking insects or fungal mycelia and also increases the resistance of cells to breakdown by enzymes. Silica content increases with the age of the plant, inhibiting growth and promoting ripening, so it is usually used late in the growing season. Silica containing rock dusts may also be used for this purpose, but rock dusts may also act as anti-feedants and are therefore possibly useful as a seed dressing.

Other minerals and chemicals, such as mineral sulphur and copper, are used against fungal diseases, but their use is restricted because of problems associated with the build-up of copper in the soil and the fact that sulphur may harm certain beneficial insects. Potassium permanganate is also allowed on a restricted basis as a disinfectant and fungal inhibitor.

In some cases, soft soap and oils derived from minerals or plants may be used to control insect pests like aphids. Diatomaceous earth is occasionally used to control ground-moving and grain storage pests. It consists chiefly of silica as remnants of microscopic diatoms laid down over millions of years. Ground into a fine powder, it scores the shells of any insects coming into contact with it. Again, because of its potential to harm beneficial insects, its use should be restricted.

Biological control

It may seem surprising that biological control, often regarded as being a key characteristic of organic farming systems, has been left until this point in the chapter before being discussed in detail. This is partly a matter of terminology. Many authors, including Jones (1974), Huffaker (1980), Lisansky (1981), Cook & Baker (1983) and Quinlan (1989), consider biological control in its widest sense covering many of the issues already discussed above such as organic manuring, rotations, sowing dates, green manures, breeding for resistance and the encouragement of natural enemies. In this chapter, the term ecological control has been used for this purpose, and biological control is used in a more restricted sense, that of the direct use of supplemental parasites, predators and pathogens against insect pests and antagonists against soilborne fungal pathogens.

The reason for leaving the discussion of biological control until now is that it should really be seen as a last resort once all other steps to create an optimal environment for plant growth have been taken. Although natural controls may frequently be all that is required, there are some instances where the population density of a pest does not need to be very high to cause economic damage and supplementary intervention may be necessary. In some cases, the predator populations may develop too slowly, a process which can be helped by the addition of supplementary predators at the time that the pest population begins to grow.

Various new developments concerning the use of added antagonists such as *Trichoderma* to control fungal pathogens have been considered earlier in this chapter. The use of compost extracts or filtrates of soil bacteria to control diseases such as mildew is another form of biological control. Scientists are also looking at the possibility of using parts of existing viruses (the so-called satellite RNA) to control other closely related viruses. Work at Rothamsted is currently focusing on the potential of a bacterium, *Pasteuria penetrans*, and a fungus, *Verticillium chlamydosporium*, for controlling root-knot and cyst nematodes respectively.

The traditional use of biological control has been in the control of plant pests, with some of the most successful examples going back to the 1920s. Two of the most important are the control of the greenhouse pests red spider mite and whitefly (Dalby, 1984; Hussey & Scopes, 1985; Gear, 1988). The reason these have been so successful is that they can be released in a controlled environment within the greenhouse and are not therefore subject to the vagaries of climate and other factors outside.

The red spider mite is controlled by another small predatory mite (*Phytoseiulus persimilis*). The primary infection starts in April-May when overwintering females emerge from the soil and walls. The predator mite lives only on the spider mite and can easily keep pace so long as sufficient food is available. It will survive at temperatures as low as 5°C and at humidity levels above 50%. The predatory mite is available commercially but, once introduced, it is important to manage both pest and predator and to anticipate attacks.

Whitefly can be successfully controlled by the chalcid wasp as well as by the fungus *Cephalosporium lecanii* and the parasitical hymenopter *Terebrantes*, although only the former is used widely as a biological control. The chalcid wasp does not survive at temperatures less than

10°C and does not overwinter in unheated greenhouses. It therefore needs to be imported and is available from commercial suppliers who raise the pest and parasite together on tobacco plants. Sections of leaf containing whitefly scales, most of which have been parasitised, are fixed to cards which can be posted to the grower and hung on plants in the greenhouse. For tomatoes and cucumbers, the wasp can be introduced at either three to four fortnightly intervals at a rate of 30,000 per ha, or six to eight weekly applications at 15,000 per ha, following the first appearance of the whitefly.

The wasps' effectiveness depends on sufficient whitefly being present and on temperatures above 18°C. Below this temperature, the predator cannot keep pace, although this changes rapidly at increasing temperatures and at 26°C its rate of reproduction will be twice that of whitefly. Where whitefly is an annual problem, the pest and the parasite can be introduced together before any whitefly is seen in the greenhouse. An immediate balance is then established which will control the outbreak when it occurs. It is also important to be careful with the use of pyrethrum and derris, leaving an appropriate gap before the introduction of the whitefly predator, or it too will be affected.

In the open field, one of the best known biological controls is *Bacillus thuringiensis*, a bacterium which has several different strains with specific applications. It is frequently used in organic horticulture as a selective microbial insecticide against many lepidopterous pests such as the cabbage white butterfly and has the advantage that it does not have a significant impact on other insects and therefore does not suffer from the disadvantages of naturally occurring pesticides such as derris and pyrethrum. However, commercially available products containing *Bacillus thuringiensis* are not effective against all pests of this type, one example being the turnip moth. New isolates of *Bacillus thuringiensis* are being developed and more powerful commercial products should eventually be available, including a strain (*var. tenebrionis*) which is effective against Colorado potato beetle.

The widespread use of *Bacillus thuringiensis* may lead to problems in the future including the development of resistant larvae. Another area of potential concern is the current attempts to isolate the toxin producing gene from the bacteria and to incorporate this directly into plant tissue, including potatoes and brassicas, using genetic engineering. In its current use, the affected larvae are not eaten, but if genetically engineered into a crop significant amounts of the toxin may be ingested, the effects of which on humans are unknown.

Other examples of biological control include the tiny *Trichogramma* wasp, which lays its eggs (available commercially) in the eggs of pests such as the cabbage white butterfly and the European corn borer, and the granulosis virus which is being tested against codling moth. Successful field experiments have been carried out in West Germany and at East Malling and Long Ashton research stations in Britain. The codling moth larvae ingest the virus and die after five to ten days. A new area for exploration is the use of chemicals contained in spider venom to paralyse insect pests (Quicke, 1988).

In New Zealand, a naturally occurring bacterium (*Serratia entomophila*) is being used for biological control of grass grub, which is one of New Zealand's major pests. The bacterium is present under natural conditions, but not usually in sufficient quantities. Work is currently underway on commercial preparations of this bacterium, which is likely to be the first New Zealand originated, commercially available biological control.

In Australia, scientists have gone a stage further and developed a genetically engineered strain of *Agrobacterium tumefaciens* which is being used to control crown gall, which is caused by the same bacterium. A benign strain of the bacterium produces an antibiotic which attacks the pathogenic strains, but the part of the gene which allows development of resistance to the antibiotic has been removed.

From time to time, biological control techniques achieve a degree of notoriety through the use of 'exotic' species which do not naturally occur in a particular ecosystem. A current example is the search for a biological control of the psyllid pest which has devastated large areas of the nitrogen-fixing 'miracle' tree *Leucaena* in South East Asia and is rapidly spreading towards India and Africa. One option is a ladybird beetle (*Curinus coeruleus*) which is the psyllid's main enemy in Hawaii. Great care has to be taken to ensure that these exotic species do not get out of control, particularly in situations where there are no natural predators. Mistakes made in the past have ensured that many countries apply stringent controls over the testing of such species, although there is now a serious risk, particularly with the use of genetically engineered organisms, that companies may attempt to test them in secret or in countries with less strict controls.

Mechanical control

A range of mechanical pest controls have been developed which include physical barriers (including slug and rabbit fences), traps and sound (e.g. birdscarers). Very fine mesh nets can be used as vertical barriers or to cover crops like carrots and cabbages during periods when the carrot or cabbage fly is in flight. Trials have shown a 50% increase in marketable yield in susceptible areas as a result of using nets, which may also result in milder temperature and better moisture conservation.

Other suitable crops for netting include radish, broccoli, Chinese leaves, kohlrabi, celery, parsley, garlic, onions and beans, although some of the more sensitive crops require support for the net to avoid yield losses.

Traps containing pheromones, the chemical substances which act as sexual attractants to insects, may also be used to entice pests away from crops, although traps containing prohibited insecticides are only allowed on a restricted basis under Soil Association standards (see Colour plate 18). Strawberry growers in California have developed a giant vacuum cleaner (the Bugvac) to remove harmful pests (New Scientist, 28/01/89).

Another form of trapping which has been developed specifically for organic farmers is the Colorado potato beetle collector (Plate 7.3). This implement, developed by Walter Kress in West Germany, is front-loaded so that the beetles and larvae are not shaken off by the tractor before it reaches them. Chains hanging down from the frame shake the leaves. A reflex action ensures that the beetles and larvae immediately loosen their hold and fall into the metal, ship-like containers, which are buffered by rubber strips to protect the potato plants from damage. The insects can then be removed and destroyed. Up to 90% control can be achieved with this approach.

SPECIFIC DISEASE PROBLEMS

The approach to the control of any specific pest or disease problem must first of all be the creation of a suitable ecological environment including all the factors discussed above:

- choice of crops and varieties appropriate to location;
- soil cultivation and choice of sowing/planting/harvesting dates;
- organic manuring and crop nutrition, including type, quantity and timing;
- spacing, rotation design and crop species/variety mixtures;
- use of resistant varieties;
- habitat management and the use of green manures/companion planting;
- use of plant extracts and minerals;
- direct biological and mechanical controls.

No single strategy is likely to be successful on its own; pest and disease control relies on the effectiveness of interactions between many factors. The control of specific pest and disease problems should be considered in these terms, rather than 'What substance can I use against such and such?'

Plate 7.3   The Kress implement for mechanical control of Colorado potato beetles

General information on specific pests and diseases (from a conventional perspective) can be found in the MAFF/ADAS advisory leaflets. Although they tend to emphasise chemical controls, some useful cultural controls are also covered and they contain important information on pest lifecycles and disease epidemiology.

Diseases of field vegetables

Blight
Of the diseases, potato blight is probably the most important one affecting organic producers. The disease is airborne and spreads from infected tubers and dumps of potatoes, particularly under warm, humid conditions. The tubers should be protected by earthing up and removing haulms before harvest or delaying harvest at least two weeks after the haulms have died. Harvest should be delayed longer than two weeks if the haulms are destroyed mechanically. The use of Bordeaux and Burgundy mixtures containing copper is allowed only on a restricted basis; the copper content is matter of concern and such mixtures should not be used routinely. Pre-chitting and the use of blight tolerant or early maturing varieties allow the crop to bulk up before blight strikes and clean seed should always be used. Compost, stinging nettle and seaweed extracts, as well as silica-containing sprays such as waterglass (sodium silicate), may help to promote growth and enhance disease resistance,

but there is little reliable evidence on this. Silica containing rock dusts have also been used, but to little effect. A sulphur containing 'herbal' fungicide Bio-S may be effective in combination with copper oxychloride. Aligning the rows with the direction of the prevailing winds and a wider row distance of 0.75 m will encourage reduced humidity and may delay the spread of blight. Excessive nitrogen levels should also be avoided.

Other diseases

Several other diseases may be significant in horticultural systems, in particular *Alternaria*, powdery and downy mildews, *Botrytis*, *Rhizoctonia*, *Pythium* and other soilborne damping-off diseases. A brief guide to vegetable disease identification is contained in Davis (1988). Appropriate rotations, especially in the case of club root, the use of organic manures, balanced nutrition and the use of resistant varieties are the most important preventive measures. Frost (1989) describes the results from a survey of organic growers on the subject of vegetable varieties suited to organic systems. In the case of club root, grazing *in situ* to prevent the spread of spores from an infested field and increasing the pH to 7.3–7.5 will reduce incidence considerably, although the latter may result in nutritional problems. In some instances, plant extracts and herbal 'fungicides' (e.g. Bio-S), waterglass (sodium silicate), bicarbonate of soda and steam sterilisation may be appropriate, with sulphur and potassium permanganate only allowed under restricted circumstances. Biological controls such as *Trichoderma* (marketed in Britain as Binab-T) may also be appropriate in some cases.

Diseases of legumes

The two main diseases of forage legumes are *Verticillium* wilt and clover rot. These are generally problems of rotation or, in the case of *Verticillium* wilt, can build up over a period of years, for example in lucerne. Clover rot results when red clover is grown too frequently in rotation (more than once every five to six years) because of attacks by the fungus and by stem eelworm. The problem is more pronounced in the south of Britain and it is not likely to be seen elsewhere. Some varieties have a degree of resistance and these can be used where a problem exists. White clover may also be severely attacked, but other forage legumes are less likely to be affected. Field and broad beans can be attacked but are more likely to be hit by a specialised strain of *Sclerotinia* which does not attack clover. If clover rot is a severe problem, red clover should be limited to once every eight or even 12 years in the rotation. Avoid over-manuring and graze in the autumn to reduce foliage.

Cereal diseases

Cereal diseases are not usually a significant problem in organic systems, due to rotational design and low nitrogen use. Even so, the use of resistant varieties and appropriate varietal mixtures can enhance resistance to disease. Varieties known to be susceptible to disease should not be grown on large areas. The NIAB recommended variety leaflets provide the best guide as to how varietal diversification strategies can be implemented. Detailed descriptions of the many cereal diseases can also be found in Gair *et al.* (1972).

Time of sowing and the avoidance of 'green bridges' which allow disease to be transferred via plants or crop residues to the new crop are also important factors. Early sowing in the autumn favours the introduction and development of diseases. Mildews and rusts survive on volunteers; crops which emerge before these volunteers are killed are particularly at risk. Warm weather in early autumn favours the development and spread of most leaf diseases. It also encourages the early development of eyespot and take-all, so that early drilled crops are at greater risk from damaging attacks by these diseases.

Foliar feeds and silica-containing sprays may be used to strengthen the resistance of leaves to penetration by fungal mycelia, although the evidence for this is not strong, and sulphur can be used under restricted circumstances. Future developments may include the use of filtrates of soil bacteria or spraying with less aggressive strains of certain diseases to enhance immunity. In any case, strategies involving direct intervention and sprays of any kind should be seen as a last resort and not as a regular part of a disease control programme.

Take-all

Much has been written in recent years about the biological control of take-all, a black-rot of the roots. In severe attacks, some stem bases may also be blackened, with plants stunted or showing whiteheads. Affected plants may occur in patches or at random. Take-all is primarily a rotational problem, although its effects may be exacerbated by high lime contents, inadequate nitrogen in the early spring, and impeded drainage. Oats are resistant to all but one strain, which does little harm in the west and north of Britain, and can therefore play an important role in achieving the intervals of at least one year between susceptible crops. There are no resistant varieties of wheat or barley.

The search for a biological control centres around the phenomenon of take-all decline, where the take-all incidence peaks and falls off each year a crop of wheat or barley is grown in succession. The take-all decline curve indicates that there may be soil microorganisms which are antagonistic to the fungus. Researchers at Bristol University are looking at soil bacteria in particular. One, *Pseudonomas fluorescens*, is currently being tested as a 'live' seed dressing and commercial products may be available in the next few years. It is questionable, however, how important such developments are likely to be in organic systems where the emphasis on organic manuring and a biologically active soil may be sufficient to encourage antagonistic microorganisms naturally.

Seed-borne diseases

Certain seed-borne diseases may become a problem where reliance is placed on homegrown seed and given the prohibition or restrictions on most seed dressings. Biodynamic farmers place particular emphasis on the use of homegrown seed in the belief that varieties will evolve over time which are adapted to the specific ecology of the farm. Increasing incidence of bunt, where the grain is converted to black, sooty spore masses with a fishy smell, has been observed on some biodynamic farms. A warm water treatment, similar to that for loose smut described below, has proved effective for many years, but in some years crops are badly affected and attempts have been made to find alternative approaches. Intensive washing of the seed, for example in a concrete mixer, proved very effective and would seem the most practicable, at least on a small scale. General preventive measures should also be adopted including the cleaning of any trailers, cleaning plant etc. which may have become contaminated and the use of clean undressed seed wherever possible.

A related problem is that of loose smut. The grains are replaced by a mass of powdery black spores, obvious at emergence, but the spores are later shed leaving a bare stem. Because the spores develop in the grain, the seed cannot be treated by the usual methods and the only effective method of seed treatment is using warm water to kill the fungal mycelia. The grain should be soaked in cold water for four hours, then put in water between 52° and 54°C for 10 minutes, after which it is spread out to dry. The difficulty is that the range of temperatures over which the mycelia will be killed and not the grain is very narrow. A thermometer is essential and hot water should be added to keep the temperature up. Both wheat and barley grain can be treated in this way.

Another disease problem which is frequently raised in discussion about organic farming is that of ergot. Hard black or dark purple fungal bodies, up to 5 cm long and slightly curved, replace a few grain in each ear, particularly in rye crops in moist years. These contain alkaloids which are poisonous to humans. There is no effective seed dressing, even in conventional agriculture. Affected grain must be cleaned, either for human consumption to avoid poisoning or for seed to avoid further disease spread. Conventional seed cleaning is usually successful, except for broken ergot grains which are a similar size to the rye grain. The European Community specifies a maximum limit of 0.5 g ergot grains per kg of rye to be used for human consumption.

Specific Pest Problems

Although the range of ecological approaches to pest control described in this chapter such as rotations, habitat management, mixed and inter-cropping, undersowing and the use of green manures should help to minimise pest incidence, there are still certain pests, particularly in horticultural systems, which can present a significant problem. Although products such as derris and pyrethrum are potentially allowable, they should be used with caution and only on an occasional basis, not as the main pest control strategy.

Pests of field vegetables

Carrot fly

Of the horticultural pests, the carrot fly presents perhaps the most serious problem. In the past, carrots have been grown in areas like west Wales which are not usually associated with horticultural crops so that carrot fly incidence has been minimal. As the organic production of carrots expands and moves into more traditional horticultural regions, the carrot fly problem is likely to become much more severe. Where infestations can be particularly heavy, it may not be possible to grow carrots organically at all. In some districts, the carrot fly may also damage parsnips, celery and sometimes parsley.

Incidence of the pest can be reduced by avoiding growing crops close to previous crops in short rotations and by avoiding growing both early and late crops close together. The flies spend much of their time in shelter around the edges of carrot fields; exposed fields and reducing the shelter available to the flies can help to keep numbers down. Hedgerow plants like nettles which act as alternative hosts to the fly should be kept under control. Infested carrots should be disposed of in a way which will ensure that the fly cannot complete its lifecycle (e.g. by feeding to livestock).

The primary method of control is through understanding the pest's lifecycle and timing sowing to avoid the first generation of the fly. The eggs are laid singly or in small groups in cracks in the soil surface near the host plant. Depending on temperature, they usually hatch in about seven days. The maggots feed on the root systems of young and mature plants. In early June, the first generation of maggots feeds mainly on the side roots near the base of the tap root; this is often the only damage showing on roots in early summer. (In Scotland, this damage is seen in late June to early July.) Early fly attack can also kill carrot seedlings directly, resulting in 'natural' thinning or a gappy crop. Delaying sowing until late May, with germination three weeks later, can help to avoid this first generation. During August and September, many of the more superficial mines caused by the first generation of maggots will heal and little damage will be apparent. Damage by the second generation of maggots becomes progressively worse during the autumn, but if the first generation has been deprived of a ready food supply, the second generation should be diminished. There can sometimes be a partial third generation during the autumn in the Fens of eastern England. The actual dates of different generations varies between different regions and there can be significant overlap.

Cabbage and turnip root flies

Similarly, cabbage and turnip root flies can be controlled to some extent by sowing out of phase with the expected egg-laying periods of the two or three generations of female flies, reducing the risk of attack at critical stages of growth. Cruciferous weeds such as wild radish are alternative hosts; hedgerows and tall weeds also provide shelter. MAFF advises control of these, but there may also be potential benefits from the use of cruciferous 'weeds' as trap crops. Natural enemies play an important role in their control in the soil. In one trial, more than 90% of all root fly eggs laid in a field were eaten by beetles (although the surviving ones can still be a problem!). The pupae are also destroyed by predatory staphylinid beetles and their larvae as well as the larvae of the cynipid wasp.

In a field experiment during 1986, McKinlay (1987) compared two of the insect control products permitted by the Soil Association (garlic and derris) with a standard insecticidal treatment, a lightweight fabric cover over the plants, sowing between the first and second generations of the cabbage root fly and an untreated control. The best treatment, even better than insecticides, proved to be the fabric cover which prevented the fly laying its eggs near the base of the plant, but this may lead to greater weed control problems. The other treatments—garlic, derris and sowing between generations—proved to be ineffective.

Flea beetle

The flea beetle can be a problem in cruciferous crops in so far as it attacks seedlings, particularly in dry springs. Damage is less after the first true leaf has appeared. A seed bed of good tilth, moist and adequately manured, will help plants to grow quickly through the susceptible stage. The use of wild cruciferous species as a trap crop may also be effective.

Wireworm

The wireworm can be a serious pest in organic horticultural systems and is associated primarily with the ploughing up of old pastures. It is the larva of the click beetle or skipjack, of which several forms occur throughout Britain. The beetle is up to 1 cm long, thin and of a uniform brown or grey-brown colour. The adults can be seen on the soil surface from March until July. During May and June, each female will lay up to 100 eggs below the soil surface. They prefer grassland for this, or any ground covered by crop or weeds, but will also lay on bare soil. The larvae hatch in July and are at first transparent white and only about 1.5 mm long. During their lives, they moult nine times, generally twice a year and then once in their fifth year, pupating towards late summer and emerging as adults in the following spring. As they moult and grow, their skins take on a distinctive golden brown colour, with darker heads, and they grow eventually to about 25 mm.

Populations in established grassland may reach 20,000,000 to the hectare. It is generally considered that less than 2,500,000 to the hectare is probably safe for most arable crops, although potatoes and many market garden crops can be damaged at populations of 250,000 per hectare. ADAS provides a wireworm count service, but it is difficult to be accurate at low population levels.

The wireworms will eat a wide range of plants, concentrating on the roots and sometimes the stem just above the ground, and eating plants at any stage from seedling to maturity, although the risk period is most acute at seedling stage. There are two main feeding periods in April/May and July to December, with a peak in September. The main symptoms are neighbouring plants wilting and dying; a worm will usually be found near a plant which has just wilted. Some plants such as linseed, flax and to a lesser extent parsnips, peas, field beans, clovers and lucerne are resistant to attack. All cereals are susceptible, but can recover. Barley is least affected. Wheat sown at the right time can escape feeding before it is well established. Spring wheat, oats and newly established leys are highly susceptible, as are all small-seeded crops, particularly when well spaced. Of the fodder brassicas, kale and rape seem to have a good chance of success. Potatoes should be avoided if there is the least chance of risk, because of the damage which can be done to tubers.

Although arable and horticultural crops following grassland are particularly at risk, ground disturbance and cultivations form the primary means of control. Cultivations directed solely at growing crops will reduce the small forms without much affecting the larger, but as these pupate, the reservoir will become steadily diminished. By keeping the ground free of grass or weed cover during midsummer, egglaying may be reduced and certainly the rate of egg survival will diminish due to desiccation. With arable cropping, a reduction of 50% after one year and up to 70% after two years can be expected, with tolerable levels being reached after three years, although the problem may persist for several years on thin chalk soils. For more rapid control, bare fallowing after ploughing in February or March could reduce numbers by 95% in time for winter wheat through disturbance leading to desiccation, although in areas susceptible to wheat bulb fly any form of fallow could create problems for a subsequent winter wheat crop. In a horticultural system, the fallow could be followed by a combination of resistant crops such as legumes, parsnips and transplanted brassicas, even if the rotation starts off with an imbalance of brassicas.

Mustard green manures, cultivated in twice in a growing season, and trap-cropping with wheat between rows of susceptible crops like sugar beet, have also been used successfully. The wheat must be hoed out before it gets too large, but not before the desired crop is established and more resistant to attack. Other cultural controls include a good firm seedbed and adequate manuring to encourage rapid growth. Seed rates should be increased to compensate and possibly top-dressed with high solubility organic fertilisers to compensate for root loss. Transplants have a better chance of survival than direct seeded crops. Further details can be found in Deane (1987) and the MAFF and Scottish Agricultural Colleges advisory leaflets.

Nematodes

Cyst nematodes or eelworms can also cause problems, but are controllable by rotational means. There are two species of potato cyst nematode (yellow and white). For these, long breaks or the more frequent use of early varieties are essential. Rotational controls can be supplemented by the use of resistant varieties, green manuring (see above) and avoiding the spread of contaminated soil, for example when clamping potatoes. Beet cyst nematodes affect not only sugar beet, but also beetroot and mangels. Although the rotational constraint clauses have been dropped from sugar beet contracts, a gap of at least three, preferably four, years between host crops represents the only economic method of control. Again, the use of green manures as described earlier in this chapter may be appropriate.

Caterpillars

Caterpillars and other larvae may cause significant damage to crop foliage. Many of these are controlled naturally by a range of predators such as ladybirds, which eat the eggs, as well as disease. Biological controls using the bacterial insecticide *Bacillus thuringiensis* and the parasitic wasp *Trichogramma*, which lays its eggs in the eggs of the cabbage white butterfly, are available commercially. These biological controls can also be effective against the European corn borer, a pest of maize crops. Although pyrethrum and derris can be used, these will harm the many beneficial insects which help to keep caterpillars etc. under control. On a small scale, hand picking of the caterpillars and eggs may be practicable, but care should be taken not to destroy the ladybird eggs.

Slugs

The control of slugs is a major headache for organic growers and can be a serious problem for farmers too, especially where reduced cultivations are practised and crop residues provide ideal conditions for slugs in subsequent crops such as cereals. Modern agricultural practices have created ideal living conditions for slugs, without providing similar encouragement (and probably, through the use of pesticides, causing direct harm) to predators and other natural control mechanisms such as birds, frogs, moles, shrews, hedgehogs and many beetles.

There are several different species of slug, not all of which are significant pests. Slugs are more active in wet seasons and do most damage in the autumn and mild periods of winter. The main agricultural crops affected are autumn-sown cereals and maincrop potatoes, particularly direct-drilled crops and crops on heavier soils, often poorly drained and with high organic matter levels, as well as peas, clover, reseeded ryegrass, root crops, vegetables and flowers, especially at their seedling stages. Damage may be caused by seed hollowing or leaf shredding, or by the cutting off of shoots from young seedlings. However, slugs play a very important role in the decomposition of plant residues and hence the cycling of nutrients and the formation of humus in the soil.

Slugs are active and feed throughout the year whenever the temperature and moisture conditions are suitable, but during very dry weather, and often when it is frosty, they stop feeding and move down into the soil or shelter under debris. Breeding cycles and growth rates also show a degree of flexibility in response to adverse weather conditions. Early cultivations in the autumn provide ideal conditions for the slugs to get under the soil and be protected during winter, or for the eggs to be buried and thereby protected. A key factor in their control is to prevent them finding shelter in adverse weather conditions. A fine, well consolidated seedbed will prevent slugs burrowing into the soil; the traditional, cloddy fields for wheat on heavy soils provides the ideal environment for slugs to survive.

Rotation and soil cultivations are also important. Damage to cereals is worst after oilseed rape and other brassica seed crops, clover, grass, cereals, beans or peas, but rarely occurs after sugar beet or potatoes, partly because cultivations during the summer deprive the slugs of the moist conditions they need for survival. Shallow cultivations during dry periods in summer and during frosty periods in winter disturb the protective environment which the slugs create for themselves under the soil and expose them to dehydration or frost. Sowing or transplanting under ideal conditions for crop germination and growth will allow the crop to become established quickly and to grow away from any significant damage. Drilling cereal seeds slightly deeper to take them out of the way of the slug activity may also help. In addition, some crops like potatoes show a degree of varietal resistance to slug damage.

Chemical controls are not allowed under the Soil Association's standards, although some exceptions such as aluminium or copper sulphate sprayed directly onto slugs when they are on the soil surface may be made (see Appendix). The use of metaldehyde is banned; the risk to wildlife and pets from this control method is significant. In any case, the implementation of chemical controls does not remove the underlying causes of slug outbreaks. One new approach being developed by scientists at Rothamsted is to isolate the chemicals in some plants which make them repellant to slugs and to use these in anti-feeding agents as seed dressings. Research into biological control using ground beetles and other organisms is also underway (Dussart & Cherfas, 1989).

Slug control in small gardens and horticultural systems will need to rely on a wide range of practices, many of which are being actively researched by organisations like the Henry Doubleday Research Association (HDRA, 1989) and the Swiss Research Institute for Biological Husbandry (Graber & Suter, 1985). Beds should be disturbed as little as possible in the autumn and residues should be put on the compost heap, not left on the soil surface to act as shelter. In the winter, when the ground is frozen, the soil can be loosened so that the slug eggs are exposed to frost. In the summer, the soil should not be loosened too deeply to avoid giving slugs shelter. Cultivation with a rotovator or similar implement in dry periods will kill many slugs buried in the soil.

Dry, absorbent surfaces prevent the movement of slugs and dehydrate them, so that mulching with dry materials like straw and leaves can help. The mulch should not be too thick, just enough so that the soil can no longer be seen. Absorbent materials such as sawdust may also help. Borders of single crop species, such as a grass strip, act as a deterrent and these can be supplemented by slug barriers made of plastic or iron or, as a temporary measure to protect seedlings, using an electrified wire as a barrier. If plants need to be watered, watering individual plants while leaving the area between plants dry will also discourage slug movement.

Slugs will migrate to the surface shortly after cultivations and can move considerable distances when searching for food. Attractants such as a chopped mustard green manure, chopped kitchen wastes or a mixture of wheat germ and cat or dog biscuits can be used to entice slugs away from sensitive seedlings and, if pursued rigorously, will also enable slugs to be collected and destroyed. Beer traps are often recommended, but may entice more slugs into the garden than are actually killed by drowning. A foul-smelling slug 'brew' can be prepared by pouring boiling water over collected slugs and allowing the mixture to stand for a couple of weeks. Slugs are actively repelled by the smell, but for hygienic reasons the material should not be spread on crop plants! Finally, it may be possible to create a slug habitat in part of the garden which is much more attractive to them than the area you are trying to protect, thus encouraging them to move off somewhere else.

Colorado potato beetle
Although not (yet) a pest in Britain, the Colorado potato beetle is a serious pest in mainland Europe and North America. Mechanical collection as described earlier in this chapter is one of the most effective control methods available to organic producers. Strains of *Bacillus thuringiensis* are now available to control this pest, and further work is concentrating on biological control using specific strains of the fungus *Beauveria bassiana* and the parasitic wasp *Edovum puttleri*. Trials using rock dusts to discourage feeding have not proved particularly effective.

Pests of grassland, cereals and grain legumes

Aphids
The control of aphids has been mentioned at several points in this chapter so that little needs to be added at this stage. Lower nitrogen availability and the increased activity of natural predators such as ladybirds, hoverflies, parasitic wasps, beetles and spiders mean that aphid problems in cereal and even grain legumes are unlikely to be significant. Fungal spores also play a role. They remain dormant in the soil until weather conditions are right, at which stage they infect passing aphids merely by contact. The fungal hyphae work their way into the aphid, which will continue to be active for several days before it dies. The dead aphid will then change to a pale colour and remain attached to a leaf for a further period. When the temperature and humidity are right, the aphid explodes, sending a shower of grey spores into the air.

Because aphids are subject to large numbers of predators, they survive by sheer force of numbers and breed prolifically. They have the ability to reproduce without the aid of another aphid. A single specimen can produce several identical copies of itself in a day, born alive and ready to start feeding. The main predators, however, cannot breed in this way and it usually involves a considerable time lag before the predator population has caught up with the pest, during which time the damage may well have been done. In wheat, this can involve reduced grain number, grain weight, specific weight and breadmaking quality. The damage is caused not only by the aphids sucking sap from the plant, but also because they produce a sugary honeydew which blocks the plant's stomata and encourages the growth of fungi. Aphids also transmit various virus diseases including BYDV.

Some cereal varieties (e.g. Rapier) are partially resistant to aphid attack and this can help to reduce the rate of aphid increase, giving a greater opportunity for the natural enemy complex to keep populations below the economic threshold. Another aspect is that aphids are forced to look elsewhere for more favourable hosts, coming into contact with ground-active predators as they travel over the ground.

In horticultural systems, dusting with rock dusts to discourage feeding and the use of stinging nettle extracts and waterglass to enhance plant resistance have been tried, but are not reliable. The use of soft soap treatments is also a possibility. Wide spectrum 'natural' pesticides such as quassia, pyrethrum and derris should be used with great care to avoid harming beneficials.

Leatherjackets
Leatherjackets, the larval stage of the cranefly, are often not paid very much attention, but they can cause substantial crop damage, particularly in spring cereals following grassland. Eggs are laid from late July to September in the soil surface amongst herbage, especially grasses, where within two to three weeks they hatch into small, legless leatherjackets. They are active during periods of mild weather throughout the winter, but grow faster as the temperature rises. They mature in late May and June, cease feeding and pupate in the soil, later emerging as adult craneflies. Cultural methods, such as top-dressing and firming the soil by rolling, will improve plant growth and restrict the movement and feeding of leatherjackets.

Frit fly
Similarly, the frit fly is a pest associated with cereal crops following grass or a cereal crop with high grass weed infestation. The frit fly prefers young ryegrasses; Westerwolds and Italian ryegrasses are particularly

prone to attack, although some ryegrasses are resistant. Close grazing by sheep can make the problem worse by encouraging new tillers. All commercial cereal varieties are liable to attack.

There are usually three generations of frit fly each year, in the spring (pupating in March—April and emerging in May to lay eggs at the base of young plants), in the summer (emerging in July to lay eggs on oat husks) and in the autumn (emerging from stored grain). An important aspect of cultural control is the time of sowing. Spring oats are very resistant to first generation attack after the four leaf stage. Early sowing into a well prepared seedbed for rapid growth is therefore important. In the case of winter wheat after a ryegrass ley, the land should be ploughed as early as possible to allow time for the larvae to die before the wheat is sown. A four week interval is important, also for establishing new grassland.

Wheat bulb fly
The wheat bulb fly is a pest primarily associated with wheat after summer fallows, early vining peas, potatoes, sugar beet and other root crops because it prefers to lay its eggs on bare soil. Despite its name, barley and rye can also be attacked, although oats are immune. It is a significant pest only in certain parts of Britain. The wheat bulb fly eggs hatch between January and early March depending on soil temperature. The symptoms ('deadhearts') usually begin to appear from late February to mid-March. Crops sown before November are usually well tillered at the time of attack and suffer less than later sown crops. Spring cereals sown after mid-March should also avoid damage. The main control is to avoid bare soil over the summer, which could be achieved using green manures or undersowing, but which also reduces the scope for weed control which bastard fallows offer. If a fallow is necessary, then a smooth tilth will reduce the number of eggs laid. The avoidance of shallow drilling can also help.

Nematodes
In Britain, cereal cyst nematodes are predominantly a pest of light textured soils in southern England. Their predominance has decreased in recent years, probably due to the decline in the area of oats grown and the parasitism of nematode cysts and eggs by soil-inhabiting fungi. These fungi take some time to build up so that their full effect is only seen after several years in a manner similar to take-all decline. Research into their use as a biological control in the future is yielding promising results. Cultural control depends primarily on rotations and the use of resistant varieties of oats such as Panema winter oats and Trafalgar spring oats. However, many 'resistant' varieties of cereals are fully

susceptible to damage caused by the young nematodes penetrating the cereal roots, and these should not be grown in heavily infested fields.

Susceptibility to cereal cyst nematode varies considerably among different cereals and grasses, with spring oats showing greatest susceptibility followed in descending order by maize, winter oats, spring wheat, winter wheat, winter and spring barley, rye and grasses showing the least susceptibility. The ability of these crops to increase infestation also varies, with winter oats worst, followed by spring oats, spring barley, winter barley, spring wheat, winter wheat, rye, maize and grasses.

Clovers are susceptible to another nematode, stem eelworm, which is primarily associated with red clover. However, there is also a distinct race which attacks white clover. If an infestation occurs, then a break of several years in the crop rotation is required, longer on heavy soils. Lucerne may be good alternative crop, and some resistant varieties are available.

Grain storage pests
Grain storage pests are not often considered to be a problem, perhaps because demand for organic produce has meant that little produce remains in store for very long. Simple hygienic precautions, such as cleaning the grain to remove the lighter, infested grains which have been dehulled, and the use of controlled, low volume ventilation of grain to keep it cool and stop the pests breeding, are important non-chemical control methods. In some cases, it may be appropriate to use carbon dioxide (dry ice) to 'suffocate' the pests. Traps are now available to monitor storage pest populations so that unnecessary treatments of crops in store can be avoided.

References and Further Reading

Pesticide problems
Arden-Clarke, C. (1988) *Pest control strategy impacts on the crop ecosystem.* In: *The Environmental Effects of Conventional and Organic/Biological Farming Systems.* Research Report RR-17. Political Ecology Research Group; Oxford
Diercks, R. (1983) *Alternativen im Landbau.* Ulmer; Stuttgart
Dudley, N. (1987) *This Poisoned Earth—the truth about pesticides.* Piatkus; London
Dudley, N. (1988) *Maximum safety: pest control and organic farming.* Soil Association; Bristol
Harding, D. J. L. (ed.) (1988) *Britain since 'Silent Spring'.* Proceedings of Institute of Biology Symposium, Cambridge, March 1988. Institute of Biology; London
Kickuth, R. (1982) *Ökotoxikologische Probleme bei der Anwendung von Pestiziden.* In: Kickuth, R. (ed.) *Die Ökologische Landwirtschaft.* Alternative Konzepte 40: 89–99. C. F. Müller; Karlsruhe
Munro, J. (1984) *What agribusiness does to you—the toxic effects of pesticides on humans.* Soil Association Quarterly Review, September

Pesticides Trust (1989) *The FAO Code: Missing Ingredients. Prior informed consent in the International Code of Conduct on the Distribution and Use of Pesticides.* Pesticides Trust; London

Causes of pest and disease problems

Chaboussou, F. (1978) *La résistance de la plante vis-à-vis de ses parasites.* In: Besson, J-M. & Vogtmann, H. *Towards a Sustainable Agriculture.* Proceedings of the 1977 IFOAM Conference. Wirz; Aarau. pp. 56–59. (Translated and summarised by Mary Langman) (1984) *Plant health is key to fighting 'enemies'.* New Farmer and Grower 3: 21–22)

Chaboussou, F. (1985) *Santé des Cultures—une Révolution Agronomique.* Flammarion; Paris. (Translated into German: *Pflanzengesundheit und ihre Beeinträchtigung.* Alternative Konzepte 60. C. F. Müller; Karlsruhe, 1987)

Cherrett, J. M. & Sagar, G. R. (1977) *Origins of Pest, Parasite, Disease and Weed Problems.* The 18th Symposium of the British Ecological Society, Bangor, 1976. Blackwell; Oxford

Coaker, T. H. (1977) *Crop pest problems resulting from chemical control.* In: Cherrett & Sagar, 1977

Eigenbrode, S. D. & Pimentel, D. (1988) *Effects of manure and chemical fertilisers on insect populations on collards.* Agriculture, Ecosystems and Environment 20: 109–125

Griffiths, E. (1981) *Iatrogenic plant diseases.* Annual Review of Phytopathology 19: 69–82

Hodges, R. D. & Scofield, A. M. (1983) *Agricologenic disease—a review of the negative aspects of agricultural systems.* Biological Agriculture and Horticulture 1: 269–326

Huber, D. M. & Watson, R. D. (1974) *Nitrogen form and plant disease.* Annual Review of Phytopathology 12: 139–165

Patriquin, D. G. *et al.* (1988) *Aphid infestation of faba beans on an organic farm in relation to weeds, intercrops and added nitrogen.* Agriculture, Ecosystems and Environment, 20: 279–288

Organic manuring, suppressive soils and rhizosphere pathogens

Baker, R. (1968) *Mechanisms of biological control of soil-borne pathogens.* Annual Review of Phytopathology 6: 263–294

Broadbent, P. & Baker, K. F. (1974) *Behaviour of* Phytophtora cinnamoni *in soils suppressive and conducive to root rot.* Australian Journal of Agricultural Research 25: 121–137

Bruns, C. & Gottschall, R. (1988) *Phytohygiene—Untersuchung zur Abtötung von Pflanzenkrankheitserregern bei der Kompostierung von Grünabfall mit hohem Holzanteil.* Ingenieurgemeinschaft Witzenhausen

Chung, Y. R., Hoitink, H. A. H. & Lipps, P. E. (1988) *Interaction between organic matter decomposition level and soil-borne disease severity.* Agriculture, Ecosystems and Environment 24: 183–193

Cook, R. J. (1984) *Root health—importance and relation to farming practices.* In: Bezdicek, D. *et al.* (eds) *Organic Farming—Current Technology and its Role in a Sustainable Agriculture.* American Society of Agronomy; Madison, WI. pp. 111–127

Cook, R. J. (1986) *Plant health and the sustainability of agriculture, with special reference to disease control by beneficial microorganisms.* In: Lopez-Real & Hodges, 1986

Fullerton, H. (1984) *How resistance is affected by the interaction of bugs.* New Farmer and Grower, 4: 7–9

Gottschall, R. C. *et al.* (1987) *Verwertung von Kompost aus Bioabfall: Aufbereitung von Frisch und Fertigkomposten.* In: Fricke, K. *et al.* 3. *Zwischenbericht Forschungsprojeckt 'Grüne Bio-Tonne Witzenhausen'.* Fachgebiet 'Methoden des alternativen Landbaus', Witzenhausen

Hoitink, H. A. J. *et al.* (1986) *Compost for control of plant diseases.* In: Bertoldi, M. *et al. Compost: Production, Quality and Use.* Elsevier

Huber, D. M. & Schneider, R. W. (1982) *The description and occurrence of suppressive soils*. In: Schneider, 1982

Lynch, J. M. (1983) *Soil Biotechnology*. Blackwell; Oxford

Lynch, J. M. (1988) *Microbes are rooting for better crops*. New Scientist, 28 April

Lynch, J. M. & Hobbie, J. E. (1988) *Microorganisms in action: concepts and applications in microbial ecology*.2nd edition. Blackwell

Nelson, E. B. & Hoitink, H. A. J. (1983) *The role of microorganisms in the suppresssion of Rhizoctonia solani in container media amended with composted hardwood bark*. Phytopathology 73: 274–278

Samerski, C. & Weltzien, H. C. (1988) *Untersuchungen zum Wirkungsmechanismen von Kompostextrakten im Pathosystem Zückerrübe-echter Mehltau*. Zeitschrift für Pflanzenkrankheiten und Pflanzenschutz 95: 176–181

Scharpf, H-C. (1971) *Die Auswirkungen der organischen Düngung auf das Abwehrpotential des Bodens gegen bodenbürtige Schaderreger im Gemüsebau*. Ingenieurarbeit, Hessischen Lehr- und Forschungsanstalt, Geisenheim/Rhg.

Schneider, R. W. (1982) *Suppressive Soils and Plant Disease*. American Phytopathological Society; St Paul, MN

Schüler, C. *et al.* (1989) *Suppression of root rot on peas, beans and beetroot caused by* Pythium ultimum *and* Rhizoctonia solani *through the amendment of growing media with composted organic household waste*. Journal of Phytopathology 127: 227–238

Stindt, A. & Weltzien, H. C. (1988) *Der Einsatz von Kompostextrakten zur Bekämpfung von* Botrytis cinerea *an Erdbeeren—Ergebnisse des Versuchsjahres 1987*. Gesunde Pflanzen 40: 451–454

Thom, M. & Möller, S. (1988) *Untersuchungen zur Wirksamkeit wässeriger Kompostextrakte gegenüber dem Erreger des echten Mehltaus an Gurken*, Erysiphe cichoracearum. Diplomarbeit, Gesamthochschule Kassel

Weltzien, H. *et al.* (1989) *Improved plant health through application of composted organic material and compost extracts*. Paper presented at the 7th IFOAM International Scientific Conference, Burkina Faso, January 1989.

Ecosystem management

Altieri, M. A. & Letourneau, D. K. (1982) *Vegetation management and biological control in agro-ecosystems*. Crop Protection 1: 405–430

Altieri, M. (1985) *Diversification of agricultural landscapes—a vital element for pest control in a sustainable agriculture*. In: Edens, T. *et al.* (eds) *Sustainable Agriculture and Integrated Farming Systems*. Michigan State University Press; East Lansing

Altieri, M. (1986) *The ecology of insect pest control in organic farming systems: towards a general theory*. In: Vogtmann *et al.* (1986)

Altieri, M. A. (1987) *Agroecology—the Scientific Basis of Alternative Agriculture*. Intermediate Technology Publications

Altieri, M. A. & Schmidt, L. L. (1987) *Mixing broccoli cultivars reduces cabbage aphid numbers*. California Agriculture 41 (11/12): 24–26

Brown, R. H. (1984/5) *The selection of management strategies for controlling nematodes in cereals*. Agriculture, Ecosystems and Environment, 12: 371–388

Cook, R. J. (1988) *Biological control and holistic plant health care in agriculture*. American Journal of Alternative Agriculture 3: 51–62

Crowder, R. A. (1987) *An organic monoculture—can it work?* English IFOAM Bulletin 1: 6–8

Dempster, J. P. & Coaker, T. H. (1974) *Diversification of crop ecosystems as a means of controlling pests*. In: Price-Jones & Solomon, 1974

Hay, J. (1987) *Natural pest and disease control*. Century; London

Horn, D. J. (1988) *Ecological approaches to pest management*. In: *Food production and natural resources*. Elsevier; London
Kahnt, G. (1983) *Gründüngung*. DLG-Verlag; Frankfurt
Mader, H-J. (1984) *Die Tierwelt extensiver Obstwiesen und intensiver Obstplantagen im quantitativen Vergleich*. German IFOAM Bulletin 50: 8–14
Martens, B. (1983) *Der Einfluss von Streifen-Anbau zwischen Hafer und Erbsen auf die Populationsdynamik der Getreideblattläuse und ihrer Antagonisten*. PhD Thesis. Botanical Institute; Heidelberg
Motyka, G. & Edens, T. C. (1984) *A comparison of heterogeneity and abundance of pests and beneficials across a spectrum of chemical and cultural controls*. Report No. 41. Department of Entomology, Michigan State University; East Lansing, MI
Patriquin, D. G. & Baines, D. (1989) *Coevolution of crops, weeds and pests*. Paper presented at the 7th IFOAM International Scientific Conference, Burkina Faso, January 1989
Pimentel, D. (1977) *The ecological basis of insect pest, pathogen and weed problems*. In: Cherrett & Sagar, 1977
Schlang, J. (1989) *Zur biologischen Bekämpfung des weissen Rübenzystennematoden* (Heterodera schachtii) *durch resistente Zwischenfrüchte*. In: 52e Congres d'hiver/IIRB pp. 249–265.
Sengonca, C. & Brüggen, K. U. (1989) *Auftreten von Winterweizenschädlingen und ihren natürlichen Feinden in unterschiedlich bewirtschafteten Ackerbaubetrieben*. Zeitschrift für Pflanzenkrankheiten und Pflanzenschutz 96: 100–106
Speight, M. R. (1983) *The potential of ecosystem management for pest control*. In: *Biological Control*. Agriculture, Ecosystems and Environment Special Issue Vol. 10 (2)
Stöppler, H., Kölsch, E. & Vogtmann, H. (1989) *Sortenmischungen bei Winterweizen*. Lebendige Erde 6/89: 430–438.
Sumner, D. R. (1982) *Crop rotation and plant productivity*. In: *CRC Handbook of Agricultural Productivity*, Vol.1, Plant productivity (M. Rechcigl, ed.), pp.273–313. CRC Press; Boca Raton, Florida
Way, M. J. (1977) *Pest and disease status in mixed stands versus monocultures; the relevance of ecosystem stability*. In: Cherrett & Sagar, 1977
Wolfe, M. S. (1985) *The current status and prospects of multiline cultivars and variety mixtures for disease resistance*. Annual Review of Phytopathology 23: 251–273

Plant defence mechanisms and plant extracts
Matthews, D. & Schearer, W. (1984) *Tansy—a potential repellent for the Colorado potato beetle*. Rodale Research Centre Report EN-84/6
Schmutterer, R. & Ascher, K. R. S. (eds) (1987) *Natural pesticides from the Neem tree and other tropical plants*. Proceedings of the 3rd International Neem Conference, Nairobi, July 1986. Gesellschaft für technische Zusammenarbeit; Eschborn.
Snoek, H. (1988) *Naturgemässe Pflanzenschutzmittel. Anwendung und Selbstherstellung*. Pietsch; Stuttgart.
Swain, A. (1977) *Secondary compounds as protective agents*. Annual Review of Plant Phytopathology 28: 479–501
Hellpap, C. (1989) *Insect pest control with natural substances from the Neem tree*. Paper presented at the 7th IFOAM International Scientific Conference, Burkina Faso, January 1989

Biological control
Baker, K. F. & Snyder, W. C. (1965) *Ecology of Soil Borne Plant Pathogens—Prelude to Biological Control*. Proceedings of International Symposium, University of California, Berkeley, 1963. John Murray; London
Biological Control. Agriculture, Ecosystems and Environment Special Issue Vol. 10 (September 1983)

Biological Control. American Journal of Alternative Agriculture Special Issue, Vol. 3 (Spring/Summer 1988)

Chet, I. (ed.) (1987) *Innovative Approaches to Plant Disease Control*. Wiley; New York

Cook, R. J. & Baker, K. F. (1983) *The Nature and Practice of Biological Control of Plant Pathogens*. American Phytopathological Society; St Paul, MN

Dalby, J. (1984) *Parasites and predators in biological pest control*. New Farmer and Grower 4: 6–7

Gear, A. (1988) Biological pest control in the greenhouse. HDRA Newsletter 112: 21–23

Hoy, M. A. (1988) *Biological control of arthropod pests: traditional and emerging technologies*. American Journal of Alternative Agriculture 3: 63–68

Huffaker, C. B. (ed.) (1980) *New Technology of Pest Control*. Wiley; New York

Hussey, N. W. & Scopes, N. (1985) *Biological Pest Control: the glasshouse experience*. Blandford Press, Poole

Jones, F. G. W. (1974) *Control of nematode pests—background and outlook for biological control*. In: Price-Jones and Solomon, 1974

Kirsch, P. (1988) *Pheromones: their potential role in the control of agricultural insect pests*. American Journal of Alternative Agriculture 3: 83–97

Lisansky, S. G. (1981) *Biological pest control*. In: Stonehouse, 1981

Lopez-Real, J. M. & Hodges, R. D. (eds) (1986) *The Role of Microorganisms in a Sustainable Agriculture*. Special Issue of Biological Agriculture and Horticulture, 3 (2/3)

Lumsden, R. D. & Papavizas (1988) *Biological control of soilborne plant pathogens*. American Journal of Alternative Agriculture 3: 98–101

Price-Jones, D. & Solomon, M. E. (1974) *Biology in Pest and Disease Control*. 13th Symposium of the British Ecological Society, 1972. Blackwell; Oxford

Quicke, D. (1988) *Spiders bite their way towards safer insecticides*. New Scientist, 26/11/88: 38–41

Quinlan, R. J. (1989) *Biological control*. Natural Farming. Summer: 24–29

Specific pests and diseases

Davis, L. (1988) *Vegetable disease identification*. New Farmer and Grower 19: 35–36

Deane, T. (1987) *Snake in the grass with catholic tastes*. New Farmer and Grower 15: 12–13, and subsequent correspondence NF&G 16: 28

Dussart, G. & Cherfas, J. (1989) *Slugs and snails and scientists' tales*. New Scientist 22/7/89: 37–41

Fream's *Elements of Agriculture*. 15th edition. John Murray; London; 1972

Frost, D. (1989) *Survey of vegetable varieties*. New Farmer and Grower 22: 18–21

Gair, R. *et al.* (1987) *Cereal Pests and Diseases*. 4th edition. Farming Press; Ipswich

Graber, C. & Suter, H. (1985) *Schneckenregulierung*. Forschungsinstitut für biologischen Landbau; Oberwil, Switzerland

Hasson, L., Stopes, C. & Davies, L. (1989) *Potato Feature*. New Farmer and Grower 25 (Winter): 16–19

HDRA (1989) *On the slug trail*. Step by Step Organic Gardening pamphlet series. Henry Doubleday Research Association; Coventry.

MAFF/ADAS and Scottish Agricultural Colleges advisory leaflets

McKinlay, R. G. (1987) *Control of cabbage root fly on culinary swedes*. English IFOAM Bulletin No. 2

World Crop Pests – a series of books on a wide variety of pests, their biology, natural enemies and control, published by Elsevier

General

Allen, P. & van Dusen, D. (eds) *Global Perspectives on Agroecology and Sustainable Agricultural Systems*. Proceedings of the 6th IFOAM Conference, Volume 2, Part 10: *Nonchemical Pest Management — Insects, Nematodes and Pathogens*. Part 11: *Soil Management and Plant Nutrition — the Relationship to Pest Control*. University of California, Santa Cruz

Carroll, C. R. , Rosset, P. & Vandermeer, J. (eds) (1989) *The ecology of agroecosystems*. Macmillan

Harris, P. & Lennartsson, M. (eds) (1989) *Undergraduate Projects in Organic Agriculture and Horticulture. Volume 1*. Henry Doubleday Research Association; Coventry.

Jackson, A. W. (1988) *Organic farming – crop protection implications*. British Crop Protection Council. Proceedings of the BCPC Conference: Pests and Diseases. Volume 1: 1043–109

Stonehouse, B. (1981) *Biological Husbandry — a scientific approach to organic farming*. Butterworths, London

Vogtmann, H. *et al.* (eds) (1986) *The Importance of Biological Agriculture in a World of Diminishing Resources*. Proceedings of the 5th IFOAM Conference. Verlagsgruppe; Witzenhausen

Recent publications

Unwin, R. (ed.) (1991) *Crop protection in organic and low input agriculture*. Conference proceedings. Monograph 45. British Crop Protection Council; Farnham.

Chapter 8

Livestock Husbandry

The Need for Change

Over recent years, farm animals have tended to bear the brunt of the effects of new, intensive systems designed to maintain or increase farm profits. The impact of these systems on farm animals has not gone unnoticed by the consumer. One of the major reasons why vegetarians avoid meat is their concern for the conditions under which animals are raised and fattened. Many people, including researchers and advisers in universities and colleges, ADAS and MAFF, now appreciate that the application of our knowledge of animal ethology, behaviour and welfare can and should come together with good livestock husbandry.

There is nothing new about this. Most aspects of traditional livestock husbandry ensured that animal ethology and production were more closely attuned than they now are, with more people attending to smaller groups of animals fed at moderate levels and living in naturally ventilated, straw-bedded housing. However, all the factors which have led to the increasingly intensive use of resources in arable agriculture have similarly moulded systems of animal husbandry with detrimental results for the health and behaviour of livestock.

At the same time, philosophers have wrestled with the question of how much animals (other than humans) should be considered to have a 'soul', or 'feelings' or suffer 'pain' (for example, Singer, 1976 and others). If animals do, then we have a moral duty to act with greater compassion towards them, and many of us are at fault. However, these are questions to which no certain answer can ever be found so there will always be the criticism that considerations of farm animal welfare are irrelevant, merely demonstrating the unerring anthropomorphic attitudes of many human beings.

Perhaps most important to organic farmers is the relationship between the health and vitality of the individual animal and the way in which that animal is kept. Much research and observation has shown that many contemporary diseases and syndromes are related to the housing, feeding or breeding methods which have been adopted to increase production or profitability. New problems have been introduced, which have required new solutions from veterinarians and animal scientists.

Some of the most significant health and fertility problems are closely related to intensity of production. Increased yields in dairy cattle are closely correlated with increased culling for infertility, mastitis and hoof ailments, and conception at first service has also been shown to be correlated to stocking rates and yields (Boehncke, 1985, 1986). Part of this is due to the use of high levels of fertiliser on grassland. A ration containing too high a potassium load can lead to fertility problems, disturbance of carotene metabolism and reduced feed intake, while a chronic high nitrate load during pregnancy has been linked to milk fever, retention of the placenta and inflammation of the uterus. Nitrates in the forage may be converted by bacteria in the rumen to nitrites, which can be directly toxic (methaemoglobin formation) or block the activities of enzymes, leading, for example, to vitamin A deficiency symptoms even though there is a good supply of carotene.

The fatty liver syndrome in dairy cows is a serious disorder which is not the result of a single pathogenic organism. The effects of the syndrome are diverse and arise from several different but related causes. High levels of concentrates in the ruminant's diet can reduce the pH in the rumen. This mild rumen acidosis can result in the development of fatty liver syndrome (Reid *et al.*, 1983). This in turn may be related to the increased incidence of mastitis (Lachmann, 1984) and foot problems (Dirksen, 1980), as well as reduced fertility. Diets with a low fibre content have also been implicated in exacerbating foot problems, which can be made worse if animals are kept on slatted floors with no bedding. Taken together, mastitis, foot problems and poor fertility are among the primary problems in the national dairy herd, and are certainly the most common causes for an animal to be removed from the herd.

Acidosis is also a significant problem in barley beef systems. A high proportion (as many as 25% according to some studies) of barley beef animals suffer from abcessed livers. The high acid content of the rumen hardens the rumen wall, causing the villi to clump together trapping hair and barley awns. These penetrate the gut wall, causing inflammation and allowing bacteria to enter the blood stream and attack the liver.

Sows confined in farrowing crates have been observed to show a higher incidence of three related health problems: mastitis, inflammation of the uterus and poor milk production (agalactia). They also appear to have longer farrowing times, a higher level of morbidity and a greater number require veterinary treatment (Ekesbo, 1981). The increase of several diseases in pigs has also been correlated with the increasing number of pigs per building (Lindquist, 1974; Fox, 1984) as well as the total size of the herd and the need for medicated feed

Figure 8.1 Effects of chronic concentrate overfeeding in dairy cows and fattening beef animals.

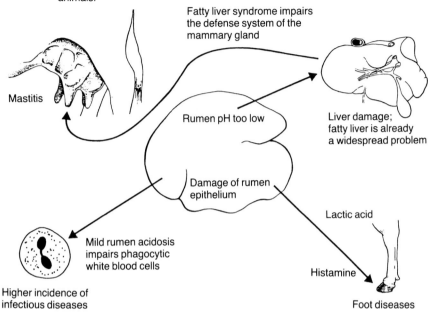

Fatty liver syndrome impairs the defense system of the mammary gland

Mastitis

Rumen pH too low

Liver damage; fatty liver is already a widespread problem

Damage of rumen epithelium

Lactic acid

Mild rumen acidosis impairs phagocytic white blood cells

Histamine

Higher incidence of infectious diseases

Foot diseases

Source: Boehncke (1985).

(Niederstücke, 1983). In addition, the breeding objectives for high feed conversion performance and leanness has resulted in modern pig breeds with weak hearts, poor circulation, susceptibility to arthritis and leg weakness, and poor quality, PSE (pale, soft, exudative) meat.

Antibiotic resistance in animal husbandry is a problem which makes it increasingly difficult to find a cure for some simple bacterial diseases, and the increasing resistance of important bacterial pathogens is attracting widespread attention. In addition, the effectiveness of the animal's own immune system may be reduced by the system as a whole. A reduction in immune competence has been related to the production of stress hormones when animals are overcrowded; the lack of opportunity for animals to develop their own immunity gradually; the presence of pesticide or heavy metal residues and even the relationship between stockperson and stock (Gross & Siegel, 1982).

Other examples of the relationship between modern livestock systems and new health problems in livestock have been described in Fox (1984). The most recent example is that of bovine spongiform encephalopathy (BSE) or cow madness. The disease, first recognised in Britain in December 1986, is thought to have arisen from feeds contaminated with the infectious agents of scrapie, a neurological disease of sheep and

Figure 8.2 Advantages and disadvantages of modern super pigs.

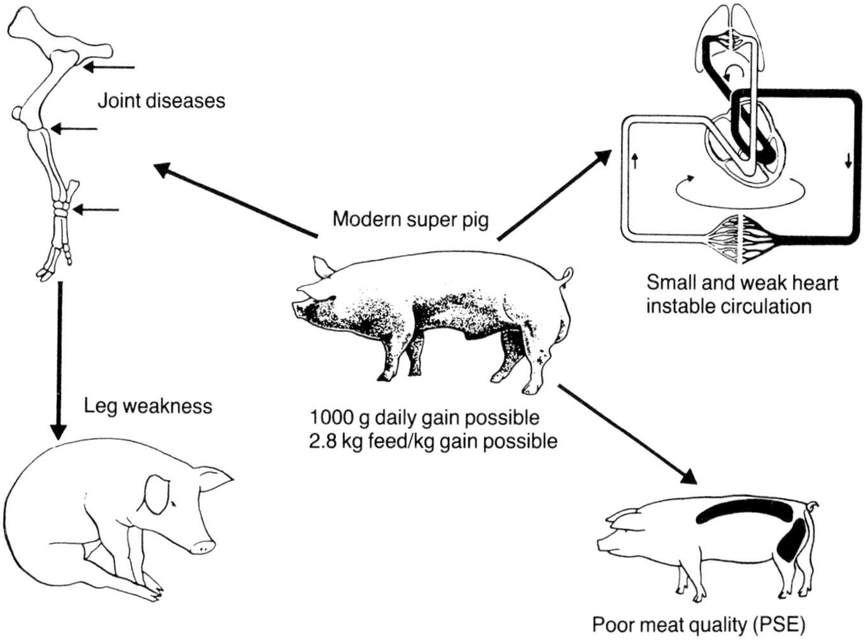

Joint diseases

Modern super pig

Small and weak heart
instable circulation

Leg weakness

1000 g daily gain possible
2.8 kg feed/kg gain possible

Poor meat quality (PSE)

Source: Boehncke (1985).

goats. Sheep brains, which contain large quantities of the transmissible agent, and other animal by-products were used as a cheap protein supplement in cattle concentrate feeds until this practice was prohibited in 1988, following the discovery of the link. Given that such additives are not even necessary, the question arises whether animal protein of any sort should be used for feeding ruminant livestock.

Such 'technological' diseases will always confront the animal scientist and veterinarian with new problems and we are at risk of digging a bottomless pit from which the only escape is to adopt a system which does not seek such high yields with only short-term financial considerations in mind. In other words, a system is needed which is closer to the animal's 'natural' capacity, however this may be defined. The wellbeing of animals must be determined with reference to their health status, disease incidence, longevity, reproductive performance and various physiological and behavioural indicators, not just the simple guideline of productivity, which is only an indicator that the animal is well fed and watered and free from clinical disease.

There are other consequences of modern livestock production systems which have implications for the soil, crops and human health and wellbeing. The divorce of livestock production from the soil not only runs counter to the nature of the animals themselves, it also creates problems in terms of massive reliance on feedstuffs from elsewhere and in terms of disposing of livestock 'wastes', usually in environmentally harmful quantities and concentrations.

The use of conventional medications may also create environmental problems: feed additives leading to the build-up of copper and zinc in the soil, drugs which under certain circumstances can actively prevent the breakdown of manures by interfering with soil biological activity, and sheep dips which are usually disposed of, however carefully, in ways which lead to their escape into freshwater and marine ecosystems adding to the general pollution burden dramatically illustrated by the deaths of seals in 1988.

For humans the direct problems range from the quality of meat and dairy products to competition between livestock and humans for food, as well as working conditions which create health problems among agricultural workers. For example, respiratory diseases such as bronchitis are prevalent among workers in intensive livestock units. Stress can also be a significant factor as fewer people are required to tend increasing numbers of animals, often single handed in relatively monotonous or unnatural conditions such as pig sweat houses (FAWC, 1988).

The contamination of milk by antibiotics is a well known problem and significant steps have been taken to deal with it, because there are also serious implications for milk processing activities. But antibiotics are also used as growth promoters, and salmonella resistance to antibiotics makes the treatment of humans as well as livestock more difficult. Poultry account for 50% of salmonella cases, which have increased from negligible levels in the 1940s to more than 20,000 cases in 1988. The links can no longer be ignored.

The salmonella in eggs scare of the late 1988/early 1989 was also linked to 'unnatural feeding practices', in particular the use of poultry by-products in feed for laying hens. The particular strain of salmonella which gave rise to the scare, *S. enteritidis*, is thought to have entered Britain in imported feedstuffs, with the problem being exacerbated by the recycling of poultry offal (New Scientist, 17/12/88). Only 392 cases of salmonellosis were attributed to this strain in 1981; by 1988 there were more than 12,000, making up more than half of all cases of salmonella poisoning.

The use of hormones as growth promoters is widely criticised because of the potential health risks. These hormones, although they may be described as 'nature-identical', are usually imperfect copies of the real thing allowing them to be manufactured commercially; the extraction of the natural hormone directly from the living animal would not be feasible. The argument that hormones used as growth promoters are not detectable is clearly false, as the tests used to detect stock which have been illegally implanted with hormones, following the European Community ban on their use, rely on their detectability. In the summer of 1988, major illegal use of hormones was uncovered in West Germany and bans were placed on the movement of cattle from several farms in North Wales because the use of hormones had been detected, shaking the complacency of farmers who had believed such tests were ineffective.

Yet serious consideration is still being given to the use of bovine somatotropin (BST) to increase milk output at a time of chronic milk surplus. A major report (Isermeyer *et al.*, 1988) for the West German parliament's Commission of Technology Assessment concluded that although most studies showed no effect on cow health from the use of BST, many showed a significant impact on fertility with a typical reduction in conception rate from 90 to 75%. It also concluded that cow health could be affected where feeding was inappropriate. In addition, the wide variability in response to BST from individual cows might mean that many farmers would gain no overall economic benefit. BST has also been criticised on the grounds that it does not confer the higher feed capacity and gut volume needed to meet the higher nutrient requirement for the increased production, except under circumstances of very high feed quality (Ørskov, Farmers Weekly 10/2/89).

With milk and grain in surplus and a population in the less developed countries who can ill afford to supply energy or protein sources and in particular soya meal (Figure 8.3) for European livestock, it behoves us to question our current approach to animal husbandry and our excessive consumption of animal products. In 1984, the UK livestock population stood at 13.2 million cattle and calves, 34. 8 million sheep and lambs, 7.7 million pigs and 129.4 million poultry. With such a population to feed, it is not surprising that so much grain needs to be produced on so much land with so many inputs.

One long-term aim must be to alter the dietary balance in developed countries. Perhaps this is a challenge which can only really be met by a sustainable agriculture, aware of the needs of animals as sentient beings and the needs of the environment as a place for sustainable food production. Fox (1984) has suggested we should aim for three Rs to achieve the welfare of animals: refinement (improvement of husbandry systems and practices), reduction (in the total livestock and poultry population) and replacement (of farm animal produce with other agricultural produce).

The potential importance of reduction and replacement can be illustrated on a global scale with one example. Simantov (1979) explored the effect of the likely increase in pork consumption in China if that country were to follow the Western approach to the development of its

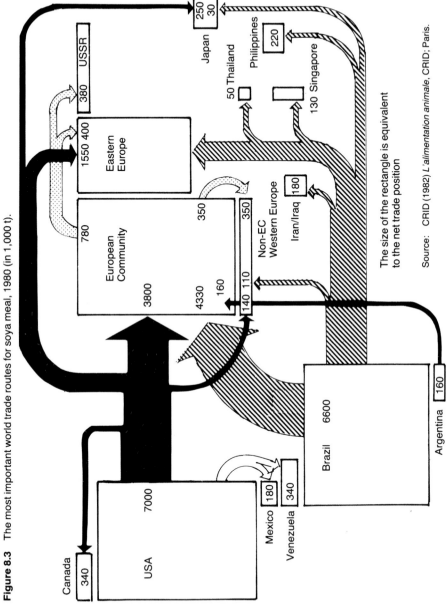

Figure 8.3   The most important world trade routes for soya meal, 1980 (in 1,000 t).

The size of the rectangle is equivalent to the net trade position

Source: CRID (1982) L'alimentation animale, CRID; Paris.

livestock industry and meat consumption. If the Chinese population were to double pork consumption in the 15 years from 1977, this would be equivalent to one fifth of the total world pork consumption in that year. Food for this large population of pigs could not be found within China, placing a demand on the world market for cereals and protein rich feeds of $7 billion, equivalent to China's total imports in 1977.

Organic agriculture seeks to ensure a more balanced system in this general sense. More importantly, at the level of the individual farm it seeks to recognise that farm animals are not just units of production but are feeling, sentient beings which are an integrated part of the whole farm.

THE ROLE OF LIVESTOCK ON AN ORGANIC FARM

An organic farm without livestock is a contradiction to many people. Livestock are usually considered to be an essential component of the farm, supplying manure and allowing a balanced rotation based on leys and arable cropping. It is certainly generally the case that organic farms are mixed farms, although there are examples of stockless farms; these may become more numerous in the future as primarily arable farms choose to convert to organic farming. The extent to which a viable system can be maintained without using ley/arable rotations is not well known, since the major benefits such a rotation gives in terms of weed, disease and pest control and fertility building are very significant.

Within a mixed ley/arable farming system, ruminant livestock are the central point around which the rotation and indeed the whole farm operates. The grass/clover ley acts as the major source of fertility within the rotation, and without cattle or sheep no profitable use could be made of this productive and fertility building crop. While cattle and sheep graze, they return nutrients to the land. During the winter, if they are housed, they eat conserved forage leaving the manure to be collected and subsequently distributed round the farm. This ability to move nutrients around the farm through the appropriate spreading and use of manures is a very important advantage of keeping ruminant livestock on farms.

Non-ruminant livestock can also play a useful role on organic farms. Poultry for eggs or meat are always a good additional enterprise and can have a beneficial effect on the pasture on which they are ranged. Pigs make light work of breaking or clearing new ground, and have the advantage of being omnivorous, making them ideal users of foods which might otherwise go to waste. Thus, skim milk from the dairy, vegetable and other food wastes may all be converted into meat. Both pigs and poultry require concentrated energy rich foods as well as a sufficiently high level of protein in their diet, so they are in competition with humans for food in a way that ruminant livestock are not (or should not be). This may sometimes make it difficult to formulate rations with organically grown food.

Finally, livestock play an important economic role on the organic farm, as they do on any farm. They may, as in the case of a dairy herd, provide a regular income; or they may be an effective additional enterprise making best use of the farm's resources. In general, livestock and their products do not attract the premium available on cash crops in the rotation. This can put financial pressure on the organic livestock farmer, particularly if investment has been made to upgrade or convert existing livestock housing or feeding systems.

A crucial element in successful and profitable organic animal production is the marketing. Some of the most successful meat producers either market directly from the farm butchery, or have a close relationship with a local abattoir and butcher so that the customer comes to know the meat and the source and is thus assured of the consistent quality of the product.

General Principles of Organic Livestock Husbandry

Livestock husbandry on organic farms depends on three major principles:

- The systems in which animals are kept must conform to the highest welfare standards.
- They must be fed in a way suited to their physiology, using food largely produced on the farm.
- Veterinary treatment should always avoid routine prophylactic drug use. Livestock health should be maintained through good preventive husbandry, animal welfare and appropriate housing and feeding systems.

The most important feature must be the attainment of the highest possible standards for animal welfare, for which the Universities' Federation of Animal Welfare Handbook *Management and Welfare of Farm Animals* (UFAW, 1988) can be taken as a basis. Thus animals must be housed and fed in a way which allows them to practise the fullest possible range of their 'natural' behaviour patterns. This requirement is not merely to satisfy the demands that animals should be well looked after because they are sentient beings, it is also one positive way in which the health and vitality of the animal may be ensured. The fact that stressed or confined animals show some health problems not seen when animals are kept in a more ethologically benign way is evidence enough for the importance of such systems. Thus, prolonged tethering and confinement in cages or in enclosed buildings without acceptable bedding or floor surfaces are all disallowed on organic farms.

Feeding should be suited to the physiology of the animal. Some of the possible effects of overfeeding concentrates to ruminant livestock have already been mentioned; organic farmers seek to ensure that the animal is fed in a way which fully recognises the potential as well as the limitations of the digestive system of the animal. In addition to this, to ensure that the system on the farm is as balanced as possible, it is important that all the forage required by ruminants on a farm is produced on that farm.

Lastly, the routine and prophylactic use of conventional drugs on organically managed livestock should be avoided. This last common feature, which is perhaps the most difficult to attain, is of great importance. Often, in modern systems of animal husbandry, there is an almost total reliance on the routine dosing of animals with anthelmintics, antibiotics, vaccines, trace elements, growth promoters and so on. Together, these help to prop up the animal's health in the face of an ever increasing challenge of disease from the environment and the farming system. They provide an excuse for and a means to cover up what is simply bad husbandry, resulting at least in part from declining labour use on farms.

FEEDS AND FEEDING

Ruminants have evolved to fulfil a specific role in agricultural systems, which is to utilise feedstuffs which cannot be used directly by humans. The microflora in the rumen convert cellulose and other plant constituents to volatile fatty acids which pass from the rumen into the blood and are subsequently used as an energy source for milk, meat and fibre production. The relatively valueless feed protein, often with only a limited range of amino acids, is converted into bacterial protein which is then available to the ruminant animal, unlike pigs and poultry which require the full range of amino acids. Thus the ruminants are unique in their ability to supplement the needs of humans without direct competition.

The ruminant's ability to digest large quantities of fibrous material has always meant that cattle and sheep have been an important part of British farms. Few climates are better suited to the production of grass, although there has been a long running trend since the second world war, heightened on Britain's accession to the European Community, towards high levels of milk or meat production through the 'efficient' use of concentrates, particularly on lowland farms. This had led to prequota dairy farmers being advised to feed up to twice the level of concentrated food compared with the recommendations given to their grandfathers.

High milk yields were the result, the average yield per cow increasing by almost 40% between the early 1960s and the 1980s with a large part of this increase being the result of increasing the levels of concentrates fed. Associated with this have been increasing veterinary bills and greater welfare problems for cows, a shorter productive life, surplus production of milk and so on.

No doubt this change, 'milk from concentrates' instead of 'milk from grass', is being slowed down if not reversed to some extent as quotas sharpen the minds of farmers, advisers and feed manufacturers towards the importance of optimising the economic production of milk from grass. No longer is high yield per cow or per lactation necessarily the most important goal in the minds of dairy farmers.

Organic farmers have always relied on 'milk from grass', and their rotations have depended on the use of long and short leys as well as break crops to provide other bulky feeds. An important aspect of this is the clover content of the clover/grass leys. Although the overall stocking rate may be reduced, the output per head is likely to be increased compared with stock on all grass swards. This is because both the protein and mineral content and total intake are increased with clover in the sward. Silage containing legumes will also give higher production than grass of similar digestibility. The MAFF/ADAS booklet 2047 *Grass as a Feed* contains useful information on nutrient values, palatability, digestibility and intake of grass, but it does not give adequate information on the benefits of clover, much of which is now the subject of considerable research at experimental husbandry farms and research institutes.

Research work at North Wyke Research Station in Devon suggests that three Hereford x Friesian steers per hectare can be finished in 16–17 months on clover/grass swards without additional feed. The 40% clover content of silage fed in winter resulted in a liveweight gain of 0. 8 kg/day without any concentrate supplement. Tight grazing at a sward height of 7 cm during the summer resulted in similar liveweight gains. Cattle offered barley did not grow faster, but ate less silage.

More recently, results from a trial comparing organically and conventionally raised beef at the North of Scotland College of Agriculture's Craibstone Farm have confirmed these results (Younie, 1989 & 1990). The beef animals were stocked at a rate of 3.6 head/ha on the grass/clover ley, compared with 4.3 head/ha on a pure grass sward receiving 270 kg N/ha. First cut silage yields from the organic system were about 10% lower at just under 6 tonnes dry matter per hectare. However, the clover content of the organic silage resulted in significantly higher winter liveweight gains with only half the concentrate use, compensating for significantly lower summer liveweight gains due to compensatory growth factors and the effects of parasite burdens late in the season. The gross margin per head was 12% lower for the organic animals, but because of lower stocking rates the difference in gross margin per hectare increased to 27%. To achieve the same gross margin per hectare after bank interest a 14% premium would be required. There may be scope for reducing this gap though further exploitation of the nutritional benefits of clover silage and reducing the use of expensive, organically produced concentrates. A similar conclusion was also reached by Lowman (1989).

In other trials, the liveweight gain of lambs fed red clover was 25% higher than when fed perennial ryegrass, and lambs fed lucerne achieved a 50% higher liveweight gain. White clover, although lower yielding than red clover or lucerne, gave the best results, increasing liveweight gain by 90% when forage was fed ad-lib, and by even more when intake was restricted.

The digestive physiology of ruminants: nutrition and health

The voracious appetite of cows and sheep for bulky green food is made possible by the action of the rumen. In an adult animal, the rumen may take up 60% of the total internal volume and contains a rich and diverse microbial population. It is worth considering what goes on in the rumen, since the balance of this living symbiotic system inside the animal is as critical as that outside the animal and above and below the surface of pastures and crop land. It is an active concern of the organic livestock farmer that this finely balanced system is protected and enhanced, since on this depends the health and vitality of the animal.

Only certain bacteria in the rumen can produce the enzyme cellulase which is needed for breaking down the long and complex cellulose molecules which are the major constituent of roughage. Through the action of these bacteria and other microbial species the complex starches and proteins of forages are broken down releasing energy, protein and the building blocks for fat. The rate at which food is broken down in the rumen depends upon the type of food, with roughage passing through the rumen at a slower rate than concentrate feed.

The anaerobic activity of the microbial population of the rumen results in the production of a range of simple organic acids (the volatile fatty acids or VFAs) which are absorbed from the rumen and the rest of the digestive tract along with any other undigested material and microbial cells. Acidity in the rumen is controlled by the production of ammonia as well as sodium bicarbonate in the saliva. If the pH falls to

Figure 8.4   The digestive system of the cow.

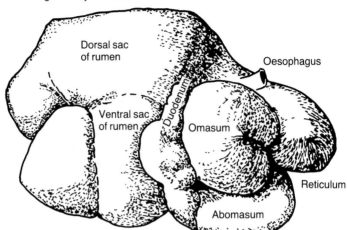

below 6.5 then several pathologies may emerge; rumen acidosis, fatty liver syndrome and foot problems are the most obvious. Apart from these physiological problems, which may be the result of high levels of concentrate feeding, there is a further problem associated with low roughage diets. This is the 'low fat syndrome' associated particularly with winter and early spring milk production when the fibre content of the diet may be reduced. There is no solution apart from the inclusion of sufficient fibre in the diet to ensure an adequate production of acetic acid, an essential VFA produced by particular bacteria whose main job is the breakdown of roughage. These microbes are put off by high acidity in favour of different, proprionic acid-producing microbes which prefer concentrates.

Diets with high levels of concentrates and low quantities of roughage should, therefore, be avoided. The feeding of ruminant livestock on organic farms involves an appropriate balance between fibrous feeds and concentrates. Under the Soil Association's standards, forage should account for at least 60% of the dry matter in the ruminant diet.

Pigs have a digestive system very similar to that of humans, and therefore cannot survive on a purely roughage based diet. Like humans, they rely on more concentrated feeds, and similarly, digestive competence and good health require some roughage in the diet, which is digested in the caecum. The omnivorous habits of pigs make them willing to eat foods of many different types, but if good growth is to be ensured, it is important to know what the value of the food given really is. Poultry also rely on concentrated food although the digestive system is very different from that of a pig. The crop, filled with grit and bacteria, allows the bird to grind food very thoroughly and ensures some preliminary breakdown before it passes through the digestive tract.

Ration formulation for ruminants

Ideally, all the food consumed by livestock should be produced on the holding, allowing the development of an enclosed and self-sustaining system. Economic considerations often mean that this is not fully possible, particularly because of the frequent necessity to emphasise cash cropping on the farm. Organic livestock farmers should always attempt to maximise their home food production, and should certainly expect to produce all the forage necessary for the livestock on the farm.

The organic standards organisations usually allow the use of limited quantities of conventional feeds, which are particularly useful for contributing to the concentrate portion of the diet for dairy cows or finishing beef cattle. The reason for this is the recognition that few farms are capable of being 100% self-sufficient, and that the current availability of suitable organically produced feeds is severely limited. At the same time, the problems inherent with conventional feeds need to be recognised, so that the conventional feed 'allowance' should only be seen as a last resort. The maximum feed allowance from conventional sources depends on the type of livestock, with the Soil Association specifying 10% for non-dairy ruminants, 20% for dairy cattle and 30% for pigs and poultry, usually calculated on the basis of daily dry matter intake. The UKROFS standards are even tighter with respect to non-dairy sheep and goats, where the limit is 5%. The International Federation of Organic Agriculture Movements' international baseline standards recommended that national standards organisations should move to 10% or less for all ruminants by 1994, and many European countries already only allow 10% for dairy cattle and 20% for pigs and poultry.

The Soil Association's standards also stipulate that although at least 50% of the ration for livestock must be organic, up to 50% may be 'in conversion' (i.e. forage produced 'organically' during the conversion period before it is fully recognised as such). If conventional feeds are used to supply 20% of the ration, this means only 30% may come from fields in conversion. The non-forage part of ration (concentrates and other feeds such as sugar beet pulp) must not comprise more than 40% of daily dry matter intake. A cow giving 5,000 litres/year will need about one tonne of concentrates or 5–6 kg/day fed flat rate for a 180–200 day winter. This falls within the Soil Association's requirements, even if all the concentrates are purchased from conventional sources.

If there is organic grain available on the farm, it will usually have a high opportunity cost since the cereals could have seen sold for human consumption where premiums are higher. If it is possible to attract a sufficient premium on livestock products to justify the use of home grown cereals or other energy or protein rich foods in the ration, this is certainly to be recommended. It may also be the case that the reduced need for concentrates because of the improved quality of the forage will result in reduced overall concentrate costs anyway. Straights produced on the farm or brought on to the farm can be pelleted readily by mobile mill and mix operators; alternatively it may be worth installing mill and mix units on the farm, but this can entail high capital investment and running costs.

Cereals will be an important component of any home mix, supplemented by suitable protein sources. Oats are preferable to wheat or barley as their higher digestible fibre, higher fat and lower starch content not only cause fewer digestive upsets, but also help to produce milk of high butterfat with lower than normal saturated fat content. Suitable sources of protein include field beans and peas, as well as rapeseed meal, dried lucerne and maize gluten. Compounded feeds made from organic ingredients are beginning to become available.

Any feedstuff of plant origin may be used in a compound feed which is part of the non-organic component of the ration, but some are preferred to others. The use of imported protein and energy supplements, particularly soya or cassava, is not recommended, partly because of the risk of pesticide residues where chemicals banned in developed countries are still in use, and partly because the production of these crops in certain less developed countries is believed to have destabled local agricultural production.

Although conventional protein supplements are allowable, care should be taken with protein overfeeding. The real problem on organic farms may in fact be insufficient energy rather than protein, especially where lucerne or red clover based forage is used. The legume content of the forage is likely to result in much higher protein levels than would be expected in a conventional system. Excess protein can lead to health problems similar to those in conventional systems with a high nitrogen load, such as excess ammonia production, liver and fertility problems, as well as high urea levels in milk. When excess protein is combined with too little energy in the ration, this can actually lead to protein deficiency in the cow and reduced protein levels in milk, because of the energy consuming process of deamination. Where necessary, high protein levels can be compensated for by feeding straw or hay.

Fodder crops other than silage or hay can form a valuable component of the ration, particularly during the winter and early spring when the availability of other foods may be poor. Root crops such as turnips, swedes, mangels, fodder beet and sugar beet are all ideal, giving fodder during the autumn, winter and spring depending on sowing date. Both cattle and sheep can benefit from these alternative fodder crops, and they will often fit well into the rotations on organic farms. Such root crops may be fed up to 3 kg dry matter per cow, with anything more than this leading to problems of protein balance in the whole diet since these foods contain abundant energy but are short in protein. Other forage crops such as kale and cabbage can be invaluable for winter feeding, although kale should not exceed 30% of the dry matter intake of the animals and can affect cow fertility. Finally, the green tops of beet, turnips and swedes can all be useful in feeding cattle, having a higher protein content than the roots. Again, intake shold be restricted to about 3 kg dry matter per cow.

The mineral and vitamin balance of the ration should be adequate on well established organic farms, where management of the forage is likely to positively influence its mineral content. If there is a known dietary deficiency in home grown feeds, or as a result of soil deficiencies (e.g. copper and selenium), then mineral supplementation in the feed or application to the land should be made to rectify the deficiency, but at the same time efforts should be made to deal with the underlying cause (e.g. poor drainage or other factors locking up soil nutrients) if at all possible. For ruminants, the only vitamins which are not produced by bacteria in the rumen are the fat soluble ones A, D and E and these can usually be provided through suitable ration formulation.

In general, mineral and vitamin rich additions to the diet should be from natural sources. Seaweed meal contains a range of mineral and trace elements, as do bonemeal, cod liver oil and yeast. If such additions cannot rectify the problem, then additions of simple mineral salts can be made, or the use of mineral licks without flavour enhancers. However, the appropriate balance between minerals is important, since surpluses of individual minerals, or relative proportions such as sodium to potassium and calcium to phosphorus, can affect the uptake or use of other minerals as well as livestock health. The diet should not be routinely supplemented with mineral salts and vitamins as is so often the case with compound feeds.

Mineral balance, and in particular the potassium to sodium ratio, has been identified as a potential problem during the period of conversion to an organic system. Under certain circumstances, the potassium content of herbage has been found to increase from a more normal level of 15–30 g/kg DM to more than 40 g/kg DM in the year following the abandonment of mineral fertilisers (Krutzinna & Boehncke, 1989). Under conditions of low sodium availability, this can result in a potassium to sodium ratio of 50:1 instead of the optimal 10:1 level needed by the animal, and can lead to fertility and other problems. Special measures may therefore need to be taken during the conversion period.

Growth promoters including hormones and antibiotics, medicated feeds and other additives, including urea, are prohibited. Feed supplements of animal origin should not be used for ruminants and, under the Soil Association's standards for pigs and poultry, fish, meat and bone meal may be used only under certain circumstances.

Naturopathic probiotics may be permitted as feed additives, but only with the specific approval of the Soil Association's Symbol Committee. Probiotics are naturally occurring bacteria which may help promote livestock growth and food conversion. This may be because they have a role in animal digestion or actively control pathogenic bacteria (Ewing & Haresign, 1989). Whether probiotics are of any advantage in an organic or even conventional system is another matter. The gut flora is determined by a wide range of factors including nutrition and environment. Introduced bacterial cultures are therefore likely to be inefficient at best and perhaps more emphasis should be placed on stabilising gut conditions through sound nutrition so that the natural bacterial flora can develop more effectively.

Ionising radiation
The Chernobyl disaster brought the problem of radioactive contamination of pasture, milk and meat into stark relief, and even in 1989, more than three years after the event, the movement of sheep from the worst affected areas in North Wales and Cumbria was still restricted. At the time of the event, the authorities seemed helpless and incapable of giving advice or determining courses of action which would deal with the problem.

Since the event, some research has been carried out into methods for reducing the radioactive contamination of meat and milk from affected stock. The main problem substances are the long-lived radioactive isotopes of caesium and strontium. Caesium behaves in a similar way to potassium, and is fairly quickly removed from the body, but in a grazing situation will be taken up again by the herbage and recycled through the stock. Moving stock to relatively unaffected grazing land will reduce the problem by spreading it over a wider area, effectively diluting it. However, it can be argued that spreading the problem in this way may not be a desirable solution. Strontium behaves in a similar way to calcium and can be incorporated into the bone structure where it may affect the bone marrow.

The feeding of substances with a highly absorptive surface, to which cations such as caesium and strontium are bound, may be a way of dealing with the problem in emergency situations. Experiments using bentonite and rock dusts have proved successful at the University of

Kassel, where 100 g/animal per day proved sufficient to reduce contamination to 'acceptable' levels. This can be supplemented by the use of low contamination forage for an appropriate period before slaughtering. Swedish research has concentrated on other minerals such as mordenite. When 5 g of mordenite per day was fed to goats for a week, 60% of the caesium was excreted.

Organic systems cannot cure the problems caused by this type of pollution and the environment and health conscious consumer is likely to reject contaminated produce even when organically produced. In the event of another such disaster in the future, the authorities should be better prepared with advice to farmers. However, some information can also be obtained on this subject through 'alternative' sources such as IFOAM, the Soil Association and the Organic Advisory Service.

Rations for non-ruminants

The formulation of rations for non-ruminants is more complex, largely because a wider range of amino acids and vitamins needs to be supplied than for ruminants. Rationing pigs and poultry for maximum feed conversion and growth rates has become a complex exercise, especially given the economic vulnerability of these enterprises. Pigs can, however, utilise a wide range of different feedstuffs, including many products which might otherwise be considered waste. Several ideas for home-grown rations for pigs are contained in the Soil Association's technical booklet *Pig Keeping*. For more commercial and precisely targeted rations, conventional text books can be referred to, but it should be remembered that every animal is an individual and careful interpretation is essential in view of the general considerations set out above and in the Soil Association's standards.

One of the biggest problems in formulating rations for non-ruminants is replacing the non-organic components, particularly those of animal origin and protein sources such as soya beans which originate from developing countries. This is particularly so with poultry, where careful attention to rationing and egg laying performance may be necessary to ensure the financial viability of the enterprise.

In an effort to find a solution to this problem, the Research Institute for Biological Husbandry in Switzerland conducted some trials to devise rations for poultry using home-produced organic feed (Züllig, 1986 & 1988). The emphasis was on replacing imported feedstuffs, which proved to be no problem as far as home-grown energy sources were concerned, but protein sources were more difficult because of the required amino acid composition. Using field beans and peas meant that the ration tended to oversupply certain protein components and was thus nutritionally inferior to imported protein sources.

The ration tested consisted of grains (maize, wheat and barley in the ratio 11 : 10 : 4) fed at a rate of 30 g per hen per day, and meal produced from home grown sources (Table 8.1) fed ad lib. The performance of hens on this ration was comparable to conventional feeds and, in fact, mortality was lower, especially in later life, and the number of eggs and egg weight per hen were marginally higher. The costs were similar, but discounts for large scale operations mean that conventional feed would work out cheaper in most commercial situations.

Another laying hen ration developed for the organic farm at the University of Kassel is intended to eliminate the use of animal proteins in the diet (Deerberg, 1989). Cereals can be used to meet the energy requirements of the ration, consisting of 40 to 60% of the total mix, although oats, rye and maize should be limited to less than 35% and in the case of younger animals rye should be restricted to less than 20%. Grain legumes (field beans, fodder peas, lupins, vetches and soya beans) can form the basis of the protein part of the ration, forming up to 30% of the total mix. The grain legumes will need supplementing to complete the amino acid spectrum; possible sources include maize gluten, brewers' yeast, skim milk powder, linseed cake, sesame meal, sunflower cake and rape meal. Minerals, mainly calcium, will form 7 to 10% of the mix; pearled feed chalk is used to avoid a dust problem. Vitamins, in particular carotinoids, are required as colour carriers to give the yolk sufficient colour; these can be obtained from herb mixtures, dried grass meal, etc. but should be restricted to 5 to 6% of the total mix to avoid increasing the fibre content of the ration above 5%.

HOUSING

The welfare of animals, and a high regard for their ethological (psychological and behavioural) needs and wellbeing should be a natural part of any livestock production system. The European Convention for the Protection of Animals states that:

> animals shall be provided with food, water, care, temperature
> and other environmental conditions which, having regard to
> their species, development, adaptation and domestication, are
> appropriate to their physiological and behavioural needs in
> accordance with established experience and scientific
> knowledge.

These requirements are reflected in the Farm Animal Welfare Council's Charter, which lists five 'freedoms' to which all animals are entitled:

Table 8.1 Composition of ration for laying birds using organically produced feeds.

Feed component	Source	Per cent of ration
Energy	Maize	25.0
	Wheat	18.0
Crude fibre	Oats	3.0
	Bran	6.0
	Dried grass	3.0
Protein	Maize gluten	8.5
	Meat meal	7.0
	Dried yeast	9.0
	Field beans	10.0
Minerals	Carbonated feed chalk	5.3
	Chalk grit	3.5
	Monocalcium phosphate	0.7
	Feed salt	0.4
Vitamins etc.	Brown seaweed	0.6
	Cod liver oil	5 litres/ tonne

Source: Züllig (1988).

Table 8.2 Standard poultry feed mixture Neu–Eichenberg.

44.5%	Feed wheat
18 %	Fodder peas
8 %	Field beans
5 %	Green meal
11.5%	Maize gluten
7.5%	Pearled chalk
2 %	Mineral mixture
2 %	Edible oil
1.5%	Molasses

Source: Deerberg (1989).

- freedom from malnutrition: the diet should be sufficient in both quality and quantity to promote normal health and vigour;
- freedom from thermal or physical discomfort: the environment should be neither too hot nor too cold nor impair normal rest and activity;
- freedom from injury and disease;
- freedom to express most normal, socially acceptable patterns of behaviour; and
- freedom from fear.

   Given adequate nourishment, the greatest influence on the welfare of animals is housing design. Intensive animal production systems have been developed to reduce labour (and other) costs per animal and to maximise output from each unit of food, but these systems have had little regard for the quality of life of the animal and have tended to ignore many ethological characteristics of the animals. Although the cows, pigs, hens and so on in these systems have continued to produce milk, meat and eggs at an economically rewarding level, this cannot be used as evidence that they are not suffering. The arguments against this 'production' approach for assessing the 'welfare status' of a system have been widely discussed (see Dawkins, 1980 (Fox); Ekesbo, 1981 (Fox); Fox, 1984); there is a generally held consensus that production alone is not an adequate determinant of the acceptability of any housing or management system.

It is necessary when considering animal welfare to avoid the trap of considering only the behavioural problems; it has been argued for example that free range poultry, while being able to fulfil their natural behavioural patterns, are subject to cold, parasites and predators such as foxes. However, behavioural aspects do play a very important role in the wellbeing of an animal and any assessment of animal welfare should include such ethological parameters as the deviation from natural behaviour patterns (the duration and frequency are as important as the activity itself), the loss of essential behaviour patterns and behavioural disturbances.

Normal behaviour patterns require adequate space to make natural movements and assume natural postures, as well as environmental enrichment with straw bedding, company, comfort and entertainment. The 'quality' of the space provided is as important as the total area, with long narrow spaces being less appropriate than ones with more equal side lengths but of similar total floor space. Extreme protection, such as is found in many controlled environment houses, can also be termed deprivation of stimulation and as such is detrimental to the wellbeing of the animal. The herd or flock, with its carefully ordered hierarchical structure related to the transfer from one generation to another, is also important in this respect.

Where livestock are prevented from exhibiting their normal behaviour patterns, then behavioural disturbances may occur. Cattle, for example, may exhibit preoccupations with the body parts of stall mates, with inanimate objects such as bars, walls and the floor, urine drinking and tongue rolling.

Kiley-Worthington (1986) has suggested a number of criteria for assessing whether or not calves are distressed and these criteria could apply to other livestock:

- the prevention of behaviours normal to that sex or age group (e.g. lying and standing, scratching, free movement, play, social contact);
- the performance of behavioural pathologies or abnormalities (for example, stereotypes—repeated, apparently purposeless behaviour);
- great differences in the distribution of time allotted to the activities that can still be performed (for example large increases in the amount of time spent licking themselves, or standing or lying);
- an increase in activities often associated with frustration or conflict (e.g. head tossing, head shaking, kicking);
- great differences in the development of behaviour as compared with animals in a field situation (e.g. when veal calves are released they are unable to walk like similar aged field-reared calves, instead their walking resembles that of calves only a few days old);

- abnormal behavioural changes (for example an increase in precocious, aggressive or sexual behaviour);
- the use of drugs and surgery (including prophylactic treatment with antibiotics, tranquilisers, sedatives or hormones to ensure reproduction).

Alternatives to intensive livestock production systems are well documented by organisations such as the Universities Federation for Animal Welfare (UFAW, 1981 & UFAW, 1988) and authors such as Carnell (1983), Kiley-Worthington (1986) and Rist & Bär (1986). General reviews of the problems of modern livestock systems can be found in Boehncke (1985, 1986) as well as Bartussek (1982), Fölsch (1982) and Hodges & Scofield (1983). Animals need to be better treated, not only because it better serves our interests as producers and consumers, but also because they have a right to be treated with respect. Many of the issues which need further investigation, however, are being denied research funding, according to the Farm Animal Welfare Council in its 1988 report *Priorities in Animal Welfare Research and Development.*

The organic approach

Livestock husbandry on organic farms is essentially different: a productive and healthy animal relies on housing which allows the expression of the fullest range of behaviour patterns normal to the species. This includes freedom of movement, the use of straw bedding (usually not sawdust, shavings or sand) and plenty of natural ventilation. The prohibition of concrete floors, slats, long term confinement and controlled environments are all obvious aspects of an organic farm. Wherever possible all stock should have access to pasture during the grazing season and this will limit the size of the herd or flock, which must be large enough to allow social contact, but not so large as to disturb the animals' behaviour.

Although it is important to allow animals access to pasture, extensive damage to the animal and the pasture results if too little space is available or if the land is not in a fit condition. Controlling many parasites of livestock depends on the adequate rotation of grazing land; housing design must allow this to happen. This means that livestock have to be able to get around the farm; for example, a flock of chickens must not be kept in an enormous fixed building with one pop hole onto a piece of ground no bigger than a small garden, otherwise the land will quickly get 'fowl sick', as it used to be called, and the birds will require the battery of prophylactic drugs.

In the same way, ground that has been poached, compacted and turned into a quagmire is completely unacceptable if the animal has nowhere else to go. Pigs can destroy a very large area in winter, and so need enough really free draining land if they are to be able to have access to pasture all the time. Poultry, though lighter on their feet, can also make a lot of mess around their house.

Housing design

Acceptable building design depends on a consideration of the basic principles of the welfare and health of the animal. This does not mean that only simple housing need be considered. If the same amount of effort had been put into designing simple but ethologically advanced housing systems as has been devoted to controlling the environment and restricting the animals' space or bedding, major advances could have been made. The development of the 'enriched' family pen for pigs (see below) is a perfect example of this. The enriched environment provides the pig with key features to allow the expression of all behaviour patterns, showing what close attention to behaviour can contribute to housing design.

An organic farm cannot have housing systems which require only the minimum of time spent with and around the animals; the relationship between domestic animals and humans is a symbiotic one where both should benefit. Livestock require human attention to notice and react to problems as well as to build up the essential sympathetic relationship between humans and animals. The work of Gross & Siegel (1982), showing the enhanced immune response in poultry when handled as opposed to those with no human contact, indicates how important this might be. It has also been observed that behavioural disturbances such as tail-biting in pigs tend to increase with reduced human attention, particularly at weekends when there are fewer visits.

The design and construction of farm buildings is usually carried out by specialists and there is plenty of information on the factors which must be taken into account in the detailed design of buildings which conform to the needs of livestock on organic farms. The pity is that far more information is available on intensive systems of slats and cages, flat-decks and environmental control.

The key factors which must be taken into account are the provision of adequate natural ventilation throughout the house and the potential for easy cleaning out. Knowledge of the relationship between slope, aspect, prevailing winds, wind speed and the air flow within a building are all essential to good design. Specialist advice is therefore usually essential. Ventilation should not, however, come into the house over manure and slurry channels, as with some building designs this may increase the amount of ammonia in the air leading to respiratory problems in both livestock and farm workers; this last point is well illustrated by the high rates of respiratory diseases among workers in intensive pig units.

The design of the house must enable each animal to have ready access to food and water all the time. There must always be sufficient space for exercise and for the animals to lie down in an area with dry straw bedding, to keep it warm and comfortable, and for an appropriate herd structure and hierarchy to become established. If the space is insufficient, then the more subordinate animals will inevitably suffer. Exercise can be encouraged by separating the areas where different functions such as feeding, drinking, resting and defaecating are carried out.

It is also important that when bedding is soiled it is easy to renew or clean out and replace. The arrangement of the housing can include slatted manuring areas, or slurry channels which can be cleaned by tractor mounted or automatic scrapers and which should be designed to ensure the best drainage possible. The Soil Association's standards set out specific requirements for different classes of livestock.

Disinfection of housing will occasionally be necessary and should use acceptable products. Similarly paint and preservatives used must conform to the requirements for the use of safe and non-toxic products.

Finally, suitable housing must always be available for the care of the individual animal when sick, calving or injured. When weaning a calf the house should be arranged to allow contact between calf and cow, since sudden and complete separation is usually very traumatic.

Appropriate housing systems for different categories of livestock

Several authors have proposed specific alternative housing systems for livestock which could make a contribution to improving the welfare and health of stock (Rist, 1987; Schaumann, 1985; UFAW, 1981). There is not sufficient space to consider all of these here, but particular systems do deserve further consideration.

Cubicles for dairy cows

The development of cubicles and kennels for dairy cows has been spurred on by the need to save labour on spreading bedding and mucking-out and to reduce the use of straw, particularly in non-arable areas. The design of cubicles, however, has not changed significantly in the past 30 years, yet cow size and breed have changed markedly, with the result that many cows are too large for the space they have available to lie down. In addition, they are liable to injury from the lower bars towards the rear end of the cubicle and sparse bedding or floor cover on concrete can lead to direct physical injuries. These factors are believed to contribute to the increased incidence of lameness and mastitis, particularly *E. coli* mastitis, in dairy herds, resulting in financial losses through depressed yields and high vets' bills.

Normal behaviour patterns are also disrupted with conventional cubicle designs. Where the front of the cubicle faces the wall, cattle are unable to stand up in the way they would normally do in the open (Figure 8.5) because there is insufficient space for the forward extension of the head and neck. In order to stand up, the cow has to adopt a horse-like position (Figure 8.6). The same applies in the case of feeding troughs in front of tie stalls. This has important implications both for stress, and the effect of stress on the animal's immune system, as well as for the willingness of cattle to use the cubicles rather than standing in the dunging passages for long periods of time, leading to soft hooves and

Figure 8.5 Normal standing up behaviour for cattle in open.

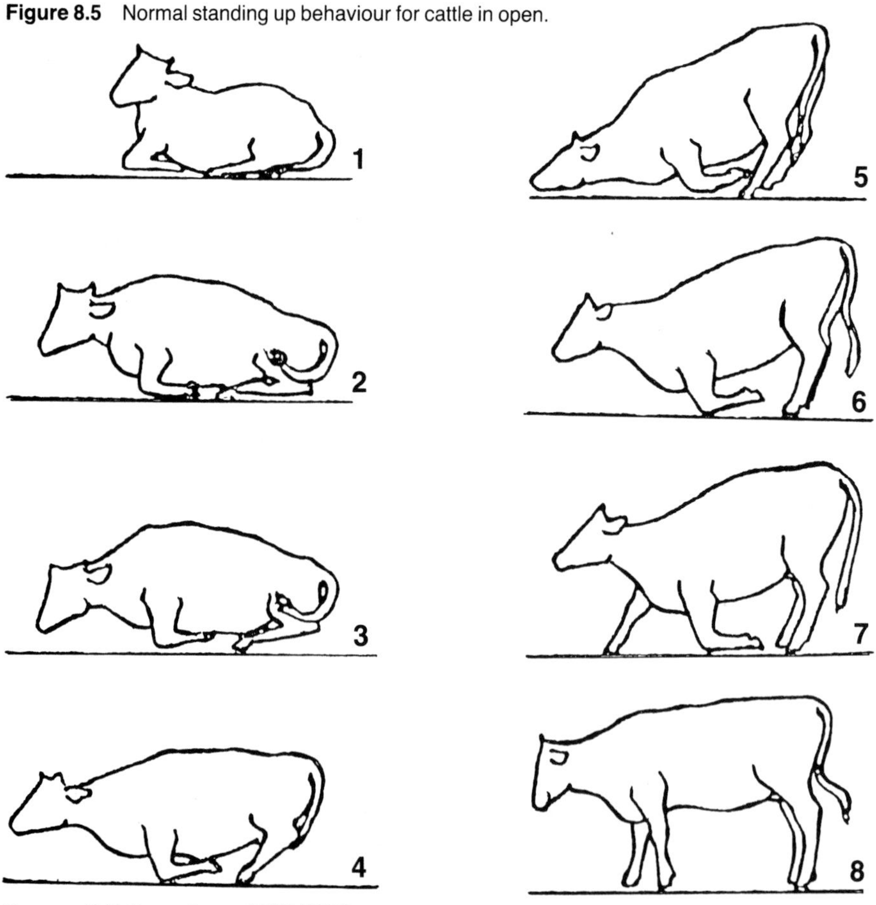

Source: Voitl, Guggenberger & Willi (1980).

lameness. This latter problem, which can also occur in loose housing, is of significant economic as well as welfare importance; the untrimmed feet of housed stock are painful and have been shown to cause a 10–20% reduction in milk/meat production.

Loose housing in straw yards would be the best solution, but the problems in terms of space, straw and labour availability are still a significant financial consideration. This has led to the development of a compromise system where the lying area is on a slope (6–8%) sufficient to cause manure to be trodden down into a cleaning passage, which can

Figure 8.6   Abnormal standing up behaviour for cattle in cubicles.

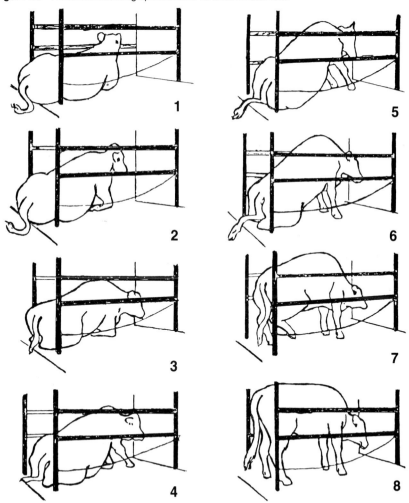

Source: Voitl, Guggenberger & Willi (1980).

Plate 8.1   New sloping floor system for cattle at the University of Bonn

be cleaned daily by tractor or automatic scraper as in cubicle housing. The straw requirements are significantly lower than in traditional straw yards. The main problem with sloping floor systems is the need to keep straw length short enough to prevent it becoming matted and immobile.

Where cubicle housing already exists, much can be done to improve the situation. In particular, the provision of additional space at the front of the cubicle (Figure 8.7), the extension of the size of the lying area to suit the breed and size of cattle housed (2.3–2.5 m for larger cows), and the use of twisted rope or flexible plastic divisions with the lower bar removed at the rear end so that large cows can 'spread' into neighbouring cubicles, will all help to enhance the comfort and the wellbeing of stock, and encourage cows to lie down as much as they would usually do in the field.

Housing for beef cattle
The same pressures for reducing labour and straw use which led to the development of cubicles for dairy cows have also led to the development of slatted systems for beef animals. These slatted systems are very

Figure 8.7   Alternative design for cow cubicle.

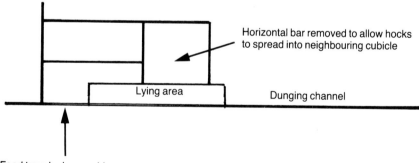

Horizontal bar removed to allow hocks to spread into neighbouring cubicle

Lying area

Dunging channel

Feed trough also provides space for lowering of head when rising

questionable on welfare grounds and are prohibited in organic systems unless the slats cover the manuring area only and not the feeding and lying areas. Cattle on slatted floors alter their behaviour patterns, with longer lying intervals, but less total lying time. Normal behaviour patterns are reduced considerably over long periods on slatted floors. Slatted floors have been shown to increase adrenalin levels in stock, although this may be due to increased stocking levels, and, in addition to the risk of direct physical injury, there is evidence for increased morbidity on slats (Hannan and Murphy, 1983). This may be due partly to increased ammonia levels in the air which counteract the beneficial bactericidal effects of acids in the air and bronchial passages.

Slatted floors are also questionable in terms of productivity. Rosemaunde EHF has found that steers which were fattened on straw had faster liveweight gain and 8% higher DM intake because more time was spent eating, and other studies have confirmed lower performance of cattle on slatted floors.

An alternative approach to the housing of beef cattle must emphasise and be adapted to the needs of the animal itself, not vice versa. This means that account should be taken of the way the beef animal would behave when at grass. Cattle are gregarious animals with a herd instinct and housing should reflect this, allowing freedom of movement and social structuring within the herd.

The herd is hierarchical; animals in a herd of up to 70 beasts will know each other and their individual place. This hierarchy is determined by the age of the animal, the length of its horns (if they still exist), as well as bodyweight, temperament etc. and is disrupted by the introduction of new animals or other changes in group composition, causing stress and hence possible detriment to the health of the animals. This very strong herd consciousness can be seen clearly when groups are mixed: given adequate space, as in an open field, they will tend to separate into individual groups again.

The herd instinct is evident not only in the hierarchical sense, but also in the activities which are usually carried out by the herd as a group, and the maintenance of natural daily rhythms such as eating, ruminating and resting. For example, although the total eating time will depend on type, quality and availability of food, the highest feed intake occurs at sunrise and sunset. Allowances should be made for this when stock are housed.

The housing of cattle has therefore to make allowance for the herd aspects, and individual penning should be avoided. Group housing allows important social contacts, including licking of areas by other cattle which the animal cannot reach itself and contact between mother and offspring. The advantages of this approach will be seen in better health, through reduced stress and the transfer of immunity, and better fertility through more synchronised oestrus. The size of the group when housed is probably optimal between ten and 20 animals.

Although group housing is an important principle, it should also be recognised that cows naturally withdraw at calving so that, if calving indoors is really necessary, calving boxes should be separate. Once the initial contact between mother and offspring has been established, young stock should again be housed in groups. Straw is particularly important as it provides young stock with material for play activities and encourages the suckling reflex in addition to the usual considerations of comfort and warmth.

Alternative systems for veal calves

The treatment of veal calves in conventional systems has rightly attracted considerable criticism, and serious efforts have been made to improve on the situation. However, some of the alternatives currently available are still not suited to the organic approach. The rearing of calves in groups on slats can be even worse than individual crates in terms of diet, discomfort and infectious disease. Group rearing in straw yards with liquid bucket feeding (the Quantock system) ensures that the calves are comfortable and have behavioural freedom, but the absence of digestible solid food can result in behavioural disturbances and an unacceptably high incidence of gut infections and respiratory diseases.

The University of Bristol has developed the Access system, where calves wearing transponders can enter one of two feeding stations and receive controlled amounts of liquid or dry feed. The calves grow as rapidly as in conventional crates, with lower feed costs and much lower mortality and disease incidence. Further work still needs to be done, but the housing systems currently being developed could help to make veal acceptable once more and a viable option for an organic farm.

Sheep housing

Housing in winter reduces the poaching of grassland and allows earlier spring grass growth on rested pastures, as well as increased access for inspection and reduced ewe mortality; these are advantages which could be of considerable benefit in an organic system. However, the higher capital cost and health problems such as *E. coli*, joint ill and pneumonia, as well as the risk of heat stress and the potential need for winter shearing, all count against this approach. If sheep are housed, then the general principles of good natural ventilation, straw bedding and avoiding overstocking must apply.

Plate 8.2   Winter housing of an early lambing sheep flock

The housing of pigs

Of all the common farm animals, pigs probably show the greatest sensitivity to housing design. Susceptibility to disease and behavioural abnormalities increase markedly under unsuitable conditions, in particular individual sow confinement and the use of concrete flooring.

The main argument for individual sow confinement is the problem of overlying, where young piglets are unable to escape and are crushed. This problem, however, usually occurs where the piglets are trapped against walls. It can be reduced by erecting a bar at a height of about 25 cm and a distance of 20 cm from the wall to give the piglets room to escape.

Concrete floored sow stalls without bedding harm the animals by making them cold and uncomfortable. Worse, they cause injuries to legs and feet, and pressure sores which prevent the sows lying down so that they adopt a forlorn dog-like position, an abnormal posture which may predispose them to urogenital infections. Normal behaviour in pigs includes a wide range of activities including rooting, wallowing and social contact with animals of a variety of ages. Where pigs are housed, the urge to root remains, even when the pig is fully fed and no food will be obtained this way. Straw, if available, will be used as a substitute, but on bare concrete behavioural abnormalities set in.

These behavioural patterns are instinctive, but they are also determined or set-off by external factors such as environmental stimulants and inner motivations or 'drives'; for example, nests are built in the evenings, not in the morning when the sow is not in the mood. In artificial situations, the need for certain behavioural patterns is removed, but the instinct and drive remain. Pigs will dung and urinate away from their bedding if given sufficient space—they are by nature clean animals. When tethered, they cannot do this, but they indicate by their behaviour that they would like to go somewhere else to lie down. Even in environmentally controlled situations, their instinct to leave the nest to defaecate remains.

There is good evidence that confined sows find it difficult to cope with life and show this by behaviour such as bar biting, head weaving, chewing on an empty mouth and tail biting. While stalled or tethered sows respond when food is given, they lose interest in whatever else is happening around them. It is known that certain proteins with an analgesic effect can be produced in the animal's brain. Apparently pointless, repetitive behaviour may be a device used by sows to switch on the supply of these substances, thereby making their confined life more bearable. Some abnormal practices, such as chewing on an empty mouth, will stop as soon as something chewable like straw is available; the mouth needs to be active even though there is no hunger.

Tail biting occurs in situations where the pigs are frustrated, such as groups which are too large or have too high a density, irregular feeding, too short a trough length so that not all animals can get there at the same time, lack of drinking water (breakdown), parasites, noise and irritating gases (ammonia). Where straw is available, the pig's frustration is taken out on the straw. If no straw is available, anything will do. The tail is a prime target, because initially it is insensitive to playing, but once blood is drawn, the taste of blood encourages more intensive biting and severe problems can set in. Where tails are removed, the problem does not go away. Anal massage replaces it, stimulating the affected pig to defaecate more frequently and causing swelling of the anus. Other pigs will eat the faeces and this is likely to be followed by further bodily injury.

Of the alternative systems currently available, free-range keeping of breeding stock and rearing of weaners comes closest to the organic ideal, allowing the animals a chance to express their natural behaviour patterns, and providing a grass break in arable rotations at low capital cost. But there are also disadvantages, such as higher feed requirements and the need to use hardy, robust breeds such as the British Saddleback. The welfare effects of the free-range approach compared with housed sows are not always positive (Table 8.3), but consideration should also be given to the increased resistance of the animal to adverse circumstances when kept under more natural and less restrictive conditions.

If the free-range approach is adopted, then rotational grazing is essential to avoid the build-up of parasites. Such a system could be based on a number of dry sow paddocks, with an appropriate number of farrowing paddocks, at a stocking density between 12 and 25 sows a hectare. Ringing may be necessary to prevent damage to pasture by

Table 8.3 Welfare effects of different housing systems on dry sows.

	Paddocks and arks	*Individual stalls (no bedding)*	*Covered straw yards*
Thermal comfort	Very variable	Fair to poor	Good
Physical comfort	Variable	Bad	Good
Injury	Slight	Feet, 'bed sores'	Fair to poor
Hygiene	Fair to poor	Usually good	Fair to poor
Disease	Some parasitism, difficult to control	Usually good	Some parasitism, control easy
Abnormal behaviour	Slight	Severe	Slight

Source: Webster, New Scientist 21/7/88.

rooting, although this form of behaviour can also be beneficial if the aim is to reclaim land or clean up docks and couch, for example.

In practice, some form of fixed housing will usually be required, even in an organic system. Attention to housing design so as to minimise the stresses which are detrimental to the wellbeing of the animal is therefore essential. The Soil Association's standards prohibit the use of slatted floors, which cause abrasions and in which piglets can get trapped, and require at least one square metre lying area per 100 kg liveweight. A system involving sloping, straw covered floors, as suggested above for beef housing, with a 'self-help' straw hopper at the front, is currently being tested by the Centre for Rural Buildings at Craibstone, Aberdeen. It represents a considerable welfare improvement and can greatly reduce straw consumption compared with traditional straw-littered pens, with lower handling and disposal costs for the manure produced compared with slurry systems.

The Soil Association's standards also emphasise the need for natural ventilation, which can be obtained with open fronted Solari-type houses (Figure 8.8) or other housing systems with access to an outdoor exercise area. It is also essential that any housed pig enterprise is treated as a

Plate 8.3   Where pigs are housed, outdoor access is important

Figure 8.8 Solari type, open-fronted farrowing pen with natural ventilation.

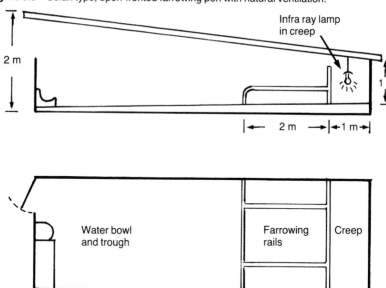

Source: Goodwin, D. H. (1973) *Pig Management and Production, Stanley Thornes*

land-using activity so that the manures can be utilised in an ecologically optimal manner.

There are several other designs for 'appropriate' pig housing currently in use, a fact which indicates the problems that exist in trying to find the right solution. Two of the most promising approaches from a welfare perspective are discussed here.

The first approach is the result of work at the Institute of Animal Production at the University of Zürich and is described in more detail in Rist *et al.* (1988) and Götz *et al.* (1986). Four distinct units are described (Figure 8.9, A–D), including a farrowing pen for breeding sows, a four-area pen for dry sows, a three-area pen for weaners and a four-area pen for fattening pigs.

The farrowing pen (Figure 8.9 A) has a heated piglet box in front of the sow's rest area (the sow naturally lies with its head away from the dunging area and towards the nesting box). The piglet nest is supplied with a curtain, straw bedding, and a thermostatically controlled air warmer with downward fan and baffle to avoid direct draughts; infra-red lamps cause upward warm air draught with colder air flowing over the piglets and are not therefore used. Pigs are bare skinned and therefore the maintenance of adequate temperatures and protection from cold draughts are crucial. The maintenance of temperature is helped by straw, high stocking rates and the covering of pens with straw

Figure 8.9   The Zürich pig housing system.

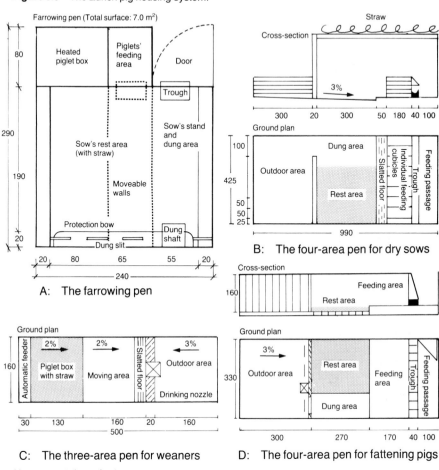

A: The farrowing pen

B: The four-area pen for dry sows

C: The three-area pen for weaners

D: The four-area pen for fattening pigs

Measurements in centimetres.
Source: Götz *et al.* (1986).

bales. These conditions encourage the piglets to spend more time in the nest, thus reducing the risk of overlying.

The sow has a rest area with straw bedding and a separate dunging and standing area and is fed separately from the piglets. The lateral walls of the pen can be removed; they serve as farrowing rails and in exceptional cases they can be put together to restrain the sow in a sow crate.

The other pens are again divided into distinct areas for resting, dunging and feeding, with a fourth area outside for exercise, except in the case of the weaner pen, where the feeder is integrated into the piglet box. The emphasis on distinct areas for different functions, space for exercise and movement and straw bedding considerably reduces stress and behavioural abnormalities and removes any need for individual penning or tethering.

The second approach, which has already received widespread attention in Britain, is the enriched-environment, family pen developed by Stolba and Wood-Gush (Stolba, 1981 & 1986). This system corresponds more closely to the organic husbandry ideal of family groups through from farrowing to slaughter. The development of the pen is a result of work by Stolba observing the behaviour of domesticated pigs in the Edinburgh Pig Park, where conditions closely resembled the natural habitat of wild pigs on the edge of forests. Stolba discovered that when the pigs were taken from their pens, the range of behavioural patterns displayed increased considerably. 103 different activities were identified, including rooting, burrowing, nesting and wallowing, all behaviour remarkably close to that of wild pigs. Several sows together formed a stable family herd, where the relationship between mothers and daughters was particularly close. A boar associated with the herd and thereby controlled the oestrus of the sows. Before the birth, the pregnant sow isolated itself from the group and built a nest in a protected place. In a few days it returned with its litter to the group. Natural weaning occurred after 12 weeks.

From this work, Stolba and Wood-Gush established three important guidelines for a housing system which meets the welfare and ethological requirements of pigs: they should have room to move freely and have separate dunging and lying areas; they should be provided with materials such as straw in which to root and burrow etc.; and they should be kept in family groups.

The 'enriched' family pen which was developed as a result allows pigs to fulfil most of their significant behavioural processes in a limited area (see Figure 8.10). Each pen contains one outer and two inner sections which imitate the forest-edge habitat. The partitions increase the movement distances and animals which do not get on can get away from each other. Eating and nesting areas are separated and head screens maintain the otherwise insufficient individual separation during eating (when eating, pigs will naturally keep some distance apart). In the rear part of the pen, the sow can build a nest protected on two sides, and can see out of the front of the stall from this position. The dunging area is situated at the front at least 3 m from the nest, again reflecting natural behaviour patterns. Various other features including rubbing posts and exercise bars allow further expression of natural behaviour. In separate work it has been shown that even a pig's desire to wallow can be fulfilled by providing a shower as an alternative; pigs learn very quickly to press a knob with their snout which gives them a shower, helping to cool them down.

Figure 8.10   The Edinburgh 'enriched' family stall.

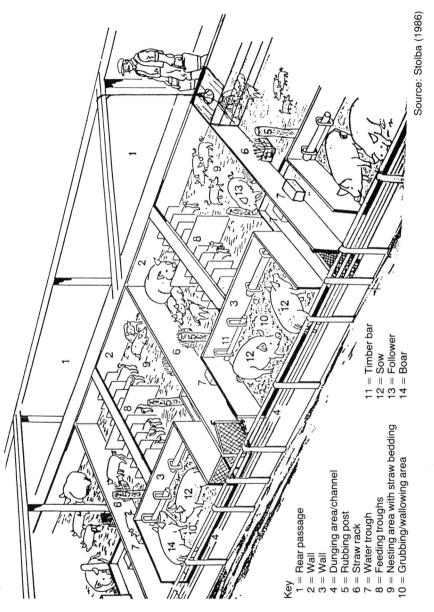

Source: Stolba (1986)

Key
1 = Rear passage
2 = Wall
3 = Wall
4 = Dunging area/channel
5 = Rubbing post
6 = Straw rack
7 = Water trough
8 = Feeding troughs
9 = Nesting area with straw bedding
10 = Grubbing/wallowing area
11 = Timber bar
12 = Sow
13 = Follower
14 = Boar

Social contact is another important feature. In the family system, stable groups of two older and two younger sows with their suckling and fattening piglets are kept in four inter-connected pens. The boar and followers also live in the herd, allowing for the same social structure to develop which would be the case in the wild.

The sows are not confined at farrowing, are covered by the boar and remain during the whole pregnancy in the same pens, while the young pigs are naturally weaned and finished without forced removal and grouping into single age groups. The sow builds her nest ready for farrowing, and the area is temporarily shut off with a gate for a few days. Afterwards, the sows are reunited. Three weeks post partum, a boar is introduced for the next six weeks, after which he is moved on to the next group. He covers the lactating sows and followers on heat, which in a stable family group are very likely to have synchronised oestrus.

The freedom for suckling sows and litters as well as dry sows with this system has avoided the stress symptoms associated with individual confinement stalls. Disease incidence in growing stock is also lower, but there have been problems with the original design such as deaths through overlying and the open front depressing growth rates during cold weather. Hardier breeds such as Duroc crosses are important to overcome these, as is tempting piglets away from the sow by providing warm, comfortable nesting conditions to reduce overlying losses. Although the area of building required (84 square metres for eight sows and their offspring) is acceptable, a major problem remains the higher capital and labour requirements, where one person is needed for every 50 sows compared with one to 80 in more intensive conventional systems.

Housing for poultry
The mass production of eggs in battery cages now dominates the poultry industry, in spite of the serious welfare concerns which exist about such systems (e.g. Dawkins & Nicol, 1989). The Soil Association's standards prohibit the use of battery cages, emphasising instead the need for access to pasture, low stocking densities in the poultry house, low colony size and housing which allows natural light, ventilation and freedom for the expression of natural behavioural patterns.

A wide range of alternatives to battery cages exist, not all of which are ideal in an organic system, including:

- modified battery or 'getaway' cages;
- deep-litter housing with a wire floored roosting area;
- deep-litter housing with outside runs for use by stock when weather conditions permit;

- free-range systems where stock return to movable arks at night;
- movable field arks in which stock are confined;
- the modified deep litter or aviary system.

The behaviour of chickens can be classified in functional groups (Table 8.4). Some of these behaviours, such as egg laying and dust-bathing, have a circadian (daily) rhythm; others, such as brooding, follow an annual cycle. Fölsch analysed the impact of selected housing systems on these behavioural patterns (Table 8.5). Various other factors which influence the results need to be borne in mind, including the method of rearing, number of birds per unit area, total number of birds per house, quality of light regime, quantity and interval of meals and climate. Hens will also incur significant costs to themselves in order to carry out their natural behaviour patterns, which is a strong indication of potential detrimental effects when they are not able to do so (Nicol & Dawkins, 1990).

Free-range systems are recommended for organic farming, but at substantially lower stocking rates and colony size than the European Community legislation permits. Details of the Soil Association's requirements can be found in Appendix 1. The Demeter (biodynamic) standards go further and specify at least two square metres per hen outside and at least one quarter hectare of land under organic management for every 100 hens, to ensure, as in the case of pigs, that the production of poultry is a 'land-based' activity and ecologically sustainable.

Organic free-range poultry is considered in greater detail in Chapter 9, so little need be said at this stage except that the free range must be on pasture, not just an area of bare soil, and that land should be rested periodically using a paddock system or mobile houses. The economics of free-range systems versus other forms have been widely reported, but Chapter 9 also gives an example of costings from an organic system.

Table 8.4 Behaviour of poultry classified by functional group.

Social organisation: the more uniform the group (age, sex), the higher the risk of losses by aggression or disease.

Locomotion: walking, running, fluttering, flying.

Feeding: search for food and water, food and water pecking, ground scratching, scraping. (Foodstuffs of different structure are preferred and should be offered.)

Courtship and reproductive behaviour: waltzing, crouching, copulation, nesting behaviour and egg laying, broodiness. (Single nests or family nests, both with loose bedding, are well accepted.)

Body care: preening, scratching, dust- and sun-bathing.

Resting: standing, sleeping – hens prefer to roost in elevated places.

Source: Fölsch (1986).

Table 8.5 Impact of different housing systems on behaviour of chickens.

		Housing system		
Functional grouping	Free-range	Aviary/ deep litter	Wire mesh	Battery cage
Social interaction +ve	–	(×)	(×)	×
Social interaction −ve	–	related to number of hens/density		
Locomotion	–	–	× qual	× qual & quant
Resting	–	–	× qual	× qual & quant
Body care	–	–	× dust bathing	×
Feeding	–	(×) choice	(×) choice	×
Courtship, reproduction	–	–	(×)	×

Key: × = problems
   (×) = problems can be solved
    – = no problems
  qual = qualitative
  quant = quantitative
Systems: free range – max 250 birds/house, 6 hens/m^2 indoors.
     aviary – more than 500 hens/house and 10 hens/m^2, 3 hens/m^3.
     deep litter – 5–6 hens/m^2 indoors.
     both indoor systems with natural light, litter, nest boxes and perches.

Source: Fölsch (1986).

Where poultry need to be housed, the following considerations should be borne in mind. Natural light, sunshine and ventilation must all be available. Sufficient nesting boxes should be provided so that not only the strongest hens get access, and the nesting areas should be darker than the rest of the henhouse. Hens should be undisturbed during the egg laying period, which takes 60–90 minutes. Facilities should be provided so that hens can exercise important behavioural characteristics such as taking dust baths, free movement and seeking out of food.

The aviary system is perhaps the most promising of the fixed housing systems from a welfare perspective and does not significantly increase labour requirements. It is currently under investigation at Gleadthorpe EHF and consists of a double decker slatted area, a deep pit below the central slatted area to collect droppings, floor area for movement, scratching etc., to allow the fulfillment of natural behaviour patterns, and individual nesting boxes with automatic egg collection (Figure 8.11). The different roosting levels allow for a pecking order to be established, but it is important that the colony size is limited to not more than 200 hens or the system will break down, and that the stocking density is not so high as to cause the same problems as occur with battery cages.

Figure 8.11   Aviary housing for poultry.

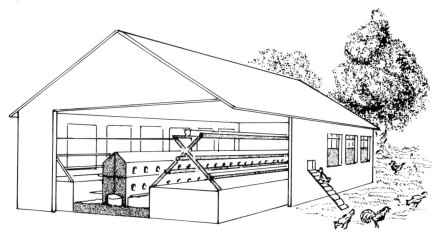

Source: Fölsch (1982).

It is possible, with poultry as with other types of livestock, to design and build housing which meets both the welfare and the ethological requirements of the animals concerned, as well as being economically viable. This can and must be a priority for the development of organic livestock systems.

ANIMAL HEALTH AND VETERINARY MEDICINE

Animal health has not markedly improved over the last 30 years (Spedding, 1982) and mortality rates have remained essentially unchanged. Wise (1978, in Fox) has shown that the costs of maintaining animal health have risen faster than the farmer's total production costs. There seems no doubt that we have created conventional intensive systems of animal production which contribute to disease aetiology (Hodges and Scofield, 1983).

Many of the diseases and syndromes which have to be coped with are the result of overcrowding in inappropriate housing with high demands placed on the physiology of the animal in terms of production (growth rate, milk yields etc.). These diseases and syndromes have largely been ignored or resolved as far as possible through the routine and extensive use of antibiotics and other drugs to destroy the pathogen and/or relieve the symptoms. Attention to the individual animal is a key to the early identification and treatment of a disease. It has been noted that as animal numbers have increased per stock person, so has the reliance on prophylactic drug use.

Making use of the animal's own resistance to disease

The ability of the animal to fend off disease through its own defence mechanisms should be the starting point for any organic approach to livestock healthcare. This involves both the immune or specific defence system utilising antibodies, and non-specific parammunity based on the action of white blood cells etc. The first step is to create the right environment for the animal so that stress and other factors do not impair its resistance. The importance of maintaining health through adopting the most appropriate forms of feeding, housing and production levels has already been emphasised; breeding and rearing are two additional aspects yet to be covered.

Young animals have the ability naturally to acquire immunity, or at least a degree of resistance, to many parasite and disease problems. The provision of colostrum is the starting point for this, because the young animal receives the antibodies already formed by the mother in response to an earlier infection from which she has recovered. The act of suckling is as important as getting colostrum, because germs transferred from the calf to the cow while suckling stimulate an immune response from the cow which is then passed back to the calf. This acquired immunity is strengthened by exposure to low levels of a disease or parasite over time. Animals which have been affected by parasites and have recovered may still carry them, albeit at low levels which will wax and wane periodically.

The degree of immunity or resistance will vary depending on a range of factors. Pregnancy and lactation in particular will cause immune relaxation due to stress and hormonal interactions, thus making it easier for pathogens to survive and develop. In addition, a whole range of other factors such as drugs, vaccinations, contact with antigens, pesticides residues, contact with humans, nutrition, stress, housing and weather conditions may also have a detrimental influence on wellbeing and immunity.

At the same time husbandry methods and the prophylactic use of drugs have resulted in previously harmless bacteria and viruses becoming more virulent and the development of resistance by bacteria to antibiotics through selection, mutation and plasmid transfer, leading to the general condition of 'hospitalism' where health problems are exacerbated by the treatment practices used (Boehncke, 1986), a clear example of which is the development of salmonella as a major health risk.

The avoidance of prophylactic drug use is an important aspect of organic husbandry, since such use merely plugs a hole in what is otherwise an unbalanced system. Regular and routine use of medication only develops problems for the future—antibiotic resistance has already been mentioned, there are other examples. The danger of cross resistance occurring in humans as a result of the veterinary use of drugs is a further reason to avoid their use when possible. Examples of the routine prophylactic use of drugs which would be avoided on organic farms include dry cow therapy to prevent mastitis, routine use of wormers and in-feed anticoccidiostats.

However, if an animal does suffer from a disease the most effective means to cure it must be found. In some cases, for example with mastitis in dairy cows, this does not necessarily involve reaching for the conventional cure. Mastitis can often be cured with frequent stripping, massage and cold water, although this obviously takes more patience than reaching for the antibiotic tube! There are many examples where alternative treatments can be used, and these should be explored in full. Some are considered in more detail below.

Whenever an animal does suffer from a disease, deficiency disorder or parasitic attack, and an alternative or 'complementary' form of treatment is not available, then it is essential to treat the animal with conventional medicines. The Soil Association's standards allow the use of conventional drugs in order to save a life, to prevent unnecessary suffering to an animal and to treat a condition where no other effective treatment is locally available.

When conventional drugs are used, the standards may require that a withdrawal period exceeding the manufacturers' recommended period is observed, and in many cases the animal treated would have to be excluded from sale as organic. This should not stand in the way of preventing unnecessary suffering. The welfare of the animal should be paramount and not subject to economic considerations of loss of premium, which in any case is likely to be far outweighed by loss of value from a sick or dead animal. Organic standards are flexible enough on this issue, but even they cannot guard against poor livestock farmers. Nevertheless, the root cause of the disease should be identified to ensure that modifications to husbandry will prevent a recurrence. To ensure that the necessary standards can be adhered to, good and effective record keeping and the clear identification of treated animals is essential.

Alternative treatments and 'complementary' medicine

Alternatives to conventional drugs are often available, although knowledge of their use and potential is often limited among veterinary surgeons. Herbal remedies, homoeopathy, traditional treatments and husbandry practices (often ignored or forgotten when a new drug is developed) are all available to the organic farmer intent on avoiding conventional drugs whenever possible. Many of these treatments can be very effective and, as experience is gained, some organic farmers achieve remarkable successes.

Herbal remedies

The medicinal use of herbs has been well documented, with plants such as garlic having great curative powers for several conditions including acting as a mild vermifuge for the control of internal parasites. Another very useful vermifuge is prepared from the seed hairs of the plant *Kamella philipensis* and is available as a proprietary preparation. Bairacli Levy (1973) provides an invaluable source book on a wide range of herbal remedies. Information on herbal remedies from a biodynamic perspective can also be found in Koepf *et al.* (1976), Sattler & Wistinghausen (1985), and Spielberger & Schaette (1983).

Homoeopathy

Homoeopathic methods are increasingly used by organic and conventional farmers, reflecting the increasing interest in homoeopathy in the treatment of human disease. The potential for homoeopathy in agriculture has been reviewed by Scofield (1984) who has characterised homoeopathy as: '. . . a therapeutic system in which diseases are treated with substances usually in extreme dilutions, which, when given to healthy individuals, produce the same symptoms as the disease being treated.'

The material used is 'potentised' by successive dilutions, the theory being that the more dilute the solution, the greater the potency of the medicine. In addition to these remedies, nosodes are also used. Nosodes are homoeopathic preparations from the bacteria or disease against which protection is desired. The difference between homoeopathy and vaccines, however, is that homoeopathic remedies produce no antibodies. Homoeopathy is an holistic method of treatment in that the whole organism is treated in an attempt to raise its level of resistance and stimulate its ability to throw off disease.

The choice of the correct remedy depends on the 'character type' of the animal and is considered to be very important in the successful application of veterinary homoeopathy. Thus the success may in part rely upon the ability of the veterinarian to correctly 'type' the animal. This will depend upon the skill of the veterinarian in observation as well as the extent to which the stockperson knows the animals.

An advantage of homoeopathy is the complete safety of the preparations which are used, since in every case they are carefully prepared from very dilute extracts of the active ingredient (which may in some cases be a poison). In fact, it is this very high level of dilution which leads many conventional scientists to consider that homoeopathy cannot work, since they say that the ingredients have become so dilute that there is nothing left. On the other hand, homoeopaths claim that it is not the active ingredient itself which is therapeutic, but the 'imprint' that it leaves on the carrier during preparation.

On the whole, circumstantial evidence for the success of homoeopathy abounds. Practising vets and farmers report success in treating or preventing a wide range of conditions. Shuttleworth (1988) claims to have successfully cured over 200 humans and several hundred sheep of orf using the homoeopathic remedy Thuja. Some farmers may, however, feel that they need more firm evidence before risking the health of their animals. A number of controlled trials have been carried out by researchers around the country (including Elm Farm Research Centre and the North of Scotland College of Agriculture). The only conclusion that can be drawn is that the evidence is contentious, some trials showing an effect while others fail to do so, and this is true for many conditions and remedies. Any farmer considering introducing homoeopathy should realise the implications of this.

In addition, although the work of vets like Christopher Day has demonstrated the ability of homoeopathy to cope with the stresses imposed by conventional livestock systems, its use must be seen as a short-term expedient and in no way alters the necessity for fundamental changes to housing and nutrition in intensive livestock systems in order to attack the root causes of disease problems.

For farmers starting to use homoeopathy, the following guidelines may be of help:

1. Do not attempt remedies to deal with specific cases until your skill and knowledge of homoeopathy is developed. Choosing the appropriate remedy, the appropriate potency, frequency of administration and the altering of these according to observed changes in the animal's condition is the key to success in homoeopathy. This requires a basic knowledge and skill that should be obtained by reading and practise in other areas first.
2. Remedies can be used by the beginner in the following areas:
   a) prevention of milk fever—dosing with Calcium Phos 30C and Magnesium Phos 30C during the dry periods affords a high level of protection;
   b) prevention and early treatment of calf pneumonia—Phosphorus 1M put in the water supply cures the condition in the early stages and prevents it spreading to other stock sharing the building;
   c) the use of Arnica as a primary treatment in all cases involving stress or shock;

d) the use of Aconite as a first treatment in all cases displaying the early signs of feverish conditions.

3. Nosodes can be used safely by beginners in the following circumstances:

a) in the prevention of mastitis. Nosodes of all the main strains of mastitis bacteria exist. In the 30C potency, they can be placed in the water supply;

b) in the prevention of husk. A husk nosode exists which is effective when given to young stock periodically through the first grazing season;

c) as an insurance when buying in young stock. A wide range of nosodes, including one for pneumonia, exists and can be given to bought-in stock as a protective measure.

Further information on homoeopathy can be obtained from the British Association of Veterinary Homoeopathy, which links homoeopathic vets around the country. There are only a few books on the subject (e.g. McCleod, 1981 and Day, 1984 & 1988), but these are by well established practitioners and can be recommended.

Other alternatives

Various other alternatives exist, such as acupuncture and probiotics, both of which have been successfully used on livestock. The mechanism by which acupuncture works is not fully understood, but it is thought that acupuncture may work by stimulating the production of opiates (natural pain killers) at freshly needled acupuncture points. Evidence for this comes from the fact that acupuncture fails to work when substances which block the production of opiates are administered. Probiotics are beneficial bacteria intended to help restore the gut flora of an off-colour animal. A development of this approach is being explored for the control of salmonella. Bacteria are cultured in the caeca of adult birds (more than 200 different kinds of bacteria grow there in a healthy adult) and the cultures are used to innoculate young chicks, giving them better resistance to salmonella due to 'competitive exclusion'—the vigorous gut flora outcompetes the pathogenic salmonella bacteria. More evidence on the effectiveness and other possible implications of probiotics is required before they can be recommended for organic farming.

The use of vaccines

Vaccines are used as a matter of routine on most farms. Under organic management their use is allowed only if there is a known disease problem on the farm, but it is normally preferred to avoid the use of vaccines (as is the case with all other prophylactic drugs) since it is considered that they can interfere with and inhibit the development and expression of the animal's own immune system.

This may seem surprising, as the intention of vaccination is to stimulate an immune response by the animal and hence give it protection from infection, in one sense creating the same effect which homoeopathy is supposed to have. There is, however, a growing body of evidence that vaccinations can have a detrimental impact on the immune system as a whole, even if protection is given against a specific disease.

It is already well known from human medicine that vaccine can have side-effects, sometimmes very serious, and that certain vaccines interact adversely with others, particularly live ones. With sheep, the use of multiple vaccines in concentrated vaccination programmes against pasteurella and clostridial diseases can result in ewes being hit with up to 16 different agents at once, representing a considerable infection challenge for the animal's immune system to respond to. The introduction of new 20-in-1 vaccines can only exacerbate the problem.

In 1989 certain vaccines, and in particular Heptavac-P which is widely used to control clostridial diseases and pasteurella, were implicated in a series of sheep deaths throughout Britain (Farmers Weekly, April/May 1989 onwards). It seems likely that interactions between a range of factors including organophosphorus dips, wormers and vaccines as well as forage quality and abnormal stress levels, are involved. Severe adverse reaction (SAR) is now acknowledged by the pharmaceutical industry to be a problem, but it is not clear whether the increased number of cases since 1989 is due to a real increase in the problem or increased reporting of incidents. However, Dr David Barbour, senior veterinary officer with the Scottish Agricultural Colleges Veterinary Investigation Centre at Ayr, has suggested that farmers should consider whether vaccinating for pasteurella is really necessary. The pharmaceutical companies have changed their advice to farmers at the request of the Ministry of Agriculture, Fisheries and Food, recommending that the period between vaccination and lambing should be doubled from 2–3 weeks to 4–6 weeks and individual companies have issued warnings that certain vaccines should not be used together on pregnant ewes.

Vaccines are never 100% effective, so the benefits and potential risks of their use have to be weighed against the likelihood of a disease problem occurring. More important, the use of vaccines does not deal with the conditions which give rise to particular disease problems and in some cases simply covers up poor husbandry. It therefore seems sensible that vaccines should only be used in situations where problems exist and not as a blanket prophylaxis.

Controlling specific disease problems

Mastitis in dairy cows

This is one of the main problems which concerns farmers when faced with the prospect of abandoning antibiotics and in particular dry cow therapy. It has been argued that the widespread use of antibiotics to control mastitis has only succeeded in diverting resources away from finding the actual cause. In addition, when mastitis does occur, by the time abnormalities are spotted in the milk, the cow's own resistance system is likely to be working and the disease may well be retreating. All antibiotics achieve at this stage is to ensure that the milk is thrown away for three to five days. Teat disinfection may also be counter-productive. Research has indicated that less important bacteria can protect cows against mastitis. These bacteria are also destroyed, leading to a higher risk of clinical mastitis where teat disinfection is practised in herds with low cell counts (Schukken *et al.*, 1989).

It is known that certain families are more susceptible than others to mastitis. This is because the speed at which white blood cells and other phagocytes are transferred from the blood stream into the mammary gland to attack bacteria is in part genetically determined and in part affected by other factors such as the stage of lactation. In cows with an efficient defence mechanism, which includes natural barriers such as the teat canal itself, coated with a wax-like substance known as teat canal keratin, the phagocytes are mobilised rapidly and migrate from the capillary blood vessels beneath the teat canal lining, through the lining surface, into the milk; the whole sequence of events occurs within four hours of infection.

Various alternative treatment methods have been tried successfully by organic farmers, some of which are reported in Woodward (1983).

Newman Turner (1954) advocated a combination of prevention and treatment. The prevention consisted of a balanced diet composed of green food, roots, hay, straw and silage grown on composted land without chemicals and devoid of artificial food as well as a weekly dose of garlic and a daily dessertspoon of seaweed meal. For the treatment, which should be implemented at the first sign of any abnormality, all food should be stopped, and only water allowed. A strong dose of garlic should be given twice a day for a week ('two whole garlic plants chopped up and made into a ball with a little molasses and bran, or four tablets of garlic fortified with fenugreek'). The affected quarter should be milked out as often as possible, at least four times daily. If there is any inflammation, alternate hot and cold fomentations should be applied. The affected quarter should be massaged three times daily and the treatment finished off with a cold water hose turned on the udder and loins for 10 minutes. This stimulates the exchange of blood and speeds the purification process. The fast should be continued for as long as three days if necessary, with a drench of cane molasses (one pint in warm water split into three doses) being given after the first day. When the abnormal discharge has ceased, then feeding can resume with green feed only. This should gradually be increased and after one week, the production ration can be reintroduced. Newman Turner had considerable success with his veterinary methods in the '40s and '50s, but there is no evidence of anyone using the full method today.

Gareth and Rachel Rowlands, who milk Guernseys near Aberystwyth (Chapter 9), use an adaptation of the Newman Turner approach. At the first sign of any problem, the affected quarter is sprayed with cold water under pressure and the quarter is stripped out. After this, the udder is dried and then massaged. This is done night and morning until any swelling or heat subsides and the udder is back to normal. In severe cases, this routine is carried out every two to three hours and in some cases, such as a cow with a high temperature, a 24 hour fast will be imposed, allowing the cow only molasses and water. The Rowlands have almost total success with this method, failing only if, because of black spot or injury to the teat, they cannot draw milk from the affected quarter, or if a case of summer mastitis is not detected early enough.

Homoeopathy has also been used successfully for the control of mastitis. The key to the homoeopathic prevention of mastitis is the use of nosodes, homoeopathic preparations made from the mastitis-causing organisms themselves. Prevention must be done on a herd basis. Nosodes are best applied weekly through the drinking water, but can also be given orally. The main types of nosodes likely to be used are *Streptococcus agalactiae* 30C, *E. coli* 30C, *Tub. bov.* 30C, *Pseudomonas* 30C (30C indicates the degree of potentisation it has undergone).

The treatment of mastitis homoeopathically is more complex because the remedies will depend upon the symptoms and the response of each individual animal. The most commonly used remedies for freshly calved cows are Belladonna 1M, Apis Mel 30C, Bryonia 30C, Urtica Urens 30C; for acute cases, Belladonna, Bryonia and Urtica can be combined. Silicea, Hepar Sulph and Phosphorus are also useful remedies. Will Best, an organic farmer in Dorset (Chapter 9), has used homoeopathic remedies including Belladonna, Phytolacca, Conium, Bryonia and Hepar Sulph with as much success as antibiotics and considerably less cost.

However, a survey by Elm Farm Research Centre of conventional and organic farmers using homoeopathy to control mastitis could not find any conclusive evidence that homoeopathy, rather than other factors such as the stockman and husbandry changes, was responsible for a decline in the incidence of clinical mastitis (Stopes & Woodward, 1990). Emphasis should still be placed on reducing stress and on good management and husbandry to control disease, rather than the routine use of medication, homoeopathic or otherwise. In particular, avoiding overstocking in straw yards, the provision of adequate bedding, attention to hygiene during milking and balanced nutrition are all factors which contribute to reduced mastitis incidence.

Milk fever

Milk fever is due to low blood calcium levels at calving. Cows are usually most susceptible when they are producing their top yields from third lactation onwards. Channel Island breeds are particularly prone. Milk fever occurs most in autumn when grass quality wanes, unless a conserved forage supplement is available. Treatment usually involves using calcium supplements, although techniques such as inflating the udder to slow down milk production and acupuncture have been effective. Prevention is possible, but difficult. The use of Vitamin D can help, as can low-calcium diets before calving which are intended to prime the body for the sudden increase in demand at parturition. This approach depends on reduced grass intake and increased cereal consumption and can lead to other problems such as fatty liver or ketosis.

Fog fever

This disease is caused by stock eating a lot of grass containing the protein L-Tryptophan. In itself, the compound is not harmful, but during digestion in the rumen, it is converted into compounds that circulate in the blood and can result in injury to and death of many lung tissue cells. It is seen mainly in suckler cows, and more recently in dairy cows turned out onto lush fertilised silage and hay aftermaths following a period on poor pasture. Where it occurs, grazing should be limited, or the pasture grazed with less susceptible stock like sheep. Stock should not be stressed.

Diseases of sheep

A range of diseases of sheep exist for which vaccines have been the primary means of control, such as the clostridial diseases (including lamb dysentery, struck, pulpy kidney, braxy, tetanus and black disease), foot rot, orf, pasteurella and enzootic abortion, although even vaccinations are not always successful where there are several strains of the pathogen, as in the case of enzootic abortion.

Where a problem persists, vaccinations can still be used, but attention must be given to reducing stress levels (for example the movement of stock the previous day may predispose the animals to pasteurella) and minimising the risk of importing infection by maintaining a closed flock. Outdoor lambing is preferable as many of these diseases are becoming more prevalent with winter housing. In the case of pasteurella, good ventilation and clean, fresh water, possibly mildly acidified (bactericidal effect) can help avoid the onset of pneumonia. For the treatment of foot rot, the Soil Association's standards allow the use of zinc sulphate and copper sulphate footbaths, but formaldehyde footbaths are only allowed under restricted circumstances. The foot rot vaccine is prohibited under all circumstances.

Abortions may be caused by other diseases such as Q-fever, listeriosis, tick-borne fever and border disease, but they may occasionally also be due to malnutrition leading to sub-clinical pregnancy toxaemia. Pregnancy toxaemia, or twin lamb disease, is the result of a lack of carbohydrate or energy intake and is seen usually in sheep on insufficient rations, but can also hit overweight ewes. It can be triggered by a change of feed, even to better quality, which causes sheep to go off food for a short period. The feeding of bulky foods in the last three to four weeks of pregnancy can also predispose ewes to pregnancy toxaemia. When a case occurs, it is an indication of a flock problem. Treatment is available, but prevention is more important, using condition scoring and supplementary feeding for thin ewes. Fat ewes should be encouraged to exercise, for example by movement between poorer and better pastures.

Pig health
Many, if not most, of the disease and nutritional problems which affect young pigs are a direct result of the production system and can be avoided by making changes to the system. Breeding is a major problem, with the emphasis on high rates of feed conversion. This leads to problems such as weak circulation and arthritis (which accounts for more than 10% of piglet mortality) resulting from an imbalance between body weight and skeletal strength.

In addition, the practice of bringing sows into farrowing crates/quarters shortly before farrowing gives them insufficient time to adjust to the microflora and thus provide protection through colostrum to their young ones. For instance, *E. coli* immunity can be obtained from high colostral protection and the active development of immunity over the first three weeks of life. The incidence of *E. coli* problems will decrease with subsequent litters as a result of increasing sow immunity, so long as contact is maintained between sows of different ages and at different stages of lactation/pregnancy, as is possible in family group housing systems.

Anaemia is another problem associated with intensive systems, due to insufficient levels of iron in the milk. In some cases, piglets will eat the sow's dung which, if it contains sufficient iron, can render injection unnecessary. By implication, access to free range and the eating of soil will have a similar effect. In less intensive systems, iron supplementation should not be necessary, and the practice is in fact prohibited under the Soil Association's standards.

The control of parasites

Parasites of livestock can be one of the most intractable health problems in both conventional and organic systems. Occasionally, preventive treatments are required by law to avoid the spread of potentially epidemic disease and these requirements must obviously be conformed with. In some cases, recently developed treatments may be more acceptable than the established ones, for example the use of pyrethroid based dips in preference to the organophosphorus dips for compliance with sheep scab control regulations. Where treatment is not required by law, the prophylactic or routine use of drugs should be avoided and preference given to husbandry and environmental management techniques such as rotational grazing and reduced stocking rates.

Internal parasites

The routine use of anthelmintics for control of intestinal and stomach worms is not allowed under the Soil Association's standards, which means that priority must be given to appropriate (probably lower) stocking rates and the use of a clean grazing system, with the use of anthelmintics allowed for treatment of infected individuals, or strategic dosing under restricted circumstances. Clean grazing systems are based on a detailed understanding of the lifecycle and epidemiology of the worms which affect cattle and sheep. Parasitic worms can be grouped into three categories: nematodes (stomach, intestinal and lung worms), cestodes (tapeworms, hydatid cysts) and trematodes (liver flukes). There are 20 species of nematode in sheep and 18 in cattle, the most important being species of *Ostertagia*, *Trichostrongylus* and *Nematodirus*. A few species are common to both sheep and cattle, but with few exceptions these are unimportant. Infections by these parasites are commonly, although not always, associated with poor growth, loss of appetite, coughing or diarrhoea; but the extent of the symptoms will depend on age, immune status and level of contamination of the herbage.

The lifecycles of *Ostertagia*, *Trichostrongylus* and *Nematodirus* are illustrated in Figure 8.12. There are two distinct phases in the life cycle of these nematodes: the host animal phase and the free-living pasture phase. The nematodes cannot multiply within the host itself. Instead, eggs laid by the adult female worms pass out with the dung. These then undergo development on pasture before they reinfect the host animal and develop into adult worms in the gut.

In the case of *Ostertagia/Trichostrongylus*, eggs deposited on pasture in early spring can take several weeks to complete their development. Eggs passed later develop faster because of higher ambient temperature, with the result that all eggs passed in spring to early summer tend to complete their development at the same time. This means that the main infection period for sheep is in late June, and for cattle in mid July onwards. Levels of herbage contamination then fluctuate during the rest of the year and can persist at a high level until the following spring, often reaching their peak during the winter months. By late spring the herbage contamination levels decline, reaching low levels by May.

Young cattle turned out in spring onto pasture carrying overwintering infection begin to pass worm eggs about three weeks later. If they remain on the pasture, they will be exposed to reinfection during the second half of the grazing season and this can result in large worm burdens and in some cases clinical disease. Where young cattle graze infected pasture in late autumn/early winter, the larvae will remain dormant in the gut until the spring, causing problems then. However, older stock will become resistant, so that adult cattle do not usually play a role in the lifecycle of the nematode. In a suckler herd, where adult animals predominate, the effective dilution of susceptible stock means that grazed pastures rarely develop dangerous levels of infective larvae and that spring born suckler calves are not usually at risk throughout the whole of the grazing season.

In sheep, the problem is exacerbated by the fact that, although the ewes do develop resistance to the worms, there is a reduction in this resistance at lambing and in early lactation which means that ewes as well as lambs contribute to the problem and there is a rapid increase in egg shedding in spring.

Nematodirus species have specific characteristics which distinguish them from other stomach/intestinal parasites. *Nematodirus battus*, which first appeared as recently as 1959, causes clinical disease (usually indicated by profuse, black diarrhoea) in lambs in April/May as result of an abrupt but short lived appearance of infective larvae on the pasture at a time when young susceptible lambs commence grazing. They are transmitted from one lamb crop to the next, with the ewe playing a negligible role. This is because the free-living development of the nematode takes place entirely within the egg shell. Hatching depends on a period of cold weather followed by warmer conditions, which means that the time of hatching will vary from year to year. This can be an advantage in that early hatching takes place when the lambs are not yet

Figure 8.12   Life cycle of *Ostertagia*, *Trichostrongylus* and *Nematodirus* spp.

Ostertagia and *Trichostrongylus* groups

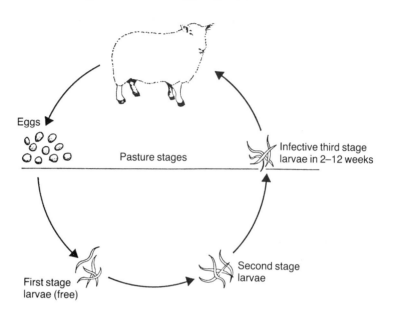

Nematodirus group

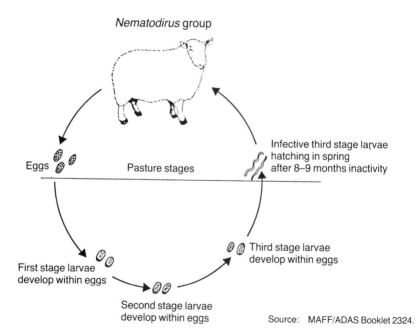

Source: MAFF/ADAS Booklet 2324.

grazing, and if hatching is late lambs may already have developed resistance. Ideally, lambs should spend the first 12 weeks of their lives on new leys which have not had sheep before, but as this is not always possible, use can be made of the MAFF *Nematodirus* forecast to assess the risk of problems occurring during the critical end April/early May period.

Clean grazing systems
An important factor in the design of any clean grazing system is the recognition that parasitism is a natural association. Balance may be achieved where the parasite causes little or no distress, but if stress or other unfavourable conditions occur, this will tip the balance. Almost all stock will be infected, but the aim of clean grazing is to keep the worm burdens to a low, economically and physiologically acceptable level, not necessarily to eliminate them.

The basis of designing clean grazing systems is the categorisation of 'clean' and 'safe' pasture as indicated in Table 8.6. The most important aspects of this categorisation are:

- the division of the grazing season into two parts reflecting the period in spring/early summer when the larvae die off and cannot serve as a source of further infection, and
- the fact that cattle and sheep are not usually affected by the same species of nematode.

Clean pasture is defined as pasture which is uncontaminated, while 'safe' pasture may be contaminated at low levels which should not affect the production of susceptible animals, but these will become a source of contamination. There is also a third category of 'acceptable' pasture, which is safe apart from in exceptional circumstances or years.

The availability of clean and safe grazing will differ from farm to farm, hence there will be a wide range of possible options. It will also be necessary to consider the next year as well as the current year, to ensure sufficient clean grazing in the following year. Detailed information can be found in the MAFF Booklets 2154 and 2324, as well as in Taylor (1987). However, the options can be categorised into three main strategy groups: preventive, evasive and dilution. These may or may not involve the use of anthelmintics, depending on individual circumstances.

A preventive strategy can be adopted where clean pasture is available in the early part of the grazing season. In the case of cattle, young stock can be turned out onto clean pasture for the whole of the grazing season. Autumn born calves which may have grazed prior to housing in winter may need to be dosed with an anthelmintic before turnout to prevent pasture contamination. With sheep, the situation is different because of

Table 8.6 Definitions of 'clean' and 'safe' grazing for sheep and cattle.

1 A classification of pastures for calves

Before 15 July

CLEAN

1. New seeds after an arable crop. If grazed by cattle in the previous autumn, these should have been dosed immediately before they were put on.
2. Pasture grazed by sheep only in the previous year.
3. Grassland used for conservation only in the previous year.

SAFE

1. Pasture grazed in the previous year by cattle in their second or subsequent grazing year.
2. Pasture grazed in the previous year by beef cows with or without calves at foot.
3. After 7 May, pastures grazed by calves in the previous year.
4. Pasture grazed by calves before mid-March or after mid-September of the previous year.

ACCEPTABLE

1. After 23 April, pasture grazed by calves in the previous year.

POTENTIALLY UNSAFE

1. Before 23 April, pasture grazed by calves in the previous year.

After 15 July

1. Aftermath not grazed by cattle earlier in the year.

2. Pasture grazed by sheep in the first half of the grazing season.

1. Pasture grazed in the first half of the season by cattle in their third or subsequent grazing season.
2. Pasture which was CLEAN at the beginning of the season and was grazed thereafter by hand-reared calves straight out of the buildings.

1. Pasture grazed by yearling cattle in the first half of the grazing season.
2. Pasture grazed in the first half of the season by beef cows with calves at foot.
3. Pasture grazed in the first half of the season by calves dosed every three weeks until the end of May.

1. Pasture grazed by calves in the first half of the grazing season.

2 A classification of pastures for lambs

Before 1 July

CLEAN

1. New seeds after an arable crop. If grazed by sheep in the previous autumn, these should have been dosed immediately before they were put on.
2. Pasture grazed in the previous year by cattle only.
3. Grassland used for conservation only in previous year.

After 1 July

1. Aftermath not grazed by sheep earlier in the year.

2. Pasture grazed by cattle only in the first half of the season.

(continued)

Before 1 July

SAFE

1. Pasture grazed only by non-lactating adult sheep in the previous year.

2. After 7 May, pasture grazed by yearling sheep in previous year.

3. After 7 May, pasture not grazed by lambs in the first half of the previous grazing season but grazed by dosed lambs in the second half.

ACCEPTABLE

1. After 23 April, pasture grazed by ewes or yearling sheep in the previous year or by dosed lambs in the second half of the previous grazing season.

2. Pasture grazed by lambs before mid-March or after mid-September of the previous year.

POTENTIALLY UNSAFE

1. Before 1 June, pasture grazed by lambs in the first half of the previous grazing season or by undosed lambs in the second half of the previous grazing season.

After 1 July

1. Aftermaths not grazed by yearling sheep, lactating ewes or lambs earlier in the year.

2. Pasture grazed by non-lactating adult sheep in the first half of the season.

1. Pasture which was CLEAN at the beginning of the season and was then grazed by ewes and their lambs, the ewes having been dosed after lambing and immediately before they were put on.

1. Pasture grazed by lambs in the first half of the current grazing season.

Source: MAFF/ADAS, Booklet 2154

the role of the ewes, which may need to be dosed with an anthelmintic prior to turnout onto clean pasture, thus avoiding the exposure of lambs in the second half of the season. Goats share parasites with sheep, cattle and horses and therefore new leys and aftermaths are the only source of clean grazing.

An evasive strategy can be adopted where clean aftermaths etc. are only available in the second half of the grazing season, and can also include the use of 'safe' pastures. Calves may be turned out on to contaminated pasture in spring and then moved to clean pasture before a new generation of infective larvae appears on the original pasture in mid-season. To keep worm burdens low, they will need to be moved to other aftermaths subsequently. In the case of sheep, dosing with anthelmintics will be ineffective anyway if the ewes are turned out onto contaminated pasture. The evasive strategy involves weaned lambs being dosed and moved to conservation aftermaths before infestation appears on the herbage.

A dilution strategy is one which can be adopted where there is no clean pasture and contamination is present. It is based on the grazing of susceptible young stock together with resistant adults, for example single suckled beef or leader-follower systems on dairy farms. The resistant animals must greatly outnumber the susceptible ones on a size/number basis. This approach is not suitable for sheep unless a large number of resistant non-pregnant yearlings are present on the farm. Dilution by mixing cattle and sheep is not usually sufficient for this strategy to be effective.

For clean grazing systems to be effective, certain practical requirements must be met, including sound, stockproof fencing, the integration of beef and sheep, the use of fodder crops to provide areas of clean grazing and possibly the housing of livestock over winter to rest pastures. There is also a range of problems which will need to be faced including the provision of adequate early spring grazing, the fact that the intervals needed for parasite control are not necessarily the best intervals for managing grass/clover swards and the fact that decisions are dictated more often by the weather than the plan, as the stock have to go where the grass is. Much of the work on clean grazing systems has been carried out in drier areas and there is some evidence that parasites are capable of surviving rather longer in the wetter western areas of the UK such as North Wales. Finally, there is always a risk that organic (and clean grazing) systems will tend to select certain parasites which may become significant in the future.

In cases where it is difficult to implement a clean grazing system, and in the case of a known farm problem or before infected animals are turned out onto clean pasture, it may be necessary to use anthelmintics. Under the Soil Association's standards, ewes may be wormed at lambing without affecting the organic status of the lambs as most will be sold before problems arise in the autumn, but where individual animals have to be treated the specified withdrawal periods must be observed[*].

In some circumstances, 'strategic dosing' may be appropriate, particularly during the conversion period. Such procedures are described in detail in sources such as Taylor (1987 & 1988) and Williams (1984). They include options such as dosing ewes with anthelmintics as they go onto clean pasture and once again as they go onto clean summer

---

[*] The New Zealand standards for organic livestock are even more restrictive and specify that ' . . . any animal treated must be clearly and permanently marked at the time of treatment and subsequently sold as non-organic; treated animals must immediately be placed in a designated quarantine area for a specified time related to the chemical used; treated stock may be returned to the main flock following the quarantine period, provided they are clearly and permanently marked'. Only drenches containing either levamisole or morantel as the active ingredient may be used, for which the quarantine periods are two and three days respectively.

grazing, as well as the treatment evolved by Professor Michael Clarkson of Liverpool University which consists of dosing the whole flock three times during the crucial mid-summer period, when infective larvae have died off and the only means of transmission is the stock. The aim is to convert sheep-sick grazing into clean pasture by breaking the lifecycle of the adult worms in sheep, thus eliminating reinfestation and the usual autumn build-up of infective eggs and larvae. This can be followed in subsequent years by a single booster dose to reduce any carry over of worms.

Where anthelmintics are used, care should be taken to use those which are most appropriate and to change the drug used periodically to avoid the build up of anthelmintic resistance, an issue which is causing increasing concern.

Although the need to use anthelmintics on an occasional basis is accepted under organic standards, many farmers, especially those with large sheep flocks on all grass farms, are concerned to find other alternatives. One such is the use of an extract produced from the finely ground seed hair of the *Kamella* plant which is a native of the Philippines. The extract paralyses the muscles of the stomach worms and produces scour, thus eliminating them. After dosing, clean grazing is practised. This treatment can be given to any lambs showing signs of worms without depriving them of their organic status.

Lungworm (husk)

Husk or parasitic bronchitis is caused by the lungworm nematode. Much less is understood about its lifecycle than other internal parasites, and therefore it has not been possible to develop a clean grazing type approach to its control. It is known that stocking intensity is a factor, and that suckler cattle, which traditionally were kept at low levels of intensity and were not usually significantly affected, are becoming increasingly susceptible. Although the disease was present at low levels in pasture, the calves picked up only small amounts of infection at a time, enough just to give them immunity rather than the disease itself. Adult animals are not usually infected by lungworm as they have already aquired a degree of immunity.

The Soil Association's standards allow the use of the oral vaccine Dictol where a known farm problem exists. The vaccine consists of infective larvae which have been irradiated so that they still go through most of their life cycle in the lung, but do not develop to the adult stage. Immunity is thus stimulated quite rapidly without any egg production. The husk vaccination should not be used in herds where the disease does not occur. Where it has occurred, use may need to be continued, because the animals will continue to pick up low levels of infection from pasture and may be subject to husk entering the farm through bought-in cattle. Although young animals can acquire immunity naturally, this is not always sufficient to meet heavy challenges of infection.

Liver fluke

The life cycle of the liver fluke is characterised by the presence of the snail as an intermediate host. The most important means of control is drainage to remove the damp, slightly acid conditions which the snails prefer and need to survive. Where this is not possible, then stock should be kept off the infected area during the high risk period between mid-September and the end of January. Black disease is a secondary problem associated with high infestations of fluke.

Tapeworms

Although tapeworms can infect humans, they are of little consequence in ruminants and no attempt is usually made to control them. However, in areas of North Wales and Cumbria in the UK, and more widely in Australia and New Zealand, hydatid disease is common in sheep. Dogs are important carriers and humans can also be affected. Control can be achieved only by ensuring that dogs are kept clear in the first place. Transmission by flies, although possible, is much less significant. Work is currently being done on the genetic engineering of vaccines to control internal parasites such as hydatids, but this raises a number of other issues relating to the release of genetically engineered organisms into the environment.

Coccidiosis

Coccidiosis is primarily a problem associated with housing of sheep and to a lesser extent the feeding of concentrates, because of the soiling which takes place at the ewe's rear end where lambs also have to suckle. Winter shearing will reduce soiling, but dagging may be a more acceptable solution if ewes have to be housed. Adult ewes are usually resistant, but will continue to pass small numbers of oocysts in their dung throughout their lives. Lambs, however, are very susceptible. Coccidiostats prevent the shedding of organisms, but the load is shed in one go as soon as the treatment stops. Although sulphamethazine is permitted by the Soil Association in restricted circumstances, good hygiene and husbandry must be the primary means of prevention.

External parasites

Many of the common external parasites such as lice and keds can be controlled reasonably effectively using derris, and even warble fly can be controlled in this way. Many farmers continue to treat for warbles using the suspect systemic organophosphorus compounds or ivermectin (which is as suspect in terms of its ecological impact) even though no warbles are present and the legal requirement for treatment applies only to those animals which are known to be affected. Some organic farmers, notably the Rowlands at Brynllys (Chapter 9) actively resisted the use of organophosphorus dressings and obtained exemption permits allowing the use of derris, which proved to be effective under close veterinary control.

It may, however, be a mistake to think simply in terms of alternative medications for the control of external parasite problems. Environmental management is also important. An example of this is ticks, which are a problem in rough grazings and can lead to tick pyaemia, hornfever, red water fever and louping ill among other diseases. In some cases, they can be controlled by grassland management, keeping the grass as short as possible, thus making life unpleasant for the free-living ticks.

Of particular concern to organic farmers is the control of sheep scab and fly strike. The former because, although only 26 outbreaks were identified in Britain in 1987, there is a legal requirement for sheep flocks to be dipped in an as yet unsuccessful attempt to eradicate scab completely, and the latter because, at least until recently, no effective alternative to systemic insecticides has been discovered, particularly with the management of large flocks.

The question of whether sheep should be dipped at all is an important one, both on welfare and on human health grounds. Sheep scab is caused by the mite *Acarus* which is just visible to the naked eye. It is a notifiable disease and dipping in the autumn is compulsory, following its reemergence in 1973 after an absence of 20 years, partly, it is believed, due to the replacement of dips by sprays and showers which also led to an increase in skin parasites such as keds, lice and ticks.

It has been argued by, among others, Jack Done, former Head of Pathology at the Central Veterinary Laboratory, that national compulsory dipping for scab is ineffective, that there is no need to eradicate scab nationally, and that scab could be eradicated at flock level, or even lived with. Scab spreads only by contact, so control can be effective if and when it occurs. The submersion of sheep is at best stressful and at worst cruel to the animals; autumn dipping disrupts mating and summer dipping, which is no longer compulsory, disrupts marketing because of withdrawal periods.

The organophosphorus dips are known to cause health problems in humans, particularly in sensitive individuals, including 'post-dipping flu', depression, weariness or dizziness, and there is serious concern about the risk from residues in meat and the effects of dips as a pollutant on fresh water and marine life (Farmers Weekly, November/December 1989).

The Soil Association's standards permit the use of flumethrin, a non-systemic synthetic pyrethroid effective against sheep scab and which complies with the legal requirements for dipping. It is not, however, effective against fly strike. Lime and sulphur dips are also permitted, but individual MAFF approval is required and preparation is very time consuming (Young, 1985).

The control of fly strike (blowfly) is a much more significant problem for organic sheep producers in Britain. Blowfly eggs are laid in the fleece and hatch in about twelve hours as sheep maggots which invade the skin and flesh of their host. An infested ewe may die in three to four days from toxaemia. The high risk season can extend, according to prevailing weather and altitude, from May to October.

In order to avoid the use of systemic organophosphorus compounds, many organic sheep producers have had either to practise dagging combined with attention to individual animals (although this is not usually feasible in larger flocks), or they have aimed for early lambing so that finished lambs can be sold before the compulsory dipping and high-risk periods, while still dipping the ewes.

Recently, the pour-on compounds Vetrazin and Cypor containing the active ingredient cyromazine have been introduced in Britain. Used for a number of years in Australia and New Zealand, where blowfly have become resistant to organophosphorus compounds, cyromazine is able to give six to eight weeks protection after a single treatment. It acts as an insect growth regulator by interfering with the maggots' moulting process. Although not totally harmless, cyromazine is non-systemic, very specific in its action and has a low toxicity to mammals. It is therefore much safer than the organophosphorus compounds and the legal withdrawal period is only three days. The product has been accepted by the Soil Association for use in Symbol lamb production in restricted circumstances where fly strike is a major problem. The New Zealand organic standards allow the use of dips or pour-on dressings containing cyromazine followed by a 21 day quarantine period. (This quarantine period also applies to the use of synthetic pyrethroids in New Zealand).

A range of non-medicinal approaches to control blowfly also exists, but these are not always effective. Dagging (the trimming of dirty wool around the crutch and tail) helps to reduce the incidence of flystrike because blowflies are attracted to wool which is damp or soiled. Grazing on stubbles after harvest may help to provide a lower risk environment. Where an individual animal is infected the wool should be clipped tightly around the infected area and the maggots combed out, followed by spot treatment with pyrethroid or iodoform products. The use of a brine solution may also be effective, but the priority must be dealing with the suffering of the animal rather than a purist approach to the use of alternative remedies.

The foregoing discussion should make it clear that there is no easy answer to some of the important animal health problems which face organic livestock producers, and that the organic standards recognise this by allowing for compromise solutions while stating the need for further research. One aspect of this has recently been recognised by MAFF with a decision to fund further research into the control of flystrike without the use of systemic insecticides. This is a step forward, but much more needs to be done before organic producers can be properly satisfied with their results.

Breeding, Reproduction and Rearing

The continuity of any livestock enterprise is dependent on the successful transfer of desirable characteristics from one generation to the next. In organic systems, such characteristics should include hardiness and longevity as well as the more conventional measures of productivity. The achievement of this covers not only the aspects of nutrition, housing and health care discussed in previous sections, but also the areas of breeding and breeding objectives, reproduction and fertility, and the rearing and weaning of the young animal. Health and fertility are determined both by the environment and by genetic disposition, but the best environment cannot produce healthy and productive animals if the hereditary disposition is unfavourable.

Breeding objectives

The current trends towards breeding, in the case of meat production, for larger animals which can be slaughtered earlier when still immature and hence less fat and, in the case of milk production, the selection on the basis of first lactation yields, make little sense in the context of an organic farming system. It is questionable even whether it makes sense in the long run for conventional systems either. In the case of pigs, where the offspring eat most of the food, the breeding of larger, faster growing animals may make sense on economic and feed conversion efficiency grounds; this is possibly also the case for beef where the parents and offspring share the food 50:50, but certainly not for sheep, where the ewes eat most of the food, although selecting for size in the sire such as Texels or Suffolks may be important.

The problems associated with breeding for leanness and growth rates are most clearly illustrated by pigs, with the Pietrain breed in particular showing high stress susceptibility as a result of a weak heart and circulatory system (although lack of exercise in inappropriate housing also plays a role). The problems of arthritis in young pigs and PSE (pale, soft, exudative) meat is also associated with these breeding objectives. In organic systems, hardier, more robust breeds such as the Duroc and the British Saddleback are likely to be more appropriate.

These problems may be exacerbated by the genetic engineering revolution, and the development of 'man-made' transgenic animals. Transgenic animals are ones which have a foreign gene for a particular characteristic introduced into them. One example related by Murphy (1988) is that of growth hormone genes introduced into pigs at the US Department of Agriculture's farm at Beltsville, Maryland, producing a severe form of crippling arthritis. Only one survived to adulthood, but was unable to stand. Genetic engineering carries with it significant welfare implications, as it is impossible to predict all the consequences of adding a foreign gene.

The pig industry also exemplifies another disturbing trend in livestock breeding—that of genetic simplification. In the British pig industry, two breeds dominate: the Large White and the British Landrace. Many other breeds have virtually disappeared, thus reducing the diversity of genetic material available to breeders and increasing the risks of specific diseases reaching epidemic proportions. These now rare breeds could have a significant role to play in meeting the new breeding objectives which organic systems would seem to require (Tudge, 1989).

Another example of breeding programmes taking the wrong direction is the double-muscled Belgian Blue. As pure-breds, Belgian Blues require routine delivery by Caesarian section because the large hindquarters of the calf do not allow for a natural birth; when Belgian Blues are used for cross breeding, this problem does not arise.

In dairy cattle, there has been considerable emphasis on first lactation yields as a measure of potential productivity, with little attention being given to longevity and lifetime yields, or to ability to utilise high forage rations. Peak milk production occurs at about seven years of age, yet the average slaughter age for dairy cows is now less than six years. The average dairy cow produces in its lifetime a yield of little more than 15,000 litres, although top quality animals can produce as much as ten times that amount.

It has been argued by some, notably Bakels and his colleagues in Germany (Bakels in Voitl et al., 1980; Bakels & Storhas, 1981; Haiger, 1986), that trying to combine meat and milk in the same animal as typified by the British Friesian has been a major error. The selection

1 *Profile of a Leicestershire fattening pasture (permanent grass), showing a deep, strongly structured soil, fissured by desiccation and cracking and granulated by earthworm activity* (V. I. Stewart)

2 *A naturally acidic marginal hill soil in Wales, showing characteristic surface organic mat and a degraded, compact mineral soil resulting from the absence of earthworm activity* (V. I. Stewart)

3 *Earthworm casts in grassland indicating good biological activity*

4 Use of natural vegetation as a green manure protects soil fr erosion in an organic vineyard in the Mosel region of Germany

5 Manure dumped in a heap creates a serious pollution risk and wastes valuable nutrients (Elm Farm Research Centre)

6 A three course stockless rotation with years cereals and one ye green fallow (red clove. field beans etc.) (T. Gips)

7   Permanent crops where no rotation is possible require a different approach to the creation of diversity, here apple trees with a floriferous understorey (R. Crowder)

Mixed cropping has ...antages even in a ...rse rotation, here a ...ture of oats and peas

9   Grain legume and cereal mixtures can be harvested together, here a mixture of wheat and field beans. The grains can be separated using seed cleaning equipment (Elm Farm Research Centre)

◄10   *Vetch and forage rye as a green manure*
(Elm Farm Research Centre)

11   Phacelia *(bee's friend) as a green man.*
(Elm Farm Research Centre)
▼

Cleavers swamping a crop of third year cereals illustrate the need for careful attention to rotation design weed control

Charlock in a spring ley reseed can be successfully controlled by topping and grazing management

14 and 15   *Common poppies in winter wheat* (above) *and spring oats free of poppies* (below) *illustrate the importance of species and variety selection and timing of cultivations for weed control. The field of winter wheat can be seen in the background*

16 Encouragement of natural predators keeps codling ▶
moth under control in orchards at the Biological
Husbandry Unit, University of Lincoln, New Zealand.
Umbelliferous plants are especially important as nectar
sources for parasitic wasps (Ichneumids). Here parsnip
is in flower at the same time as the apple tree
(R. Crowder)

◀ 17 Competition between the floriferous understorey and
the crop needs some control, but this orchard has had no
pest problems in the eight years since conversion and no
biological control techniques other than natural
management have been used during this period
(R. Crowder)

18 Pheromone traps in
fruit production (Elm
Farm Research
Centre)

19   Cockle Park type mixture shut up for silage, with red clover, white clover and chicory in flower (Brynllys Farm, Dyfed)

20   Medium-term white clover based leys are best suited to grazing or a mixture of grazing and forage conservation (Luddesdown Farm, Kent)

21 *Lucerne is a high yielding crop for forage conservation, best suited to dryer areas and soils with high pH levels (Manor Farm, Dorset)*

22 *Short-term red clover leys provide high yields for forage conservation (Neu-Eichenberg, Hessen)*

23   *Winter wheat, Kite's Nest farm, Worcestershire*

24   *Winter oats, Kite's Nest farm, Worcestershire*

25   *Rye, Luddesdown Farm, Kent*

26   *Field beans, Luddesdown Farm, Kent*

27   The non-use of herbicides in organic systems results in a larger number of non-crop species in organic fields, with consequential benefits for insect and bird life and a reduced risk of soil erosion

28   Many organic farmers place special emphasis on the creation of conservation habitats such as this one at Brynllys Farm, Dyfed

for beef characteristics has led to dairy cows which do not have a milk producing constitution. Milk production is a secondary sexual characteristic of the female, while meat production is a secondary sexual characteristic of the male; trying to combine the two represents a contradiction which cannot adequately be resolved. Bakels *et al.* argue that the maxim 'milk from the dam, and meat from her sons' should be re-established as a breeding objective.

Bakels and colleagues argue that historically, dairy cows were characterised by length and depth and that recent selection policies have tended towards smaller animals. Yet it is only large animals which, they argue, have the ability to take in large quantities of forage in order to produce large quantities of milk. In addition, longevity can be attained only by using late maturing animals. A long growth period means a long youth and a long immature stage delays the ageing process, which, it is argued, is a necessary precondition for a long life. The selection of immature but relatively well proportioned bulls will tend to select against a long lifespan, as will the selection of early maturing cows producing high first lactation yields at the expense of lower lifetime yields.

On the basis of their analysis, Bakels and his colleagues concluded that the American Holstein-Friesian came closest to being the ideal dairy cow. Unlike European breeders, the Americans have concentrated on milk production characteristics and this was reflected in the fact that it was still possible to find Holstein cows with lifetime yields of over 150,000 litres. They set up a breeding programme based on German cows and North American bulls, using as a basis family lines within which individual cows had given lifetime yields of up to 100,000 litres. The aim of the programme is to produce cows which will achieve average lifetime yields of 30,000 kg on a high forage ration, a target which they are achieving with a considerable degree of success proving that it is possible to breed for constitution.

Breeding objectives is an area to which little thought has been given by organic producers and much more work is required, although it is fair to say that biodynamic researchers have considered these issues to a greater extent (Sattler & Wistinghausen, 1985 and Schaumann, 1988). The general requirements of health and a strong constitution, high forage intake capacity for ruminants, regular cycling/fertility and longevity are easily stated but perhaps more difficult to achieve.

Reproduction and fertility

Fertility in livestock is clearly a husbandry issue, although a large range of factors are involved, including nutrition, housing, stress and the presence of male animals. The presence of a boar amongst sows has a marked influence on when they come into oestrus and the same may also be true for ruminant livestock. Biodynamic farmers set great store by the presence of a bull on the farm, believing that artificial insemination (AI) has significant disadvantages in this and other respects. Most organic dairy farmers, however, make full use of AI, but other techniques such as embryo transfer and the use of hormones for oestrus synchronisation are not permitted under organic standards.

The technique of embryo transfer is questionable on the grounds of welfare and genetic diversity. Artificial insemination, which allows a few bulls to be used on a large number of dairy cows, already contributes significantly to the reduction of genetic diversity in the national dairy herd. Embryo transfer, if used on a wide scale, threatens to exacerbate the problem. The use of hormones for oestrus synchronisation, like the use of other routine medication, is merely an excuse for ignoring the impact of the husbandry system on the health of the animals themselves.

The link between nutrition and infertility is still ill-defined, although researchers have described the conditions which are least likely to encourage optimum fertility in dairy cattle, such as high peak yields or low energy diets in early lactation causing large energy deficits because the cow is unable to take enough food on board, with a consequential rapid decline in body condition after calving. Important factors in maintaining fertility are therefore the avoidance of high levels of concentrates and dramatic body weight losses post calving, as well as allowing sufficient feeding space at the silage face and plentiful access to forage under low concentrate regimes.

There is evidence that the organic management of forage can promote better fertility in livestock. Lotthamer (1982) has shown that concurrent high potash and low sodium levels in the diet, a problem associated with the use of soluble potash fertilisers, can lead to irregular cycling and other fertility problems. In addition, controlled experiments with animals fed organically and conventionally produced foodstuffs have shown significant differences in semen quality, fertility, birth rates and morbidity (Chapter 15).

The rearing of young stock

The production of healthy and long-lived replacements depends as much on the care of the animal from birth as it does on breeding and the management of the mother.

The feeding of colostrum is now widely recognised as essential to the wellbeing of the young animal through the provision of immunity against common diseases. Surplus colostrum can be stored by freezing, or at ambient temperatures using natural fermentation. Freezing is ideal because it maintains the nutrient content. Fermented colostrum should be stored below 20°C and should be fed within a few weeks of collection because nutrient content declines through storage. It needs to be stirred daily to counteract separation and should be handled hygienically to prevent contamination by pathogens.

When rearing dairy replacements, the minimum requirement in an organic system should be full milk (no replacers) over a period of at least five weeks with at least two to three days opportunity to suckle, a healthy housing environment for the calves and the feeding of as few concentrates as possible, because of potential damage to the rumen epithelium. The rumen should be trained to cope with high levels of forage intake by feeding high quality hay etc. supplemented later on by 1 to 1.5 kg concentrates per day, but it is important that any changes made should be gradual.

For dairy cattle the preferred, if more expensive, approach is the use of multiple-suckled nurse cows with weaning at three to four months. The Rowlands at Brynllys (Chapter 9) practise this approach, as do many biodynamic farmers, in the belief that it gives the heifer a much better start in life and enhances the longevity of the animal. Efficiency of milk utilisation is higher in calves that are suckled compared with calves which are bucket fed, due to the fact that psychological stimuli, rather than physical factors, control the closing of the oesophageal groove which directs milk past the rumen to the abomasum, which functions as a simple stomach. If this mechanism does not work, milk spills into the rumen where fermentative digestion reduces the quantity and quality of nutrients available to the calf, in particular through the coagulation of casein, and can lead to ulceration problems. If bucket feeding is necessary then it should be done using buckets with teats fitted and placed so that the calf has to tilt its head upwards.

Natural weaning, with no forced removal of the dam from its offspring, is a further important consideration. The Youngs at Kite's Nest Farm (Chapter 9) have shown that it can work well for suckler beef, and Norwegian research has shown that it also works well with pigs. If the sow is allowed an escape route into a separate stall for feeding (such as a counterweighted sliding door which she can push aside, but which prevents the piglets following), she will initially leave them only to get food for herself, but later on (after about four weeks) she will leave them for increasing periods of time, encouraging the piglets to move gradually onto solid feed. The time of natural weaning will vary, but is usually nine to fifteen weeks. In addition, the social hierarchy established among piglets means that normally each acquires its own particular teat or two, and that milk output from each teat will adjust to the requirements of the individual piglets.

BEES—THE FORGOTTEN LIVESTOCK

This chapter and the following chapter deal in some detail with the main domesticated farm animals such as cattle, sheep, pigs and chickens, and Blake (1987) covers some of the other livestock suited to the smaller organic holding. But any consideration of livestock in an organic system would be incomplete without reference to the humble bee.

Bees not only provide a product for human consumption, honey; they are essential for the pollination of many crops and often a major influence on yield and the creation of a synchronised, early crop. Most perennial legumes depend on insects such as bees for cross pollination so that they can set seed, as do many horticultural and fruit crops. Crops like cereals, which are mainly wind pollinated, still attract bees and even in crops which are largely self pollinated, visits from insects such as bees can increase yields.

Honey is not only nice to eat; it has been shown to be very effective at helping the rapid healing of wounds (British Journal of Surgery, 75: 679). The properties of honey (slight acidity, extreme viscosity and high water absorption) enable it to absorb water from wounded tissue, clean the wound and protect it from further infection. It also contains the bacteriocidal agent inhibine and can be used to remove dead tissue from persistent wounds without the need for surgery. Its use for the treatment of livestock appears to be as yet untried, but there may be possibilities.

Yet bee populations have been decimated by the use of insecticides in conventional agriculture, particularly on oil seed rape, and by the use of aphicides on cereals later in the season. The decline in the range of flowering species in crops and the number of farmers keeping bees has played a significant part in this problem.

In biodynamic agriculture, the role and importance of bees has been recognised from the outset. The bee colony is conceived of as a single organism whose organs have acquired a certain spatial freedom, but which functionally remains a single unit. The management of the hive should aim not to disrupt this organism. The bees build hanging combs parallel to each other and spaced about 4 cm centre to centre. Eggs are laid from the centre outwards in an hierarchical structure, then from the centre again once the bees have developed. Inserting new combs, a practice common in conventional beekeeping, disrupts this natural progression and should, it is argued, be avoided.

It is not only the honey bee which is the farmer's friend. Bees native to the UK, such as the bumble bee, are important for certain crops, because honey bees are not always the best pollinators (Corbet & Prŷs-Jones, 1987). For example, with field beans, plants that are the products of self pollination produce few seeds unless they have the benefit of cross pollination. Long-tongued bumble bees are the most effective pollinators of field beans. Honey bees can pollinate the field bean when they collect pollen, but when seeking nectar, they often 'steal' it through holes bitten at the base of the corolla by short-tongued bumble bees.

Bumble bees tend to work faster, for longer hours and in worse weather than honey bees and they willingly patronise some crops which honey bees avoid. They are likely to work better on crops in which the flower has a long corolla tube and relatively dilute nectar, such as the field bean and red clover, and hence are an important element in the persistence of clovers in pastures.

The native bees of the United Kingdom have suffered serious decline in recent years, but there is much which farmers can do to encourage their revival. The provision of patches of rough ground and hedgerows where bees can nest and have access to wild flowers is a starting point which can be supplemented by the deliberate sowing of non-competitive wild flowers, such as blue cornflowers, in and around cereal crops. The provision of stacks of straw bales or wooden nest boxes where bees can set up home and the creation of a mosaic of crops in small fields suited to distances which bees can fly will also help.

INTEGRATION – THE KEY TO ORGANIC LIVESTOCK HUSBANDRY

The integration of different species, or stock of the same species but kept for a different purpose, and of crops and livestock is the key to many of the husbandry changes needed to make an organic system work. The integration of beef with sheep and of arable crops with livestock is a key factor in making clean grazing systems work and improving grassland management. The integration of sheep with poultry allows poultry to range over much larger areas of land, reducing the potential for pest and disease problems while making effective use of the forage produced. The integration of bees with crops provides benefits in terms of higher crop yields.

Integration runs counter to the trends in conventional livestock production towards specialist dairy and beef herds and sheep flocks, although a strong link between milk and beef production has long been a characteristic of British livestock farming. Preston & Vacarro (1989) have argued that the demand for milk and beef, which is usually in a ratio of about 4 : 1, could be more efficiently met by restricted suckling of milking cows. Increases in the biological efficiency of such an integrated system arise from increased milk production of the cows, found with *Bos indicus* (Zebu) cows to be about 27% due to an increase in lactation length, and increased health, in particular less mastitis incidence and reduced stress to cow and calf. Increased incidence of silent heats does occur with restricted suckling, but this can be compensated for with natural service.

The further development of integrated livestock production will be a major feature in improving organic livestock husbandry and in the development of sustainable livestock productions systems (Phillips, 1989).

REFERENCES AND FURTHER READING

General

Blake, F. (1987) *The Handbook of Organic Husbandry*. Crowood Press

Chamberlain, A. T., Walsingham, J. M. & Stark, B. A. (1989) *Organic Meat Production in the '90s*. Chalcombe Publications; Maidenhead.

Haiger, A. , Storhas, R. & Bartussek, H. (1988) *Naturgemässe Viehwirtschaft*. Verlag Eugen Ulmer; Stuttgart

Koepf, H. , Petterson, B. & Schaumann, W. (1976) *Biodynamic Agriculture—An Introduction*. Anthroposophic Press, Spring Valley, NY

Phillips, C. J. C. (1989) *Sustainability in cattle production systems*. Proceedings of 3rd Egyptian-British Conference on Animal, Fish and Poultry Production, Alexandria, October 1989. pp. 23–31

Preston, T. R. & Vacarro, L. (1989) *Dual purpose cattle production systems*. In: New Techniques in Cattle Production (ed. C. J. C. Phillips) pp. 20–32. Butterworths Scientific; London.

Sambraus, H. H. & Boehncke, E. (eds) (1986) *Ökologische Tierhaltung*. Alternative Konzepte, 53. Müller; Karlsruhe

Schaumann, W. (1985) *Gesichtspunkte zur landwirtschaftlichen Tierhaltung*. In: Vogtmann, H. (ed.) *Ökologischer Landbau*. Pro Natur Verlag; Stuttgart. pp. 69–77

Vogtmann, H. *et al.* (eds) (1986) *The Importance of Biological Agriculture in a World of Diminishing Resources*. Proceedings of the 5th IFOAM Conference, Witzenhausen, 1984. Verlagsgruppe; Witzenhausen

Problems of intensive systems

Bartussek, H. (1982) *Probleme der Massentierhaltung aus ganzheitlicher Sicht*. In: Kickuth, R. *Die ökologische Landwirtschaft*. Alternative Konzepte, 40. Müller; Karlsruhe

Boehncke, E. (1985) *The role of animals in a biological farming system*. In: Eden, T. *et al.* (eds) *Sustainable Agriculture and Integrated Farming Systems*. Michigan State University Press

Boehncke, E. (1986) *The role of animals in biological farming systems*. In: Vogtmann *et al.*, 1986, pp. 317–331

Ekesbo, I. (1981) *Some aspects of sow health and housing*. In: Sybesma, W. (ed.) *Welfare of Pigs*. Martinus Nijhoff; The Hague

Fox, M. W. (1984) *Farm Animals: Husbandry, Behaviour and Veterinary Practice*. University Park Press; Baltimore

Guillot, J. F. *et al.* (1983) *Anti-biotherapy in veterinary medicine and antibiotic resistant bacteria in animal pathology*. Rec. Med. Vet.159 (6): 581–90

Hodges, R. D. & Scofield, A. M. (1983) *Agricologenic disease—a review of the negative aspects of agricultural systems*. Biological Agriculture and Horticulture 1: 269–325

Isermeyer, F. *et al*. (1988) BST—*Technologie, Zusammenhänge und Folgen, insbesondere ökonomische, agrarstrukturelle, soziale und ökologische Folgen*. Institut für Agrarökonomie, Universität Göttingen

Lachmann, G. *et al*. (1984) *Einfluss einer chronischen metabolischen Azidose auf die Phago-zytoseaktivität neutrophiler Granulozyten beim Jungrind*. Archiv für experimentelle Veterin-ärmedizin, 38: 75

Lindquist, J. O. (1974) Acta Vet. Scand. Suppl.51: 1–78

Niederstücke, K. H. von (1982) *On the economy of health promoting measures in pig production*. Deutsche Tierärztliche Wochenschrift, 89: 370–373

Reid, I. M. *et al*. (1983) *Immune competence of dairy cows with fatty liver*. In: Swedish University of Agricultural Sciences—Proc. 5th Int. Conf. on Production Diseases in Farm Animals

Simantov, A. (1979) *Outlook for world agricultural markets—a general view*. In: Tracy, M. & Hodac, I. (eds) *Prospects for Agriculture in the EEC*. College of Europe; Bruges

Animal welfare and ethology

Dawkins, M. S. (1980) *Animal Suffering—The Science of Animal Welfare*. Chapman & Hall; London

Fölsch, D. W. (ed.) (1978) *The Ethology and Ethics of Farm Animal Production*. Birkhäuser; Basel

Fölsch, D. W. (1982) *Probleme der Nutztierhaltung und Aufgaben der angewanden Verhaltens-forschung*. In: Kickuth, R. (ed.) *Die ökologische Landwirtschaft*. Alternative Konzepte, 40. Müller; Karlsruhe, pp. 139–150

Kiley-Worthington, M. (1986) *Ecological ethology and ethics of animal husbandry*. In: Vogtmann, H. *et al*. (1986), pp. 339–358

Rist, M. (1978) *Müssen tiergerechte und wirtschaftliche Nutztierhaltung einander wiedersprechen?* In: Besson & Vogtmann (eds) *Towards a Sustainable Agriculture*. Proceedings of the 1st IFOAM Conference, Sissach, 1977. Verlag Wirz; Aarau, Switzerland, pp. 39–48

Rist, M. & Bär, M. (1986) *Basic principles of species-appropriate animal production and its ethical, ethological and constructional consequences*. In: Vogtmann, H. *et al*. (1986), pp. 332–338

Sainsbury, D. (1986) *Farm Animal Welfare*. Collins

Singer, P. (1975) *Animal Liberation*. Random House; New York

UFAW (1988) *Management and Welfare of Farm Animals*. The Universities Federation for Animal Welfare Handbook. Bailliere Tindall

Nutrition

Deerberg, F. (1989) *Fütterung von Legehennen*. Bioland 1/89: 10–12

Ewing, W. & Haresign, W. (1989) *Probiotics UK. The guide to probiotics in the United Kingdom 1989*. Chalcombe Publications

Ferrando, R. (1968) *World Review of Animal Production*, 4: 16–27

Krutzinna, C. & Boehncke, E. (1989) *Zur Problematik der Natriumversorgung von Milchkühen in der Umstellungsperiode*. Bioland 2/89: 29–30

Lowman, B. (1989) *Organic beef production*. In: Chamberlain *et al*. (1989): 19–32

Younie, D. (1989) *Eighteen-month beef production: organic and intensive systems compared*. In: Chamberlain *et al*. (1989): 41–53.

Younie, D. (1990) *Organic beef in perspective – a comparison with conventional methods*. New Farmer & Grower 25: 22–23.

Züllig, M. (1988) *Der Einsatz inländischer Futtermittel in der Legehennenfütterung*. German IFOAM Bulletin 66: 4–9 and Schweizerische Landwirtschaftliche Monatshefte, 64: 288–303 (1986)

Housing

Carnell, P. (1983) *Alternatives to Factory Farming.* Earth Resources Research; London

Dawkins, M. S. & Nicol, C. (1989) *No room for manoeuvre.* New Scientist 16/09/89: 44–46

FAWC (1988) *Priorities in Animal Welfare Research and Development.* Farm Animal Welfare Council

Fölsch, D. W. (1986) *Ethological aspects of the behaviour of hens in relation to different housing systems.* In: Vogtmann, H. *et al.* (1986), pp. 396–403

Götz, M. *et al.* (1986) *Ethologically sound pig production.* In: Vogtmann, H. *et al.* (1986), pp. 380–386

Hannan, J. & Murphy, P. A. (1983) *Comparative mortality and morbidity rates for cattle on slatted floors and in straw yards.* In: *Indicators relevant to farm animal welfare.* Martinus Nijhoff Publishers; Den Haag. pp. 139–142

Nicol, C. & Dawkins, M. S. (1990) *Homes fit for hens.* New Scientist 17/03/90: 46–51

Rist, M. *et al.* (1987) *Artgemässe Nutztierhaltung. Ein Schritt zum wesensgemässen Umgang mit der Natur.* Verlag Freies Geistesleben; Stuttgart

Stolba, A. (1981) *A family system in enriched pens as a novel method of pig housing.* In: UFAW (1981)

Stolba, A. (1986) *Ansatz zu einer artgerechten Schweinehaltung — der 'möblierte Familienstall'.* In: Sambraus & Boehncke, 1986, 148–166

Thornton, K. (1988) *Outdoor Pig Production.* Farming Press

UFAW (1981) *Alternatives to Intensive Husbandry Systems.* Proceedings of the symposium at Wye College. Universities Federation for Animal Welfare; 8 Hamilton Close, South Mimms, Potters Bar, Herts.

Health, homoeopathy and herbal medicine

Bairacli Levy, J. de (1973) *Herbal Handbook for Farm and Stable.* Faber; London

Day, C. (1984) *The Homoeopathic Treatment of Small Animals — Principles and Practice.* Wigmore Publications; London

Day, C. (1988) *A Guide to Homoeopathic Treatment of Beef and Dairy Cattle.*

Farrant, T. (1989) *Lameness.* Farmers Weekly, 26 May: 58–60

Gross, W. B. & Siegel, P. B. (1982) *Socialisation as a factor in resistance to infection, feed efficiency and response to antigens in chickens.* American Journal of Veterinary Research, 43, 2010

Johnson, P. (1989) *Keeping (sheep) disease at bay.* Farmers Weekly, 27 January: 56–61

Macleod, G. (1981) *The Treatment of Cattle by Homoeopathy.* Health Science Press; Saffron Walden

Schukken *et al.* (1989) *Intramammary infection and risk factors for clinical mastitis in herds with low somatic cell counts in bulk milk.* Veterinary Record, 7/10/89.

Scofield, A. M. (1984) *Homoeopathy and its potential role in agriculture — a critical review.* Biological Agriculture and Horticulture 2: 1–50

Shuttleworth, V. S. (1988) British Homoeopathic Journal.

Spedding, C. R. W. (1982) *Concern for the individual animal.* Vet. Rec. Oct. 9 1982, p. 345

Spielberger, U. & Schaette, R. (1983) *Biologische Stallapotheke.* Verlag Lebendige Erde; Darmstadt, West Germany

Stopes, C. & Woodward, L. (1990) *The use and efficacy of a homoeopathic nosode in the prevention of mastitis in dairy herds – a farm survey of practising users.* English language IFOAM Bulletin 10: 6–10

Turner, F. N. (1954) *Herdsmanship.* Faber; London

Wise, J. W. (1978) J. Am. Vet. Med. Assoc. 171: 1064

Woodward, L. (1983) *Alternative medicine offers hope for mastitis problem.* New Farmer and Grower 2, Winter

Parasite control

Kettle P. & Boland, C. (1989) *Sheep solutions* Soil and Health (New Zealand), Winter 1988: 28–33

MAFF/ADAS (1981) *Clean grazing systems for sheep.* Booklet 2324

MAFF/ADAS (1982) *Grazing plans for the control of stomach and intestinal worms in sheep and in cattle.* Booklet 2154

Taylor, M. (1987) *Parasitic worms in ruminants and their control.* Paper presented at BOF seminar on parasite control in ruminants, July 1987. British Organic Farmers; Bristol

Taylor, M. (1988) *Anthelmintics: The choice and the strategies.* Farmers Weekly 6/5/88: 58–61

Williams, E. D. (1984) *Economics of controlling worms in sheep by prevention of pasture reinfestation.* Welsh Studies in Agricultural Economics 3: 41–44

Young, R. (1985) *Alternative sheep dip is allowed.* New Farmer and Grower No. 7, Summer, p. 8

Breeding and fertility

Augstburger, F. *et al.* (1989) *Vergleich der Fruchtbarkeit, Gesundheit und Leistung von Milchkühen in biologisch und konventionell bewirtschafteten Betrieben.* Landwirtschaft Schweiz 1: 427–431

Bakels, F. & Storhas, R. (1981) In: *The Research Needs of Biological Agriculture in Great Britain.* Elm Farm Research Centre Research Report No. 1. 54–62

Haiger, A. (1986) *Breeding dairy cattle for high life-time yields.* In: Vogtmann *et al.* (1986), pp. 359–368

Lotthamer, K. H. (1982) *Umweltbedingte Fruchtbarkeitsstörungen.* In: *Fertilitätsstörungen beim weiblichen Rind.* Verlag Paul Parey; Berlin

Murphy, C. (1988) *The 'new genetics' and the welfare of animals.* New Scientist, 10/12/88

Sattler, F. & Wistinghausen, E. v. (1985) *Der landwirtschaftliche Betrieb—biologisch-dynamisch.* Ulmer; Stuttgart. pp. 229–268

Schaumann, W. (1988) *Tierzucht und Tierhaltung am Beispiel der Rinder.* Lebendige Erde 6/88: 348–355

Tudge, C. (1989) *Variety in vogue.* New Scientist, 18 March: 50–53

Voitl, Guggenberger & Willi (1980) *Biologischen Land und Gartenbau.* Orac Pietsch; Vienna

Bees

Biodynamic Beekeeping. Star and Furrow 57, Winter 1981, pp.17–23

Bienen. Zum Beispiel, 11 June 1986

Corbet, S. & Prŷs-Jones, O. (1987) *Bumblebees.* Cambridge University Press

Recent publications

Boehncke, E. and Molkhentin, V. (1991) *Alternatives in animal husbandry.* Conference proceedings. University of Kassel, Witzenhausen.

Part Two

Organic Farming in Practice

The principles of organic farming set out in the first part of this book form the basis for a wide range of different farming systems. There is no one true organic farming system; each will be unique and adapted to the particular environmental, physical and financial constraints of the individual farm. In the following chapters, aspects of different organic farming systems in Britain and Ireland are described, illustrating the varied ways in which the principles and theory behind organic farming are applied in practice.

(New Farmer and Grower)

Chapter 9

Livestock Systems

It would be fair to say that the 'organic management' of livestock has been ignored by many organic producers, even though livestock form the backbone of many organic systems, because of the complications and uncertainties involved, particularly over health care. A number of producers, however, regard the organic approach to livestock husbandry as a major priority and have successfully overcome many, if not all, of the problems involved.

Dairying

Gareth and Rachel Rowlands, together with their sons Mark and Johnny and daughter Sian, farm 100 ha of land overlooking the coastal resort of Borth, near Aberystwyth. The Grade 3–4 land rises from sea level to a little over 100 m, and annual rainfall is about 1,100–1,400 mm. The farm has been run organically since 1948 and recognised as such since 1952. Before that, it was a traditional Welsh farm with few external inputs. The main enterprise on the farm is a 70-cow Guernsey herd, complemented by smaller beef and sheep enterprises. Forage comes from long-term Clifton Park and Cockle Park type leys, with short-term Italian ryegrass/red clover leys providing bulk forage for conservation. Cereals, mainly oats, are also grown and meet most of the supplemental energy requirements of the various livestock enterprises.

Management of the dairy herd starts with the calf from the day it is born. In summer, weather permitting, calvings take place outside, increasing the calf's ability to make use of colostrum and making suckling easier. In the field, when the cow is managing on her own, she will rest for a few minutes before rising to attend to her calf. During this time the blood flow through the umbilical cord declines naturally. These various factors contribute to increased resistance to disease on the part of the calf and fewer problems later on. Calving indoors, although usually necessary in winter, affects the health and vigour of the newborn calf and is avoided whenever possible.

Plate 9.1 Brynllys Farm, Dyfed – an organic dairy farm for over 40 years (A. Martin, courtesty of DBRW)

Colostrum is fed to the calf as soon as possible, as its ability to absorb it decreases every hour. The calf will remain with its mother for three to four days, although sometimes, when several calvings take place in a short space of time, calves may be grouped together on one cow, again after three to four days with the dam. The milk would not be tanked at this stage anyway and the care and attention which the calf receives from the cow during this period is seen as important to its future wellbeing, in much the same way as nurturing is to humans.

Afterwards, the calves move on to fostering using Welsh Black x Channel Island nurse cows, which have a good temperament and produce high value calves themselves. The young calves receive about one gallon of milk/day, sufficient to avoid scouring through gorging. The process of sucking is important to the development of the calf's digestive system, which is specially adapted to avoid milk going directly into the fourth stomach. This requires that the calf's head is at the right angle, tilted upwards, but most bucket feeding systems involve the calf putting its head down into a bucket. A raised bucket fitted with a teat could provide an acceptable alternative, but emphasis is also placed on the use of warm, fresh milk. The use of reconstituted or dried milk is considered unacceptable, because of the altered nutritional quality of the milk and the impact which this can have on the health and wellbeing of the calf.

Gradually, the calves are introduced to good quality hay, which is always available in front of them, although they only start eating significant amounts at about three to four weeks. Rolled oats are introduced at the three to four week stage, at a rate of about 0.1 kg per calf per day, sometimes with sugar beet pulp and seaweed meal as a mineral supplement. The calves are kept in groups of three or four in loose boxes, where they have plenty of space for movement and ample straw for bedding. Suckling continues until about 14–16 weeks, when the calves are weaned and new calves introduced to the nurse cow. There is usually some overlap at this point, with calves being weaned and new calves being introduced individually, rather than in batches.

During the weaning period, the calves are given ad lib, good quality hay or silage and cereals at about 1 kg per calf per day. The aim is that they should receive sufficient cereal, but not so much that it is constantly available to them. The emphasis is on fibre intake through hay and the development of a good capacity for forage intake in later life.

The first four months is the critical period in the rearing of replacements for the Brynllys herd. Although it is expensive, there are several justifications. There is a firm belief that the use of natural rearing results in the incidence of disease being kept to a minimum and increased longevity. Longevity is particularly important in a pedigree breeding herd as it allows good families to be defined. In addition, cows only really mature at three to four years of age. The Rowlands argue that the impact of this approach can be seen through their vet's bill for the whole farm, which in 1986/7 was only £540 (including sheep and beef as well). The cost of veterinary treatment per cow was little more than £7, compared with a national average nearer £15. Apart from Dictol, which is used for husk control, there is no worming or other medication for the calves and the Rowlands would like to put more emphasis on the use of homoeopathy to replace Dictol.

After weaning, the system of feeding aims to maintain a steady rate of growth until calving at 27–30 months. In some cases, the period may be longer or shorter than this, but the main determinant is whether the maiden heifer has sufficient growth on her to be ready for putting to the bull. Summer calvers receive no additional concentrates before calving, although some are fed in winter.

The calving policy is reverting to an all year round system, with a slight emphasis towards spring and autumn calving. This is largely due to the expansion of the retail sales side of the business and the need for a constant supply of milk. In the first lactation, the aim is to avoid pushing the heifers too much, because they are still maturing at this stage. They are served to calve again at about 13 months.

Winter feeding during lactation is based on about 7–9 litres from big-bale silage, fed at a rate of 55–60 kg per cow per day, with DM levels over 25% and metabolisable energy (ME) values between 10.2 and 10.3 MJ ME/kg DM. The aim is that the cattle should get the same amount of conserved forage on a daily basis throughout the winter, from the beginning of October through to the end of April. The uncertain conditions in the spring in recent years have created a situation where, although there is good growth in March and April, the regrowth does not materialise. As a result, the decision has been taken to rely on late turn-out after the winter. Formerly, forage rye was used as an early bite, but the emphasis is now on making additional silage, which is considered to be a more effective use of land than producing early forage crops.

The conserved forage is supplemented with a home-mix, consisting of about 40% purchased (conventional) protein balancer and 60% home-grown cereals. The aim is a concentrate feed containing 14–16% protein. The source of protein is usually English linseed, but may be some other protein balancer depending on relative cost. In some years, it is possible to feed silage ad lib. This allows fewer concentrates to be fed and for the homegrown supplies to last longer into the summer. The home-mix is supplemented with sugar beet pulp. It is fed at a rate of 0.5 kg/litre of milk over that produced from silage but, at a certain point in the lactation, concentrates are cut back. This depends to a large extent on the likelihood of producing over quota, although generally the aim is to feed towards expected yield and in this sense the system is no different from that practised on other farms.

Grazing in the spring is controlled and limited to 0.4 ha of fresh grass per day. This is supplemented with in-parlour feeding, again at a rate of 0.5 kg/litre for cows yielding over 16 litres. Keeping both fat and protein levels high is important, in view of the retail outlets for milk and milk products. As the homegrown cereals run out, more reliance is placed on purchased cake, but still within the Soil Association's 20% limit.

The cattle are loose housed in straw yards with ample bedding during the winter. Health problems are minimal. There are no problems with

Plate 9.2   Guernsey cattle at Brynllys Farm provide the right quality milk for Rachel's Dairy produce

hypomagnesaemia, largely due to the management of forage and the restricted use of fertilisers. Milk fever does occur from time to time, but it seems to come in bouts and appears to be associated either with forage from particular fields or with certain families in the herd. Occasionally, there are problems with pneumonia, but these are not significant.

Mastitis does occur from time to time and is treated using a combination of stripping out the affected quarter and hosing it down with cold water to stimulate blood circulation and the animal's own defence mechanisms in the affected area. The more virulent strains of mastitis require more intensive care, involving the withdrawal of concentrates and a change to high quality hay for feeding, as well as stripping up to four times a day to prevent heat building up in the udder. In the 35 years as a recognised organic farm, penicillin has never been used in the udder and no dry cow therapy is practised. Iodine teat dips are not used, although fly lesions are more serious in some years and cases of summer mastitis can occur. The cell count is usually around 300–350 thousand. Counts lower than this are regarded as an indication that mastitis problems are likely to occur. The cell count is as much a measure of the antibodies present and the ability of the animal's immune system to respond to an attack as it is an indication of the presence of the disease.

This non-conventional approach to health care has also been seen in the controversial area of warble fly control in the past. Organo-phosphorus compounds were never used; warbles were successfully cleared with derris. Records of infected animals were kept at the time and show a decrease on a monthly basis. The farm was clear by the time statutory notification was introduced and has been clear ever since.

The calving index for the dairy herd is under 380 days and the average yield (including heifers) is 4,400 litres/cow/year at 5.1% butterfat and 3.8% protein. Yields are lower now than they used to be due to a deliberate policy decision to maintain cow numbers and reduce individual yields, following the introduction of quotas. This decision was made so that output could later be expanded to make whole milk yoghurt, which is not subject to quota.

Clifton and Cockle Park type mixtures form the basis for the longer-term leys. The newer leys are generally used for forage conservation, and grazed at a later stage, although some of the oldest leys may also be used for conservation. After grazing in June and early July, the fields are topped to stimulate good regrowth. Big-bale silage is produced, because it fits better into an organic system where it is unusual for all the fields to be ready for cutting at one time and so produce sufficient material for clamping. In addition, both quantity and quality control are much easier. Big-bale silage is also well suited to the short-term Italian ryegrass/red clover leys which are used primarily for conservation, with three to four cuts possible in a season. These are grazed in the autumn/ early spring by sheep, which do not seem to have a significant impact on the red clover content of the ley. The red clover seems to thrive on the treatment and has even reached 80% in some years.

Plate 9.3   Big-bale silage is the preferred option for forage conservation at Brynllys (A. Martin, courtesy of DBRW)

Sheep and other stock are subsidiary to the dairy herd, but they play a very important part in the overall success of the system. The main role of the sheep is in keeping the pastures clean through tight grazing over winter. The flock consists of 125–130 ewes, producing 180 lambs of which 80% are sold to Soil Association symbol standard, i.e. before dipping in August, although they are not actually sold as such. In recent years there has been a move from buying in to rearing replacements, so that now more than one third are home reared. Routine worming is not necessary, because of the rotation, but conventional dipping is used to counteract the fly strike problem. In addition to the farm's own flock, 200 tack sheep are kept over the winter.

About 85 other cattle are kept, including dairy replacements and a few sucklers and cross-breds. The dairy herd replacement rate has not yet stabilised following quotas and is considered to be high at 15–18%. There is a gradual shift towards more beef calves and fewer replacements, with the emphasis on Limousin crosses.

The farm has a very diverse range of enterprises. In recent years, considerable emphasis has been put on processing, marketing and distribution of milk products under the Rachel's Dairy brand name. All the milk produced meets Soil Association standards. Milk which is not processed on the farm is supplied to the Welsh Organic Foods cheese plant in Lampeter, where it is turned into Cardigan and Pencarreg cheese. A horticultural enterprise has been developed on the farm on a share farming basis. The expansion of the marketing side of the business has given rise to opportunities for the Rowlands' two sons, Mark and Johnny, and daughter, Sian, to make a living from activities on the farm as well. At the same time, the complexity of the operations which have evolved has also demonstrated the need for considerable flexibility in any farming system. The days of a hard and fast approach to rotations and a rigid style of management are past and no longer appropriate to a dynamic business such as this.

A Lowland Beef System

Kite's Nest Farm, run by the Young family, is on the scarp slope of the Cotswolds overlooking the Vale of Evesham. The soil type ranges from blue clay at about 100 m, through fine silty soils, to Cotswold brash at 250 m above sea level. The predominant aspect is north-westerly, but because much of the farm is in the form of a valley a few fields are south facing.

Plate 9.4   Kite's Nest Farm, Worcestershire

The farm consists of 89 ha of arable land, half of which is fairly steep, and 67 ha of permanent pasture, made up of 8 ha of reclaimed quarried land which is too irregular to plough or even mow, 6 ha which are not ploughed because they are rich in rare and semi-rare wild flowers, and 53 ha which are too steep to plough. The farm is lucky enough to be surrounded on three sides by a large block of woodland, 34 ha of which are owned by the Youngs, protecting them from the risk of spray drift.

The farm is in two blocks, one of which has been farmed organically since 1974; the other was bought in 1980 and conversion to Soil Association Symbol Standard was effected on a field by field basis, with the whole farm having qualified by the middle of 1986. The principal enterprises are cereals and beef, the aim being an appropriate balance between cattle and cereals. The beef enterprise consists of a 65 cow suckler herd and about 40 ha of cereals are grown annually.

At its simplest level the rotation is a four year ryegrass/white clover ley, followed by two crops of winter wheat and one of winter oats, with the land being direct reseeded in early September. Considerable variation from this is now emerging as the cleanest and most fertile fields are identified. These, it is hoped, will be able to cope with the inclusion of a red clover/Italian ryegrass ley and come close to the theoretically possible level of five cereal crops in a ten year rotation.

Total cattle numbers are usually around 200, giving a relatively low stocking rate by today's standards. The major limiting factor as far as stocking density is concerned is the amount of grass which can be conserved as hay or silage and this is restricted by the fact that most of the permanent pasture fields are steep. In addition, the farm is critically short of buildings and with current uncertainties, despite organic premiums, there has been a reluctance to increase fixed costs. The cattle are out-wintered as much as possible, but usually spend most of the three winter months under cover.

Suckler cow systems are one of the easiest livestock enterprises to manage from the point of view of organic standards. Their strength comes from the fact that each calf is looked after and monitored by an animal precisely designed by nature to carry out such a task: the mother. Organic beef systems based on multiple suckling and even bucket rearing of calves are possible, but the further one moves away from rearing any animal 'as nature intended', the more difficult it becomes to do it organically.

After experimentation with both traditional and continental beef breeds, reflected in the variety of animals in the herd, the Welsh Black appears to do best on this grass system. There are no calving problems; the calves are hardy and thrive in the extensive conditions, producing lean carcasses of good conformation.

Looking after cows and calves can be quite taxing, especially in February and March, but the Youngs feel one should never be afraid to let the cattle decide what is best for them. It is vital to eliminate stress and keep the animals happy; that way they grow faster and become ill less often. Whenever possible they should have the run of more than one field, so that they can choose the most sheltered spots. This applies particularly in the spring and autumn.

Plate 9.5   The suckler beef herd at Kite's Nest Farm

Careful consideration needs to be given to calving dates; if large numbers of calves are born between the end of October and the beginning of April they will be very susceptible to virus pneumonia and scouring problems. Spring born calves will thrive, but cows which produce too much milk on spring grass require hand-milking of individual quarters. For this reason late spring calving is favoured for heifers, who will rarely have too much milk, and late summer calving for the cows. This also allows the heifers a two to three month recovery period after their first calf has been weaned before they join the rest of the herd on a regular 365 day calving index.

It is also found that there are immense benefits to be gained from allowing the cows to decide for themselves when to wean their calves. This is usually no more than three or four weeks before they calve again. The extra 'free' milk not only results in as much as an additional 55 kg liveweight gain compared with calves forcibly weaned at nine months. It also means that the calves have a high degree of 'finish' and only require time to grow to be fit for slaughter and to qualify for the variable beef premium.

Occasionally, a heifer may find it difficult to wean her calf without help. Where this is the case, the calf is weaned by separating it from the cow in a situation where the cow can still see and lick her calf. The cow and calf may be either side of a gate in a cattle yard or either side of a suitable fence in a small paddock. This helps to eliminate the stress that is otherwise caused, preventing the check in growth rate which usually occurs at weaning and avoiding the problem of teats ripped on barbed wire fences where cows make a determined and often desperate attempt to find their calves.

Slaughter weights for the steers are in the region of 350 kg dead weight and 285 kg for the heifers. Slaughtering takes place when the cattle are between 26 and 30 months of age. With only a few exceptions, the cattle receive no cereals at all. Each animal is taken individually to an abattoir at Evesham, where slaughtering is done by appointment to avoid stressful waiting around. The sides of beef are then returned to the farm, where they are jointed by a local butcher.

The cattle have been entirely free of stomach worm problems for the past seven years, even though no specific clean grazing system is operated. The main reason for this is almost certainly the fact that cows, calves and finishing cattle all graze together as part of a herd.

After four months of age, calves begin to develop resistance to stomach worm infestation, but cows and cattle over 12 months old have a high degree of immunity. Because of the relatively small amounts of grass grazed by the calves in their first four months, over 90% of the active larvae being ingested do not complete their lifecycles, and those few eggs that do hatch out are sufficient to stimulate the calves' immune system, but insufficient to cause clinical symptoms. The cows and older cattle act like mobile cleaners, mopping up infective larvae and keeping the grazing safe for the calves.

Other factors which have helped this farm to use no wormers since 1981 are the relatively low stocking rate and the fact that some arable land is reseeded each year, providing a completely clean area which some of the cattle can graze after hay-making during July, which is a peak month for stomach worm infestation.

Because of the way beef production generally and beef cows in particular have been seen to become unprofitable over the last few years, it is worth looking at why the Youngs have retained their suckler cows and how they fit into the farming system and contribute to its success. To do that it is necessary to look at the arable and livestock systems as interrelated parts of the whole farm. Exclusively arable systems may have a place in the future of organic farming, but mixed farms form the basis of the vast majority of successful holdings, with the livestock enterprises forming the most indispensable part. Most farmers understand the importance of this as far as soil fertility is concerned, but what often seems to be missed is the important role the grass/clover ley plays in weed control. The Youngs aim for a minimum of three years, even longer if weeds are a problem or fertility is low. Under organic conditions, white clover becomes very vigorous and prolific; mowing and grazing prevents weeds seeding and when ploughed up for cereals or vegetables fields can be almost entirely weed free.

For many years, the profit from the cattle has been fairly modest and their main justification has been their role in the overall system. More recently, however, the marketing of the meat has improved through the establishment of a farm butcher's shop. Because of the butcher's profit and organic premium, gross margins for the cattle are beginning to look quite healthy at around £430/ha for the sucklers and £600/ha for the fattened cattle, although the difference between the two is partly due to the somewhat arbitrary point at which the organic premium is applied. Even so, the Youngs' system is a clear illustration that the extensive, organic management of suckler beef can be profitable and rewarding.

LOWLAND SHEEP AS AN INTEGRAL PART OF AN ORGANIC SYSTEM

Sheep frequently form a vital and integral part of mixed organic farming systems; in this respect the management of the sheep flock by the Bests and the Chapmans at Manor Farm in Dorset is not unique. In other ways, particularly their commitment to the use of homoeopathy, their flock is a good example of the potential of an organic approach to sheep husbandry.

Manor Farm covers 105 ha of land. The main enterprise is a 70 cow dairy herd, together with both milling wheat for sale off the farm and feed cereals for use on the farm. The role of the sheep on the farm is to complement the dairy cows in the grazing rotations by:

- keeping the pastures healthy and grazing out certain weeds such as ragwort and docks;
- alternating the worm burden with those of another species;
- applying the effect of the 'golden hoof', particularly to reseeds, through aeration and dunging; and
- grazing down the autumn growth of forage crops, such as red clover, established leys and the proud corn crops, before the winter frosts, but without damage from poaching.

The sheep enterprise consists of about 100 ewes, selling off fat lambs, wool and occasionally some organically reared breeding ewes. The stocking rate is 12 ewes and 18 lambs to the hectare during the summer grazing season.

In order to avoid the use of anthelmintics, attention has to be paid to the grazing rotations throughout the year. Feet need to be regularly checked and trimmed and lameness dealt with as and when it appears. Where sheep are being folded over different crops of varying heights and consistencies there is a likelihood for scald to develop, a condition caused by abrasion between the claws, and also for damage to feet from the brittle stubbles. Particular attention has to be paid to fly strike during the summer as the non-organophosphorus dips offer no sustained effect against blowfly.

The aim is to operate a closed flock. Where the buying-in of stock is necessary, sheep are not bought in the market place so as to minimise the risk of importing disease. If at all possible, one farm (such as a hill farm) is found which offers a single source for incoming stock.

The sheep are not vaccinated as there is no history of clostridial diseases on the farm. Reliance is placed instead on the extensive use of homoeopathy, but if pulpy-kidney or another soil-borne clostridial disease did occur, then the sheep would be vaccinated. In 1988, 15 lambs were lost from pasteurella, due primarily to inadequate nutrition during pregnancy. The level of nutrition at this stage has since been increased. Acute diseases such as pneumonia, mastitis, eye inflammation and extreme foot troubles and abcesses can also be treated with homeopathy and remedies such as pulsatilla lycopodium, euphrasia, hepar sulph, pyrogen, apis and phytolacca have all proved useful for sheep ailments.

A considerable amount of time has to be allocated for fencing and a flexible electric system provides the possibility of moving the sheep around any part of the farm.

In the early autumn the lambs are taken from the ewes and weaned on to the aftermath of the red clover (grown as a forage crop). This provides excellent fattening and at the same time benefits the clover. The lambs are folded over it, tidying up each paddock before moving on. In a normal autumn in Dorset this can be gone over twice.

Plate 9.6   Lowland organic sheep production at Manor Farm, Dorset   (W. Best)

Manor Farm lambs are selected for slaughter as they grade and sold off direct to the consumer. There is considerable advantage in selling the lambs even as late as February, when the larger animal will fetch a good price and the lamb subsidy is high. However, this is not readily achievable with the Clun ewes currently kept on the farm because they tend to finish lighter and earlier. It may therefore be necessary for the breeding policy to be reconsidered.

The ewes are put onto a clean pasture to dry off and to bring them into a suitable pre-tupping condition. They will be checked through for suitability for breeding (feet, udder, teeth, general health and lambing records). The flock will then be gathered up and moved onto the flushing pasture four to six weeks before tupping, which is usually in early November. The flushing pasture can be an undersown stubble, a lush ley, or the red clover following behind the lambs.

The rams are also brought into good condition before tupping, with special attention being paid to feet. If a ram is hired, the same source is preferred each year to minimise the risk of introducing disease. Colour raddle is used to identify groups for lambing and infertile sheep. In early November the flock is divided according to breed into suitably sized groups, allowing one ram per 40 (approx.) ewes, and the rams are introduced. Grazing red clover has not proved to disrupt the oestrus cycle of ewes here and tupping groups folded over this crop have shown no fertility problems.

It is difficult in a growthy autumn to maintain the pregnant ewes in a good condition and to avoid them becoming too fat. The balance has to be found between the needs of the farm, in terms of grazing and treading, and the needs of the sheep. It is not necessary to give supplementary feed until the winter weather prevents access to grazing with snow and deep frost. This is a time to pay particular attention to the accessibility of water and to introduce silage or hay, depending on which is available. Silage is more popular with the sheep and less bulky in the rumen.

Where the sheep have done well through the winter, the silage or hay are of good quality and lambing is late, it may not be necessary to supplement feed, but the condition of the ewes is watched carefully and those with a low condition score are grouped separately and fed accordingly. Concentrates are fed to the ewes at a flat rate of 600 g per head per day for 6 weeks prior to lambing. The concentrate mixture consists of three parts homegrown oats to one part of a commercial high protein mix.

Preparation is made for lambing and the flock is brought in one or two weeks before the expected date. The lambing area allows plenty of room for daily exercise, to keep the animals in good condition and to avoid the build up of muscle toxins and the tendency to twin lamb disease. The use of homoeopathy for difficult lambings, with remedies such as arnica, aconite, veratrum album, caulophyllum and sabina, makes antibiotics unnecessary. Ample bedding straw is used to prevent the build up of disease, and pure iodine is applied to each lamb's navel immediately after lambing.

The pastures which are to be grazed after lambing are divided into six paddocks and the ewes and lambs rotated around these, spending approximately one week per paddock. A creep gate allows the lambs to graze the paddock one step ahead of the ewes and to benefit from an early bite of the clover. Paddock grazing encourages clover growth, resulting in increased weight gains in the lambs compared with set stocking. However, the six week rotation does not ensure clean grazing and it may be necessary to shut some paddocks for conservation and direct the flock to alternative grazing to avoid the build up of worms in July. If anthelmintics can be avoided, a natural resistance to worms develops.

Until 1989, all the sheep had to be dipped twice during the summer and autumn against sheep scab. Organophosphorus dips are not used on Manor Farm as they are considered to be too hazardous for the sheep, the shepherd and the consumer. Instead, Bayticol, a synthetic pyrethroid alternative is used. This offers no sustained effect against fly strike and close observation of the flock is essential during the summer so that any fly strike is detected and dealt with immediately. The wool is cut back beyond the area affected and the affected area is soused in an iodine product. This helps to prevent infection and deters further strike because of its strong smell. Where there is much fly strike it may be necessary to machine dag all lambs for easier inspection.

Although there are many advantages in including sheep in a mixed organic farm system, the Bests and the Chapmans acknowledge that they need considerable time and attention. Their presence necessitates extra labour on the farm, both for the actual shepherding and also for fencing and the maintenance thereof and the conservation of grass into silage or hay.

AN UPLAND BEEF AND SHEEP SYSTEM

Many upland farms find themselves trapped on the one hand by their unsuitability for tillage and crop diversification on any scale (because of topography, climate and soil type), and on the other by their relatively high rates of grass production and stock-carrying capacity. Walkerhill, a 36 ha upland farm in Galloway, combines a comparatively high proportion of good grazing land (about 30 ha or 84%) with a rather low proportion which is accessible to combine or forage harvesters (about 12 ha or 33% of the total). The farm covers four small hills and is composed entirely of fields sloping in at least two directions, often quite steeply. Some part of every field is not negotiable by tractor.

Currently the emphasis on stock-rearing farms is on ever-increasing stocking densities in an attempt to maintain margins. Given the difficulties of practising any meaningful form of clean-grazing system, when perhaps only 30% of the land is suitable for harvesting machinery, there is among conventional farmers an increasing dependence on anthelmintics and other drugs for animal health. There is also the tendency to get as much as possible out of the small amount of good land, which is cropped year after year for hay or silage. So two of the first questions to be tackled in setting about farming Walkerhill organically were what types and proportions of stock should be carried and what is the most appropriate tillage programme. In tackling these questions, specific aims were set down to work towards. Those which seemed foremost were:

- to achieve increased output from permanent pasture with a minimum of cultivation;
- to reseed all potential cropping fields with a grass/clover/herbal mixture;
- to minimise reliance on bought-in feeds;
- to explore ways of avoiding the use of drugs and feed supplements for animal health.

The farm lies between 90 and 130 m, right on the 1,250 mm annual rainfall contour. The soil is shallow, a light sandy silt but with some deeper pockets due, it is believed, to glacial deposits. The land is very stony. The soil temperature reaches 6°C at the end of March.

For many years the farm had carried only dairy cattle. Regular applications of $20:10:10$ had been made but little else had been applied. Consequently, the levels of available phosphate were very low in every field, although the insoluble phosphate reserve levels were satisfactory. The same was true for a number of important trace elements, particularly copper and magnesium. In recent years all available FYM had been used on two fields lying close to the farmyard, which had been cropped every year for hay. Soil pH, at around 5.5, was surprisingly uniform across the farm.

All of the fields which were not scheduled for reseeding within the first five years underwent a programme of lime, rock phosphate and calcified seaweed, spread over several applications. As a result of various trials it was found that smaller, regular applications of lime (not more than 2.5 t/ha at a time) are preferable to large, single applications, especially in light soils with little buffering capacity. This, coupled with careful grazing management, has resulted in a visible, steady improvement in the composition and vigour of the pasture. Unfortunately the Scotch thistles now reach waist high every summer, and they bristle with health too.

The decision to carry sheep and suckler cows as the predominant types of stock was made early on, but the numbers of each depended on many factors, which included:

- the amount of winter feed which could be grown (cereals and fodder);
- the grazing systems used; and
- the relative profitability of sheep and beef cattle.

In the event, an even split was decided upon, counting one cow as equivalent to five sheep, and there are now 18 suckler cows and 90 breeding ewes.

After experimenting with different calving dates over five years, the traditional calving months of April and May were chosen as being the most satisfactory, for the following reasons:

- Cows are more easily settled to the bull in July.
- Late spring calving conforms closely with the animal's natural nutritional and feeding cycle. Requiring the cow to calve and begin milk production in the autumn, at a time when its own system is trying to build up mineral reserves for the winter, places additional stress on the cow.
- It worked best in practice. In 1986, 16 of the 18 cows calved within a three week period and, in 1987, 15 did. In 1988, all the cows calved within four weeks. This gives very even batches of calves, which can be managed efficiently and which sell well as stores at 12 months old in a strengthening market. Also, there has not yet been an identifiable case of hypomagnesaemia, something which claims several cows per year from some of the neighbours.

After calving the cattle are set-stocked in small batches for two months, until they are due to be put to the bull in July. A balance between sheep and cattle in each field is aimed for at this stage. Once all the cattle are together, and the sheep are no longer suckling their lambs to any great extent, they begin a rotational grazing sequence—cows and calves, followed by ewes and lambs, followed by ewe hoggets—spending about two weeks in each field. During the autumn the cows and calves together graze each field in turn quite heavily before they are weaned and settled into the winter routine in November. This gives them an enormous range of plant varieties to eat, and it leaves the fields tidy, with just enough time for some late season growth to carry the sheep through the winter.

After trying several different systems, all cattle are now wintered outside in two fenced, well sheltered enclosures of a bit under 1 ha each. Both of these open onto a concrete court at the edge of the farmyard where fodder and concentrates are fed. The courts are scraped down into a 'dungstead' which is large enough to hold the whole winter's FYM. The dungstead is not yet covered but will be as soon as finances permit.

From weaning until early March the cows are fed only on silage and straw. Four to six weeks before calving, depending on condition, they also get 1 kg per head per day of a home mixed, barley based concentrate. This continues through calving until turnout in the second week of May, when only straw is available until they no longer eat it.

The young stock also are fed on silage with barley straw and 1.5 kg concentrates daily from November onwards. If there is enough barley this provides the basis for their concentrates too, but often it has to be supplemented with a proprietary beef nut. The stirks are sold in two or three batches during March to May. The aim is to have none of the previous generation left by the time of turnout, mainly because of grazing pressure.

Lambing begins in the third week of March and batches of ewes and lambs are moved and set-stocked thereafter, along with the cows. Mixed stocking seems to make much better use of the lush spring growth, sheep being very selective feeders in times of plenty. The ewes are fed up to 0.5 kg of a whole-barley based concentrate, and hay ad lib, from the beginning of February until well into June. Although it seems unnecessary to feed them so late, it does ensure a good start for the lambs and it seems to pay off later in the season.

The lambs are weaned in the first week of August with the first few being sold for slaughter at that time. They are weighed regularly thereafter and sold as they reach condition, aiming for a dressed carcass weight of 19–20 kg. So far, all of the lambs each year have been sold for slaughter rather than as stores.

For six weeks after weaning, the ewes are held back on rather thin grazing. This is both to assist their drying off and to permit some good

Plate 9.7 Sheep gathering for feeding in winter at Walkerhill Farm   (S. Biggar)

regrowth ahead of them. They are flushed on grass alone and put to the ram at the end of October. From then until supplementary feeding begins in February they eat just about everything there is.

As a sideline, outdoor pigs have been by far the most successful. Berkshires have been 'strip-rotated' along a number of steep bankings and overgrown corners. Electric netting keeps them in well at all times, except when one comes into season the day before expected. They usually farrow with no intervention and often make a nest of bracken or reeds if the weather is mild; otherwise they use their portable wooden huts. The male piglets are castrated, but they get no other treatment whatsoever. The litter is creep-fed from six weeks and weaned at eight, at which time they are housed. If feed has to be bought for them, then sow-meal with no hormone additives is bought. Almost without fail the sows come back into season three days after weaning and hold to the first service.

At one time in this part of Galloway there was a great deal of arable cropping (principally of wheat, oats, potatoes, swedes and turnips) but now there is practically none. Even the single field of spring barley is becoming rare, and reseeding is done by herbicide and direct reseed with no break whatsoever. Various different systems have been tried in the four fields which have so far been reseeded, but the one that has been chosen is the simplest of these.

During September, while the land is still quite dry, the field is sub-soiled with a home-made machine, aiming for a depth of 45 cm, although this is sometimes impossible to achieve. Cultivation in February is followed by 2 t/ha of lime, and spring barley drilled in the first week of April. On emergence this is undersown with 20 kg/ha of Italian ryegrass and late flowering red clover. After harvest the field is kept for flushing the ewes, then grazed lightly throughout the winter. In early February it is grazed bare by the sheep then ploughed again. Sheep achieve a very even dung distribution, but if conditions permit, 12–15 t/ha stockpiled FYM is spread on it too. Ploughing is as shallow as possible, certainly not more than 15 cm, and the severe grazing beforehand helps to minimise any regrowth of the old crop.

Rock phosphate is applied at a rate of 0.5 t/ha before a second crop of spring barley. The variety used is often Patty, which has never lodged nor been prone to disease, and which seems to yield well without artificial fertiliser. Growing barley in this way is not economic, by any common standard. It does provide the basis of the home mixed concentrate, and it is certainly cheaper than buying organic grain from hundreds of miles away. This second crop of barley is again undersown, this time with the grass/clover/herb mixture which has been developed on the farm (Table 9.1).

Table 9.1 Ley mixture used on an upland organic farm in Galloway.

kg/ha	
3.5	Frances Perennial Ryegrass (or e.g. Premo)
3.5	S23 Perennial Ryegrass
3.0	Melle Perennial Ryegrass
4.0	Whisper Italian Ryegrass (or e.g. Combita)
3.0	S26 Cocksfoot
2.0	S37 Cocksfoot (or e.g. Cambria)
2.0	Kampe II Timothy
3.5	S215 Meadow Fescue
1.5	Late Flowering Red Clover
2.0	Kent Wild White Clover
3.0	Alsike Clover
2.5	Burnet
1.0	Chicory
0.5	Yarrow
35.0	

The reseeded field has usually been grazed lightly in the first winter and conserved for silage the following summer. Experience has shown, however, that 'unfertilised' seeds are rather slow to tiller, and cropping in the first year may be a bad practice. Grazing the field in its first year and cropping it only in the second came up with rather better results. Now that four fields have been reseeded, and there is therefore greater choice about which ones to use for (organic) hay and silage each year, it is a practice that will be followed in future.

It would appear that the natural fertility of a cultivated field takes rather longer to recover than that of a permanent grass field to which chemical fertilisers are no longer applied. This may be due to the low humus levels of regularly fertilised and cropped land, the length of time taken to develop a good root system with mycorrhizal associations, and consequently poor plant nutrition, low plant density and slow tillering. Rebuilding the biological activity in the soil must be a gradual process using composted FYM and small, regular applications of lime and rock phosphate which do not dramatically disturb the balance of the soil. On the other hand, the experience shows that after perhaps five years a grass/clover/herb ley with no inputs is at least as productive as a conventional ley receiving perhaps 150 kg N/ha per year.

The effect on animal health of any particular aspect of organic practice must be almost impossible to measure, except in very specific cases. At Walkerhill, a principal reason for outwintering all stock, for example, was to avoid the possibility of pneumonia and the consequent need to use penicillin. This seems justified by the fact that over the period 1984–1987, penicillin has been used on a bovine animal only once, and that was administered by the vet to a heifer after a difficult first calving.

Other factors which support the idea that a balanced soil will support healthy livestock are:

- good conception rates in both cattle and sheep, and
- the diminution of a trace element problem which was inherited.

The tightness of calving pattern has already been mentioned, and this pattern is reflected by the sheep also. Perhaps a more significant indication of the beneficial nature of good pasture is the high conception rate achieved by the sheep. Flushing on grass and clover in October has given rates, as measured by scanning in February, of 170–175%, which is high for Scottish Blackface sheep. On this pasture, they also seem able to maintain their condition later into gestation.

The suckler herd was begun by rearing Angus x Friesian heifers alongside beef store cattle. All of these were bought in as young calves. By the time the first batch of these had been on Walkerhill for one year and the store cattle were approaching market, they began to show signs of severe copper deficiency. Soil analysis from Elm Farm Research Centre had shown that copper reserve levels on the farm were in fact high, so the problem was obviously one of availability. Bloodtests are still done every year on two or three cows, as the only way properly to monitor copper levels, and it is still found necessary to administer copper directly to both cattle and sheep, but thankfully the problem is diminishing year by year. 'Solway pine' was a condition well known in Galloway in the past and seems to have been due to a combination of copper and cobalt deficiencies. It is possible that it will take a long time to make mineral supplements unnecessary altogether.

Drenching and dipping remain the two greatest problems in aiming for a whole-farm organic system. The problem in young stock is particularly serious. It is known that severe worm infection at an early age can lead to physical damage to the intestine and stunted growth, the effects of which are never overcome. A stock-rearing farm survives or fails on the health of its young stock and it has had to be accepted that there are times when anthelmintics may be necessary. This relates also to the difficulties in practising a clean-grazing system which were mentioned earlier. Neither the sheep nor the cattle are drenched routinely but the lambs have to be dosed at some stage of the late spring or early summer and the calves in early autumn. There are patches of comfrey and garlic in each field which have to be transplanted regularly from the garden because the animals keep grazing them out, but the use of these alone seems inadequate to deal with what is basically a problem of overstocking with too few kinds of animals. Its resolution may involve a marked change in the types of stock which are carried.

Walkerhill is now in one of the Environmentally Sensitive Areas declared in 1987, and this may provide opportunities for developing the proper management of the woodland on the farm. Overall, there is no doubt that the farm is on a much more secure footing than it would have been had the predecessor's previous conventional system been continued. It would be possible to reduce all inputs other than those for maintenance and machinery almost to zero without seriously affecting the farm's income for a number of years. The farm consists of a group of systems, for reseeding, for grazing management, for forage conservation and for winter feeding, none of which is ideal in purist, organic terms but which go just about as far in that direction as it is possible for an upland Galloway farm to go.

Free-range Organic Pig Production

The organically run free-range pig unit at Sir Julian Rose's Path Hill Farm was established in 1985 to expand the 'mixed farm' foundation of the 105 hectare holding.

The farm is mostly stony, grade 3/4 light sandy loam over chalk on the southernmost point of the Chiltern Hills. Conversion to a fully organic enterprise started in 1975 and was completed in 1985. The principal features of the farming system base themselves around the overall need to build high levels of fertility into this hungry land and consequently boost the overall productivity per unit.

The farming system consists of a 50 head Guernsey milking herd producing organic unpasteurised milk and cream for a daily local delivery round; some 35 beef cattle (per year) resulting from crossing Sussex, North Devon and Hereford on those Guernseys not bringing on dairy replacements; 600 hens annually ranging over medium term mixed grass leys and 1000 table poultry per year raised from day old chicks and taken on to 2/3 kg weights some 5 months later. Free-range turkeys are also raised for the Christmas and Easter trade. A recent addition is a 100 head flock of ewes which gives a boost to the overall fertility building programme while producing a crop of lambs for the late autumn organic meat trade.

Broadly speaking, the farm comprises some 28 ha of permanent pasture with the other 76 ha consisting of 3 to 4 year grass/clover leys followed by 2 years of cereals and a break crop of roots or legumes. All livestock rations are based on home produced wheat, barley, oats, roots, hay and straw with the addition of some bought in dried lucerne nuts. A mobile miller mix unit is used to make the rations up for the various requirements of each enterprise. Skim milk, from the cream separating process, provides a valuable boost for fattening the pigs, table poultry and turkeys.

The free-range pig unit consists of 20 Canbarough Blue and Gloucester Old Spot sows and gilts producing 2.25 litters a year each, usually comprising an average of 9 piglets per litter. These run with the mother to a separate electrically fenced paddock or, in poor weather, to strawed down, covered housing.

Rations are based on a combination of oats, barley, beans, wad of straw and skim milk when available. Soya is also included as 8% of the ration. Weaners are fed on an ad lib basis starting at about 0.5 kg a day and rising to 1.5 kg a day as weights reach 45 kg. Porkers will be ready for killing at 63–68 kg within approximately five months. This compares favourably with indoor systems. Farrowing sows receive 2.3 kg per head plus about 0.25 kg per piglet.

The entire outdoor unit (at present 2 ha) will move on once every 18 months, after which 2 years of crops will be grown on the original site. The animals also have autumn access to a 1.6 ha beech/oak wood adjoining the paddocks. They pan for acorns and root amongst the leaf mould, thus picking up additional beneficial foods. Paddocks are electrically fenced using a mains supply. This has proved largely effective. But from time to time, piglets have got out between wire strands or, where flexinet was used, through the netting.

Plate 9.8   Organic free-range pigs at Path Hill Farm, Berkshire
(M. Carpenter)

Standard portable shelters are used, with the addition of individual farrowing huts which incorporate a metal bar to deter the sow from rolling onto her piglets. Drinking troughs, hooked into the nearest outlet, provide a constant supply of fresh water. None of the pigs are ringed, but their rooting ability is dependent on ground conditions. In a drought season hard ground conditions prevent rooting and only grazing is possible. In these conditions some supplementary minerals are provided. The paddocks are sown with a mixed ley comprising perennial ryegrass, cocksfoot, timothy, meadow fescue and white clover. This provides a rich diet although a simpler ley of red clover and Italian ryegrass would suffice.

The marketing aspect of the enterprise centres around selling into the organic meat trade and, as such, has a number of variables. One of the main difficulties is to develop a steady and consistent production pattern which the buyer can rely upon for continuity of supply. This is hard to fulfil, and would be considerably easier should other mixed organic farms come on stream in the area and thereby enable the development of a co-operative system for meeting requirements with round the year production.

At present, sales are concentrated on shops selling organic meat and individual customers. Premiums are set at approximately 15% above quality Meat and Livestock Commission prices. Pigs going to shops are usually sold whole or split, whereas for local customers jointing and packing is carried out on the farm premises using existing staff. The main difficulty, which faces all producers of organic meat, is the lack of a developed marketing infrastructure. This does place severe limitations on the profitability of otherwise cost effective units. Porkers reaching prime weights must be caught at the right moment; otherwise the farmer will lose out through unwanted weight gains. Having to sell organically reared pigs through local livestock markets at standard prices severely limits the economic position.

To sum up, it is perfectly possible to raise consistently well proportioned, lean and healthy pigs on the outdoor system at Path Hill. The herd is disease free and no programme of worming is required.

Customer reaction nearly always rates them superior in flavour in comparison to conventionally raised animals and this must be largely due to the free-range conditions and the variety of general food intake. Aggression is very rarely expressed in the herd, and the sows, young stock and boars are quiet and easy to handle. Rooted out paddocks require cultivating or ploughing to get them back into suitable condition for re-sowing, and overall fertility gains are reflected in the yields of subsequent crops. So far, these have indicated that wheat, barley and oats when sown behind pigs will return good yields (from 4.2 to 5 t/ha), bearing in mind the quality of land on the farm.

FREE-RANGE ORGANIC EGG PRODUCTION

Egg production was a sideline at the Raymont family's Garlands Farm in Berkshire for many years, with around 50 hens being kept to supply the family and a few local customers. In 1983 it was decided that the farm needed to have a second enterprise in addition to the small Jersey herd, and free-range organic egg production was chosen. In the autumn of 1984, 460 birds were purchased and the numbers were increased to 730 in the spring of 1985. Based on the experience which was gained from this operation, it has been possible to draw up guidelines for producers potentially interested in organic egg production.

Much of the knowledge required to run a successful organic egg production enterprise is the same for any free-range unit. Since free-range production has increased markedly over the last few years, the Ministry of Agriculture has devoted considerable efforts to determining the requirements for profitable production, and full use should be made of this information. The Ministry Code for the welfare of domestic fowls (leaflet 703) was extensively revised and republished in 1987. It should form the starting point for any producer who desires to keep poultry in humane conditions, which is, of course, a cornerstone of organic livestock production. There are, however, special features of organic egg production, and these deserve attention.

A flock of 1,000 hens will consume the grain produced on some 7 ha of land, while themselves occupying only 1.5–2.0 ha of grassland (depending on stocking rate). If they are to justify this consumption of valuable organic grain, they must clearly play their part, not only in the economics of the farm, but also in its nutrient budget. There is no place for overstocked, over manured permanent chicken fields.

1,000 hens will produce 95 t of fresh droppings in a year, and these must be utilised efficiently. If the hens are kept in large buildings with over 1,000 birds, as is increasingly common, they will spend most of the time inside, and so most of their droppings will be inside and these can be composted and used where most needed. This system, however, surely contradicts the principles, though not the regulations, of free-range production which should be based on access to and consumption of good quality green food. This can only be achieved through the use of smaller units; for this reason and because a smaller colony size suits the social behaviour of the hens, the organic standards specify a maximum of 200 birds and a preferred unit of 100 birds.

With these numbers per house, good grass over a large part of the year can be achieved in two ways, either by a system of paddocks or by using mobile houses, as has been practised at Garlands Farm. With the former, the area around each house which will initially become denuded of grass can best be treated as a straw yard (Figure 9.1) and the separation of the individual yards will greatly lessen the mixing of birds, even where the grazing area is undivided. The latter is by far the most flexible. For example, the hens could spend the summer on the last year of a grass ley, and the winter on an undersown stubble, boosting its fertility before a spring sown grain crop or vegetables. The permutations are endless and highly beneficial, but the capital and labour costs are obviously higher than in a static system.

Figure 9.1   Illustration of layout for static free-range poultry system.

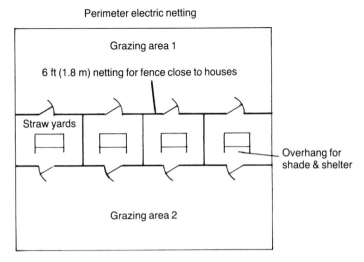

Adequate fencing is essential except in the rare instance of a fox-free area where loose dogs are never encountered. It can be achieved either by six-foot high wire netting or by electric netting. This must be of a smaller mesh than the normal sheep netting and a little higher, unless the hens' wing feathers are clipped. Electric netting is effective against foxes. If cattle are to be grazed nearby it is essential that wire netting fences are protected against them with electric or barbed wire guard fences, otherwise an extremely expensive investment will soon be in ruins. Electric fencing is obviously the only solution on a mobile system, and if houses are frequently moved, a relatively small length of electric fencing will be required and a higher stocking density can be employed.

Provision of clean water is essential and this means either a drinker inside the house or troughs designed so that the birds cannot stand on the edges. A frequently moved system will probably require a mobile water tank to supply the drinkers.

*Plate 9.9   Free-range
poultry at Garlands
Farm, Berkshire
(N. Rebbeck)*

The housing type will depend on decisions made as to how the laying flock is to be managed in the farm context as discussed above. Organic standards recommend 15 kg/m^2, but allow up to 25 kg/m^2, given additional shelter against wind and rain. The houses must be big enough for the birds to feel happy staying indoors in bad weather, as otherwise egg production will suffer, food intake will increase and the eggs will become dirty. 25 kg/m^2 is adequate for sleeping quarters, but extra shelter outside the house becomes essential as the birds will not stay willingly in such close proximity in daylight hours.

If houses are moved regularly, perhaps twice weekly in winter and once in summer, they may have slatted floors without dropping boards, the chore of cleaning out is avoided and litter will not be required except for the nest boxes. The houses will have to be as light as possible while being strong enough to withstand being moved about, and will need to be mounted on sleds or wheels, with a simple device to allow a rapid connection and disconnection from a tractor. For seasonal movement only, the houses will still have to be strong structurally and will also need either solid floors or a slatted or weldmesh area and dropping boards.

All houses should have at least 15 cm perch space per bird, and at least one nest box for every eight birds. These should be lower than the perches to reduce the tendency of birds to roost in the nest boxes. A roll-away system which could be based on old battery cages, can help very much with the wet weather problem of dirty eggs. Feed (and preferably water) should be available in the houses in feeders that avoid the possibility of it becoming fouled.

There are many chicken houses available for smallholders and also for large unit sizes, but choice is limited for the 100–200 colony size that is advocated in organic standards. There are probably none to fit the requirements for a readily moveable house. On-farm design and construction, preferably after taking a good look at what is commercially available, may be the best answer. Minimising ledges and cracks on the inside facilitates cleaning and helps to keep down dust, parasites and disease. Creosoting before restocking a house is usually the most cost-effective way of disinfecting and prolonging the life of a timber house. Plenty of ventilation must be available, with shutters so that it can be restricted in bad weather.

If the hen is to produce economically she must be fed as near as possible to her ideal diet. Since, in an organic system, her feed will be based on valuable organic grain, there can be no place for casual diets low in energy, protein or minerals. This presents free-range producers with a problem as the hen will need a varying ration according to the quality and quantity of the grass and invertebrates that she will be eating on range. It could be as wasteful to feed a constant 18% protein ration at all times as it would be to feed a low protein ration leading to excessive consumption or low production.

There are two ways round this problem. The normal approach is to feed an 18% ration except when grass is freshly growing, when 16% is adequate. The alternative is to allow the hen to be her own feed chemist and provide her with two or more different feeds from which to balance her diet.

Choice feeding of hens has been scientifically investigated over many years, much of the evidence being summarised in the British Journal of Nutrition 34: 363–373. This shows major advantages to be gained from choice feeding, both through better feed conversion and reduction of milling costs as one feed choice can be whole grain. The second choice available should be a high protein mix, preferably based (in an organic system) on home grown peas and beans with a little fishmeal, but otherwise on soya. Some milled grain could be included, especially if the mix is based on soya, to bring the protein level down a little. Minerals and vitamins would also be included, but the calcium supply could best be given as a separate third choice to avoid overeating of protein at times of high calcium requirement.

This system allows each hen to optimise her feed requirements over time as one is no longer fitting the ration to 'the average hen', and to take account of green food quality. The hen will select approximately 70% grain in her diet and this is typically wheat, though oats and to a lesser extent barley can be used. The ideal is probably to provide a mix of grains. With organic grain contributing 70% of the ration, some of the protein balancer should be of organic origin to satisfy more than just the letter of the standards, but a bit under half of the grain could be conversion grade.

A modern hybrid hen will almost certainly be required to make the best use of expensive organic feed. The heavier types such as the Shaver

ORGANIC FARMING IN PRACTICE

and Warren are preferred and are extensively used by free-range producers. The feathering on Warrens is usually softer and therefore sheds rain less well, but this is also affected by whether the pullets have been reared outdoors. The Arbor Acre, a large black hybrid originally bred for free-range use, is very suitable but less widely available than the other two strains.

Organic standards permit on a restricted basis the buying in of point of lay pullets so long as they are fed organically for ten weeks before their eggs are sold as organic, but the purchase of growers from symbol standard sources or, in restricted circumstances, from any source under ten weeks of age is obviously preferable. Producers who wish to rear their own replacements from day-old must be aware that this requires a high level of stockmanship. MAFF leaflets on small scale rearing and even natural hatching are still available, and should be consulted. Anecdotal evidence strongly suggests that pullets reared outdoors make better use of pasture and may well make healthier adults. If the routine use of coccidiostats is to be avoided, pullets must be raised on clean pasture and great vigilance must be used to spot the first signs of disease. Mixed choice feeding can be practised as soon as chicks are able to cope with whole grain, but to start with a chickmeal of conventional formulation is required, preferably based on organic grain.

As with any organic product it is especially important that the market should be well researched before commencing production. It is currently quite hard to persuade buyers that they should pay an additional premium over and above that for free-range eggs, but sales are expanding with the general recognition of the value of organic produce. There are many regulations covering the sale and grading of eggs which can be obtained from the Egg Marketing Authority. With the recent increase in free-range production, the government and EC have created standards and to use the term free-range it is now necessary to register with the Authority as a free-range producer. To summarise, hens must be stocked at not more than 1,000/ha, and the ground must be covered mainly with vegetation. A maximum of 7 hens/m² is permitted in deep litter type houses, or up to 25 hens/m² in houses with at least 15 cm perch space per hen. As can be seen these standards are somewhat lower than those for organic production.

Any eggs not sold retail at the farm gate or to a licensed packer must be graded; and if the free-range designation is used, must be in retail boxes with the size, class, packing date and packing station number of the farm on it. To register as a packing station the farm will have to have its grading and packing facilities inspected. With free-range eggs having to be packed in retail boxes anyway, the small additional cost of having a box designed for the farm with the organic symbol will soon be paid for in increased sales and customer loyalty. The latter also depends on grading being carefully carried out, rejecting poor quality shells which can still find a good market with caterers looking for the good flavour of organic produce.

The main difference between the costs of free-range and organic production lies in the cost of feed. The organic producer is unlikely to produce a feed for less than £180 per t compared with £145 for a conventional layers' mash. It is difficult to give precise figures, but Table 9.2 gives an idea of the variations possible while showing that the crucial factors in the equation are the level of production achieved and the price received. The figures are adapted from ADAS average production costs for free-range units. These figures lead to the profit figures in Table 9.3 for a 1,000 bird unit at various selling prices and laying percentages.

Table 9.2 Organic free-range egg production – costs per dozen eggs (p).

Feed cost (£/t)	145	180	180
Average laying percentage	70	70	60
Feed	32	40	47
Labour	20	20	23
Livestock depreciation	8	8	9
Deadstock depreciation	8	8	9
Packaging	10	10	10
Misc.	2	2	2
Total	80	88	100

Table 9.3 Profit on free-range eggs for a 1,000 hen unit under different % lay and price assumptions.

% lay	70	70	60	60
Price/doz. (£)	1.00	1.20	1.00	1.20
Costs (£)	18,740	18,740	18,250	18,250
Revenue (£)	21,290	25,550	18,250	21,900
Profit (£)	2,550	6,810	–	3,650

OVERVIEW

Although the emphasis in this chapter has been on individual livestock enterprises, their context within the whole farm system is just as important and should not be ignored. In particular, the need to balance different types of livestock for optimum nutrition and health care and to balance crops and livestock on the farm has to be borne in mind. In most cases, it should be possible to derive a system which is suited to the particular requirements and limitations of the individual farm, but that system may well be very different from those described here.

FURTHER READING

Chamberlain, A. T. *et al.* (eds) (1989) *Organic meat production in the '90s.* Chalcombe Publications.
New Farmer and Grower – numerous articles describe organic livestock production systems on farms including those covered in this chapter.

Chapter 10

Grassland and Fodder Crops

GRASSLAND

For most organic farmers in Britain, the clover/grass ley forms a central part of the farming system. On farms where the emphasis is on arable production, the ley will probably be seen primarily as a fertility restorer, whereas on all or mainly livestock units, it will be the cornerstone of the production system. In either case, it is desirable that the grassland achieves its maximum potential under organic management.

Grassland swards may be managed as short- or medium- to long-term leys or as permanent pasture. In each case, the constituent plants will be different, but in practice the resultant swards will usually be utilised by ruminants, whether by grazing or cutting. In some cases they may also be used for free-range pig and poultry production.

The role of legumes in the ley and the role of the ley in the rotation have been discussed in some detail in Chapters 3 and 5. The aim here is to concentrate on the actual management of the different types of grassland, including consideration of the component species, establishment, manuring, and grazing and conservation management.

The composition of the grass/clover ley

There has been a trend for many years (which is only now being challenged) toward pure grass swards including only one or two types of grass. In these situations, productivity depends primarily on the supply of soluble nitrogen from the bag. On organic farms, the ley not only provides food for livestock, it actively contributes to the fertility of the soil through the nitrogen fixation of forage legumes included in the more complex mixtures.

A typical ley on an organic farm will include a range of different species and varieties of grasses, clovers (or other forage legumes) and other species, generally referred to as 'herbs'. These complex mixtures, including the traditional 'Clifton Park' and 'Cockle Park' types (Tables 10.1 and 10.2), may often include seven or more components, but this need not necessarily be the case (see Colour plate 19). Simpler mixtures, discussed in more detail below, can also perform well in organic systems.

Table 10.1 Traditional 'Clifton Park' type ley mixture (Hunters, 1984).

Species	Variety	Quantity (kg/ha)
Cocksfoot	Roskilde	3.4
Cocksfoot	S 26	4.5
Meadow fescue	S 215	2.2
Tall fescue	Alta American	1.1
Rough-stalked meadow grass		0.6
Smooth-stalked meadow grass		0.3
Italian ryegrass	RVP	4.5
Perennial ryegrass	Reveille tetraploid	3.4
Perennial ryegrass	S 23	4.5
Timothy	Canadian Climax	0.6
Chicory		1.1
Salad burnet		2.2
Sheep's parsley		1.1
Ribgrass (plantain)		0.6
Yarrow		0.3
White clover	Grasslands Huia	2.2
White clover	Pertina wild	0.6
Alsike clover		1.7
Red clover	Altaswede	1.1
TOTAL		36.0

Source: Woodward & Foster (1987).

Table 10.2 'Cockle Park' type mixture with herbs.

Species	Variety	Quantity (kg/ha)
Cocksfoot	Cambria	5.00
Italian ryegrass	Trident	2.50
Perennial ryegrass	S 23	7.50
Perennial ryegrass	Frances	5.00
Perennial ryegrass	Melle	5.00
Timothy	Kampe II	3.75
Chicory		1.25
Ribgrass		0.60
White clover	Huia	0.60
White clover	Kent wild	0.60
Red clover	Altasweed	2.00
Alsike clover		1.25
TOTAL		35.00

Source: Arthur Evans, Llandysul.

Herbs are included for two reasons: firstly, they are thought to
contribute to the health of livestock by increasing the mineral content of
the sward (incidentally also a major benefit from the inclusion of
clovers); and secondly, the deep rooting habit of certain species (for

instance ribgrass and chicory) is considered to bring minerals into the upper part of the soil, thereby contributing to soil fertility. That neither of these characteristics is well supported by hard evidence does not deter the farmer from including such species in the ley mixture, even though they may represent 15–20% of the cost of the seed. Furthermore, both establishment and persistence of herbs in sown leys can be poor, and observations on farms have shown that many herb species, if they establish at all, will only be present in any quantity in the first year of a ley. This has led some researchers to consider that herbs should be sown and managed separately in strips. Newman Turner, author of a classic ley farming book, *Fertility Farming* (Faber, 1951) and a devotee of the inclusion of herbs in ley mixtures, often claimed that what many people felt were weeds in grassland should be encouraged and used positively to contribute to fertility of the soil and health of livestock.

The choice of a suitable mixture of species and varieties of grasses, legumes and herbs can be very important in determining the success of a ley, and must be made in the light of a number of contrasting and often conflicting factors. These include the intended lifetime and purpose of the ley, the environmental conditions, the compatibility between species and the cost of seed.

Grasses

Improvements in the productivity of grassland have in part been the result of the efforts of breeders, although the breeding priorities of the 1960s and 1970s have not been consistently advantageous to organic farmers. For example, the aggressive and nitrogen demanding growth of modern highly productive ryegrasses may mean that they outcompete other grasses in the more complex organically managed leys, although clovers are also capable of outcompeting some of these grasses. Furthermore, little emphasis has been placed on good early season growth in the absence of applications of nitrogen. Compatibility with clovers, improved persistence and higher quality are now major objectives of breeders, and all three could play a major part in providing varieties which are better suited to the organic ley.

There are essentially two types of ryegrass, short duration and perennial. Of the short duration ryegrasses, the biennial Italian ryegrass is now one of the most significant grasses used. It will establish rapidly, with early spring (for grazing in June/July, early autumn and providing early bite the following spring) and late summer (for early spring grazing) both being good times for establishment. A well managed sward will persist for two years; the greater yield of Italian ryegrasses compared with perennial ryegrasses is obtained at the expense of reduced winter hardiness, persistence and increased susceptibility to disease.

The annual Westerwolds ryegrass behaves essentially like a cereal. It will establish rapidly and is often used as a component of a catch crop mixture, sown after an early harvested crop. This will provide autumn and early spring grazing.

Other hybrid varieties suited to short duration leys have been bred from Italian and perennial ryegrass. The tetraploid varieties are very popular in conventional leys due to their high productivity under intensive management, as well as their higher palatability, better ability to withstand lax grazing, and higher digestibility in the first conservation cut. They are, however, wetter and slower to dry so they should be used only in small proportions in conservation leys. They do have a place in organic leys, but their aggressive growth may lead to heavy competition with some of the other species and varieties included in a mixture intended to perform well under organic management.

The short duration ryegrasses may usefully be included in mixtures for longer leys, since they provide a balance between good productivity in the first two years of the ley, with the longer term productivity being provided by other varieties and species.

Perennial ryegrass varieties are bred to be suited to both grazing and conservation, with varieties for early, mid and late season production according to time of heading. The late perennial ryegrasses have the advantage of persistence, winter hardiness, and resistance to diseases, although the greater persistence means they tend to out-compete clovers. Intermediate perennial ryegrasses are believed to combine the advantages of persistence with earlier spring growth, but not all the varieties in this class show early spring growth. They may be higher yielding for first cut, but overall yield is likely to be similar to the later perennials. The early perennial ryegrasses do give early spring growth, but at the expense of lower persistence and being prone to stemminess, and their performance in organic systems will be less good because of lower nitrogen availability in early spring. All varieties will persist in leys for several years, although first season growth and productivity will not be as high as the shorter duration ryegrasses.

Timothy can be an important perennial component of ley mixtures, being particularly leafy and productive in early summer when most other grass species will tend to start setting seed, reducing their feed value. Timothy is well suited to wetter and heavier land of higher fertility and is one of the most palatable grass species. It is very winter hardy, remaining green in winter, but spring growth is slow.

Meadow fescue is another perennial species which, though not particularly productive in the year of establishment, will perform well in subsequent years. Production of leaf is good after cutting, making it an ideal component of a conservation ley with aftermath grazing. It is a deeper rooting grass than timothy, and performs well on heavy land. Both timothy and meadow fescue can contribute well to early and late season grazing. Meadow fescue combines well with timothy and white clover for low-input systems.

Cocksfoot is one of the perennial species which has been lost from many modern simple leys. It can be an important component of leys on organic farms and it is claimed that its strong and deep rooting characteristics make it particularly beneficial to soil structure and restoring fertility to tired soils as well as conferring the advantage of drought resistance. It is a very productive species, although the palatability is not as high as that of the ryegrasses and tight management is necessary to maintain herbage quality. Cocksfoot is particularly well suited to mixtures for silage production since if too mature when cut it may out-compete clover.

In the first year of a series of NIAB trials, timothy, cocksfoot and tall fescue were lower yielding than the average of six perennial ryegrass varieties, but were higher yielding in the second year, especially in the east of the country where the difference was as high as 18%.

White clover

In the longer-term leys, white clover is the most significant of the legume species, and a variety should be selected which is suited to cutting or grazing. Careful management to maintain white clover in the sward is very important. The contrasting growth habits of grass and clover, the susceptibility of clovers to diseases such as *Sclerotinia*, and their high demand for potassium and phosphate, while being sensitive to and easily out-competed by grass in high nitrogen conditions, mean that swards based on white clover can be variable.

The smaller leaved, late, prostrate varieties, such as Kent wild white and S-184, are well suited to hard grazing with sheep in both lowland and upland conditions. The earlier, less prostrate varieties such as Blanca, Olwen, and Kersey are distinguished by having relatively long leaf stalks, which raise the leaves towards the light, and larger leaves. They are generally less persistent than the smaller leaved varieties, but the long petioled varieties do offer the option of including conservation in the management strategy and are better suited to more lenient grazing (i.e. cattle with the occasional silage cut) or shorter-term leys. (See Colour plate 20.)

There is a close correlation between leaf size and yield, although under hard defoliation (such as intensive sheep grazing) the smaller leaved varieties still give the highest output. The true performance of a clover variety, however, depends on how it reacts with its companion grasses, the nature of defoliation, climate at the particular site and the *Rhizobium* strains present as well as the tendency of a particular soil to harbour *Sclerotinia*.

On organic farms, which do not benefit from the use of nitrogen to stimulate spring growth, the development of newer varieties which show better early growth is likely to be of considerable importance in the future.

Ley mixtures

Selecting the right combination of grasses and clover for a ley is a complex process—there is no universal recipe. The issue of compatibility between grasses and clover is particularly important, but still poorly understood. Trials at the Welsh Plant Breeding Station (WPBS) have shown that compatibility can mean a difference of as much as 30% in dry matter yield, which is largely due to the contrasting growth habits of the different components and, in particular, complementarity between erect and horizontal leaf types. The WPBS plans to publish league tables of the compatabilities of different grass and clover varieties which should make assessment easier.

In general, though, clover is favoured by perennial ryegrass varieties with low persistency and the open nature of the tetraploids. Varieties with upright growth habits make it easier for clover stolons to spread across the soil surface. The growth habit below ground is also important; where grass roots develop rapidly, they compete with clover for resources. Early flowering grasses may be better suited as companions to clovers because the aggressiveness of grasses declines rapidly after flowering, allowing the clovers to make earlier progress. Timothy and perennial ryegrass allow better establishment of clover than Italian ryegrass or cocksfoot.

Another important factor is the differing ability of similar-yielding clover varieties to fix nitrogen and transfer the nitrogen to grass. In trials at the WPBS, the variety Nesta, which is now outclassed, increased grass yields by nearly 2 t/ha compared with other varieties. The choice of variety also influences the residual nitrogen, with the smaller leaved varieties Gwenda and S-184 leaving greater quantities of nitrogen in the soil.

The proportions of clover seed in the ley mixture will have a bearing on the establishment of the ley. Best clover establishment is achieved with high seed rates of clover and a relatively low seeding of grass. Researchers now advise that seed mixtures should include 4 to 5 kg/ha clover to ensure adequate clover establishment.

Two contrasting ley mixtures are illustrated in Tables 10.3 and 10.4. The first is one which is better suited to grazing by cattle on reasonably fertile clay loams where lack of soil moisture is unlikely to be a problem. It is characterised by the medium-leaved, persistent white clover variety Menna and the high yielding, early spring growing, but less persistent large-leaved variety Alice. Kent wild white might also be a suitable variety, to give bottom to the ley. The perennial ryegrasses include two late varieties, Perma and the tetraploid Condesa, both of which have an erect growth habit and are clover compatible, as well as the intermediate variety Talbot. The timothy variety Motim has been included because of its high palatability, relatively early growth and persistence.

Table 10.3 Ley mixture for cattle grazing.

Species	Variety	Quantity (kg/ha)
Perennial ryegrass	Perma	7.50
Perennial ryegrass	Condesa	7.50
Perennial ryegrass	Talbot	5.00
Timothy	Goliath	3.75
Timothy	Motim	3.75
White clover	Menna	2.00
White clover	Alice	2.00
TOTAL		31.50

Source: Elm Farm Research Centre.

Table 10.4 Ley mixture for sheep grazing.

Species	Variety	Quantity (kg/ha)
Perennial ryegrass	Parcour	5.00
Perennial ryegrass	Perma	7.50
Perennial ryegrass	Condesa	5.00
Cocksfoot	Sparta	3.75
Meadow fescue	S-215	3.75
White clover	Menna	2.00
White clover	S-184/Kent wild white	2.00
TOTAL		29.00

Source: Elm Farm Research Centre.

The second example is a ley intended for grazing by sheep in drier conditions on lighter land. The small-leaved white clovers S-184 or Kent wild white which are better suited to grazing by sheep are included this time, although the high cost of varieties like Kent wild white may be a factor to consider. Of the perennial ryegrasses, the late variety Parcour has a good degree of persistence and winter hardiness, while Condesa is valuable for its drought resistance. Drought resistance is also a feature of the cocksfoot and meadow fescue. The variety Sparta is the most palatable of the cocksfoots, but even so cocksfoot may not always be an appropriate choice for sheep grazing.

The choice of ley mixture will inevitably depend on the circumstances of the individual farm and the purposes for which it is intended, so that it is not possible to prescribe suitable mixtures here. Further information on the choice of herbage varieties should be sought and can be obtained from the various NIAB recommended varieties leaflets as well as other sources such as Ingram (1985).

Herbs

The main 'herb' species usually sown in herbal leys include chicory, ribgrass (plantain), yarrow, salad burnet, sheep's parsley, cat's ear and caraway. A detailed review of these herbs and their use in leys can be found in Foster (1988) as well as various articles in the New Farmer and Grower magazine based on the same work. The main herbs are described briefly below.

Chicory is a deep rooting plant noted for its drought resistance and its mineral content. Although it has a low germination rate, once established it can be very productive, especially in early spring, and it provides important leafage in winter. The usual sowing rate in a mixture is about 1–2 kg/ha.

Ribgrass is a very palatable herb, again with a high mineral content, which does relatively better under poor conditions than ryegrass, although it can also be highly productive under better conditions. Its disadvantages lie in its spreading growth habit and its high water content, which can cause problems with haymaking. It would usually be included in a ley mixture at a rate of about 0.5 kg/ha.

Burnet in its native form is most commonly found as a component of calcareous grassland and it is often recommended for poor or chalky soils. It is deep-rooting and drought resistant, but does not compete well against more competitive species. Although not particularly rich in minerals, it is well liked by stock and is regarded by some sources as having medicinal properties. In modern ley mixtures, it is normally used at a rate of about 2 kg/ha.

Sheep's parsley is considered important for its contribution to the health of sheep, which may be due to the high iron and vitamin content. It lacks persistence, which may be due to overgrazing, or to the fact that it is largely biennial in nature. It is generally sown at a rate of 1–2 kg/ha.

Yarrow is a very persistent herb which can rapidly dominate the sward unless kept closely grazed. It has an extensive rooting system which makes it drought resistant and, as with sheep's parsley, medicinal properties are attributed to it. Sowing rate is usually 0.5 kg/ha.

It is worth reiterating that the cost of including these herbs in the ley is significant, and that their value is closely related to their persistence, which is generally low. In field trials at Elm Farm Research Centre, as well as fields surveyed on other farms, salad burnet and sheep's parsley often failed to establish and rarely persisted beyond the first year. Yarrow was persistent at low density in many fields, while chicory was the most frequently recorded species. However, though it rarely failed to establish, it seldom survived more than two to three years. Consideration should therefore be given to herbal strips in a field, sown with less competitive grass species such as meadow fescue, which would allow the benefits to be obtained under more carefully controlled conditions.

Sowing and establishment of the ley

There are two principal approaches to sowing and establishing a new ley. The first, undersowing a cereal crop, involves the selection of a mixture which will establish well and not too competitively with the cereal crop. Although the method has the advantage that the costs of seed-bed preparation are borne by the cereal crop, there may be less certain establishment of the ley, and the lower seed rate of the cereal may reduce yields of the cash crop. However, it has the advantage that late summer grazing is assured, with good early spring grazing and/or subsequent conservation.

Mixtures with a high clover content can be very well suited to undersowing, although there is a danger that vigorous species such as red clover may out-compete and swamp the cereal crop, making harvesting impossible, particularly in areas with relatively high rainfall. The timing of undersowing can be critical, and can vary between before germination of the cereal to some time after establishment. In general, earlier undersowing is to be preferred in spring sown crops, with the undersowing of winter sown cereals being carried out as early as possible in the spring. The undersowing of winter cereals is not usually recommended, however, except in the case of short-term red clover based leys.

Because timeliness is critical and ground conditions are not always ideal in the spring, some thought needs to be given to alternative ways of distributing the seed. On small areas of land, the hand-held seed fiddle could still be a viable option; it is cheap and allows the operation to be carried out when soil conditions would prohibit the use of more conventional mechanical alternatives. The sowing rate is about 1.3 ha/ hour with this method. On the more usual larger areas of land, two interesting options have been developed by farmers wishing to combine undersowing with other operations such as weed control. One of these is to attach a seed distributor to spring-tined harrows; this system has the disadvantage that, in order not to put too much pressure on the harrows, only a small amount of seed can be carried at any one time, and distribution of the seed may be affected adversely by wind drift. The second idea is to attach a seed box above a set of Cambridge rolls, so that the seed is spread at the same time as rolling the crop.

The alternative to undersowing cereals is direct seeding into a seed-bed prepared especially for the establishment of the ley, without the benefit of a cover crop. This can considerably improve the chances of establishing a successful and productive ley in comparison with under-sowing. If reseeding is carried out in the early spring with a mixture showing early and vigorous growth, then grazing and/or conservation will be available in the year of establishment. The late summer reseeding allows an early harvested crop to be taken off the land and it will also ensure strong spring growth the following year.

It is most important, however, not to sow too late for clover establishment before autumn frosts and rain. This date will vary with different climates and soils, but reseeding after the middle of August may risk winter kill of clover, which has very serious consequences for organic management.

For successful establishment of clover, shallow sowing (not more than 1 cm) into a very firm seed bed is essential. Although clover seed can be drilled, it is difficult to ensure that they are drilled sufficiently shallow. The ideal solution would be to drill the grass seed first and then broadcast the clover seed, ideally onto a ring-rolled surface which should then be lightly harrowed and followed by flat rolling to conserve moisture. This process is, however, labour intensive so some compromise solution may need to be found.

An improved method of sowing developed by the East of Scotland College of Agriculture and the Scottish Institute for Agricultural Engineering involves an over-sowing attachment to the seed drill allowing clover and timothy seed to be broadcast at the same time as other seed constituents are drilled. This combined drilling and broadcasting technique resulted in a 50% improvement in white clover establishment rates by minimising competition between grass and clover seedlings.

Once drilled, care during the establishment of the crop is also essential. Ideally, the new ley should be grazed lightly in autumn and the following spring, then shut up and a light cut taken to encourage establishment. Prolonged autumn grazing by sheep will delay spring growth, although recent evidence from Wales suggests that clover establishment is favoured by fairly intensive grazing in late September to allow light into the sward and control competition from the grass. This should be followed by a rest period until as late as November/December where sheep are being wintered.

There is increasing interest in the development of techniques for introducing clover into existing leys. Many of these require the use of herbicides, which would not be acceptable in organic systems. Slurry seeding techniques being developed at the Welsh Plant Breeding Station and in Ireland, however, may provide an alternative way forward. The technique usually involves mixing the seed with slurry before spreading onto a field which has just been cut for silage, but the seed can also be broadcast first and then covered with slurry. The slurry provides moisture for the seed to germinate and the seedling can get a good start while the ley is recovering from the silage cut. The Irish researchers have found no difference between the two approaches, but the WPBS research favours mixing the seed with the slurry before spreading. The seed is added at the same time as the slurry is pumped into the tanker. A period of rainfall before and after the operation will help to ensure successful establishment.

Direct drilling techniques which do not involve ploughing or complete destruction of the old ley are also being explored, usually involving some form of narrow band cultivation and slot seeding. The problem is giving the seedling enough space to grow without being out-competed by other, already established species. The most promising approach is the rotary strip seeder developed by the SIAE, which consists of free floating rotors which, unlike rotary cultivators, are capable of passing over rocks without breaking and can therefore cope with the worst conditions.

The best results are achieved by this method if the pasture is grazed down hard before sowing, and any drainage or nutrient deficiencies rectified. After sowing, stock can be returned after three days to keep down the original sward until the seedlings begin to show. Once this happens, 'mob' grazing of stock for short periods helps to prevent preferential grazing and keeps the old sward down allowing the seed-lings to establish.

In a tight sward, tillering can be restricted and the wide spacing between slots may mean that the improved grasses and clover remain there without spreading and preferential grazing will then become a problem. Trials in Scotland have shown, however, that this machine is second only to full cultivation and reseeding, compared with other methods of reseeding, and that clover rhizomes do, over a period of years, meet between the rows. Even so, where there is the possibility of this problem occurring, cross drilling can help to overcome it.

The major advantages of the strip seeding approach are that the initial loss in yield usually associated with establishing a new sward can be avoided, and the risk of environmental damage to sensitive land, such as soil erosion on steeper slopes, can be avoided. In addition, there is a wide range of other applications, including the introduction of forage rye into permanent pasture for early spring grazing, and the direct drilling of swedes and turnips into an existing grass sward as a pioneer crop. The implement is now available commercially as the Rotary Strip Seeder from Hunter's of Chester.

The disadvantages of the strip seeding approach include the fact that the cultivated rows provide an ideal habitat for slugs which then eat their way down the new rows of seedlings, and the need to rely on a contractor for the machinery. A cheaper alternative is the technique of sod seeding. In April, or immediately after the first conservation cut, the sward is harrowed hard to expose the soil. The clover can then be spun on and rolled in. The ground should then be grazed until the clover begins to germinate, at which point stock should be taken off for three to four weeks, followed by light grazing to allow the clover to establish without excessive competition from the grasses.

Management of grass/clover swards

Clover/grass swards on organically managed livestock farms will be required to produce high quality forage by utilising and developing existing soil fertility, relying on the legumes for nitrogen fixation and minimising the need for brought-in mineral or organic fertilisers. The management of the sward will have important and long lasting effects on its composition and productivity. To these ends, the organic farmer has certain management techniques available to him in common with conventional farmers:

- physical treatments such as harrowing, rolling, topping, moling, subsoiling, etc.;
- fertiliser and manurial inputs;
- grazing and conservation management.

The timing and effectiveness of these operations will affect soil structure and fertility, sward composition, output and persistency. When successful, the persistence of sown and desirable species can be such that long-term leys can be maintained indefinitely. White clovers need particular care and attention during the first 18 months of a ley, after which they enter their second, stoloniferous stage of growth, where they are much more resistant to overgrazing and pest and diseases and are able to grow back much more readily.

Physical treatments
After spring establishment of a direct seeded new ley, the first grazing should not take place until there is sufficient root establishment to avoid 'pulling'. The first grazing should be followed by topping, which will help to control both annual and perennial weeds (which can look

alarming in a spring reseed, but should not be a cause for panic) and allow the young grasses to thicken without competition. The first topping must be carried out before the weeds have seeded. Topping may also be required for a summer reseed if significant numbers of weeds become established.

Chain harrowing in spring is an essential operation on established swards. It will level mole hills and dung pats, rip out dead grasses and aerate the soil surface. Swards intended for mowing should also be rolled.

In addition to the regular grazing and cutting regime, mid-season topping will control weeds and remove rank, ungrazed material, thus permitting new growth. Whenever manure is applied, chain harrowing is helpful to hasten the incorporation of the material.

Soil conditioning under grassland

Subsoil treatments, such as mole ploughing and the use of the paraplow and other sub-surface cultivations which do not invert or destroy the sward, are an important feature of grassland management in organic systems, particularly where heavy soils impede drainage. These practices break up compacted soil and create fissuring, allowing better drainage (where drainage systems exist), and they increase the availability of oxygen to the soil biomass and plant roots, allowing deeper penetration by plant roots and better exploitation of soil mineral reserves.

Various 'spiking' machines have been developed for grassland aeration, among which is the Grass Master from Broadwater Machinery in Suffolk, which consists of 45 or 90 blades fixed to a horizontal rotor in a spiral pattern. The blades will cut a constant slit pattern into the soil up to a depth of 180 mm if weights are added. In Welsh trials with another implement, the Groundhog produced by Urry Agricultural in Shropshire, spiking increased early season yields by 45% and mid season yields by 30% in Denbigh series soils compacted by poaching.

These treatments are best carried out early in the season, provided there is a sufficient soil moisture deficit. For instance, after mole-ploughing in May, growth of the sward roots, increased earthworm activity and subsequent dry weather will enable the restructured soil to be stabilised during the growing season before the soil becomes saturated again in the autumn.

Careful treatment of difficult soils will reduce the need for regular moling. Stock should never be allowed on heavy land swards in very wet conditions and heavy machinery such as muck spreaders and silage trailers should only travel on these soils in good conditions.

Grazing management

Grazing management, as in conventional systems, has to be directed towards making optimal use of the grassland available throughout the grazing season. In part, it involves flexibility in adapting to and making best use of the changing pattern of herbage growth during the year, which is determined by a range of factors including rainfall, soil temperature and the growth patterns of the individual components of the ley.

For pure grass swards, growth will follow the characteristic pattern shown in Figure 10.1. Under organic management, however, the early growth of leys in the spring will often be less good than under conventional management. This is because nitrogen fertilisers are not used and soil temperatures need to increase before sufficient biological activity takes place to make nutrients available. In a grass/clover sward, the low growth habit of clovers means that they are generally less active in the spring, but more active in summer when grasses enter a rest period. As a result, the grass:clover ratio will vary during the growing season; but this also means that the mid-season dip in production will be less severe than with pure grass swards. Although suitable ley mixtures can be chosen which will grow faster and earlier than others, it will usually be essential to take precautions against the shorter grazing periods.

Figure 10.1   Grass growth during the growing season
(average 1970–76, using 375 kg N/ha annually).

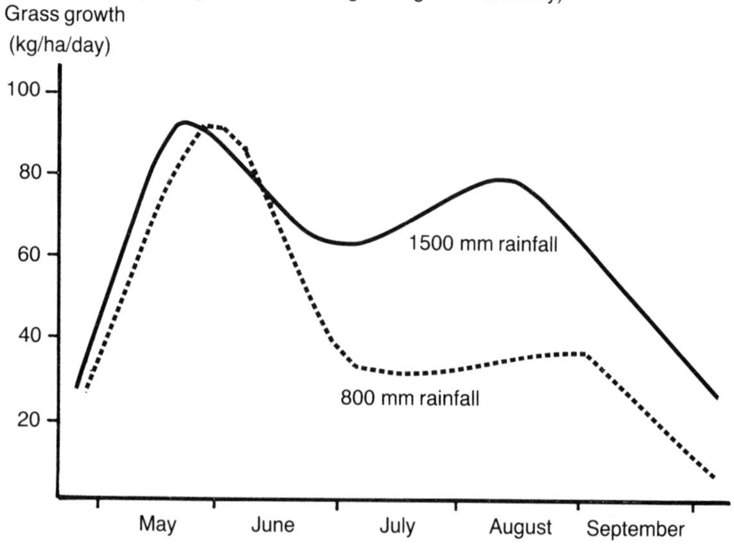

Source: MAFF/ADAS Booklet 2051, *Grazing management for beef*, 1982 ed.

Permanent and older pastures in good condition may be better for earlier grazing, their more resilient structure reducing the likelihood of damage through poaching. A vigorous summer reseed which has established well will put on good early growth, although this should be grazed carefully to avoid damage and reduced subsequent growth. In general, it is better to provide for sufficient conserved forage to take the livestock through to a more realistic turnout date, and the overall farm stocking rate should be adjusted accordingly.

Late season grazing may also be limited on organic farms. The traditional method of grassland management to provide winter forage by closing up some pastures in late summer to graze through into November could be practised; however, the tendency will be to conserve enough grass to start feeding silage or hay earlier. An alternative to extending the grazing season or conserving more forage is the use of alternative fodder crops which will be available during the autumn, winter and early spring.

Grazing systems

There are as many grazing systems as there are farms, and the complex rotational or paddock systems which have been developed through the extensive research efforts directed towards maximising grassland productivity could all be applied to organic farms. These are fully described in the MAFF/ADAS series of booklets on Grassland Practice. However, most farms do not operate ideal systems, due to the difficulty of justifying capital costs and labour requirements for the fencing on which the more complex and 'efficient' systems depend. There is no doubt that a good grazing system is as important to the organic farmer as to any other. In fact, without the use of soluble fertiliser the features of a good system are perhaps even more important. These may be summarised as:

- the maintenance of a balanced sward;
- good regrowth after grazing;
- optimum stocking rates for returning manure to the sward;
- provision of an even and sufficient supply of food to the flock or herd throughout the year.

Grazing systems which contribute to these objectives must be used, since only this way can the best levels of production be achieved. The management of any grazing system, however, should take into account that under grazing the clover content may be depressed, with levels typically less than 25% of total herbage, especially under continuous grazing systems for sheep. This does not necessarily mean that output will be poor, because recycled nitrogen from the grazing stock will reduce the contribution needed from clover. However, the general principle that moderately intensive grazing favours clover while severe grazing favours grass, should be remembered.

Grazing systems which require a high degree of management as well as labour and capital inputs for fencing can be justified in the case of intensive livestock enterprises such as dairying and lowland lamb finishing. Such systems may include:

- strip grazing in the early spring and at the end of the season when forage is scarce and needs to be rationed;
- rotational field or paddock grazing, which would normally form the basis of mid season grazing;
- leader-follower systems for grazing cattle, which contribute to the performance of the higher yielding cows by allowing them access to the best grass;
- forward creep grazing systems for sheep, which fulfil a similar role by allowing lambs access to new grass ahead of the ewes.

The advantages of rotational grazing are illustrated by data from North Wyke Research Station in Table 10.5. Rotational grazing systems in particular allow for the resting periods necessary to ensure the maintenance of clover in the sward and are likely to encourage clover to a greater extent than set stocking. Rotational systems have the additional advantage that in periods when there is surplus grass, paddocks or fields can be shut up for conservation.

Table 10.5 The effect of grazing management on the overall productions from grass/white clover swards (all treatments at 14 ewes and 25 lambs per hectare)

Treatment	Lamb growth rate from day 0 to sale (g/hd/day)	Silage made (kg DM/ ewe)	Supplement fed (t DM/ha)	Productivity (GJ/ha)
Rotational grazing (6 paddocks)	225	—	0	—
Set stocking	142	—	1.4	—
Rotational grazing (6 paddocks)	236	46	0	110
Continuous grazing	209	7	1.4	98

Source: Newton & Laws cited in Newton (1989).

The rotational requirement will affect the size and number of paddocks and the time allowed for grazing individual paddocks or fields; but in general the regrowth period for clover needs to be about ten days longer than pure ryegrass stands, with a gap of at least four weeks, and according to some practioners five to six weeks, being essential. Rotations shorter than this are likely to have adverse effects on the clover content of the sward and may result in occasional herbage shortages, although this can be compensated for by the use of conserved forage for buffer grazing.

Increased grazing intervals, particularly when the late season flush of clover occurs, will help to minimise the risk of bloat; the increased maturity and stemminess of the grasses counteracts the formation of gases in the rumen from certain proteins which occur in significant quantities mainly in legumes. The increased grazing intervals may, however, need to be supplemented by strip-grazing to counteract selective grazing, or the feeding of hay and straw as another source of fibre. In spite of the concerns which are often raised, bloat occurs only rarely on organic swards.

Rotational paddock and field grazing systems are also important in the control of internal parasites in both sheep and cattle, the key features of which are described in Chapter 8.

Set stocking or continuous grazing systems give less efficient utilisation of grassland, but also involve less interference with the stock and reduce the management and labour requirements, although ideally the management input and in particular the monitoring of sward height, should be maintained. Experience in organic systems has shown that they can provide the benefit of leaving tight swards and allowing clover to flourish without competition from grass. Set stocking is probably better suited to more extensive enterprises such as beef and sheep, but the potential future market requirement for finished animals all year round means that management should still aim to achieve a good level of liveweight gain throughout the season. The labour-saving advantages of continuous grazing, however, need to be set against the problems of parasite control particularly in sheep.

Sward height
Wherever possible, grazing should be controlled so that the sward is never overgrazed. The ideal height of the sward during grazing will vary according to conditions, and can be critical in ensuring the optimum production of the ley (Figure 10.2). Sward height is an important concept because it is related to:

- the leaf area available for trapping sunlight and hence herbage growth rates;
- the proportion of leaves which age and die without being utilised;
- the total amount of herbage available; and
- the amount of herbage which can be eaten by either cattle or sheep and hence individual animal performance.

Figure 10.2   The importance of sward height for grazing management.

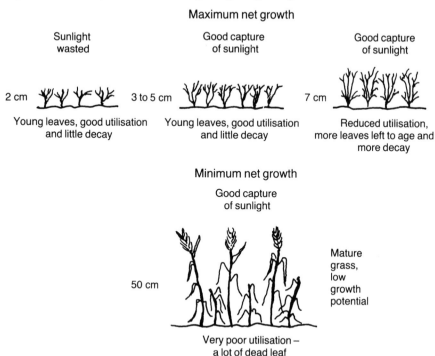

Source: *Farmers Weekly Supplement*, March 15 1985, p.29.

In practice, sward height is a measure of the average height of the sward surface above ground level, without extending the leaves or flowering stems. It can be measured with a ruler or tape measure and with practice can be recognised by eye. Changes in sward height can give an early indication as to whether a field is under or overstocked.

Research results from the Hill Farming Research Organisation show that growth rates may be doubled at grass heights of 8–10 cm compared with 3–4 cm, where stocking rates are low on set-stocked swards. More recent work at Hurley (Parsons *et al.*, 1987) has shown that increasing grazed sward height from 3 to 6 cm considerably improves animal performance (Table 10.6) with grazing at 6 cm giving the best performance for sheep on grass/clover swards. Grazing at 3 cm supported greater animal numbers, and so a greater dung and urine return, but performance per animal was reduced. At 9 cm, too few animals led to poor performance per hectare, despite increased performance per animal. All three intensities led to low clover contents and apparently low rates of N fixation. Despite this, output on the 6 cm grass/clover sward was only 20% less than that of an all grass sward receiving 420 kg/ha fertiliser nitrogen and the system had the potential to exceed the financial returns of the top third of MLC recorded lowland sheep flocks, which use an average of 180 kg N/ha. In Scotland, where the growing season is shorter, a higher sward height (up to 9 cm) may be preferable to achieve faster lamb growth.

Table 10.6 **Performance from lambing to weaning of ewes and twin lambs continuously grazing grass/clover compared with fertilised grass only swards.**

	Grass/clover			*Grass + 420 kg N/ha*
Sward height (cm)	3	6	9	6
Stocking rate* (ewes/ha)	20	14	9	18
Lamb growth rate (g/day)	223	268	295	260
Lamb live weight gain (kg/ha)	1,054	920	630	1,148
Clover content (by weight)	22%	15%	16%	

* summer stocking rate, average 1.6 lambs/ewe.
Source: Parsons *et al.* (1987).

For pure grass swards, the recommended target heights for different types of livestock and at different times of year are given in Table 10.7. These are based on research at the AFRC research institutes and the ADAS experimental husbandry farms. For grass/clover swards which are able to produce similar growth rates at lower intake rates per animal, the optimum sward heights should be a little lower. This is because the clover needs more light for optimal growth, although the fact that stock such as sheep graze clover preferentially makes it difficult to generalise about grazing height. The target heights given assume dense and leafy swards in which little bare ground can be seen. For grazed swards with more bare ground, the target heights should be increased by 1–2 cm.

At turnout, grass height is usually more than 8 cm, with rapid growth taking place. If it is not possible, due to the risk of poaching, to achieve the optimum sward height immediately, stocking rates should be increased later to bring the sward down to optimum height as soon as possible. With set stocking, lower sward heights tend to restrict intake and depress performance, but this can be of advantage if the aim is to produce store stock. With rotational or paddock grazing, the aim should be to keep the grazing period as short as possible, leaving the correct stubble height to encourage rapid regrowth. If understocking or above average growth in the spring lead to stemminess, then early topping is advisable. This should take place before the end of June at a height of 6–8 cm depending on the weather and the height of grass required in the grazed sward.

Table 10.7 Summary of target heights for all-grass swards.

Conservation			
First-cut silage	45		
Cut stubble	6		
	Sheep	*Beef*	*Dairy*
Grazing:			
Continuous (set stocking)			
Spring	4	5	6
Summer			
– fattening	5	7	8
– store	4	6	–
Autumn	7	8	10
Rotational (paddock)			
Stubble height	6	8	9
Winter	below 5 to prevent winter kill		

Source: Farmers Weekly Supplement 15/3/1985: 31.

In the autumn, when feed quality is lower, the target heights should be increased slightly so as to encourage increased intake. In early winter, it is important that pastures are grazed off to below 5 cm to prevent the possibility of winter kill. Hard grazing by sheep at this stage also means that the clover stolons are trodden into the ground, giving them added protection.

Permanent pasture

There has tended to be a prejudice which has suggested that permanent pastures are not very productive or efficient. This is not well founded; well managed permanent pasture can be productive for both conservation and grazing, and can also act as an important habitat for wildlife. From the point of view of wildlife, so called upgrading of pasture is perhaps not the best thing, although it may be preferable to reseeding and it may be important to introduce clover, grasses and herbs which are absent from a permanent pasture. Surface seeding and sod seeding can both be usefully applied.

Maintaining a variety of species in permanent grassland and traditional meadows is important and is closely related to management. Late harvesting of conservation crops allows many of the species to flower and set seed, so that they can survive, and short periods of intensive grazing followed by long recovery periods also give the slower growing species a chance to survive. Application of manures should be closely related to intensity of use, while bearing in mind that many important species are not able to compete against fast growing grasses under high fertility conditions. The maintenance or re-establishment of traditional hay meadows may also require the cutting and removal of forage without returning manure so as to reduce the fertility status of the land.

Forage conservation

High forage, low concentrate diets are an important part of the organic approach to feeding ruminant livestock. The Soil Association's organic standards require that at least 60% of ruminant diets must be forage, which gives added importance to good quantities of well conserved silage and hay for winter feeding.

In most cases, an organic farmer will choose to produce both hay and silage, since it will always be useful to have good quality hay available for young and nursing stock. However, the major emphasis will be on silage production, even though this will mean that traditional, species-diverse hay pastures may not be used on the farm. Big bale silage has considerable potential and is in many ways better suited to the range of leys with differing maturity dates which are to be found on organic farms, as well as providing a supply of long fibre material which is necessary for optimum conditions in the rumen.

Organically managed clover/grass leys have certain inherent advantages for conservation. Dry matter levels at cutting will generally be higher than grassland receiving artificial fertilisers. Thus wilting to bring dry matter down to acceptable levels for ensiling or hay-making will be easier and quicker, and effluent problems in silage will be less severe.

If the clover content of the sward is 25% or higher, this will provide a significant buffer against declining digestibility (D) after flowering in grasses. Although white clover gives relatively low yields, it retains a high D value of about 75 over long periods with very slow rates of fall. Thus the quality of hay and silage from organic swards cut after flowering is always likely to be higher than all-grass material, with higher protein content and increased palatability and intake providing additional advantages.

In silage making, the unacceptability of additives means that greater attention has to be paid to the conditions during cutting and making the clamp. It will almost certainly be necessary to wilt the cut herbage to achieve acceptable dry matter levels, and the clamp should be carefully constructed to ensure an even, well compressed and anaerobic heap. The legume content, and therefore the protein content, of the silage will be higher and management will need to reflect this.

The need for some form of additive when making silage with high protein content is often commented on, and it is sometimes suggested that the newer 'biological' additives based on bacteria or enzymes may be suitable alternatives in organic systems. There is still insufficient evidence on the effectiveness of enzyme-based additives to be able to recommend their use even in conventional systems, and bacterial inoculants are highly influenced by the bacteria already present on the grass. In practice, experience on organic farms with potential problem crops such as lucerne has shown that even these crops can be successfully conserved without additives. Good wilting and higher soluble carbohydrate levels in organic swards will ensure satisfactory fermentation in most silages, the exception being very wet material harvested in autumn when sugar levels are lower. In such circumstances, a carbohydrate additive such as molasses may be necessary, but this is the exception rather than the rule.

Higher cutting (8 cm instead of 6 cm) can help to achieve more rapid drying, as a result of the longer stubble promoting air circulation underneath the cut herbage. There are additional advantages to this which include cleaner silage and less damage to machinery, as well as greater tiller survival, faster regrowth and better persistence. These benefits are not necessarily achieved at the expense of overall output.

Ideally, fields intended for cutting should be rotated with grazed swards. This will help to maintain potash and phosphate levels and encourage species diversity in clover/grass swards. When fields are cut twice a year every year, there is a tendency for clover to become dominant late in the season, and for some grasses to become less vigorous. In these circumstances, the manuring strategy must provide for the replacement of phosphate and potash removed. Where rotational cutting and grazing is possible, this will also assist weed control, as thistles do not survive regular cutting, whereas docks are more discouraged by grazing.

In practice, some fields will be unsuitable for cutting, but the combined benefits of maintaining sward diversity, weed control and fertility provide a strong argument for cutting/grazing rotation. In fact, the ideal situation for clover maintenance is one where cutting and grazing by both cattle and sheep are employed.

Leys for forage conservation

Short-term leys based on red clover and medium-term lucerne leys are ideally suited to the production of high quantities of good quality bulk winter feed in organic systems. Both these crops can produce substantial annual yields of high protein silage for several years, and leave behind up to 200 kg N/ha for the following crop. Lucerne is primarily a crop for conservation, being better suited to ensiling than hay making due to its fragility when dry. For the same reason, conservaton may be difficult in high rainfall areas and overconditioning while wilting may result in considerable losses through leaf shatter. In general, red clover is better suited to the wetter, western parts of the country and to soils with lower pH levels.

Lucerne

Lucerne is a deep-rooting, drought-resistant and persistent crop, but is a hungry feeder, fussy about soil type and can be hard to establish and a poor competitor against weeds (see Colour plate 21). It needs a well drained soil with a pH of at least 6 and preferably 7, with adequate phosphate and potash levels (ADAS Index 2). If supplementary phosphate is required, then sources soluble at relatively high pH levels will be necessary. It is not advisable to try to grow it on heavy soils or in high rainfall areas where soils are prone to waterlogging, and any acidity in sandy or leached light soils must be corrected ahead of sowing.

April is the best month for seeding, July the latest recommended. It can be sown alone, or undersown in a cereal nurse crop such as oats. If undersowing, the cereal should be sown late, just before the lucerne, at a low seedrate. Care over establishment is essential. A good method is to prepare a level, rolled seedbed, drill the lucerne about 1 cm deep and roll again. Either the drill should have narrow spacing or the crop should be drilled two ways to create a dense stand using a seedrate of 12–15 kg/ha.

Since the strain of bacteria which colonises lucerne's roots and fixes nitrogen is not usually naturally present, it is advisable to inoculate the seed. This sounds elaborate, but is in fact a simple process. The bacteria are supplied as a powdery peat-like culture medium, from which a paste is made with water. This is then mixed with the seed. Shovelling on the floor is fine, but it is also possible to shake about 10 kg of lucerne seed at a time with the correct proportion of paste in a sack or to use a concrete mixer for large quantities.

The lucerne will grow away slowly, and the best that can be expected in the establishment year is one light cut yielding not more than about 4 tDM/ha. It is very important in the establishment year and throughout its life not to cut lucerne after the end of September, so that reserves can be built up in the roots for the winter. When the first frost comes, the greenery will begin to die back. This can be grazed as required, or just left; no harm wil ensue. At the end of the winter it will look quite dead.

Sometime in the following April, as the ground warms up, growth will suddenly start, and should continue steadily until autumn. Grazing is a bad idea at any time in the growing season, as bloat will almost certainly result and future plant growth may be adversely affected. If it is grazed, then sufficient time should be allowed for the crop to recover.

With lucerne and red clover, digestibility falls in the same way as grasses, but these crops are usually harvested at a D value of 60–63 rather than 67 for grasses, because intake is higher at similar digestibility and lower D values are therefore acceptable. In principle, lucerne should be cut first at the very early bud stage to obtain reasonable quality. In the southern half of England and Wales this is the first week of May. In practice, however, it may not be advisable to take the first cut before June. This is because when leafy, lucerne is very sappy and may need two to three days wilting to achieve a decent fermentation.

The high protein and low soluble carbohydrate content of lucerne is traditionally believed to make it unsuitable for ensiling without the use of additives. In practice, experience shows that additives are not necessary if the crop is wilted adequately.

Up to three further cuts can be taken as required which should either be well wilted or molasses used as an additive. The silage looks coarse and stemmy but makes a good high dry matter, high protein feed for livestock and feeds very well, either chopped and clamped or in the very firm round bales it makes. The big bale bags may be punctured by the lucerne stems, so a second old bag can be used to avoid spoilage.

The main problems are weeds and potash supply. There is little that can be done in an organic system for weed control. If the field is dirty when the lucerne is sown it will be considerably more so by the time it is broken up. A solution may be to grow a companion grass, which should fill up the gaps in the sward, smother weeds, utilise free nitrogen and help the fermentation process. Traditionally this was cocksfoot, which is really much too early-heading not to pull down the D-value of a June cut. A tetraploid ryegrass would possibly be more suitable, as long as it is grazed down well at the end of the year. Lucerne can also be combined with both timothy and meadow fescue which are less competitive varieties.

Potash supply is critical, because large amounts are removed by frequent cutting of green material. The types of light soil which lucerne favours have less ability than clay soils to release potash and the normal recommendation would be to apply 125–250 kg/ha of potash every year. This figure could probably be halved in an organic system, but would still require large quantities of muck or slurry, which could make lucerne a non-starter on many farms. If it is well fed and managed, however, it is reliable and, with its deep roots, almost independent of rainfall.

Lucerne can be troubled with eelworm and *Verticillium* wilt, but these are unlikely hazards in a sensible rotation with a break of at least four years between successive crops.

Managed carefully, however, lucerne should give four productive years after the establishment year, yielding 8–10 tDM/ha per year.

Red clover

Red clover (see Colour plate 22) is altogether easier to handle. It is normally sown (drilled or broadcast) with a companion grass such as RVP, Augusta or Condesa, and may be sown direct in spring or early autumn, or undersown. A typical seed mix would be 7.5–10 kg of clover with 25–30 kg ryegrass per ha, depending on the varieties used. It seems to be not very particular about soil type or conditions, and grows aggressively for two to two and a half years. The tetraploid varieties such as Hungaropoly perform well; they are quite unlike the older types, being late-heading and erect.

Like lucerne, red clover can be sappy and should not be cut too early or ensiled unwilted. It is usually cut at early flowering stage in mid June, when digestibility is about 63D. It makes excellent silage and can be grazed with care, although not by cattle in the spring. It is ideal for finishing lambs, as the autumn growth needs to be controlled but not grazed down too hard. Concern is often expressed about varieties which contain high levels of oestrogens and which can affect fertility in breeding stock. With careful management, problems such as this rarely occur in practice.

Two cuts of 2.5–4 tDM/ha plus grazing are possible in the first and third years, with three cuts and an aftermath grazing in the second. Again, care must be taken that soil potash and phosphate levels are kept up and pH levels should be around 6.5. It is said that winter grazing pigeons can be a problem, but experience has shown that once established the clover suffers little harm.

Manuring of grassland

On a mixed organic farm, the major fertility building period of the rotation is the ley and it is at this stage that the majority of available manure and slurry should be applied for the rotation. Applications should not be made in the late autumn, as is often the case on conventional farms eager to empty manure and slurry storage space, because nutrients will be wasted and the risks of environmental pollution are high. Manures and slurries will contribute to growth best during the spring and summer, and a top dressing in early spring to the leys intended for the first grazing will certainly improve their growth. Dressings will also be beneficial after conservation cuts. Normal application rates of between 10 and 30 t/ha will be sufficient. These should be distributed around the leys depending on how much is available. A programme of applications of sewage sludge (provided heavy metal contamination is within acceptable limits) could also be considered as a useful source of nutrients, particularly phosphate.

Nitrogen supply is not a significant problem. The clover will provide up to 200 kg N/ha when the population is sufficient (i.e.25–50%). Grazing livestock will return significant quantities to the sward and small additional quantities may be returned through applications of manure and slurry.

In order to ensure optimal conditions for soil biological activity and grass growth, pH levels should be maintained at around 6.0 or higher by the addition of ground limestone, chalk or calcified seaweed.

Phosphate and potash deficiencies, where indicated by soil analyses, should be corrected within the context of the Soil Association's standards. Basic slag, if it can be obtained, or Gafsa rock phosphate fertiliser can be used to provide slow-release sources of phosphate to correct deficiencies, but must be seen as part of a longer term fertility building programme. On chalk soils, alternative sources of phosphate which are soluble at higher pH levels may be used. Where extreme potash deficiencies exist, the current Soil Association standards permit the addition of sulphate of potash under licence; although this is readily soluble it is less harmful to the soil ecosystem than the chloride containing salts. Mineral fertilisers should generally be applied in dry conditions during the growing season, or at reseeding in the seed bed.

If levels of potash and phosphate are adequate, there will be little or no need to provide these elements on grazed swards. Dung and urine from grazing stock will replace most of the nutrients removed by the animals. Where significant quantities of herbage are removed by cutting, attention must be paid to regular fertilising with acceptable forms of phosphate and potash. The cutting of grass/clover leys makes considerable demands on these elements and although the power of a biologically active soil to release nutrients is considerable, there will almost certainly be a case for a rolling programme of rock phosphate and potash applications through slurry, manures and other sources. A dressing of 15 t/ha FYM, for example, will provide about 50 kg phosphate and 100 kg potash over the rotation (not all will be available in the first year). One application of FYM at about 30 t/ha in spring will generally supply sufficient nutrients for two cuts of silage.

Weed control in grassland

Organically managed long-term leys will generally consist of a wide variety of sown species, including ryegrasses (perennial and Italian), cocksfoot, timothy, fescues, several varieties of white and red clover and, in some cases, sown herbs. When weeds and non-sown species begin to take over a sward, remedial action will generally be successful only if the causes of the invasion are understood and corrected.

The persistence of sown species is clearly of great importance as this will affect output, the nutritional quality of the sward and the health and productivity of the livestock consuming the forage. The invasion of non-sown species may therefore be seen as a serious problem, but this is not necessarily so and in some cases, such as dandelions, they may even be beneficial. In addition, well managed swards containing a high proportion of weed grasses in typical organic conditions are likely to do as well as relatively pure perennial stands. The threat of weed problems is probably overemphasised because of their relatively greater importance in intensive, high nitrogen fertiliser systems.

The control of undesirable species in the sward will depend on an understanding of the environment preferred by the particular species. To take the simplest example: the presence of rushes usually indicates a drainage problem and the appropriate treatment will be to remove the rushes by creating an environment in which they cannot survive, i.e. by improving drainage.

In general, then, undesirable species can be seen less as isolated weed problems and more as indicators of soil management and environment. This approach can be of great value to the organic farmer, enabling him or her to take effective action to remove or encourage the particular plant species. Table 10.8 summarises the main desirable and undesirable components of grassland and the measures which can be taken to encourage or discourage their survival.

The two most significant weed problems in grassland are thistles and docks. Their control is considered in some detail in Chapter 6, but it is worth reiterating some of the points here.

For thistles in grazed pasture, a regular topping programme over two to three years will usually be successful in eliminating creeping thistle. The best time to top is at the early flowering stage in late June or July. This should be followed with a second cutting in August in the case of severe infestations. Thistles do not survive long in fields that are regularly cut for hay and silage.

Docks are the most difficult perennial weed to control in organically managed swards. They thrive in poached or compacted soils which have high applications of slurry or other undigested manures, and will survive a silage cutting regime. With severe infestations, the only effective control is repeated cultivation during the summer, as in the bastard fallow, but avoidance of poaching, reduction of slurry applications and prevention of seeding will help to prevent severe dock build up.

Other undesirable weeds such as rushes can be overcome by providing a good soil environment and sufficient fertility for the sown species to thrive.

Apart from these undesirable species, the presence of some unsown species should be welcomed. Wild white clover will often establish by self seeding and is quite capable of becoming a substantial percentage of

Table 10.8 Components of grassland, their environmental preferences and control measures.

Species	Preferred conditions	Remedial action
A: SOWN SPECIES		
Ryegrass	pH above 5.5, high fertility, regular defoliation.	None necessary.
Timothy	Heavier soils, tolerates lower pH.	None necessary.
Cocksfoot	Dryer, lighter soils. Does not thrive on wet land. Deep rooting, drought resistant.	If cocksfoot predominates and grows coarse, increase cutting/grazing intensity.
Red clover	Short term leys, will not persist. Very effective at N fixation (up to 300 kg N/ha).	Allow to seed for persistence. Regular cutting is tolerated.
White clover	pH above 5.5. Regular defoliation.	Avoid overgrazing in winter or undergrazing in summer. Rotate cutting/grazing management.
Chicory	Deep, well structured soil. Cannot tolerate frequent cutting.	Cut for hay not silage (allow to flower).
B: UNSOWN SPECIES		
Thistle	Undergrazed, fertile pastures.	Top at least once per season. Silage/hay cutting will eliminate thistles in two to three seasons.
Dock	Rank fertility, high nitrogen compaction, regular silage cutting, slurry applications, poaching.	Fallow with repeated cultivations. Row crops/veg. Avoid heavy machinery and poaching.
Rush	Low pH, waterlogging, compaction, poor under drainage.	Drainage, moling, subsoiling. Avoidance of grazing, tractor access in autumn, liming.
Buttercup	Low pH, poaching, waterlogging, under grazing.	As above.
Daisy	Low fertility, low pH, open swards.	FYM, lime.
Moss	Low fertility, low pH, waterlogging.	FYM, lime, drainage.
Annual meadow grass	Open swards, high fertility.	Avoid poaching, reseeding.

the sward under the correct management, leading to very significant nitrogen fixation. Dandelions will often colonise silage fields but not grazing swards as cattle and sheep will usually graze them out. The dandelion can make a positive contibution to dry matter output and its deep tap root helps to bring up useful minerals from the lower layers of the soil. In grazed swards, dandelions do not necessarily reduce herbage quality and may actually increase intake. Indigenous plantain and yarrow will also thrive in some soils and can be regarded as beneficial to the sward.

Managing upland grassland

Large expanses of the hills and uplands still managed using traditional techniques could be considered to be 'organic' although not necessarily through the conscious decision of the farmer. However, in recent years, overstocking encouraged by headage payments on ewes has led to dramatic environmental degradation. Heather is giving way to bilberry in the moorlands, bracken is taking over large areas of the countryside, broad-leaved woodlands are degenerating because grazing sheep eat the young saplings and man-made features such as dry stone walls are disintegrating because there is insufficient labour available to maintain them.

Hill and upland areas are characterised by a harsh climate, with excessive rainfall and low temperatures; thin, acid soils with low organic matter turnover and hence peat formation; and by inaccessibility and market isolation. Organic management under these conditions will therefore have to have specific features which bear little relationship to crop rotations, intensive organic manuring and clean grazing systems for livestock.

Upland grassland is generally characterised by poor grass species such as *Molinia* and *Nardus*, as well as, in some areas, heathers, heaths and bracken. On these unimproved areas, stocking may need to be reduced to allow for the survival of indigenous species and to assist with the maintenance of parasites at an acceptable level. The traditional practice of burning to rejuvenate heather may be justified in organic systems (unlike straw burning) because of its contribution to the preservation of heather moorlands.

The improvement of grassland in the hills and uplands is technically feasible, even under organic management, but the environmental and landscape consequences are significant and it is questionable whether further improvement should take place. The maintenance of improved grassland can, however, make the difference between viability and non-viability of an upland holding, so there can be no rigid right or wrong position on this.

As far as the management of improved grassland is concerned, there is little difference between organic and conventional approaches. The basics include:

- restoration and maintenance of pH and phosphate status using lime and phosphate (phosphate enrichment will also help with bracken control);
- the use of white clover with inoculants to supply nitrogen;
- possibly drainage and seeding with earthworms, once the right pH has been established, to encourage organic matter turnover, soil improvement and herbage production.

One such approach to grassland improvement used successfully in Wales involves the application of 5 t/ha ground limestone and 2.5 t/ha basic slag. In late winter, following frosts, the *Molinia* and *Nardus* grasses are burned off, to be followed in early spring by cultivation with a light set of discs to produce slits without raising clods. Second quality seeds (perennial ryegrass, timothy, red fescue, white and alsike clover) are then lightly harrowed in, then rolled and dressed with fertiliser to encourage establishment. In an organic system, suitable brought-in organic fertilisers would need to be used as there is unlikely to be sufficient, if any, farm yard manure available.

Under this system, early seeding gives the sown grasses a head start over *Molinia*. The new seeds are encouraged by heavy stocking with non-selective grazers such as cattle, but stock need to be moved in wet weather to avoid poaching which can encourage the germination of rushes. An alternative to this approach is to use no cultivations at all, to broadcast the seed and allow sheep to tread it into the ground. This approach reduces the potential problems with rushes, but establishment is not as good. The Hunter strip seeder described above, and the use of pioneer cropping such as direct drilled turnips, can also be used, but it should be remembered that the potential for soil erosion is greatest on some improved upland sites.

The lack of early spring grazing characteristic of organic systems is likely to be exacerbated in the colder upland areas and it may be necessary to use brought-in organic fertilisers to ensure that ewes have sufficient feed in the period before and during lambing. Hay should also be available to supplement the diet; the average ewe will need 100 kg hay in the winter and 1 hectare of well managed improved hill sward can produce 6 t hay.

The maintenance of the improved sward will be greatly helped by the presence of both sheep and beef, with their different grazing habits. This is likely to be easier on upland farms with a greater proportion of improved swards. Although it is unlikely that organic upland systems will be able to produce finished stock, there is certainly potential for the sale of organically reared stores to other lowland organic producers for finishing.

FODDER CROPS

Fodder crops have an important role to play in organic systems, as an energy source to balance the high protein, legume based forage, as a cleaning crop for weed control, and as an opportunity for thorough soil cultivations. There is a wide range of possible crops, from catch crops such as stubble turnips to main crops like fodder beet or kale. Table 10.9 lists the principal fodder crops along with their respective sowing and utilisation dates.

Table 10.9 Sowing and utilisation periods for different fodder crops.

Crop	Apr	May	Jun	Jul	Aug	Sep	Oct	Nov	Dec	Jan	Feb	Mar
Swede		★★★★★						++++++++++++++++++				
Turnip (yellow)		★★★★★★★					+++++++++++++++++++++++++++					
Thousand head kale		★★★★★★						++++++++++++++++++				
Marrow stem kale		★★★★★★★★★★			++++++++++++++							
Hybrid kale		★★★★★★★★★★				+++++++++++++++++++++++++++++++++++						
Turnip (white)		★★★★★★★★★			+++++++++++++							
Stubble turnip		★★★★★★★★			+++++++++++++							
Giant rape		★★★★★★★★★			++++++++++++							
Mangel	★★★★★★★★						Roots lifted		+++++++++++++++++			
Fodder beet	★★★★★							Roots lifted	+++++++++++++++++			
Peas	★★★★★★★			++++++++++								
Maize	★★★★★★				+++++++++							
Rye/triticale	+++		★★★★★★★★★★★★★★★★★									+++++

★★★★ Sowing period.
++++ Utilisation dates.
Source: Jarvis (1985).

Fodder root crops

Potential fodder root crops include fodder beet, mangels, turnips and swedes and can be grown on a wide range of soils providing they are well drained. It is unwise to grow any of them in succession because of the risk of club root and beet cyst nematode. Swedes and turnips provide a useful opportunity for keeping grass weeds in check; growing on ridges allows for inter-row cultivations for weed control.

Fodder beet
Fodder beet is a high yielding, high energy crop which grows well on loamy, free draining soils with a pH of about 6.5. On heavier soils, the

deep rooting habit of fodder beet makes it difficult to harvest. Fodder beet often follows grass in the rotation, but there may be a problem from wireworms and leatherjackets. Fodder beet needs to be topped and handled carefully at harvest to avoid wounding, which could lead to rotting in storage. The tops are a useful feed and can be fed fresh or ensiled.

Plate 10.1   Organically grown fodder beet being harvested at Hindford Grange, Shropshire. This crop won the Oswestry and District Agricultural Society's award for the best field of fodder beet   (E. Goff)

Advances in breeding have produced over 20 bolt-resistant varieties, with dry matter varying from 13 to 23%. These vary in the propagation of root in the ground, which has implications for the type of harvester to be used, and also vary in the amount and spread of leaf growth, which is very relevant to the organic producer looking for a weed smothering effect. Monogerm varieties are available which can be precision drilled and do not require thinning.

The crop is usually sown in 50 cm rows in April. The dilemma for the organic grower is that while earlier sowing gives greater yields, in a cold spring germination will be very slow, allowing weeds to establish before the rows can be seen to facilitate inter-row cultivations. Blanket pre-emergence thermal weeding would not be cost effective. It is therefore better to delay sowing until warmer weather will allow more rapid germination, which may not be until the second week in May. Weeding should begin as soon as possible and continue until the leaves meet in the rows and shade out the weeds.

The crop will grow strongly through to November. Considerations for maximum yield must be tempered by the need for a clean harvest, which will be dependent on weather conditions, soil type, method of harvesting and whether winter wheat, rye or a green manure of vetch is to follow immediately afterwards.

Satisfactory storage of the roots depends on minimising handling damage, soil and top contamination, and poor tapping. The clamp should be well ventilated and cool before covering with straw. In wetter areas a polythene sheet could cover the straw, providing the apex of the clamp is left uncovered to provide for better ventilation. It is recommended that the clamp should be no more than 5 metres wide and 3 metres high to avoid over-heating (Figure 10.3).

The beet should be fed as clean as possible. It can be fed to cows whole, but unless very low dry matter varieties are fed to sheep and young stock, it is better chopped. If cows are given the choice, they will often prefer fodder beet to silage. To ensure satisfactory silage intake, the amount of fodder beet is usually restricted to 15–20 kg/cow/day. Work at Crichton Royal Farm has shown that fodder beet in dairy cow rations can increase milk yield and improve milk quality, but care must be taken to balance the low protein, fibre and mineral levels in the roots (Table 10.10).

If the tops can be harvested cleanly, they can be ensiled, but more usually they are grazed after two or three days by sheep or young stock, or ploughed in as a green manure.

Figure 10.3   Construction of a clamp for fodder beet.

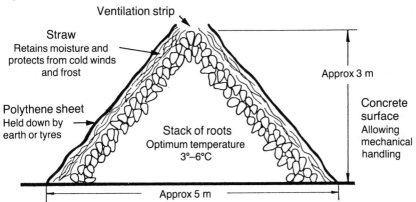

Table 10.10 Comparison of feeding values and yields of different fodder crops.

	ME (MJ/kgDM)	CP (g/kgDM)	DM (t/ha)	ME (GJ/ha)	CP (kg/ha)
Fodder beet					
– roots	13.2	60	11	145	660
– tops	10.0	60	3	30	180
Grass silage					
– 3 cuts	11.0	160	9.5	105	1,520
Barley					
– grain	12.9	110	5	65	550
– straw	6.0	40	2	12	80

ME = Metabolisable Energy; CP = Crude Protein; DM = Dry Matter.
Source: New Farmer and Grower, 22:15.

As sowing and harvesting can reasonably be left to contractors, the only equipment required is for weeding. However, if the beet cannot be harvested cleanly, then a beet cleaner becomes essential; these are not cheap.

Other fodder root crops
Mangels are a very high yielding, palatable and succulent feed, but with lower total dry matter than fodder beet. They are easier to harvest, because the beet grows above the ground, and can therefore be grown on heavier soils, but are still easily damaged and bleed if the leaves are cut off too low. As a crop, they are better suited to the south and east of the country. In traditional arable rotations, fodder roots usually follow and are followed by a cereal. Mangels can, in some cases, be cropped continuously, because they are relatively free from pests and diseases, as long as the field is manured regularly.

Turnips and swedes are better suited to the wetter areas of the north and west than fodder beet or mangels. Traditionally, they are used as a pioneer crop after a ley, where cultivations help to control weeds and feeding *in situ* helps to build up fertility on marginal land. They are frequently followed by cereal crops.

Leafy forage crops

Brassica crops also usually occupy the root break in the rotation, permitting cleaning operations until leaf cover smothers weeds. They are usually grown as catch crops. They can be grown as mixtures, but show little advantage over crops grown singly. In conventional systems, increasing use is being made of direct drilling techniques. These are not acceptable in organic systems because of the use of herbicides. Further work needs to be done to assess the possibilities of sowing these crops into cultivated strips in existing leys, which may help to reduce weed control problems, but would certainly help to avoid the problem of soil poaching which is usually associated with their utilisation.

Kale

Kale falls between being a main crop and a catch crop. In southern parts of the country, kale may follow early potatoes, forage conservation, or forage rye/triticale. There are two groups—true kales, including marrowstem, thousand head and their hybrids, and 'swede' kales, including hungry gap and rape kales, but which are not widely grown. The marrowstem types are not frost hardy and need to be used in autumn. Thousand head kale can be used in January—March; the dwarf types are hardier and leave less stem.

Kale grazed *in situ* is losing popularity, largely because of the problems of labour intensity, dirty udders and poaching, as well as the risk that when fed in very large quantities, goitrogens can interfere with iodine use by cows. Kale does not fit well into the rotation, because if it is to be followed by a ley, a spring reseed is necessary. This may be a problem as far as cultivations on heavy land are concerned and means that production will be lost while the ley is becoming established. In dry areas, soil moisture deficits during the establishment of the ley may also be a problem. However, in certain situations, kale can still have a role for supplementing winter feed and represents a valuable opportunity for stock to get some exercise outside during the winter months.

The crop can be drilled either on the flat or into ridges, depending on the weed control practices to be followed. If drilled on the flat, precision drilling is advisable. Mechanical weed control during the growing

Plate 10.2   Strip grazing organically produced kale at Manor Farm, Dorset   (W. Best)

season will be necessary to control problem weeds, particularly docks and fat-hen. As kale is a hungry crop, manure application at a rate of about 20 t/ha is advisable.

Other forage brassicas
Forage rapes are widely grown, both as pioneer and catch crops. They are characterised by rapid growth and high yields and are better yielding than kale over short periods. They are not very hardy and need to be used before Christmas. They are susceptible to club root, so their place in the rotation with respect to other crops needs to be considered carefully.

Fodder radish has very rapid growth and is club root resistant. It will outyield rape over an eight to ten week period, but quickly deteriorates in palatability after the onset of flowering. Its utilisation by stock is therefore limited to two to three weeks, but it is often used as a green manure. It is very frost susceptible and should be grazed off by the end of October.

Stubble turnips can be used as a catch crop for grazing in late summer/autumn and, as their name implies, can be sown immediately following an early harvested cereal crop. Stubble turnips have good club root resistance. They are grown for the leaf rather than the root and should be fed before the end of the year, as the roots become woolly and unpalatable and the leaves can be damaged by frost.

Cereal-based forage crops

The range of cereal-based forage crops including forage rye, triticale, arable silages and forage maize, can have a useful place in an organic rotation. They have a particularly important role as a catch crop to avoid nitrate leaching and, in the case of forage rye or triticale, as an early bite in the spring for cattle and sheep.

Forage rye
Forage rye is the most important of the forage cereal catch crops in organic systems. It can act as a winter cover crop for a field at the end of an arable rotation before a spring reseed, as a winter cover crop after early harvested vegetables such as potatoes and as a break crop before direct reseeding, where the old ley will be ploughed in in the summer and the new ley reseeded the following spring.

In horticultural and predominantly arable systems, forage rye will tend to be used primarily as a green manure for soil protection and nitrogen retention, whereas livestock producers place more emphasis on its ability to provide an early bite from late February onwards, depending on the sowing date. To get the most benefit from forage rye as a catch crop for whichever purpose, early sowing is essential. The crop must be drilled by mid September or even late August, although there can be problems obtaining seed this early unless one-year-old seed is used.

The seed may be broadcast or drilled at rates upward of 190 kg/ha as appropriate, and rates as high as 250 kg/ha are quite frequently used to get dense establishment. Although both grain and forage varieties are potentially suitable, the forage varieties such as Lovaszpatonai are usually preferred.

Forage rye is normally strip grazed. The nutritional value of forage rye is not equal to that of spring grass, but it can provide considerable bulk and is a green food, representing a useful supplement for late winter feeding to postpone turn out onto pasture. The heading date can be as early as late April and rye becomes considerably more fibrous and less digestible after this time. Where it has become too mature, some producers have considered allowing the remaining crop to proceed to harvest, but forage rye gives low grain yields, and the crop may not recover well from grazing and poaching unless it is grazed only lightly very early in the season.

Other forage cereals
Of the other forage cereals, triticale fulfils a similar role to forage rye, but there is less experience with its use in organic systems.

Arable silages are often used as a summer catch crop, particularly where another crop or reseed has failed, so as to ensure a certain level of production during the growing season. Arable silages tend to consist of a mixture of species such as oats and vetches and may include field beans, forage peas and other species.

Forage maize is widely grown on organic farms in central Europe and and can be grown in southern parts of Britain, providing full use is made of techniques such as undersowing with legumes as described in Chapter 5.

REFERENCES AND FURTHER READING

Herbal leys
Foster, L. (1988) *Herbs in Pastures. Development and Research in Britain, 1850–1984.* Biological Agriculture and Horticulture 5 (2): 97–133
Turner, N. (1955) *Fertility Pastures.* Faber; London
Woodward, L. & Foster, L. (1987) *Herbs in modern organic systems.* New Farmer and Grower 14: 22–25

Woodward, L. & Foster, L. (1985) *The possibility of herbs in ley farming*. New Farmer and Grower 8: 12–14

Herbage varieties
Ingram, J. (1985) *Grass and clover varieties—the current scene*. In: *The Grass Ley Today.* Proceedings of the 18th NIAB Crop Conference, 1984. National Institute for Agricultural Botany; Cambridge
NIAB Booklets on recommended varieties, NIAB; Cambridge

Potential of grass/clover swards
Holliday, R. J. (1989) *White Clover Use on the Dairy Farm*. FMS Information Unit Report No. 65. Milk Marketing Board; Thames Ditton
Holmes, W. (1989) *Grass—its Production and Utilisation.*2nd edition. Blackwell Scientific Publications; Oxford
Newton, J. E. (1989) *Organic sheep production*. In: *Organic Meat Production* (edited by A. T. Chamberlain *et al.*). Chalcombe Publications; Maidenhead
Parsons, T. , Penning, P. D., Orr, R. & Jarvis, S. (1987) *Are grass/clover swards the answer to nitrate pollution?* In: Hardcastle, J. (ed.) *Science, Agriculture and the Environment*. AFRC; London
RASE (1988) *The grassland debate: white clover versus applied nitrogen*. Royal Agricultural Society of England; Kenilworth

Grassland management
MAFF/ADAS Grassland Practice series of booklets
MAFF/ADAS *Lucerne*. Advisory Leaflet 67
NIAB *Let's look again at lucerne*. NIAB Fellows Conference Report No. 1. NIAB; Cambridge

Fodder crops
See Chapter 5

Recent publications
Newton, J. (1992) *Organic grassland*. Chalcombe Publications, Ashford.

Chapter 11

Arable and Horticultural Crops

Arable and horticultural crops have traditionally been the most profitable parts of organic farming systems, with the products often achieving high prices and attractive gross margins. Though tempting, higher returns on these crops need to be balanced against lower returns from the fertility building phase of the rotation and the higher fixed costs of a diversified farming business. Even so, many farmers have been able to create successful organic systems based on arable and horticultural cropping.

CEREALS

Cereals are grown by the great majority of organic farmers, especially milling wheat and oats which are in demand and can attract substantial premiums. There is no single system for organic cereal production. Here, three farms are described, with an emphasis on the role and management of cereals within the different farming systems.

Kite's Nest Farm

At Kite's Nest Farm, the Young family grows about 40 ha cereals on soils ranging from heavy clay to Cotswold brash. These crops are integrated with the suckler beef enterprise described in Chapter 9. This plays a major role in the success of the system, which relies on the balance between livestock and arable enterprises for adequate supplies of manure, weed control and straw for bedding.

Methods of pre-drilling cultivations for cereals can vary according to the availability of implements. The system at Kite's Nest is based on the rotovator mainly because it has been found to be a very positive machine on banky land where traction is a limiting factor with trailed-type cultivators. However, in fields where perennial weeds are a problem, ploughing and dragging type cultivations are used to avoid chopping and proliferating deep-rooting weeds.

Coming out of the grass ley, the first pass with the rotovator takes place in July following a cut of hay or silage. If the field is known to be fairly clean, however, cultivation may be delayed until September when the field is ploughed. Rotor speed is as fast as possible and depth of tilth as shallow as ground conditions allow. Even so, tractor forward speed is little more than 3 km per hour with a 1.75 m rotovator. This gives a work rate of just over 0.4 ha per hour. It is important to restrict speeds on this first pass because turf particles will be reduced in size only marginally by further passes. This leaves a fairly fine tilth with most weeds and grass seeds retained in the top 2 cm or so of soil, ready to germinate when conditions allow. In practice it is usually necessary to wait for rain before maximum growth takes place.

Within three to four weeks the field will have 'greened up' with a mixture of weeds, germinated grass seed and regrowth from the original ley. At this point a second pass is made, again with the rotovator. Cultivation depth is inevitably greater at 5–6 cm, but tractor forward speed can be increased greatly, depending on available power, to 10 km per hour or even more in some circumstances. With the same machine this gives work rates of up to 1.5 ha per hour. The effect is to masticate all the green material. This usually kills a high proportion of the germinated weed and grass seeds but only in exceptionally dry conditions will the old ley be killed at this stage. Regrowth again takes place, with usually a small number of additional weed seeds being released from locked up soil particles and some being brought to the surface from greater depths. A third pass is needed some three weeks after this; this time a power harrow is preferred as it is faster, in most conditions as effective, and preferable in a wet season.

In ideal conditions it might be possible to plant into the seedbed obtained by these cultivations, but due to the far from ideal nature of most British summers and the tenacity of many weeds and grasses, a 'belt and braces' approach usually has the best results. This requires that the land be ploughed to a high standard.

Large manure-type skims are essential and, because of the compact nature of reversible ploughs, all the normal adjustments are just that bit more crucial. Exactly equal tyre pressures and equal left and right arm length are required, and inside wheel width must match the plough width. Depth of ploughing should be sufficient to bury all surface trash completely but not be so deep as to bring up subsoil or stones. Depending on soil type this is usually between 15 and 20 cm, although with models designed specifically for shallow ploughing it may be possible to achieve this at depths of less than 15 cm. Once ploughed, ideally before mid September, the land is best left until the seedcorn is ready to be planted.

Most of the land involved is heavy and this gives rise to two conflicting considerations: the land is often difficult to work down and the rough seedbed produced is less than ideal for maximum germination. One solution would be to work the land a few weeks before planting, even if that meant fewer pre-ploughing cultivations. The false seedbed produced would also help to control weeds. The conflicting argument to this is that so often in the autumn there are not sufficient drying days and ploughed heavy clay soils, once cultivated, can retain moisture for a long time. Therefore, unless the weather is set fair, the land is worked down, planted and harrowed-in all on the same day, even if that means someone tractoring through the night.

On light soils just one pass with the power harrow will prepare an ideal seedbed. On heavier land the use of a flat roll followed by one or possibly two passes of the power harrow will achieve the desired effect. The roll can only be used, however, if conditions are dry, otherwise there is compaction. Alternatively disc harrows can be used, but if ground conditions are at all sticky, wheel slip can cause deleterious effects.

The idea of planting a green manure crop such as mustard into the tilth from the first rotovating, and ploughing it in before sowing the wheat is attractive. The bastard fallow may seem like an expensive and time consuming way of controlling weeds, but it has to be remembered that in the case of Kite's Nest Farm, it is carried out only once every seven years, and if effected properly can ensure a succession of weed-free organic crops.

It has been found useful to have an idea of the wireworm population. When ploughing up old pasture this pest can present a serious problem, although after three or four years of ley damage is usually insignificant. Where a problem does exist, drilling both ways with an increased seeding rate is good practice. If there is evidence of wireworm in the spring, rolling is an effective way of reducing damage.

Seed corn is totally untreated and recleaned only. Cleaning standards must be high since there is little point taking great trouble to kill all existing weeds if more are introduced via the corn drill. Sowing rates are kept somewhat higher than average. It can be assumed that some loss to pests will take place. Wireworm, slugs and birds may take a toll. Also, tillering during the following spring is likely to be less than where artificial nitrogen is applied so that plant density needs to be greater in order to achieve optimum yield. A dense crop also has the ability to suppress any weed seeds that may have managed to survive previous cultivations. Sowing rates which have been found to be satisfactory on various soil types are: winter barley 190 kg/ha, winter oats 125 kg/ha and winter wheat from 225 to 235 kg/ha. For spring wheat, rates as high as 380 kg/ha are often preferable.

Standard 18 cm spacings between rows produce quite adequate results. Use of a narrow-spaced drill with spacings of 12 cm is also acceptable and gives a slightly more even crop with greater weed suppression ability. In theory, too, it should be possible to reduce sowing rates when using narrower spacings.

For first cereals after a ley, sowing is done as early as possible. This is in order to utilise as much of the available nitrogen as possible. The earlier the crop is sown, the more extensive will be the root system and less leaching will occur over the winter. Careful thought needs to be given to the choice of variety; at present only one milling variety, Mercia, is suitable for sowing in early September although others are available for sowing from 20th September onwards. Harrowing in the corn in both directions helps to reduce damage from birds, but seedbeds should not be allowed to become too fine.

If the cultivations have been carried out successfully no more needs to be done to the field until harvest time, unless docks were present in the original ley. If so, any that survive need to be pulled by hand in the early spring taking care to extract all the root.

The use of the rotary cultivator to clean the land does not necessarily have any advantage over the traditional 'bastard fallow'. However, in the average season land needs ploughing and cross ploughing at least three times to obtain a successful weed-kill, with many other cultivations in between. This is more expensive and time consuming.

The method of stubble cleaning adopted at Kite's Nest is exactly the same as that used in breaking up a ley. The only difference is that there is usually barely time to fit in two passes with the rotovator. However, as there is not such a surplus of nitrogen at this stage, it is acceptable to plant second cereals somewhat later. One problem which can be encountered in a wet autumn is rotovated land retaining moisture, with attendant delays in planting; for this reason the land is ploughed immediately after the second pass with the rotovator.

As far as disease control is concerned there is little that can or needs to be done after the corn is sown. With trusted, high-quality, high-resistance varieties, disease is rarely a problem. The lack of stress and excess sappy growth in the organic crop generally results in levels of disease no higher, and sometimes lower, than those found in commercial crops which have been sprayed. The planting of two or more varieties together is a technique which is becoming more popular especially where higher yielding varieties with lower disease resistance are being used.

Where things have gone reasonably well, it has been possible to grow three cereal crops in a row. However, in this system using no inputs during the cereal part of the rotation, nitrogen gets a bit low by this stage and unless there have been two successive good autumns, certain weeds can be gaining an upper hand. In this situation, winter oats have been found to be a suitable crop. Not only do they thrive in conditions of lower fertility than wheat, but their remarkable tillering ability shuts out all the light and does not give the weeds a chance. There is also a strong premium market for the oats, and the large amount of straw makes a superior cattle feed.

A ready market exists for all types of truly organically grown cereals, with strong competition between millers for the best types and premiums for both milling and feed grain. Premiums in 1989 were up to 100% over commercial prices for equivalent grades of non-organic grain, which has given an ex-farm price of up to £240/t for bread making wheat.

Yields on the farm in 1984 and 1985 averaged over 6 t/ha for wheat on 36 ha, 4.8 t/ha for oats on 5 ha and 4.2 t/ha for barley on 8 ha. The wheat yields have ranged on a field by field basis from 4.7 to 7.7 t/ha. More recently, as the arable area has rotated to the less fertile land, winter wheat yields have dropped to about 5½ t/ha. Yields from the small area of spring wheat grown are much more disappointing at around 2.2 t/ha. Growing spring cereals successfully on this type of land, which is cold and wet in the spring, is more difficult. Early attempts proved almost total failures. More recently, limited success has come through the use of a weed strike and later sowing.

Establishing grass leys after the cereal should present no problem, although in practice it often does, due to pressure of work at harvest and/or difficult weather conditions. There is just time to establish red clover leys after winter oats, if everything goes according to plan; but it is safer, if large acreages are involved, to undersow wheat or barley. In the latter case, the clover can sometimes do too well at the expense of the barley; it is advisable to plant a long strawed variety. With leys based on white clover, direct sowing after harvest usually works well, with those planted by mid-September almost certain to succeed.

Rushall Farm

Barry Wookey, who farms the 670 ha Rushall Farm in Wiltshire, is well known for his approach to organic farming, both in organic circles and further afield. The story of Rushall Farm and its 15 year conversion to organic farming is told in Barry Wookey's book: *Rushall—the story of an organic farm.*

In many ways, Wookey's system is similar to that operated by the Youngs at Kite's Nest Farm and other large-scale cereal growers, many of whom have adopted and adapted Wookey's system to suit their own farms and, in particular, the differing soil types. In contrast to the heavy clays at Kite's Nest, Wookey farms lighter, thinner calcareous soils on the edge of Salisbury Plain, which has significant implications for ease of cultivation and nutrient availability, particularly potash. As a result, the yields achieved by Wookey, averaging 4.5 t/ha, are somewhat lower than at Kite's Nest.

The most striking impression one gets when visiting Rushall comes from the virtually weed free cereal crops, achieved by a combination of preventing annual weeds from seeding, timeliness of operations and attention to detail. To prevent weeds from seeding while the crop is growing, they have to be controlled before the crop is planted. This is done using a false seedbed, where the ground is prepared for planting ten to 14 days before drilling actually takes place. Given normal weather conditions, the weeds should germinate to a stage where they are very susceptible and easily killed by the drilling and harrowing operations. Timeliness is also important; crops sown too early may suffer from infestations of weeds like blackgrass. The control of problem perennial weeds is achieved using a bastard fallow following the ley in the rotation. Although this operation is both machine and labour intensive, and involves considerable working of the soil, aerating it and encouraging possibly excess mineralisation of nitrogen, the benefits in terms of weed control are considered to be justified.

To meet the nitrogen requirements of the rotation, a break crop of one year red clover/Italian ryegrass is used after the second cereal, into which it is undersown. In addition, a pure red clover stand follows the third cereal crop as a catch crop/green manure.

Ballybrado—making stockless rotations work

Ballybrado Farm in County Tipperary, Ireland, is a 93 ha estate based around a large Victorian mansion with extensive cottages and outbuildings and occupying a fertile valley. Another nearby farm of 79 ha has been rented and is run in conjunction with Ballybrado, bringing the total farmed area to 172 ha with 91 ha arable, 40 ha permanent grassland and the balance in parkland and woods. The soil is a sandy loam and rainfall is about 1,000 mm.

Ballybrado is unique for two reasons: the arable land is farmed largely without stock; and it is a joint venture between a group of people, Richard Auler, Helmut Borchers and Josef and Marianne Finke, who moved from Germany bringing machinery and implements with them. The work on the farm is divided into distinct areas of responsibility which reflect the previous experience of the individuals concerned. Richard Auler is responsible for farming the arable part of the farm, helped part-time by an agricultural student. Josef Finke is responsible for the animals and the marketing, including all the paperwork such as export documentation which is required. Helmut Borchers maintains all the farm machinery, manages the timber and fencing operations and takes care of grain movement. Marianne runs the household, looks after the book keeping and, when time allows, works in the vegetable garden, helped by one of the students. In total, there are 12–13 people to be fed and the group tries to be self-sufficient, which requires a large amount of time for preserving and storing food.

The move from Germany was prompted by the problems of working and living in a polluted environment; Ballybrado was chosen because of the relatively unspoilt environment in Ireland and the suitability of the soils and climate. Considerable work was needed, however, to get the farm into the condition it is now. Half the tillage land is too stony, so a stone rake and elevator were purchased. The fences were poor or non-existent, and no one knew anything about cattle. It was decided after much discussion that it would not be possible to farm a 50:50 stock/arable system and so a stockless system was chosen, on the basis that manure from earthworms would be as good as manure from cattle.

The stockless rotation at Ballybrado is designed to sustain fertility without grass leys by substituting green manure crops and producing cereal crops for milling and marketing. In planning the rotation, many things had to be taken into account: the needs of the soil; the labour and machine power and its availability at certain times; weed control; and last, but not least, the marketing of the produce.

At present (and it may develop differently in future) the crop rotation is of three years duration and consists of:

- 1 year spring field beans;
- 1 year spring or winter wheat (50:50) undersown with clover;
- 1 year winter rye or spring oats (50:50) followed by a mustard catch crop over winter.

This rotation has been chosen for a variety of reasons. Wheat is the best crop for the market, but is very nutrient demanding. Field beans make the nitrogen available and are deep rooting. They are a hoeing crop and give the best weed control. Oats are not too demanding on the soil. They are able to absorb silica, which is used as a seed dressing because of the alkaline conditions it produces, and are generally very healthy. They can be marketed on the home market or exported to Germany. Mustard was chosen because of the considerable organic matter production even when late sown. The rotation also allows for a more balanced distribution of labour requirements, with stubble

cultivating, winter oats, wheat and mustard in the autumn and field
beans, wheat and undersowing of clover in the spring.

As much organic matter as possible is incorporated by chopping the
straw from long-straw varieties and using catch crops for green manures.
The rotation has been practised successfully for five years without
utilising the green manure crops for grazing (the farm carries a flock of
sheep on the permanent pasture and free range pigs in part of the
woodland), but some of the land is now being fenced and the green
manures may be utilised for grazing in future. The sheep are, however,
allowed on some areas of the stubble for a short period. Root crops such
as turnips for folding are also grown, but the high stone levels in some
fields have been very punishing to inter-row weeding implements.

Richard Auler has had some success using autumn-sown spring
varieties of wheat which are better suited to the market than winter
wheats and survive the relatively mild winters. A mixture of hard and
soft varieties (Sicco and Kleiber) is used, but the intention is to increase
the number of varieties in the mixture. Pennal is the preferred variety of
winter oats, producing a 1.5 m high crop with plenty of straw for incorporation. Whenever possible, home-grown seed dressed with silicate
powder is used. For undersowing, both white clover and subterranean
clover have been tried, but the latter really prefers a warmer climate so
now only white clover is used.

For field beans, the early maturing variety Troy is used. Trials were
carried out at 50 cm and 20 cm row spacings and it was found that the
wider row spacings produced a heavier crop which had a better
smothering effect on weeds. Because of the importance of legumes for
nitrogen fixation, various trials have been conducted using cereal grain/
legume mixtures such as peas and barley, vetches and oats and lupins
and naked barley. The use of mixtures proved difficult because in wet
years the legumes produce too much leaf which finally overpowers the
crop.

The Ballybrado soils have pH levels between 5.5 and 6 and potash
and phosphate levels are increasing. Neither lime nor phosphate and
potash have yet been applied. There are, however, plans to use a
Dolomite by-product to supply some lime and magnesium, and a silica
by-product to supply other minerals. Nitrogen from the field beans and
undersown clover and the incorporation of straw and green manures is
expected to maintain adequate levels of soil nutrients.

The weed control techniques have also developed over time. At first,
20 cm rows were used to allow for the hoeing of cereals. Now cereals are
sown at 7.5 cm because the time available when the field can be entered
for hoeing is too short. The competition and smothering effect with 7.5
cm spacing is better. Mechanical control is used after drilling. The first

Plate 11.1   Marketing is critical to the success of Ballybrado's near-stockless system
(J. Finkle)

step is blind harrowing using a Rabe spring-tined harrow or a light draw harrow four to five days after sowing, followed by harrowing at the four to five leaf stage. The later harrowing covers the seeds and increases the tillers. At the same time, the clover seed is incorporated.

The field beans are machine hoed until they are 20 cm high, but pre-drilling stubble cultivations are also very important to get the weed seeds to germinate. The mustard green manure smothers the weeds and protects the soil by producing a good mulch.

Richard Auler has also been working on a new machine to control docks, consisting of a subsoiler with four legs connected by a chain, which has been successful in bringing dock roots to the surface.

At Ballybrado ploughing is not carried out, unless necessary for weed control, except on stony ground. Richard Auler prefers the chisel plough for primary cultivation. He has developed a one-pass cultivation system on the more stone-free land incorporating sub-soiler, rotary harrow, drill and roller (based on the Weichel system described in Chapter 2). A heavy disc harrow has also been used, but proved less suitable under dry conditions.

Even under a stockless system, the cereal yields are good, with the aim being to produce 5 t/ha of wheat and 4 t/ha of oats or rye. The premiums available on these crops are sufficient even to cover a fallow period of a green manure crop.

The wheat and rye are marketed as flour. Bakeries in Ireland were slow to pick up the idea of bread made from organic flour. So Ballybrado started its own test market and baked bread twice a week, labelled it with the Ballybrado Bread label and sold it to 20 shops around the local town. After five months, 10,000 loaves had been sold and this was a breakthrough. Today the flour is sold to bakeries all over Ireland and to health food shops in 25 kg bags and now in prepacks as well. The business has become so successful that flour is brought in from other organic farms to meet the demand.

Oats are sold as oatmeal in Ireland and exported to a breakfast cereal company in Germany. Some of the rye is also exported to Germany, where premiums are higher. The field beans are used on the farm and sold on the conventional market at a good price.

Although the arable side of the farm is run successfully as a stockless system, livestock are an integral part of the farm to make the best use of the permanent pasture. The lambs are marketed to Superquinn and Quinnsworth in Dublin, who pay a 30% premium but this also has to cover transport costs. There is, however, also a market for the untreated wool in Germany and potential for the development of a sideline in the hides as well.

GRAIN LEGUMES

Grain legumes are becoming an increasingly popular crop on organic farms as the demand for organically produced stockfeed develops. They are a nitrogen fixing crop and make a good one year break between cereals, although much of the nitrogen fixed may be lost if the grain is sold off the farm.

Field beans

Grain legumes, and field beans in particular, form an important part of the arable-orientated and stockless rotations being developed by Gerry Minister at Luddesdown Organic Farms in Kent. Beans act as a break between cereal crops and have two primary functions: boosting nitrogen fertility and control of weeds, particularly of broad-leaved weeds; beans have little impact on the spread of couch. At Luddesdown, spring beans are preferred because they minimise the risk of chocolate spot disease and to reduce the problem of volunteer weed beans in the subsequent cereal crop (Plate 11.2), a problem which can be exacerbated with winter beans under certain winter conditions. This can be particularly serious in crops such as Maris Widgeon where the straw is to be sold for thatching.

Plate 11.2   Volunteer field beans can become a significant weed problem in winter wheat

A mustard green manure usually follows the cereal crop. In mild winters, the mustard may not be killed off by frost and therefore needs to be disced or cut to kill it. The mustard is allowed to wilt for ten days to two weeks before being ploughed in at a depth of 15 to 20 cm. This process usually takes place in January. On the light soils at Luddesdown, spring-tined cultivators are then used to prepare the seedbed. No other manures are applied.

The crop is sown between mid February and mid April at a rate of about 250 kg/ha and at a depth of about 7–8 cm. The later the crop is sown, the finer the seedbed needs to be. Although the later sown crops will emerge faster because of higher soil temperatures, the later cultivations lead to increased weed germination and tend to exacerbate the weed problem. The early sown crops, however, will be at risk from rooks for a longer period so March has been found to be the best time for sowing at Luddesdown. The earlier sown crops also tend to have a slight edge on yield.

The beans have been sown on both narrow rows (10 cm) and wide rows (45–50 cm) to allow for mechanical weed control. The crops sown on narrow row spacings have proved more successful at suppressing weeds because of the dense canopy which is formed. No yield difference between the two approaches has been observed. After drilling, or as soon after emergence as possible, the land is ringrolled to help with weed control and to create a fine surface to help moisture conservation without compacting the lower soil layers excessively.

Narrow spaced rows can be weeded with a spring-tined harrow at any time up to 15 cm tall. It does not seem to matter if the plants are damaged by the weeder, they recover rapidly; more damage is done by the tractor wheels. The crop can be gone through as many times as is necessary, but once the stem starts to hollow out and become less flexible, the damage will be more severe and an uneven crop will result. Wide spaced rows can be cultivated with a hoe or inter-row cultivator until the crop is too high for the tractor; but canopy formation is less effective, so that once cultivation is no longer possible, the weeds can develop relatively unhindered.

The main pests are the pea and bean weevil and the black bean aphid. The weevil causes notching of the leaves, but the crop seems to grow away from it successfully and no action is necessary. The black bean aphid sits on the underside of the leaf and causes it to curl and wither. Action against the aphid has also not been found to be necessary, which may be partly due to the lack of spindle trees in the vicinity. The main disease, chocolate spot, is usually present but is not a significant problem with spring beans.

Harvesting usually takes place in the first or second week of September. The variety Corton is preferred; it has an oval shaped bean which attracts some interest for human consumption as well as animal feed and has yielded an average of 4.5 t/ha in 1987 and 1988. The yields have been remarkably consistent over these two years, despite very different growing seasons. It would usually be appropriate to expect yields upwards of 3.5 t/ha in organic systems given reasonable conditions.

Field peas

Peas are grown by very few organic producers in Britain, largely because weed control and harvesting problems have not been balanced by a sufficiently high price to make their production worthwhile. John and Mary Wakefield-Jones have been producing peas for nearly ten years on their farm at Bromyard in Herefordshire, for the majority of that time organically, and can therefore be seen as pioneers in this field. Their approach is to grow peas following the grass ley, primarily for reasons of weed control. The ley is ploughed up in late winter, so as to avoid excessive nitrogen losses while still gaining some benefit from frost action to get a good seedbed. No fertiliser is used, although some basic slag (just over 1 t/ha) is applied to the grass ley in its last season.

The crop is sown in late March or April. Because undressed seed has not been available, seed with a dressing against damping-off diseases is used. The seed is drilled with a conventional cereal drill at 10 cm row spacings and a seed rate of about 250 kg/ha, depending on individual seed size and weight and the supplier's recommendations. The seed is harrowed in after drilling, but not rolled.

After drilling, nothing further is done until harvest. Weed control depends primarily on the rotational effect of the grass ley. If full use is to be made of the nitrogen fixed by both the peas and the ley, however, the peas would be better located at another point in the rotation, e.g. as a break between cereals or after a cleaning root crop such as potatoes. In these situations some form of mechanical weed control such as spring-tined harrowing may be necessary.

Pests and diseases are not usually a significant problem, although in the wetter years mildew can be serious. Pigeons are also a nuisance, but can be allowed for by a higher seed rate and dealt with to some extent by using a bird scarer.

As the crop ripens, lodging occurs. This creates difficulties for the combine, which needs to be suitably adapted. The use of spring barley broadcast at a rate of 30 kg/ha can help to provide support for the crop. The peas and barley can be harvested together and separated at the seed cleaning stage. The crop is harvested in September, ahead of field beans, which is one of its advantages. Yields range from 2.5 to 5 t/ha and are typically about 3.5 t/ha. Apart from lodging, a further problem at harvest is uneven maturity and seed shedding at the cutter bar if the crop has ripened too far. Once harvested, the peas need to be dried immediately because they can heat up very quickly when damp, even in the trailer. Augurs need to be adapted to handle peas, as they jam easily.

The pea haulm has a market as a feed for goats and can be sold for up to £1 per bale (1989 prices), so that if the weather is suitable, baling may be worthwhile. If not baled, the haulms can be incorporated into the soil. Peas leave a particularly good tilth for the following crop.

Field-scale Vegetables

Many farmers on the larger mixed organic farms like Bwlchwernen Fawr near Lampeter, farmed by Patrick Holden and Nick Rebbeck, Pentood Uchaf near Cardigan, farmed by Jeremy and Valerie Harding, and Mangreen Hall Farm in Norfolk, farmed by Graham Hughes and Edward Howard, will consider the possibility of devoting part of the rotation to field-scale organic vegetable production and even making vegetables the main enterprise on the farm.

The integration of a field of vegetables into the rotation has many advantages, not least being the opportunity to cash in on the build up of fertility during the grass ley. In addition, if the rotation is seven years or longer, then pest and disease problems should be minimal and weed control may also be less of a problem. Income per unit land area can be substantially increased, with possible gross returns of £3,500/ha or more, giving attractive net returns even after allowing for expensive hand weeding and harvesting costs and comparing well with dairying.

There are, however, drawbacks, and these should be considered carefully before launching into field vegetable production. Vegetable growing is an exacting business as any horticulturalist will tell you. Management skill and time must be available to oversee the operation. Timing will be critical for primary cultivations, planting, weed control and harvesting. Specialist advice may also be required for selecting varieties, choosing plant densities and achieving uniform produce at the correct time of year.

At several points during the growing and harvesting season, extra labour will be required, in particular for hand weeding and harvesting. If there is an insufficient local labour pool available, this could cause serious difficulties. In addition, some vegetable operations may conflict with other seasonal farm work, for example hay and silage making conflicting with cabbage planting. Finally, to grow vegetables without having already arranged market outlets could result in poor prices for produce. Membership of a producer co-operative is regarded as essential by many growers.

Soils suitable for vegetable production should generally be easy working with a reasonable depth of lightish, fairly stone-free soil, not too sloping, especially if the slope is in two directions, and should contain adequate reserves of phosphate and potash and a pH value above 6.0.

Vegetable crops can be grown successfully at most points in the rotation, but if the crop follows long-term grass, potatoes should be avoided because of the risk from wireworm. Certain crops like carrots will not thrive in heavy soils and brassicas will perform better if the pH is closer to 7.0 or above. The range of vegetables now being stocked in organic retail outlets is considerable, but when grown on a farm scale ease of production, soil type and rainfall, timing and length of harvest period, market outlets and storage potential of the crop will all influence the choice of crop to produce. The main crops likely to be grown are potatoes, carrots, parsnips, beetroot, onions, brassicas of all types and some legumes.

Fertility is obtained primarily from the use of grass/clover leys being ploughed up and cropped for two to three years and then returned to grazing. Farm yard manure and compost are not used in large amounts and then only for one year in a three course rotation. Correction of major deficiencies is through the use of rock minerals, nitrogen is supplied through the clover ley and cultivations aimed at oxidising organic material built up during the ley period. When time allows green manures are grown, particularly over the winter period, to trap nutrients and preserve soil structure and to provide a bulk of organic matter for working in during the spring to feed subsequent crops.

A bed system is used for most row crops, the field being marked out after ploughing using either 1.5 or 1. 8 m centres; all subsequent cultivations are then done from the same wheel marks. This reduces problems of soil compaction, especially on difficult soils, and makes for far easier inter-row working. Seedbeds are made using either a single pass machine, such as a reciprocating power harrow, or a combination of lighter equipment such as spring tines and harrows. On some soils rolling may be necessary prior to seeding or planting. It is usual to adopt a standard toolbar, such as the Accord, for use as a seeder/planter unit and for inter-row hoeing and ridging operations. This removes the need for a wide variety of different machines. Various combinations of hoeing equipment are used, the most common being L blades and goose-foot hoes with plant protection discs. Thermal weeding using propane gas burners has become more common recently, especially so for the growing of weed-problem crops such as carrots and parsnips. Many of these practices are described in detail either in the first part of this book or later in this chapter.

The larger grower of organic vegetables is, however, able to specialise more and use modern methods of harvesting, such as complete harvesting, grading equipment and bulk bins, which are popular for use either to transport the crop to a central packhouse or for longer term storage. Scale of operation reduces the production costs and also the labour inputs.

The market for organically produced vegetables is crucial to their success as an enterprise. There has always been a strong demand, although it is not always consistent or reliable. Supermarket demand is increasing and becoming more important, but grade-out costs are high and large quantities are required to bring down unit costs. As the distribution of organic produce becomes rationalised by better transport systems and bigger marketing outlets, the problems facing organic vegetable producers are changing. Appearance and uniformity of produce are now much more important.

The pricing of organic vegetables is always a complicated question, but the rationalisation of marketing and distribution has resulted in a 'market price' becoming established as is already the case with organic grains. Organic fresh produce is attracting a significant premium over conventional produce, perhaps 25% or more. It is likely that for the foreseeable future the demand for organic vegetables will be sufficient to ensure an adequate return for producers willing and able to grow to organic standards. However, before rushing in and producing a glut in the early autumn, every grower should consider whether to sell the crop, and if possible make arrangements to hold a good proportion back until the late winter when demand is much stronger. With this kind of orderly marketing, plus plenty of communication and co-operation between producers, there will be plenty of room for expansion.

Potatoes

For farmers preferring a relatively mechanised crop which can be sown, cultivated, harvested and stored using machinery as opposed to hand labour, potatoes are an obvious choice. There is now a growing demand for crisping varieties of potatoes and a strong year-round market for ware potatoes, particularly from the supermarkets. The Potato Marketing Board has started to award quota specifically for organic production, which is helping to make quota restrictions less onerous.

It should be noted, however, that potato production is not necessarily the most profitable organic vegetable option and there are specific management problems which have yet to be overcome.

In a survey of organic potato production in Britain undertaken by the Potato Marketing Board in 1986, covering 72 farms producing between 0.51 ha and 17 ha, the frequency of potatoes in the rotation was every two years on three farms, every three years on three farms, every four years on 23 farms, and less than once every four years on 41 farms. Clearly the longest of these intervals represents by far the safest option. The most common main crop varieties were Desiree and Wilja, with Pentland Javelin and Maris Bard the preferred early varieties.

Weed control in potatoes under organic management should not present any serious problems. Most producers adopt the approach of chain-harrowing and re-ridging the field at least once and often twice between planting and emergence, the second treatment being carried out when the potato sprouts are just emerging. Subsequent cultivations between the ridges, followed by re-ridging, should eliminate the need for any hand weeding.

Pest and disease control is more of a problem (Figure 11.1), the main problems being wireworms and nematodes (particularly for those producers who risk potatoes immediately after the ley) and blight, for which there is as yet no satisfactory organic solution. In areas where early blight is prevalent, many growers expect blight to appear in late July or early August, and to proceed right through the crop at considerable speed. Blight is probably the main reason that organic potato yields are considerably below conventional levels, and for average tuber size of organic samples tending to be relatively small. Preventive measures include spraying of Bordeaux mixture, but Soil Association restrictions

Figure 11.1   Problems experienced during one or more stages of organic potato production.

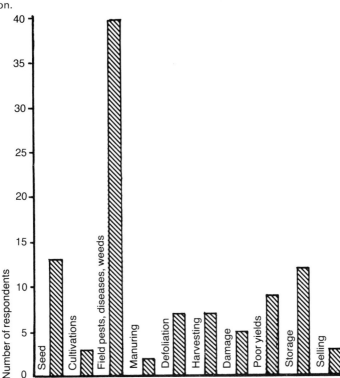

Source: Ginger & Aspinall (1986). Results from a PMB survey.

on routine applications of copper present problems for those producers relying on this method of blight control.

Recent developments at the Pentland field station of the Scottish Crops Research Institute (SCRI) include new varieties with significantly improved blight resistance. Sheilagh and Tina are showing particular promise and could well help to overcome the most serious problem that organic growers face.

After blight has become established, it is common practice among organic growers to defoliate mechanically, particularly where the variety is susceptible to tuber blight. However, nearly 50% of growers in the PMB survey allowed their crop to senesce naturally. This practice is relatively safe with Desiree, the most popular maincrop variety, as it is fairly resistant to tuber blight.

Yields of organically grown potatoes are usually 30% or more below those of conventional crops with 25 t/ha being the acceptable average for a maincrop variety (Table 11.1).

Plate 11.3   Potatoes at Neu-Eichenberg, Kassel, West Germany

Storage conditions are very important. Ideally, temperature control and ventilation should be available, particularly for long-term storage into late winter, as sprout suppressants are not permissible and potato samples showing sprouting are not acceptable to most buyers.

Organic premiums range from 30% to 100% according to year, season and quality, with farmgate and local sales attracting highest prices. Large quantities will require wholesale and multiple buyers. Quality specifications for these categories will be more exacting and premiums lower.

Table 11.1 Average total and marketable yields (t/ha) for popularly grown potato varieties in 1986 (conventionally grown crop yields are included for comparison).

| *Maturity class* | *Organic* | | *Conventional* | | *Relative yield (conv=100)* | |
cultivar	*Total*	*Marketable*	*Total*	*Marketable*	*Total*	*Marketable*
First earlies						
P Javelin	19.7	18.7	24.0	20.0	82.1	93.5
Maris Bard	17.6	15.9	27.0	25.0	65.2	63.6
Second earlies						
Wilja	26.6	25.9	40.0	33.0	66.5	78.5
Main crop						
Desiree	23.8	22.3	38.0	34.0	62.6	65.6

Source: Ginger & Aspinall (1986).

Carrots

Organic carrot production in Britain is expanding quite rapidly as demand increases, but techniques for producing consistent, weed-free crops have not been widely researched and there are still significant weed control problems to be overcome. On a field scale, therefore, farmers would be well advised to experiment with a small acreage first before growing carrots on a large scale.

Carrots can be introduced at several points in the rotation. They can successfully follow grass, or take their place after cereals or other arable crops. They have a medium to high potash and phosphate requirement, but do not require high nitrogen levels. Carrots are particularly sensitive to soil type, preferring lighter, free draining soils.

The two principle systems employed are ridges at 55 cm to 75 cm spacing and bed systems, where three or more rows are drilled on 1.5 m spaced beds. The advantages of ridges include good soil structure and depth for uniform rooting and maximum nutrient availability, a raised profile for easier hand and machine weeding, and easier harvesting, both manual and mechanical. The disadvantages are the need for specialist ridge drilling equipment, yield limitations due to a smaller number of row metres per hectare and a tendency to green top due to crown exposure during the bulking phase, which can also lead to frost damage later in the winter.

In good, fertile soil conditions where there is an adequate depth of light, stone-free loam, the bed system has much to recommend it. New mechanical weed control implements such as the brush weeder are also helping to overcome the weed control disadvantages of beds. Where soil depth is limited, or in generally more marginal conditions, the ridge system should be adopted.

Whether the ridge or bed system is adopted, it is important that the crop should be drilled with the slope of the field to avoid implement drift during subsequent mechanical weed control operations, although this will also increase the risk of soil erosion under susceptible conditions. In such situations, row crops are not appropriate anyway. The seedbed should be fine and level to achieve maximum germination and present a uniform soil profile for steerage hoe work.

Pelleted seed drilled with a traditional swede drill will give reasonable results, but precision drilling is preferable, as plant population is crucial in determining final carrot size and uniformity. A Hestair drill with a K27 or K42 cell wheel can be used in conjunction with graded seed to achieve appropriate plant populations. A Stanhay belt-type drill will give similar results. If the carrots are drilled on ridges, one way to achieve a higher plant population is to drill two rows on the top of each ridge using an appropriate twin culter. Plant populations of less than 26 plants per metre will produce large and sometimes unmarketable carrots. The best compromise between size and yield will be achieved between 26 and 65 plants per metre, depending on market requirements.

Berlicum and Autumn King varieties both perform well under organic management, the latter providing a more acceptable shaped root for retail sales. There are many modern F1 hybrid varieties, some of which produce excellent uniformity and yield, but performance under organic management seems to be more variable than with standard varieties. If the carrots are to be packed for multiple retailers there will be exacting size and weight requirements and F1 varieties may be the only choice. In any case, it is wise to have a clear idea of the intended market before choosing a variety.

The drilling date for early field carrots could be as early as late March, but for main crop varieties the best period is between late April and early June. Although the earlier dates will produce earlier crops, two significant advantages will be lost. Pre-drilling cultivations in late April and early May will dramatically reduce weed populations if the seedbed can be prepared beforehand. In addition, late May or early June drilling will avoid the first hatch of carrot root fly which is a serious problem for organic carrot producers. If the sowing date is any later than early June, the crop will have insufficient time to reach maturity, except in the early areas of the South.

Weed control is the most difficult and important aspect of field-scale organic carrot production. Stale seedbed cultivations, pre-emergence flame weeding, mechanical cultivators, hand weeding and, in a ridge system, ridging are the chief techniques available.

Stale seedbed cultivations in the spring can deal very effectively with early germinating weeds. This can take place either before or after seedbed preparation. After drilling, the carrots will not emerge for between seven and 14 days, depending on soil warmth and moisture and depth of drilling, which should be between 0.5 and 1 cm. During this period, and preferably as late as possible before emergence, flame weeding can take place. A propane flame weeder, usually tractor mounted, can be driven through the crop at around 5 km/h and cover up to 3 metres depending on its size and gas capacity. Good weed control is achieved using this method, leaving only the late germinators as a problem. Inevitably, some of these will come up between the carrots or too close to be controlled mechanically. Since the carrot plant is a slow grower when newly emerged, hand weeding for this last category of weeds is difficult to avoid except in very weed free soils. The cost of this final operation will vary greatly depending on the effectiveness of pre-emergence control, and of brush or steerage-hoe weeding.

Plate 11.4   Carrots produced on a bed system at Eichwaldhof, Darmstadt, West Germany

Rotary brush weeders with splash guards, or effectively set steerage hoes, are capable of passing very close to the carrot rows, as long as the drill lines are straight. In these circumstances, hand weeding costs can be below £125/ha, but where control measures are less effective, the cost of hand weeding can increase to more than £600/ha.

Hand weeding, although slow and labour intensive, can be very effective when carried out properly and, with organic carrots at least, can definitely be cost effective. It is best carried out with a hand hoe in the early stages. The aim is to take virtually every weed out in one pass or to weed twice, the first time very quickly and the second taking out the few weeds left after the first pass. The hand weeding stage ought to be completed by mid August at the latest. It would be unwise to consider organic carrot growing without a hand weeding operation unless the crop is grown on remarkably clean land.

At a later stage, tractor cultivations with an inter-row cultivator can be very effective in controlling weeds if carried out on a regular basis and combined with ridging up the soil against the carrot root crowns. This operation will also prevent green top and protect against frost damage later on. The final ridging operation should be carried out in the early autumn, before the tops drop too low to ridge soil up round the roots.

Yields can vary with soil type, fertility, weed control, row spacing, planting date and plant population. With light soils, well fed and watered (drought is obviously a factor, especially in the east), and with adequate control of weeds and correct population, yields of between 20 and 50 t/ha have been achieved on a 70 cm ridge system. On bed systems, some German growers have achieved up to 100 t/ha.

Plate 11.5
Transplanting vegetables
at Pentood Uchaf,
Dyfed (A. Martin,
courtesy of DBRW)

On a ridge system, hand harvesting is a possibility as the carrots will 'pull' quite easily. This method might be appropriate for smaller acreages or for harvesting for direct sales, but if large quantities of carrots are to be lifted for storage, a mechanical system will be necessary. The simplest solution is a vibrating undercutter which simply loosens the crop. In light soils, or dry areas, elevator lifters can also be effective, but neither of these methods deal with the tops, which must be removed by hand. Combined root harvesters are now being used by organic growers on a co-operative basis, but can run into difficulties in wet conditions.

To avoid the problem of an autumn glut and low prices, the carrots will have to be stored in some form. They will store for long periods at 100% humidity and 1°C without deterioration, but are much harder to wash after storage and grade-outs are also higher. Handling costs are increased and there is a need for capital expenditure on storage space. In addition, stored carrots are not as fresh as they would be if newly harvested. On the other hand, there are advantages from being able to lift the carrots under ideal conditions, with less vulnerability to bad weather and no risk of frost damage. Harvesting costs are also reduced.

Field storage is an alternative to a purpose-built store. One system involves ridging up the carrots in the autumn. This will be effective if the soil completely covers the crowns. Another traditional solution is earth clamps. In practice, however, the British organic carrot growers have failed to achieve continuity throughout the winter months and Dutch and Spanish organic carrots are imported at the end of the winter period.

Brassicas

The starting point with successful production of brassicas is the selection of a suitable area of land. Brassicas need plenty of moisture, but the land must be free draining to avoid water-logging. The soil should have a high degree of fertility and pH level above 6. If the crop is following a ley, manuring should not be necessary, but after any other crop where residual fertility is reduced, manures can be used at a rate of 25 t/ha. Seaweed preparations may be used as a foliar feed when the plants have reached a reasonable size and as a stimulus to soil microorganisms.

On farms where a ley is included in the rotation, a six to eight year gap between successive brassica crops is common and ideal for pest and disease control. The brassicas can follow directly on from the ley, but they could also follow cereals, field beans or a forage rye catch crop. If the forage rye can be established early enough, it can be grazed hard by sheep so that when it is ploughed in there is little green vegetation left and this makes the soil easy to cultivate.

The land should be ploughed or prepared early, in late January or February, and manure can also be applied at this stage. The soil should be cultivated thoroughly to get a really good, friable tilth, although in dry years cultivations may need to be restricted to conserve soil moisture. This work usually needs to be done a fair bit before planting to allow for a weed strike. The land is cultivated again just before planting, again depending a bit on the weather and the need to avoid losing too much moisture in a dry season or compacting ground in wet conditions.

The plants are usually bought in and transplanted. Good, even plants are important; they may be large or small but all must be of a similar size in order to obtain an even crop. In most cases, the plants should be planted as deep as is reasonably possible, with only the growing points showing, although one or two of the brassica crops do require to be planted on the surface.

Varieties should be selected which sit flat on the ground and smother weeds; varieties with leaves like spoons on a long stalk tend to favour weeds. Varieties also need to be selected to suit the market; beautiful cabbages which are too large are unsaleable. Adequate spacing between plants is important, but they should not be too far apart so as to maintain ground cover for weed control and to ensure that cabbages do not get too large for the market.

The field should be hoed or cultivated for weed control just before the first weeds show. Timeliness is critical. If they are already showing, it is almost too late. Cultivations need to be repeated any time more weeds appear—control is faster and more effective if they are not allowed to get beyond seedling stage. Regular examination of the crop to assess weed status is therefore essential.

Once the plants have reached a reasonable size, e.g. about 25 cm across, a deeper cultivation should be carried out. The intention of this cultivation is to push soil under the leaves so as to bury weeds within the rows. This operation can be carried out quite late, as superficial damage to the leaves is not usually important so long as the roots are not disturbed; the plants will recover quite rapidly. If these tasks are carried out properly, hand weeding should not be necessary, although where weather conditions and other factors do not permit cultivations at the right time, some form of hand weeding may become essential.

Plate 11.6   Brassicas at Pentood Uchaf, Dyfed   (A. Martin, courtesy of DBRW)

Although caterpillar pests can be controlled biologically using *Bacillus thuringiensis*, and other problems such as club root can be dealt with through the rotation, it may not actually be necessary to do anything against pests and diseases. In order for pest predators to become established, the pest has to be present in certain numbers and so complete control can be counter-productive. The pest—predator balance changes over time as the organic system becomes established; in the early stages of conversion, active pest control may be more critical.

ORGANIC HORTICULTURE

Many organic producers operate small-scale, commercial horticultural units, the management of which has not been dealt with in detail in this book. The intention here is to provide a general overview of organic horticulture rather than to attempt detailed descriptions of individual crops.

The serious growing of organic horticultural crops in Britain started during the mid 1970s, with the majority of the area being in the western extremities of the country, particularly west Wales. Many of the sites were on less than suitable land, with fairly unfavourable climatic conditions; many were established by people who generally had little experience, knowledge or capital, but were fuelled by strong ideals about establishing viable holdings aimed at being fully organic and sustainable in the long term.

These pioneers of organic growing have largely been responsible for generating the rapid increase in interest in organic methods and strong demand from the consumer. Despite the disadvantages they faced, such as site, soils, lack of capital, little research and development back-up, no market structure and lack of experience, many of the holdings were able to produce good crops at economic costs. During the last few years a sound marketing structure has been developed and this has encouraged more growers to make the change to organics.

Organic horticultural units can be roughly divided into two groups. The first is the small intensive holdings, often less than 4 ha, specialising in a large range of crops being sold to a fairly local market. Many of these holdings also grow a range of crops in polytunnels. There are few fully organic glasshouse growers, although there are several now converting their holdings. The second main group is the larger scale outdoor vegetable units, growing up to 15 ha of vegetables on a field scale, in systems such as have been described earlier in this chapter. These systems are usually part of stock farms with horticultural crops being

Plate 11.7   Cucumbers in polytunnels, Bad Nauheim, West Germany

grown on fertility provided by the grazing stock. They tend to specialise in a smaller number of crops, but in larger quantities making use of more specialised equipment.

Fruit production is almost non-existent, particularly so with top fruit. The reasons are the problems of achieving a high cosmetic finish on the produce (virtually no research or experimentation has been carried out in this area) and the fact that existing organic growers of vegetables are often on sites totally unsuited to fruit producion. Most of the best sites in the UK are already occupied by conventional growers who are loath to make a conversion to organics due to lack of technical development and advice. There is a limited quantity of soft fruit grown, usually on quite a small scale. There are several self-pick units operating with sales also of organic vegetables through their own farm shops. There is no significant production of hardy nursery stock or flowers, except for the production of herbs for sale in pots or for use dried.

The smaller intensive growers of 4 ha or less tend to concentrate much of their production on salad crops early in the year and field vegetables in the later part of the year. Fertility is provided by the use of animal manures brought in from approved sources and the use of green manures. Grass/clover leys seldom play more than a very small part in the fertility of the smaller units; the problems of having to graze off the material and the keeping of the necessary stock are often too much of a burden.

Since the tightening of the Soil Association standards with respect to brought in manures from intensive livestock units, some of these smaller holdings have had to rethink their long-term fertility position. This means a greater use of green manure crops and compost making and the thorough composting of any brought in manures. Many of these smaller units also rely on fertility imported in the form of organic fertilisers such as hoof and horn meal, seaweed products and various forms of rock dust materials. Some of these products are quite expensive and can only be justified on the higher-valued cropping programmes carried out by the very smallest growers.

Rotations are much more flexible than on the larger holdings due to the large variety of produce that is usually grown. For the smallest holdings this often means no crops of the same family for four seasons or more being the main criteria, with crops being divided up into roots, brassicas, legumes and others. Intercropping is often practised as a means of utilising available land and as a method of pest and disease control.

Cultivations are carried out initially by tractor equipment, the usual method being shallow ploughing, sometimes preceded by a light disc-harrow or rotovator to break up previous crop residues or green manures. This is followed by spring-tined implements and harrows to produce a seed or planting tilth. On the very small holdings much of the secondary cultivations may be carried out by pedestrian operated machines such as rotovators. These small machines are particularly useful where many crops are being grown as their light weight prevents any serious damage occurring to soil structure. Their relative cheapness and versatility is particularly relevant to the smaller producer who is unlikely to be able to justify the expense of a tractor and its equipment. Greater use can also be made of the land area as little space is required on headlands for manoeuvring. A rotovator may even be used for primary cultivations, mixing in manures and chopping in green materials, as land for cropping may become available in small areas due to the diverse nature of the cropping programme.

Weed control on these holdings will be by several methods. Rotation may play an important part where only a few crops are grown, less so as the number of individual crops increases and weed control through rotational practices becomes less feasible. For the very small intensive grower, much will be done by hand methods, such as hand hoeing and inter-row cultivations by small machines. Mulching with materials such as straw, compost, mowings, peat, FYM and any other available material is particularly useful for long-term or perennial crops but is seldom used for annual crops.

Black polythene mulches laid on the soil are becoming more common, particularly for longer-term crops such as strawberries. They have definite advantages in weed control and also promote more rapid, even growth, better quality produce, retention of moisture and help to overcome some pest and disease problems.

The more diverse the cropping programme, then generally the more work there is in weed control. Much will depend on the soil's weed seed bank, soil fertility, the nature of crops grown, suitability of equipment, available labour and weather, which can exacerbate weed problems if wet conditions persist. To overcome weed problems in a crop's early stages, reliance is placed on the use of either bare root or module/block raised transplants. These plants can either be raised by the grower or, more commonly, bought in; modules/blocks tend to be bought in as these require specialised raising equipment. Most of the modules used are in the block form, being grown by conventional nurseries using a proprietary compost, but excluding insecticides and pesticides. However, conventionally raised module/block plants are due to be phased out by January 1991 under the Soil Association's standards; conventionally grown bare-root transplants are already prohibited. Conventional liquid feeds are generally used to provide the necessary nutrition. For smaller areas, marking out and planting is often done by hand, two people being able to set about 1,000 plants an hour if well organised. Onions are grown using multi-blocks or sets, which can give high plant populations and a good head start over the weeds.

Pests and diseases are less of a problem on well developed organic holdings that have good soil structure and fertility. The main problem area is with brassicas, especially swedes and turnips. The main pests are cabbage root fly, flea beetles and caterpillars. Cabbage root fly is a particular problem in seedbeds, often causing the grower to discard a high percentage of plants. The only really effective method of control is the use of a crop protection material such as Growtect or Argyll P17 laid over the seeded beds and left on until the plants are removed to their final position. It is often necessary to remove the material once to allow for hoeing. Drilling can be delayed until after the main pest egg laying period for swedes, but more protection may still be desirable in some seasons and some parts of the country until quite late in the season. Flea beetle damage can largely be prevented by the same method and also by the use of irrigation. Another method is to create a dust of fine soil over the plants early in the day when dew is present; this makes them less palatable to insects. Both these pests are generally only a problem in seedbeds; once the seedlings are planted out damage, if it occurs at all, is not of economic importance.

Other pests are generally of little consequence. There are approved materials for use against a wide variety of problems, such as Savona, derris, pyrethrum, etc., which are usually effective if properly applied. Outbreaks of disease are unusual; there are few approved materials for their control and good husbandry methods are an essential element.

The harvesting of small scale crops is generally the most labour intensive part of the operation. Few small growers have the specialised equipment needed to harvest efficiently the wide range of crops often grown on the smaller holdings. Crops such as brassicas are the easiest, these being cut and packed direct into crates on the field. Legumes require hand picking. For this reason few are grown, except the higher value types such as sugar-peas. There are no organic growers producing large scale pea and bean crops for the freezer or process market due to the expensive nature of the equipment needed. Root crops are lifted with undercutting equipment and hand picked, usually into bags direct, and weighed in the shed later, grading being carried out in the field. Salad crops are hand picked, cleaned, trimmed and packed in the field, being put into outer containers at the same time. Few small growers possess much storage equipment, most crops are sold direct from the field.

The very earliest and latest crops of salad vegetables are grown in polytunnels. These are usually grown exclusively with the help of brought-in manures or compost made on the holding from waste materials. Weeds are controlled by hand or poly-mulch, and pest problems tackled by biological control methods wherever possible. Rotations in tunnels tend to be somewhat slower than those employed outside. Fertility is maintained by the use of organic fertilisers where appropriate. Several crops can be grown in one season; the higher value of these makes the increased labour element justifiable.

OVERVIEW

A wide range of crops, both arable and horticultural, can be and are being grown successfully in organic systems. There are notable exceptions, such as sugar beet and oil seed rape for which there is no organic market, as well as crops for which there is a market such as top fruit, but where the husbandry problems have not yet been surmounted. In future years, as the potential for organic systems is increasingly recognised, it is to be hoped that greater research efforts will be made to find ecological and husbandry solutions to these problems. Meanwhile, it is still frequently the case that producers have to find such answers for themselves by trial and error, although the increasing availability of technical information and advice will mean that this experience can be more readily shared with others.

REFERENCES AND FURTHER READING

Blake, F. (1987) *The Handbook of Organic Husbandry*. Crowood Press
Ginger, M. & Aspinall, L. (1986) *A survey of organic potato production in the United Kingdom*. Potato Marketing Board; Sutton Bridge
New Farmer and Grower. Numerous articles on different arable and horticultural enterprises
Wookey, B. (1987) *Rushall – the story of an organic farm*. Blackwell

Chapter 12

Marketing and Processing

Close attention to marketing is an integral part of successful organic farming. The development of modern agriculture in Britain has suffered to a large extent from excessive attention to production and little concern about what the consumer actually wants. Organically produced food offers an opportunity to remedy this.

Marketing is not about selling, or persuading the consumer to buy goods which are not really wanted; the consumer will probably not be persuaded a second time. Marketing is about identifying products which the consumer does want, and supplying them at the right price, where and when and in the form which the consumer wants them.

At the same time, the demands of the consumer are not always compatible with protection of the environment or the development of sustainable agricultural systems. The marketing concept needs to be prominent, but it cannot be allowed to dominate totally. It has to be carefully integrated with the other objectives of the organic approach and, if necessary, the consumer needs to be educated as to why external visual appearance may not tell the full story and why continuous wheat production (even organic wheat) is harmful for both the environment and for the long-term future of the farm system. The consumer does, increasingly, understand the need to protect the environment and is willing to pay for it.

STANDARDS

The environmental and food quality benefits of organically produced food (Chapter 15) are not always immediately identifiable in the end product, at least not in the same way as external appearance. If the consumer is to support environmentally sound production through purchasing decisions, then some other form of identification is necessary. The *bona fide* producer also needs protection, otherwise what is to stop conventionally produced food being repackaged and sold as organic? There have been many cases where this has happened.

444

The deception can be more subtle. Potatoes 'grown with organic manures' are not organically grown potatoes, yet the retailer may not realise the difference, deliberately or otherwise. Many conventional farmers use farm yard manures on their potatoes and could sell their produce (at a premium in some cases) using the 'grown with organic manures' label. But, as should be clear from this book, the use of organic manures is only a small part of an overall organic system. Such deceptions, legal or otherwise, are harmful to the genuine organic producer because they distort the market and they confuse the consumer, creating a lack of confidence in genuine, organically produced food.

One way in which both consumer and producer interests can be protected is through production standards laid down by independent, competent bodies without direct commercial interests. In Britain, the Soil Association's standards (Appendix 1) are the prime example, although there are others. Some, such as the Demeter and Biodyn standards, are genuinely independent; they cover biodynamic farming which has certain additional characteristics (Appendix 2). Others are less independent, because they are associated directly with commercial interests and the risk exists that they can be adapted to meet the needs of what is a seriously undersupplied market. As a result, governments are intervening to bring more control to the market, either through legislation or the establishment of voluntary schemes such as the United Kingdom Register of Organic Food Standards.

The development of production standards is closely linked to the development of organic farming since the 1940s. The Soil Association, founded in 1946, provided a focus for a relatively small group of dedicated producers; but organically produced food was to have little impact in the commercial market place until the mid 1970s, with a few notable exceptions such as the Mayalls at Lea Hall in Shropshire, trading under the Pimhill brand name. By the end of the 1970s the proportion of agricultural land farmed organically in Britain was less than one twentieth of one per cent.

During the 1970s, the only established market in Britain was for organic cereals and in particular for milling wheat. Organic Farmers and Growers Ltd. (OF&G), a marketing co-operative set up in 1975, was the main supplier, selling grain for flour milling on behalf of its members. Since milling wheat was the only readily saleable organic product at that time, considerable pressure came to bear on OF&G members to specialise in wheat production. This led to problems with unbalanced rotations, resulting in poor weed, pest and disease control and low yields.

Rather than correct the rotational problems by introducing new crops for which there was little demand in the organic marketplace, the easiest solution was to create a second, half-way house grade: OFG2, which was later superseded by BFG2 and more recently by the Guild of Conservation Food Producers' 'Conservation Grade'. The second grades allowed selective inputs. Chilean nitrate, superphosphate, potash and a series of selective herbicides, including glyphosate and asulam, were all justified as it was claimed that these inputs had less environmental impact and effect on food quality than other more concentrated or toxic chemical inputs.

Other organic organisations, notably British Organic Farmers (BOF), the Organic Growers Association (OGA), the Soil Association and Elm Farm Research Centre, have always believed that this departure from genuine organic standards was a serious error, not just because they considered that the consumer was being misled, but more importantly because the inputs allowed had no relationship to the development of a sound sustainable agriculture and created as a by-product disillusioned arable farmers in eastern England who were unable to sustain economic production using these methods.

During the 1980s, there was a dramatic expansion in both the demand for and the supply of organic produce, in terms of both quantity and range of produce. By the end of the 1980s, the Soil Association's standards were adhered to by about 450 producers, many of whom were growers, but including an increasing number of larger scale farmers. The Soil Association's standards have also become a requirement of several of the major multiples.

Organic Farmers and Growers Ltd. organic standards have been carefully distanced from the so-called Conservation Grade standards and brought into line with the government's UKROFS standards. Apart from a few, arguably important details, they are similar to those of the Soil Association. Some 200 producers are members of and market their cereals through OF&G and OF&G Scotland and are required to meet the OF&G standards, although many are also Soil Association symbol holders.

The Conservation Grade standards are now operated by the independent Guild of Conservation Food Producers, but can still not be considered to be close to an organic standard, nor an acceptable standard for the transition or conversion phase.

This brings us to an increasingly important problem: standards for production during the conversion period. Most organic standards require that land has to be managed organically for two years before it can qualify as organic. There are two main reasons for this. The first is to reduce the risk of pesticide residues contaminating organic produce. The second, more important, reason is to ensure commitment to organic management of the land on the part of the producer and to give a chance for the ecological balance which is necessary for the proper functioning of an organic system to become established. It would not be in the genuine organic producer's interest for other producers to alternate annually between conventional and organic production, thus obtaining the benefits of both chemical aids and the premium market, while yielding no significant benefits to the consumer in terms of environmental protection or improved food quality.

During this conversion period, producers may suffer lower yields, but are unable to benefit from premium prices and may therefore suffer financial difficulties. In order to support these producers, many organic organisations have introduced conversion standards. In Europe these include several of the major organic groups, such as Bioland in West Germany and the biodynamic organisations. Biodyn is the conversion grade (as distinct from 'conservation' grade) for the Demeter standards of the biodynamic movement. These conversion grades generally require a statement on packaging to the effect that the farm concerned is in the process of conversion to full organic management. Produce sold under these conversion grades generally qualifies for a smaller premium than produce from fully organic systems.

In Britain, the Soil Association has hesitated over introducing a conversion grade, although it does allow livestock consuming a limited proportion of forage from land under conversion to qualify for its symbol. This is in recognition of the fact that otherwise it would in fact be five or more years before livestock producers could qualify for the Soil Association's symbol, assuming a step by step conversion over a period of years, and this is a major disincentive to the development of organic meat production. The Soil Association does now permit produce from farms in conversion to be sold with the wording 'Soil Association Approved Organic Conversion' but not the Soil Association's symbol. Organic Farmers and Growers Ltd operates an 'in transition' grade, and in some cases, management of land under the now unconnected Conservation Grade could also qualify for time off the transition period to full organic status.

More recently, some firms have started to accept produce allegedly 'grown to IFOAM standards'. IFOAM is the International Federation of Organic Agriculture Movements. Its standards, last revised in 1989, are intended as a baseline from which national organisations can develop their own standards. The IFOAM standards are also used in some cases to decide whether produce grown to one set of national standards is acceptable to importers in another country. Individual producers cannot 'produce to IFOAM standards', because IFOAM does not, and never will, operate its own inspection and control scheme, although IFOAM is establishing a system for assessing national control schemes.

The plethora of standards is not helpful to the consumer or to retailers and causes confusion in the market place. The scope for fraud or inaccurate labelling is greatly increased in the absence of a universally accepted definition. Recognising this, there have been two important policy developments. In the European Community, a regulation defining organically produced food is in its final stages, which should improve the situation for intra-community trade. In the United Kingdom, the Government has established the UK Register of Organic Food Standards (UKROFS) under the aegis of Food From Britain. These standards were launched in May 1989, and are broadly similar to the Soil Association's standards.

There is at least one important outcome from all the discussions on standards which have taken place, and that is the recognition that standards cannot hope to define the end product in terms of chemical residues or other quality characteristics, as some national governments have tried to do. Standards for organic produce must be based on the production system—it is that which is being guaranteed, not the end product as such. This also means that accurate labelling is essential; descriptions such as 'produce from an organic farming system' and 'organically grown food', although less snappy than 'organic food', are less likely to cause confusion and misunderstandings.

The Market for Organically Produced Food

Consumer concern over high levels of saturated fats, sugar and salt in foods, as well as the risks from food additives and pesticide residues, has stimulated the demand for health foods and led to significant changes in the food sector, including the active promotion of additive-free foods. Furthermore, an increasing awareness of the environmental damage associated with the use of modern agricultural techniques has become linked, in the mind of the consumer at least, with the use of agro-chemicals. At the same time there is considerable public dissatisfaction with the consequences of European food surpluses.

These concerns have contributed to the development of the market for organically produced food. It is noteworthy that there have been no major promotion campaigns centring around organic food, neither has there been rapid and innovative product development, nor any development within the retailing trade to entice the consumer to buy these products in favour of others. The development of the market for

Figure 12.1 Organic standards organisations in Britain (1990).

Organisation	*Full organic status symbol*	*In-conversion status wording or symbol*	*Number of farms in Britain (1989/90)*	*Symbol(s) used on imported produce?*
Soil Association	SOIL ASSOCIATION · ORGANIC STANDARD ·	SOIL ASSOCIATION APPROVED ORGANIC CONVERSION	450	Yes
Organic Farmers & Growers Ltd	ORGANIC FARMERS & GROWERS	OF&G symbol plus wording 'IN TRANSITION'	200	Yes
Bio-Dynamic Agricultural Association	DEMETER	Biodyn	15	Yes
United Kingdom Register of Organic Food Standards	UK REGISTER OF ORGANIC FOOD STANDARDS UKROFS ORGANIC	None	Not fully operational	No

organically produced food has been largely consumer led, often in the face of industry indifference.

The market for organic food appears, therefore, to have developed as a result of exogenous factors relating to the wider arena of public concerns such as health, the environment, and resource use. These concerns have hitherto been largely ignored within the food industry, leaving discerning consumers to search out for themselves food that satisfies their selection criteria. Despite their lack of resources, organic producers have been remarkably successful in attracting the attention of these and other consumers. Undoubtedly, media coverage which has been antagonistic to conventional agriculture and relatively sympathetic to organic alternatives has been important in attracting this market. This has to some extent offset the lack of product promotion through commercial advertising channels.

Consumer demand and perceptions

A number of surveys (e.g. DMB&B, 1986; Fallows & Gosden, 1986; Presto, 1986; MAFF, 1987; NOP, 1987) have been carried out in this country and in Europe which indicate that there are three important areas of consumer concern with regard to food:

1. The 'healthiness' of food in general is now a significant attribute contributing to the consumer perception of quality of diet.
2. There is a concern over the risks of contamination of food by residues of agrochemicals.
3. There is a widespread concern over the quality of the environment and the negative impact of modern agricultural systems on the countryside.

There is undoubtedly a general concern about the health attributes of food. 78% of respondents in a survey of 500 food shoppers in three Berkshire towns, carried out by Elm Farm Research Centre in 1985 (EFRC, 1988), agreed that people should be more concerned about what they eat, while only slightly fewer (69%) felt that this concern should be reflected among food manufacturers when formulating their products.

65% of respondents in the EFRC survey felt that residues of agrochemicals were increasingly appearing in food. In a survey for the Ministry of Agriculture, Fisheries and Food (MAFF, 1987) fears over the use of food additives, and the presence of agrochemical residues on fresh produce came second only to a general concern over diet and health for the majority of consumers questioned. The Henley Centre for Forecasting's regular *Measures of Health* survey in winter 1987/88 found nearly 60% of respondents concerned or very concerned about the effect of pesticides on food, with a further 20% slightly concerned.

Turning to the impact of agriculture on the environment, 60% of respondents in the EFRC survey agreed with the statement: 'the countryside and the environment are being destroyed by today's farming methods'. Although 50% of respondents had heard of organic agriculture, perceiving the potential of it as an alternative, only 28% were able to give a reasonable definition of it as 'farming without chemicals, sprays or fertilisers'.

In a National Opinion Poll survey (NOP, 1987), 58% said that farmers should avoid using modern methods of farming and be compensated financially, with a further 28% saying farmers should be paid to avoid using them. A Gallup survey for the Daily Telegraph in 1988 found 42% of respondents felt that on the whole, modern farming was a bad thing for the environment, with 25% undecided. In a subsequent poll (Daily Telegraph, 14/8/89) 79% of respondents considered the British countryside to be in danger, with 66% prepared to spend 'a good bit more' to buy free range livestock products and 62% favouring policies to reduce intensity of land use. Clearly there is considerable concern among consumers about environmental quality.

In April 1989, a survey conducted by NOP for the marketing organisation MINTEL (MINTEL, 1989) found that 26% of consumers (21% of men, 30% of women) believe artificial chemicals should be banned with a further 42% (52:37%) believing their use should be significantly reduced, although 20% don't know. There was a marginal bias in terms of extreme concern among the middle-aged, although concern was relatively evenly balanced over the whole age spectrum and by region. Opinions were less extreme in the AB class compared with others.

Several surveys have been carried out in West Germany which support the conclusions of UK consumer research. In one survey (Vogtmann et al, 1984) 99% of the respondents considered that agricultural land was environmentally damaged. In a survey of 2,206 consumers throughout Germany (Allensbacher Institute, 1985), 86% of the total population associated the words 'environmental protection' with organic agriculture, while 30% of the respondents were frequently concerned that foodstuffs contain substances which are harmful to people (with a further 45% occasionally concerned).

The Allensbacher Institute survey also explored the respondents opinion on a number of positive and negative attributes of organic food. 57% of the total population (87% among respondents with a strong interest in produce from organic agriculture) considered organic food to be more healthy, while 45% considered that there were higher levels of nutrients and vitamins in organic as compared to conventional food (71% with a strong interest). 34% considered that organic food was more tasty. However, the response to the negative statements showed that 29% of respondents considered that organic food is no better than other foodstuffs while 24% considered that they are only a fashion or trend. One of the most important conclusions that emerged was that consumers did not necessarily expect organic produce to be 'residue-free' and were well aware of the problems created by general environmental contamination by pesticides.

In a survey of 250 Edinburgh consumers conducted by the Edinburgh School of Agriculture (Dixon and Holmes, 1987), 31% of consumers who purchased organic food did so for health reasons, 24% for the better taste and 16% because of the fear of pesticide residues. Only 10% gave environmental reasons, but interestingly, 7% stated opposition to agricultural surpluses as a reason for buying organic food.

The Henley Centre's *Measures of Health* survey (Winter 1987/88) found that 50.3% of the British population claim to eat at least some organic food, with 13% claiming to eat organic food frequently. Among the frequent eaters of organic produce, 59% gave health related reasons. The environment was not mentioned specifically as a reason, although concerns about the use of pesticides (mentioned by a further 29% of frequent eaters) includes both health and environmental aspects.

The Consumer Association (Which?, 1990) surveyed 1,477 adults who took a 'great deal or quite a lot' of responsibility for choosing food for their household in October/November 1989. More than 20% had purposely bought organic food. Of these, half did so for health reasons, with nearly one in three saying this was their main reason, with nearly one in three saying this was their main reason. One in three purchased organic food because they believed it to be free from chemical residues, with one in five giving this as their main reason for buying it. One in five mentioned better taste and one in six environmental reasons, with only 4% mentioning animal welfare.

It is not only in Europe where demand for organically produced food is very strong and health and environmental concerns score highly. In California, nearly 60% of regular organic food purchasers (23% of those surveyed) regarded food safety and environmental impact as the most important reasons for purchasing organic food, followed by freshness, general health benefits and nutritional value (Jolly *et al.*, 1989 a & b). Flavour and general appearances were rated relatively low. 62% of those surveyed reported purchasing organic food at some time, while 40% believed organic food to be better than conventionally produced food. The results were independent of gender, age and income. In New Zealand, Lamb (1989) found in a survey of 410 households in Christchurch that awareness of organic produce had increased from 46% in May 1987 to 65% in December 1988, with 1.2% of households identifying themselves as committed purchasers of organic produce, although 24.5% of households regarded themselves as health conscious and 8.6% as vegetarian or liberal vegetarian (limited meat eating).

A study carried out by Presto (Argyll Stores Ltd) on healthy foods and healthy eating (Presto, 1986) included a survey into the attitudes of a nationally representative sample of 1,013 British housewives to organic food. 75% of respondents agreed with the statement that "all food should be grown organically without chemical fertilisers". Almost the same proportion (72%) accepted that food produced without chemical fertilisers would be more expensive, although only slightly fewer (69%) agreed that it would be worth paying a lot more for food that is free from chemical fertilisers.

In the EFRC survey 61% of respondents said they would be willing to pay some premium, while 19% claimed to be prepared to pay a premium of more than 10%. In the Scottish survey over 19% of organic food consumers interviewed were willing to pay more than a 10% premium, while 42% were prepared to pay up to 10%.

In a survey commissioned by the NFU and carried out by Research Surveys of Great Britain in March 1988, 28% of the 1,831 people interviewed in England and Wales were 'definitely interested' in organic food, with a further 23% 'possibly interested'. Members of environmental/countryside organisations showed an even greater interest, with 71% of RSPB members and 75% of National Trust members questioned showing some interest in organic food. The NFU survey also showed that almost 50% of the people expressing an interest in organic food were prepared to pay 10% more for organic food, while 17% were prepared to pay a 10-20% premium.

A more recent survey, by the Henley Centre for Forecasting (Henley Centre, 1989), found that although most consumers say they are not very willing to pay premiums of more than 10%, and less than 2% were prepared to pay a 35% premium, only 17% of current consumers believed they were paying more than 10% extra, even though the retail price for organically produced food is 30-50% higher on average. The Henley Centre report concludes that consumers are relatively price insensitive at the point of purchase and that, if price premiums are kept below 30%, price is not a crucial influence on future prospects for the organic market.

There was a significant upsurge in demand for organic produce in 1989, as reflected in surveys by Here's Health and MINTEL. In the Here's Health survey one third said they would be prepared to pay 10% more for organically produced food, and a quarter 20% more. MINTEL found that only 20% would not pay extra for organic food with 14% don't knows and little regional variation except in Scotland (33% would not pay more and 13% don't know). Opposition to premiums is greater in the 65+ age group (27% would not pay more) and in the E social class (25% would not pay more). Interest in obtaining organically grown food was greatest in the 25–44 age group and was confined primarily to less processed produce such as fresh vegetables, which may be a factor of availability and familiarity with the concept. 67% of consumers were interested in fresh vegetables against only 24% for tinned vegetables. Around 50% were interested in fresh meat, eggs and milk and more than 40% were interested in bread and butter. Only 21% of the consumers surveyed said they were interested in wine and beer. More significantly, the proportion of consumers prepared to pay a premium of 50% or more increased from 1% in most previous studies to nearly 20% in the UK and 25% in regions like Wales and the South

West (Table 12.1). The shift in demand can also be seen from the quasi-demand curves in Figure 12.2.

Poor availability and lack of supply emerge as more important factors restricting purchases of organic food among consumers. In the EFRC survey, 77% of respondents stated that they would purchase organic food if it were more freely available, while in the Scottish survey, only 28% of all respondents stated that poor availability restricted their purchases of organic food, although over 60% of organic food consumers in the sample claimed that the product was not sufficiently available. The Henley Centre study found that among those who did not eat organic food, 42% mentioned problems of availability as a reason for not eating it while only 29% claimed simply to be indifferent to organic food. Of the 78% who had never purposely bought organic food in the Consumer Association survey (Which?, 1990), one third said they had never seen it or it wasn't sold where they shopped, while one quarter blamed high prices. Only 2% blamed poor appearance. Many were unaware of the existence of organically produced food, with only 38% aware of the existence of organic cheese and milk and 22% aware of tea, compared with 76% awareness of organic fruit, according to unpublished data from the Which? study. California consumers face similar problems with location of stores, time required for search and availability being cited as major reasons for not purchasing organic produce (Jolly et al. (1989a & b)).

Because of the high prices, organic produce is often thought to be of interest only to the higher income groups. Little research has been done on this in Britain, although the Henley Centre study did find interest in

Table 12.1 Analysis of entry into the organic food market as price premiums reduce, by TV region, 1989, base: 933 adults.

	All (%)	London/ TVS (%)	Anglia/ Central (%)	Harlech/ TSW (%)	Yorks/ Tyne Tees (%)	Granada (%)	Scotland (%)
100+% premium	1	1	1	3	1	3	–
100% premium	5	6	2	5	7	7	3
50% premium	13	14	12	17	13	13	6
25% premium	21	24	21	25	17	19	12
20% premium	27	33	24	32	22	27	16
15% premium	32	38	29	41	25	30	18
10% premium	49	60	48	57	40	41	28
5% premium	67	79	62	73	56	60	54
No premium	87	90	82	85	85	82	87
Balancing don't knows	13	10	18	15	15	18	15

Source: Mintel (1989).

Figure 12.2 Demand for organic food as indicated by willingness to pay premium

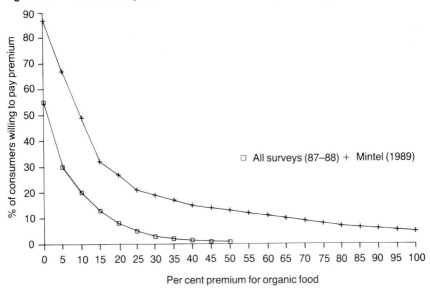

organic food biased toward the more affluent and more health and environment conscious ABC 1 classes. In West Germany, where more detailed market research has been carried out, various surveys (e.g. Böckenhoff & Hamm, 1983, Werner & Alvensleben, 1984 and Altmann & Alvensleben, 1986) have shown that purchasers of organic produce are not necessarily those in high income groups. If anything, the West German organic market is patronised more by low income groups. It is more likely that the purchaser will be young, highly educated, more socially critical and sufficiently highly committed to pay premium prices even though on a low income. The older purchasers tended to purchase more for health reasons and be more strongly committed to 'health foods' as such. According to the MINTEL study, the willingness to pay more for organic produce is remarkably constant across the full A to E social class and age spectrum, although the older groups and the DE classes tend to be less willing to pay the higher premiums.

The demands and perceptions of consumers within any market sector are inevitably hard to determine and quantify. However, it is clear that there is a significant demand for organic food which would appear to be more than a passing fashion or trend, and which is likely to increase in the future. Demand is not necessarily limited to a few specific social or income groups, although the committed purchaser tends to have a higher level of education. Generally speaking, the more altruistic, environmental concerns are tending to become increasingly important reasons for buying organic produce, especially among younger people, while the more self-centred health motives may be more important for older age groups and for opportunistic or occasional purchasers.

Structure and size of the current UK retail market

The market for organically produced food developed relatively slowly until the mid 1980s, resulting in a fragmented marketing structure. Individual organic farms scattered across the United Kingdom would sell their produce at the farm gate, through local retail outlets or perhaps by local delivery rounds. The quality of the product may have been variable, but the direct contact between grower and consumer led to a trust that developed a firm marketing base.

With the recent expansion of interest in a wider market, there has been dramatic development in the channels for marketing organic produce, especially vegetables. The small individual grower could not offer year-round continuity of supply, nor the packing and marketing skills which a larger market demands. With these problems in mind, several growers throughout the country have established cooperatives to pool resources for more effective marketing. National collection and distribution networks have been developed by two companies, Organic Farm Foods (Wales) and Geest.

Organic produce is now sold by most of the big supermarket chains in Britain (Figure 12.3). The major demand of multiples is for continuity of supply and, to meet this, Britain has become the main European importer of fresh organic produce. Although a proportion of these imports are 'exotics', staple products are also imported. The shortage of supply of British organic produce is illustrated by the fact that almost 70% of organic produce is imported. In spring 1989, a rapid surge in demand led to more than 95% of fresh vegetables being imported as homegrown stocks ran out. This clearly demonstrates the potential for expansion of the production base in the UK, since a major aim of this expansion must be import substitution.

A ready market exists for a wide range of produce from vegetables through to top fruit and cereal products. The market for vegetables is strongest among the winter crops and potatoes, particularly brassicas, carrots, leeks, onions, swedes, turnips and parsnips. These have already been widely marketed and do not present a problem from the point of view of visual quality. Animal and dairy products are also now being produced to recognised organic standards, and the potential in these fields is considerable. Although national marketing has been achieved for only two types of cheese so far, there is every likelihood that this could be developed.

Figure 12.3   Availability of organic produce in British supermarkets

	Fruit	Vegetables	Milk	Cheese	Yoghurt	Flour	Bread	Cereal	Rice	Meat, eg beef	Wine	Fruit juice	Tea	Coffee	Jam/marmalade	Crisps	Biscuits
Asda 50 stores	●	●		○	●		●	●		●					●	●	
Co-op ‡	○	○				○	●	○	○					○			○
Gateway 100 stores	●	●		●	●	●		●		●							●
M & S 30 stores	○	○					○										
Safeway All 270 stores	●	●	○	●	●	●	●	●	●	●	○	●	●	●	○	●	●
Sainsbury's 140 stores	●	●	●	●	●	●	●	●		●			●	○		●	
Tesco 100 stores	●	●	●	●		●	●	●	●		●		●			●	
Waitrose 88 stores	●	●	○	○	○	●	●	●			●	●	●			●	●

● indicates where organic food is available
○ indicates where organic food is only available in selected stores, subject to availability
No organic eggs, butter, pasta or poultry are available in supermarkets.
‡ Organic fruit and vegetables are available in 4 superstores only

Source: Which? (1990).

There are few reliable estimates of the current size of the UK market for organic food. Claude Hill in his report commissioned by Food From Britain (Hill, 1986) suggested that in 1985, sales of the limited range of vegetables available in the supermarkets were valued at £1 million, or less than 1% of supermarket vegetable sales.

In a survey of multiple outlets by Elm Farm Research Centre in 1987 (EFRC, 1988), only four of the groups already selling organic produce were prepared to state the value or proportion of their sales. Total sales of fruit and vegetables were valued at over £2 million, with bread and cereal products at £370,000 and milk and dairy products valued at £219,000. The latter two lines were sold by only one of the multiples contacted. Finally one multiple (with 112 shops in the group) stated that organic produce represented 0.5% of their total produce sales, although they would not put a figure on this.

EFRC's investigations of organic food processors, wholesalers and retailers produced estimates suggesting that the total market for all types of organically produced food may have amounted to over £34 million in 1987 (Table 12.2).

The Henley Centre survey at the end of 1988 (Henley Centre, 1989) estimated the market for organically produced vegetables alone at over £110 million, more than 1.5% of all vegetable sales. Other sectors of the market, however, were less well developed (Table 12.3). The figure of 1.5% is considerably higher than other studies have found and probably overestimates the real size of the market, although by how much given the recent growth in demand it is difficult to say.

A recent paper by the Fresh Foods Trading Director of Safeway plc (Murdy, 1989), which now sells a wide range of organic products including cheese, wine, yogurt, bread, prepared salads as well as fruit and vegetables, reported that despite premiums for the processed lines of between 33% and 45% and for fresh produce of nearer 90%, the total organic sales in 1989 by Safeway would amount to £4 million, or just over 1% of total group produce business.

Table 12.2 Market sales of organically produced food, 1987.

	(£ million)
Domestic produce	
Cereals	6.6
Fresh produce	16.8
Imports	
Processed foods	10.0
Dairy produce	0.5
Meat	0.25
TOTAL	34.15

Source: EFRC (1988).

Table 12.3 Size of 1989 organic market as % of total market for different commodities.

Vegetables	1.56
Fruit	0.63
Bread	0.57
Cereals	0.30
Dairy	0.25
Meat	0.21

Source: Henley Centre (1989).

Market potential

Various estimates have been made of the potential penetration of organic produce into the retail food market, and a potential of 5-10% of vegetable sales has been widely quoted. Although it is hard to determine potential consumer demand, the surveys reviewed above do indicate a very strong and sustained demand. One clear factor which currently limits market penetration is the lack of availability of produce.

Claude Hill's report for Food From Britain (1986) concluded that there was considerable potential among all retail outlets to increase sales of organic produce. The study was restricted primarily to sales of vegetables, and it was reported that two multiple stores predicted that the potential sales of winter vegetables and potatoes, (particularly brassicas, carrots, leeks, onions, swedes, turnips and parsnips) could reach 10% of present sales of this produce. Among the supermarkets alone this could represent sales of up to £13 million, and of £34 million for all outlets. These figures have clearly been exceeded already if the Henley Centre's results are accurate.

Multiple retailers
Elm Farm Research Centre surveyed 14 multiple supermarket groups in 1987. Seven of them were already selling organic produce and several of the others were thinking about it. There was general interest in the possibility of selling organically produced fruit, vegetables, bread, cereal products, milk and dairy products. None of the respondents claimed to be currently selling meat and meat products, although nearly half of the supermarket chains contacted were considering doing so. By 1989, virtually all the major multiples including Marks and Spencer were stocking organic produce, either on a regular or a trial basis. However, multiple retailers still cite poor appearance and shelf life of produce and inconsistency of supply as major problems (Henley Centre, 1989). This point is also emphasised by Murdy (1989) who argues that despite average premium of 90% at Safeway, the gross margins on the organic lines are 7% lower than conventional due to extremely restricted availability which results in rationing between stores and the expense of low volume transactions. Murdy comments that 'much of the produce is either deformed, non-standard or extremely small, appears very much to

Plate 12.1
Supermarkets are an important outlet for organically grown produce (A. Martin, courtesy of DBRW)

resemble rejected outgrades from conventional packhouses', attributing much of this to production on second-rate soils.

Multiple bakeries, greengrocers, butchers and dairies were also contacted by EFRC. There was no response from the dairies, perhaps demonstrating that the established dairies do not see any role within their operations for organically produced milk or dairy products. However, Unigate has started retailing organic milk in south-west England (Anon, 1989), which may reflect the increased general level of interest in organic food. Of the four bakeries who responded, three already claimed to be selling organic food although none would specify the value of their sales.

As in the case of the supermarkets, none of the multiple butchers were selling organically produced meat, but some were considering doing so, but this would depend on the level of supply. Clearly there is potential for retailing meat through multiple butchers, although the marketing challenges are perhaps greater in this than in any other sector.

The comments received from the EFRC survey of multiple outlets demonstrates that there is some confusion over the potential for development of the market for organic food. Supermarkets currently offer a significant outlet for organic produce. The claimed intentions of consumers to buy organic food if it were more available suggests that the projected figure of 10% of sales by the multiples who have trialled such produce may be an underestimate. The potential market may be much greater if the product were available and appropriately promoted. This challenge is still waiting to be met.

One approach which could be explored is the potential for emphasising the availability of organic and health food products by developing a 'store-within-a-store' concept in the supermarkets. This would enable the consumer to identify and purchase a range of products which are distinct from the normal foods available on the shelves. In one multiple store, this has been developed and appears to be very popular.

It is important to both multiple buyers and the consumer that an effective guarantee scheme is available, so that the authenticity of origin of produce is assured. This must be coupled with developing increasing supplies of produce through expanding the production base. It would also be beneficial to develop a generic advertising campaign for organic food, supported by high quality point of sale material. Unless these steps are taken, it is possible that organic produce could be removed from the shelves of supermarkets in what is a volatile and competitive marketing environment.

Potential demand from other retail outlets
The upsurge in interest in organic food shown by the multiple retailers has resulted in these outlets offering the most significant market to

wholesaler distributors. It is possible that this has resulted in less attention being paid to other potential outlets than they deserve. There are a large number of smaller retail outlets for organic produce, as indicated by the Soil Association's Symbol Holder and Organic Meat lists as well as Thorsons Organic Consumer Guide (Mabey *et al.*, 1990).

Health food shops offer an excellent example. There are more than 1,000 outlets in Britain, some of which have attempted to stock fresh produce with relatively little success, although they may stock some organically produced dry grocery lines such as flour, rice, pasta and pulses.

These outlets have traditionally concentrated on dry goods, and their lack of experience in handling, storing and retailing fresh produce does pose problems. However, with appropriate support and encouragement it would seem likely that these outlets could be developed to offer an important and consistent market for organic foods.

They have the particular advantage of the food being presented within the context of healthy foods and healthy eating to customers who are already committed to making such purchases. It is difficult to estimate the future market in this sector, since considerable obstacles must be overcome before it is possible to realise their potential.

Health food shops are just part of the spectrum of retail outlets which offer potential for the sale of organic produce. This ranges from local grocers and greengrocers to delicatessens, health food shops and whole-food shops, the latter often trying to combine environmental and health awareness. These outlets can be very important to the individual producer, because dealing directly with the retailer often allows a bigger margin to be obtained, although this must, of course, offset the additional costs of so doing.

At the end of this spectrum is the 'all organic' shop. These are usually run by committed individuals who may or may not be producers. Prominent examples include the Edinburgh Organic Shop and Philip Haughton's all-organic shop, Real Food Supplies, in Bristol. Real Food Supplies attracts regular customers from as far afield as Birmingham and Swindon, and aims to supply a wide range of organic produce including fruit and vegetables, meat, dairy produce, baby foods, grains and pulses. Philip Haughton is particularly keen to develop organic meat sales and to expand by opening additional outlets in the West Country and farther afield.

As far as the farmer is concerned, marketing directly to the consumer can also be a viable option. This can take the form of pick your own enterprises or a farm shop, either specialising in produce from the farm or importing organic produce from elsewhere to increase the turnover and viability of the operation.

One example is that of the Raymonts' farm shop at Garlands Farm,

Plate 12.2   Another popular organic farm shop is that at Lea Hall Farm, Shropshire, well known for its Pimhill brand flours.   (G. Mayall)

near Reading. The Raymonts' farm is situated within easy access of Reading town centre and the aim is to supply customers with as complete a range of produce as possible, so that they can do the bulk of their shopping from the farm shop. In their experience, high value goods with a long shelf life sell well in a farm shop and a commodity like frozen meat is bound to be successful. Accessibility to passing motorists is an important factor. Local advertising proved to be less successful at attracting customers compared with advertising in national magazines such as Here's Health. The most important aspect of a farm shop is to present produce really well and to promote organic food by talking to the customers and using good point-of-sales material.

Direct marketing can also take the form of a market stall in local towns and villages, which can, as in the case of Frost's Fresh in Aberystwyth (described in detail later in this Chapter), or Peter Redstone's ice cream shop in Torbay, become an established business in its own right. Mail order for non-perishables or delivery rounds may also be a possibility in certain circumstances.

Direct marketing, although time consuming, allows direct contact with the consumer, which can be invaluable for ensuring a solid market base and close attunement to customer demand. This has to be balanced against the additional legal and regulatory requirements, as well as all the other costs such operations entail.

The consideration of local retail outlets is not merely an issue of practicality, it can also become one of principle. Many of the more committed purchasers of organic produce are also committed to the concept of an environmentally sound lifestyle, involving decentralisation and reduced use of resources such as packaging. In some countries, there has been considerable debate as to whether organic produce should be sold through multiples at all, particularly vegetables which need to be highly packaged to keep them separate from conventional produce which is sold loose. Producer-consumer cooperatives, where producers and consumers together decide cropping programmes for a particular farm, and consumers carry some of the financial risks of a poor harvest, are becoming increasingly popular in certain countries, notably Japan, Switzerland and West Germany. Such cooperatives allow consumers and producers to reconcile their interest in organic food with wider social objectives. In most situations, however, the availability of supply exceeds the ability of the local, 'alternative' outlets to deal with them; supermarkets and other multiples which control most of the conventional market appear to be the only way forward. The major multiples in practice offer the only way of making organic produce available to a much wider range of people, many of whom are likely to be less committed and only occasional purchasers.

The debate over decentralised versus centralised marketing of organic produce has not been significant in Britain, but this does not mean that the issues raised are irrelevant. If the production system is meant to be environmentally benign, then every effort should be made to ensure that the processing, distribution and marketing systems are equally so.

Demand forecasts

Future demand for organically produced food is particularly difficult to estimate. The food scares in early 1989 saw demand shoot up, with Geest reporting a 400% increase for some of its lines and with demand for onions increasing from 10 to 40 t/week and potatoes from 3 to 80 t/week (Farmers Weekly, 2/6/89).

It is in this context that the Henley Centre's forecasts for future demand (Table 12.4) should be seen. An estimate of a 30-50% increase between 1989 and 1990 and growth continuing steadily but not exponentially for the next five years is probably conservative. Even assuming that organic produce accounted for only 1% of retail sales across all commodities at a premium price of 50% by 1994, this would represent a retail market worth £300 million in total, a seven-fold increase on Elm Farm Research Centre's 1987 estimates. A similar rate of increase has been predicted for the Californian market (Franco, 1989).

Table 12.4 Size of future organic market as % of total market for different commodities.

	1989 (%)	1990 (%)	1994 (%)
Vegetables	1.56	2.31	5.31
Fruit	0.63	0.91	2.03
Bread	0.57	0.80	1.72
Cereals	0.30	0.39	0.75
Dairy	0.25	0.36	0.80
Meat	0.21	0.32	0.76

Source: Henley Centre (1989).

THE ORGANIC MARKET FROM THE PRODUCER'S PERSPECTIVE

Much of the previous discussion of the market for organically produced food has concentrated on the consumer and retailing aspects. From the producer's point of view, the availability of intermediaries such as wholesalers, distributors and processors is often more relevant. The organic market divides fairly neatly into four sectors: fruit and vegetables, cereals and cereal derivatives, dairy produce and meat.

Fruit and vegetables

The fruit and vegetable sector is clearly the most developed part of the organic market. All the major supermarkets now offer organic vegetables in at least some of their stores. These supermarkets are supplied primarily by six packer/distributor companies: Organic Farm Foods (Wales), Geest, Donald Cooper (Kent), Sunnyfields Organic (South of England), Organic Farm Foods (Scotland) and Polytun in Scotland. There are also a number of general wholesalers including Geest, OFF (Wales) and OFF (Scotland), as well as several based in London (e.g. The All Organic Company, Forest Organic Produce and Tony Smith) and regional wholesalers such as Glen Broughton in Nottingham and Patrick Evans in Somerset. In addition to marketing through wholesalers or through packer/distributors, many individual producers market their own produce through local market stalls, farm shops and pick your own enterprises.

Increasingly, however, membership of a producer cooperative is being regarded as essential by many growers. This is because as supply increases, the local outlets tend to become saturated and individual growers may find themselves competing against each other. The work of marketing produce becomes more time consuming and is arguably better left to specialist marketing agents who are also better placed to assemble quantities of produce sufficient to bring about significant economies of size in terms of preparation and distribution.

There are currently four grower cooperatives: Organic Growers West Wales, Somerset Organic Producers, Green Growers (Herefordshire) and Eastern Counties Organic Producers. In addition, Organic Farmers and Growers Ltd markets a limited quantity of vegetable produce.

Organic Growers West Wales, described in more detail later in this chapter, is the oldest, but not the first, of the cooperatives. Started in 1985, OGWW had grown to about 30 grower and farmer members with more than 100 ha of vegetables between them by 1989. The first cooperative was in fact Cornish Organic Growers, set up in early 1985, but this failed primarily because of lack of grower commitment and because the cooperative attempted to market all its own produce rather than use another company as marketing agent. OGWW chose to follow a different path and adopted Organic Farm Foods (Wales) as its marketing agent.

Eastern Counties Organic Producers was formed in late 1986 by an initial seven growers in Lincolnshire with 30 ha (Bowers, 1988). A year later they had expanded to ten members with nearly 60 ha in vegetable production. The formation of ECOPs was helped considerably by the collaboration of Geest, and Geest are used by ECOPs as marketing agents.

Somerset Organic Producers started in early 1987 with nine producers and expanded in the first 12 months to 15, with about 50 ha of vegetable production between them (Stewart, 1988a). Somerset Organic Producers have not opted for one sole marketing agent, but rely on a range of outlets including OFF (Wales) and other wholesalers.

Green Growers is the most recent cooperative to be formed, originating from SHACS/Ledbury Farmers in Herefordshire (Hunter & Dart, 1988). It was founded in late 1988 with 15 members and also uses Organic Farm Foods (Wales) as its marketing agent.

All four cooperatives insist on Soil Association symbol standard produce. Since in most cases the actual marketing is the responsibility of the marketing agent, the cooperatives' role is in the coordination and planning of production to meet the needs of the buyers, particularly the multiples. This has often required growers to reduce the number of lines they produce and to change the varieties and timing of production to suit the needs of the market.

The supermarkets can also be very demanding in terms of quality (Clackworthy, 1988); rejected produce can have a significant impact on small producers. Membership of a cooperative therefore requires a high degree of discipline to meet the necessary quality standards and production targets. Full commitment is also essential, as trading outside the cooperative weakens the ability of the cooperative manager to plan and coordinate production and sales effectively.

Although grants from the Cooperative Division of Food From Britain have been available to help with the establishment of these cooperatives, members also have to be prepared to commit some of their own capital in the form of a joining fee. In addition, both the cooperative and the marketing agent will take a percentage of the price the grower receives in order to finance their activities. This can be as high as 15%.

In this brief review of the vegetable market, it is impossible not to mention OGA Packaging, the trading arm of the Organic Growers Association. OGA Packaging has played a crucial role in the development of a uniform, generic image for organic produce, supplying retailers with millions of paper bags and degradable plastic carrier bags since the mid-1980s. In addition, OGA Packaging supplies growers with a range of packaging for produce in bulk.

Cereals and processed products

The market for organically grown cereals was the earliest to establish and, along with vegetables, is currently the best developed. Premium prices of 50% and in some cases as much as 100% can be obtained, but as supplies increase and more companies move into the market, quality factors are playing an increasingly important role. The prices quoted for organically produced cereals need to be treated with caution. Transport and storage costs are a significant factor, with the unit costs for small loads being considerably higher and the distances to millers specialising in organic cereals longer. The result is that the net price eventually received by the producer can be considerably lower than the £170–£250 often quoted in the farming press.

Wheat
High quality milling wheat is at a premium, because millers have to import expensive French and North American wheats for blending. With very strong demand in 1989 as a result of three major companies entering the market to produce organic bread (UAM supplying Marks & Spencer, Allied producing the Allinson organic loaf and RHM producing the Hovis organic loaf) and imported wheat costing £300–400 a tonne, prices for domestic milling wheat reached £230 a tonne, more than twice the price of conventional wheat. 1989 was also a good year for organic wheat quality. In bad years, such as 1988, poorer quality wheats may be heavily discounted, not achieve a premium at all or even be rejected.

The important quality factors for milling wheat are bushel weight, protein content, moisture content, Hagberg level, impurities and pest damage. A bushel weight target of 76 kg/hectolitre is typical. If the consignment falls too far below this, it may be rejected because it will contain too many shrivelled grains and too high a proportion of bran to flour.

The minimum protein level required is usually 11%, but 10% is acceptable for organic samples. The quality of the protein is also important. When the protein is mixed with water, it forms gluten. A strong elastic gluten is needed for bread making, so that the gas bubbles are retained, whilst for biscuit making, a less elastic, more extensible form of gluten is required. The protein quality is usually determined by variety (see NIAB Recommended Varieties list for details), but the protein may also be destroyed by overheating when drying grain.

The Hagberg level should be at least 250 for bread-making wheats, although 200 may be acceptable in some seasons when supplies are short. The values for biscuit making flours are about half this, but the market is not so well developed and premiums will be lower. The Hagberg falling number test is a measure of the activity of alpha-amylase, an enzyme involved in building up the starch granules during grain filling and breaking them down to simple sugars during germination. When too much alpha-amylase is present (low Hagberg), there is too much sugar in the grain, which results in bread with a wet, sticky crumb structure and less volume, since starch forms the bricks from which the structure is made.

To avoid alpha-amylase build up, milling wheat needs to be given priority in both harvesting and drying, so that conditions likely to lead to the onset of sprouting are avoided. In wet summers, Hagberg values are likely to be low and prices paid will reflect this. Blends do not give a simple average; a 50:50 mix of wheat with Hagbergs of 80 and 250 will result in a value of only 102, which explains why millers are reluctant to accept low value samples.

Most contracts specify a moisture content of less than 16%, but 15% is becoming more common and ideally the grain should be dried to 14% in order to reduce pest problems and better meet the millers' requirements. Pest damage in store can also be minimised by careful attention to store hygiene and by keeping the store cool and well ventilated. The grain should be cleaned to remove impurities. A limit of 2% admixture is usual; dirty grain can be rejected.

Further details on quality requirements can be found in the Home Grown Cereals Authority Booklet *Marketing Guide for Milling Wheat*, published annually, and in a special feature on cereals marketing in the Summer 1987 issue of New Farmer and Grower.

Oats

The market for milling quality oats is also strong, with premiums as high as those for milling wheat. Bushel weight is the most important factor, the limit usually being defined as 50 kg/hectolitre. The other quality factors will include moisture content and impurities, but the market for milling oats is less sensitive to these factors than in the case of milling wheat.

Other grains

There is a limited demand for other grains such as malting and feed barley (for flaking), rye and fields beans and, apart from field beans, these also command a substantial premium. Samples which are rejected for milling may find a market at a lower premium for stock feed. In 1989, however the quality of and demand for cereals of any type for human consumption was so high that virtually no cereals were available at a reduced price for livestock feed. This created severe problems for livestock producers attempting to meet the new demand for organically produced meat.

Outlets

There is now a wide range of food companies with interests in the organic cereals sector, so that most producers should have no difficulty finding a market for suitable quality cereals.

The earliest entrant to the market was the producer cooperative Organic Farmers and Growers Ltd whose marketing arm, OF&G (Marketing) Ltd, is now 60% owned by Edward Billington and Sons of Liverpool. OF&G Ltd is one of several organic producer cooperatives in the UK, although it and OF&G (Scotland) Ltd are the only ones currently specialising in cereals. OF&G Ltd was established by David Stickland in 1975 and was the first cooperative attuned specifically to organic production. Its members are primarily arable producers and, although the cooperative has become involved in small scale trials with both vegetables and meat, the vast majority of its business is confined to cereals marketing. Membership of OF&G Ltd is currently around 200, farming 4,000 ha organically, which represents some 10% of the total area farmed by its members. In addition to the membership fee of £50, members must contract to sell all their produce through the cooperative and pay a commission of 4% on sales. OF&G Ltd operates its own standards and inspection programme.

Several producers, notably Michael Marriage, Barry Wookey and the Mayalls, have milling operations attached to their farms, but they differ in the extent to which they rely on wheat bought in from off the farm, and in the relative importance of the milling business as part of the farm operation.

The Marriages' milling business, Doves Farm Flour, is now the largest miller of organic cereals in the UK. Set up by Michael and Clare Marriage in 1978, Doves Farm Flour has grown from a small family enterprise into a major company with a multi-million pound turnover. To cope with its rapid expansion, the company has recently commissioned and moved into a new mill. Set in a one hectare site, the mill is the second largest stone-grinding site in the country and boasts large warehouse and grain silo facilities. More than 60 lines are produced with organic wholemeal flours being the most important.

Doves Farm Flour buys much of its organic wheat direct from farmers who must hold the Soil Association symbol, and currently purchases almost half the organic wheat available on the open market in Britain. In 1987, 2,000 t of organic cereals were processed. The flour is supplied to all types of users, including bread and biscuit manufacturers, baby food and dairy companies, caterers and a wide range of retail outlets. The company has developed a small and successful range of biscuits and also bread baked nationally under licence by a wide range of bakeries, including Goswells for supply to supermarkets in the south of England. Doves Farm Bread is the only loaf widely available to hold the Soil Association symbol.

Unlike Doves Farm, where the milling business now completely overshadows the farm operation, the milling operation at Barry Wookey's Rushall Farm is an integral part of the farm business, described in his book *Rushall—the story of an organic farm*. Barry Wookey grows organic cereals on about half his 670 ha farm in Wiltshire. The wheat is milled on the farm and sold as flour (at least 500 tonnes annually) and as bread and other products baked on the premises.

The Mayalls' operation at Lea Hall farm in Shropshire is the longest established of the three examples given here. The 270 ha farm was converted in 1949 and grows about 100 ha cereals each year to Soil Association standards. These are all processed through the farm mill and the finished products sold under the well known and respected Pimhill brandname. Some of the flour is turned into Pimhill brand bread at a local bakery. 700 tonnes of grain goes through the mill annually, about half of which is bought in from other farmers. A number of other farm-based or specialist mills process organic flour, including Shipton Mill in Gloucestershire, which supplies some supermarket instore bakeries as well as Cranks and a number of wholefood shops, and York Grounds Farm in Yorkshire, which produces flours under the Hider, Raywell and Shepcote brandnames. Breads are also produced by Sunnyvale (formerly Springhill) in Aylesbury.

It is not only producers who have taken a major stake in the organic market. Several national cereal processors and traders are also involved. W. T. Jordan (Cereals) Ltd have had an interest in organic cereals for some years, but initially found it difficult to obtain sufficient volume of organic grains and therefore assisted with the halfway house Conservation Grade standard in order to increase the volume of cereals available to them for processing. This move was widely criticised by the organic movement and Jordans have now committed themselves to a policy of developing their organic range using only Soil Association approved produce. In summer 1987, Jordans launched their 'Farm Organic' programme to encourage more farmers to grow organic cereals with the promise of contracts and advisory back up from Elm Farm Research Centre and the Organic Advisory Service. Their aim was to obtain 400 t of milling wheat and 1,000 t of milling oats in the first year to form a base for a range of organic products including a rolled oat breakfast cereal. Jordans continue, however, to use Conservation Grade cereals in a wide range of products not marketed as organic, but still attracting a small premium. This label may, in future, also be applied to 'in conversion' cereals which currently do not qualify for the Soil Association symbol but are managed organically. As such, they would more than meet the Conservation Grade standards.

Apart from Jordans, Morning Foods is a cereal processing company specialising in breakfast cereal products including Mornflake oats. They purchase both from OF&G Ltd and Soil Association symbol holders but all Morning Food products carry the OF&G symbol. More recently, the milling company Marriages (not connected with Doves Farm Flour) has also moved into the organic cereals market.

On the trading side, W. M. Gleadells is a large independent grain merchant based in Lincolnshire which has developed a specialist interest in trading organic grain, largely through the enthusiasm and commitment of Bill Starling, who manages the organic section. Gleadells can help to place parcels of grain with buyers for a modest commission. Bill Starling is also associated with the Lincoln Green Lager Co. which has started brewing organic beer and is seeking supplies of malting barley.

As the market for organic livestock products expands, the demand for organically produced feed grains is increasing. One company which has responded to this is T. E. Jones and Sons. The Joneses are organic farmers diversifying into a small milling business specialising in organic feedstuffs for animals. At present, they supply limited quantities of Soil Association standard barley (rolled or ground), mixed corn (containing wheat, oats, peas and barley) and 'in conversion' bean meal, processing 200 t of feed annually at their Batchley mill.

Overall, an estimated 10–15,000 t of organic cereals are traded annually in Britain, including about 4,000 t of imported wheat. The demand for wheat is estimated to be at least five times current supply and 5,000 t of barley are being sought in 1990. The market is so undersupplied that even screenings are achieving £140–150 per tonne as livestock feed and in-conversion standard grains are also being bought for feed at about £150 per tonne.

Milk and milk products

The dairy sector is still small compared with cereals or fresh vegetables, but it is much further developed than the meat sector. Even so, the majority of the milk produced on organic farms is still sold through conventional channels. At the end of 1989, there were 20 farms with the Soil Association symbol for milk production, producing more than 3 million litres of organic milk between them.

The majority of the marketing and processing initiatives which do exist for dairy produce have come from individual producers making and selling yogurt, cream, cottage cheese etc. on a small scale, or in some cases farmhouse cheeses and speciality sheep and goats' milk cheeses. Some producers have also explored ice cream as an option. One example, Rachel's Dairy, is described in detail later in this chapter.

Significantly, very few milk producers have explored the option of raw milk sales. The requirements for pasteurisation run counter to the philosophy of a 'wholefood' product and the difficulties inherent in setting up a delivery system for a scattered group of consumers are almost prohibitive. In spite of this, Path Hill farm, Old Plaw Hatch farm and Busses farm (which also takes milk from Elm Farm Research Centre), all in the south of England, have built up successful businesses supplying green-top milk at up to 40p/pint as well as yogurt and other products. The additional margin, even without a premium, over selling milk direct to the MMB may still make liquid milk sales worthwhile, but the ideal situation is likely to be where farms are situated close to centres of population where single dropoff points can cater for significant numbers of discerning consumers willing to pay a premium.

The increasing involvement of the major dairies, in particular Unigate and County Dairies, is transforming the market for fresh milk. Unigate takes milk from farms in the south-west and retails the milk locally via doorstep deliveries. County Dairies supplies three major supermarket chains. However, premiums remain low at between 15 and 25% (2.5 to 5p/litre) because of the problems associated with and the high cost of distribution from a small number of widely scattered farms. In the case of Unigate, the transport cost is about 16p/litre.

A number of organic farms produce farmhouse cheeses such as Penbryn, Llangloffan and Tyn Grug from Wales, Staffordshire Farm

Cheese and Danbydale and Botton cheeses from Yorkshire. Prices are usually low, with extra income coming from direct sales rather than an organic premium. On-farm processing of milk is, like direct marketing, labour intensive and plagued by other pitfalls. Hygiene control is essential, especially when using unpasteurised milk. Unpasteurised milk allows for the production of cheeses and other milk products with distinct individual characteristics, but contamination with the wrong types of bacteria can cause problems with the cheese making process. Careful attention also has to be paid to compliance with local authority health and trading standards regulations and a range of other relevant legislation. The perishable nature of many milk products means that storage, distribution and marketing need to be carefully organised. Considerable capital investment at all the different stages is likely to be involved; many producers have found that a gradual entry to processing using second-hand equipment considerably reduces the risks when things do not quite work out as envisaged. Many of these problems can be avoided with good advice and training. The revival of interest in on-farm milk processing following the introduction of EC quotas means that ADAS and agricultural colleges are now much better placed to provide this information than a few years ago.

An alternative to on-farm processing of milk is centralised processing by groups of producers. This approach is ideally suited to cheese production, because the quality and flavour of cheeses is determined by local conditions and each group of producers can produce a cheese or cheeses with unique regional characteristics in the same way that genuine farmhouse cheeses are unique to the individual farm. So far, the only example of such producer cooperation is Welsh Organic Foods in West Wales, described in more detail below.

Plate 12.3   Farmhouse cheese making, such as this Tyn Grug cheese from Dyfed, is popular among organic milk producers.

Meat

Meat is the least developed sector of the organic food market, but even in this area the outlook is changing rapidly. At least two major supermarkets in Britain have started stocking organic meat, and other parts of the retail trade are also keen to get involved. A survey conducted for one of the supermarkets showed 70 to 80 producers who could potentially supply organically produced meat. However, the supermarkets have specific requirements which need to be met before they will enter into the organic meat market (Hunt, 1989). These include guaranteed volume, with individual stores able to stock organic meat 100% of the time, guaranteed quality including slaughtering at an approved abattoir, and an acceptable price.

The prices paid to organic meat producers are adversely affected by a number of factors. Firstly, the customer is much more price sensitive with respect to meat than to vegetables, so that a significant premium is achievable only on the less expensive cuts. Secondly, not all the parts of the carcass are in equal demand, so that some parts of it will have to be sold into conventional channels, bringing down the overall value of the individual animal. Thirdly, the transport costs for bringing together scattered supplies and transporting them to abattoirs acceptable to the multiples, such as Lloyd Maunder in Devon and City Meat Wholesalers in Shropshire, who both supply Sainsbury, can be very high. This is further complicated by the fact that only three abattoirs – in Bristol, Tregaron (Dyfed) and Scotland – had been approved by the Soil Association by early 1990. The result is that premiums paid to the producer were less than 15%, but looked likely to increase to 20% by the end of 1990.

Another problem is the wide variety of 'alternative' meat production systems currently competing for the consumers' attention. A report by the Meat and Livestock Commission (MLC, 1987) identified several categories including hormone-free, real, conservation grade and organic. The consumer does not yet have a clear perception of how organic meat differs from all the other categories. The strengthening of organic standards with respect to meat production by the Soil Association and the inclusion in the UKROFS standards of special processing standards for meat, may make a significant difference in this respect, particularly if the European Community ban on the use of hormones stays in place and continues to render the description hormone-free superfluous as well as inaccurate. The MLC report estimated, however, that organically produced meat, which they inaccurately define to include conservation grade, would still account for considerably less than 0.1% of the meat market by 1992.

Even so, a number of initiatives are developing which could change this situation dramatically. In Britain, producers like Greenway Organic Farms, Peter Mitchell and Evan Owen Jones in Wales, and Ian Miller in Scotland (trading as Organic Meat and Products (Scotland) Ltd) are actively trying to obtain supplies of organic meat to meet retailers' requirements. Farm shop sales of organic meat and meat products are also important at Lea Hall farm in Shropshire and Path Hill farm near Reading. The formation of the Organic Sheep Society (affiliated to British Organic Farmers) by over 60 sheep producers in early 1989 also heralds a new spirit of producer cooperation, further strengthened by the launch of an organic meat producers' cooperative, Welsh Organic Livestock Farmers, in March 1990.

Kite's Nest Farm
One major success story in the marketing of organic meat by individual producers is that of the Youngs at Kite's Nest Farm. The lack of outlets led them to establish their own farm butcher's shop selling meat produced to Soil Association standards. Initially, interest was very limited and the confidence did not exist to charge more than conventional prices. The Youngs did, however, find that a very high proportion of the people who came once returned, either out of curiosity or because they were genuinely looking for organic produce, and became regular customers. They have also received considerable free publicity from television, radio and the farming and national press which has resulted in an enormous increase in customers.

The main things that had to be learnt were connected with butchering techniques. The 'Meat Trades Journal' and 'Butcher and Processor' became required reading. For a long time, certain cuts would sell out very quickly, but it was almost impossible to clear others. This was partly due to the fact that the majority of people who had sought them out were relatively well off and so tended to go for the more expensive joints, but they were also unaware of the full range of alternative ways that the meat could be prepared. For example, rolled top rib, which did not sell, can be prepared as two different braising steaks: London fillet and feather steak, which sold well and could be priced considerably higher. As trade increased, a premium price structure was established together with a scale of discounts for wholesale purchasers.

The licence to operate a butcher's shop has been granted by the Local Health Authority under the favourable regulations which apply to farm shops in general. About £3,000 was spent on converting some basement rooms of the house, including cold store, freezers, plumbing and equipment. The drawback is that only produce from the farm is allowed to be sold and cattle may not be bought in from elsewhere. However, scope for further expansion is limited anyway, with sales reaching a fairly consistent three sides of beef a week.

Retail outlets

Retailers like the Salad Shop in Aberystwyth, Phil Haughton of the Bristol-based Real Food Supplies and Charlotte Mitchell of Real Foods (not connected) in Edinburgh, as well as the specialist Land and Food Company, are also expanding into organic meat retailing. Even the more traditional 'alternative' meat suppliers such as the Real Meat Co. and the Pure Meat Co. are showing an interest in organically produced meat, and Billingtons are marketing organically produced meat in the North West of England under the name of the Carousel Meat Company. Superquinn and Quinnsworth in Ireland both stock organically produced lamb.

A further increase in the availability of organically produced meat will depend critically on the level of premiums paid and on the availability of reasonably priced, organically produced feed cereals. The serious shortage and high price of organic feed cereals in 1989/90 has proved to be extremely restrictive on meat and milk producers alike, particularly in the case of pig and poultry production. This may leave the door open to producers in countries like Denmark, who benefit from subsidies to convert to organic farming, to exploit the British market before domestic producers have had a chance to get established.

Alcoholic beverages

A review of the organic food market would not be complete without a brief mention of wines, cider and beer. These beverages can be and are being produced organically. The raw materials must meet specific production standards and the end product must also meet processing standards with regard to the use of additives and preservatives. There are a couple of organic cider producers in Britain, notably Aspalls and Dunkertons, and at least one producer of English organic wine.

Large quantities of wine are being imported from France, Spain and Germany and are being sold in Britain by a variety of specialist organic wine suppliers, including Vintage Roots in Reading, Vinceremos in Leeds and Haughton Fine Wines in Cheshire. Organic beer is being produced in Britain by the Lincoln Green Lager Co.; good quality organic beer is imported by Vinceremos and some of the other organic wine suppliers. The Henry Doubleday Research Association also supplies organic wines and ciders. Further details can be found in Mabey *et al.* (1990).

THE WEST WALES EXPERIENCE

In many respects, West Wales provides a model for the development of an integrated processing and marketing infrastructure which has considerable potential for producers in other regions.

Several organic farms and smallholdings were established in the 1970s, mainly by English immigrants who were attracted to West Wales by the relatively unpolluted environment and a generally less intensive approach to agriculture which together made organic farming an attractive proposition. West Wales is also the home of one of Britain's oldest organic farms, Brynllys, run by the Rowlands family. More recently, other indigenous Welsh farmers have started to convert their holdings.

The area now boasts a number of complementary processing and marketing operations, all of which have developed since the early 1980s. Three of the initiatives described below (Organic Growers West Wales, Organic Farm Foods (Wales) and Welsh Organic Foods) are based in Lampeter and have a strong emphasis on producer cooperation, making a substantial contribution to job prospects in an area of high unemployment. The other examples (Rachel's Dairy and Frost Fresh) are businesses based on or closely connected to individual holdings in West Wales.

Organic Growers West Wales

Organic Growers West Wales Ltd (OGWW) is one of the four grower marketing cooperatives in Britain. Formed in November 1985, it has grown from the original nine members to nearly 30 in 1989, with 100 ha of produce available and turnover estimated at £1 million.

In the process, some hard lessons have been learnt about marketing, in particular the need to use a marketing agent and the importance of a disciplined cropping plan. After an initial phase selling prepacked produce directly to Sainsbury and Safeway and marketing other produce surplus to the multiples' requirements through Organic Farm Foods (Wales) (OFF), it became clear that the cooperative was better equipped to specialise in production and members decided to engage OFF as marketing agents. Since 1986, all the cooperative's marketing has been undertaken by that company.

The cooperative has enabled growers to pool their resources, giving them access to new markets and allowing them to plan their activities as a group without the problem of competition between individuals. Members of the cooperative must hold the Soil Association symbol and growers are required to commit virtually all their produce to be marketed through the cooperative, the only exception being a small allowance for farm gate sales.

The scale of the cooperative's operation has made tight cropping and marketing strategies essential. Growers have had to adjust both the types and the varieties of crops they grow and their management systems including planting and harvesting dates. They can no longer always choose exactly what they would like to grow and when they harvest, but have to fit in with the overall cooperative plan.

The larger, more extensive holdings tend to grow the high volume, lower value crops such as swedes and potatoes, while OGWW aims to support the incomes of the smaller growers by encouraging them to produce higher value crops like salads.

One of the most important lessons that OGWW has learnt is that production should be market-led and it is essential to be prepared to tailor growing to fit in with market requirements. After consultation with the major customers, the marketing agents may advise the cooperative to opt for particular varieties of vegetable, although they may not always be the easiest to grow.

A major consequence of these detailed cropping plans is that growers have had to reduce the range of crops they produce. Although this makes sense from a cooperative point of view, it can be hard for individuals to accept if they have traditionally derived enjoyment and satisfaction out of growing many different crops. It can also be a factor which makes some growers unsuitable for cooperative membership.

OFF packs produce six days out of seven; growers have to harvest and deliver crops accordingly, aiming at all times to avoid any stockpiling, particularly of the highly perishable crops, although there are adequate cold storage facilities at the packhouse.

Selling to the supermarkets is by no means easy; buyers are highly demanding, specifications are tight and they will reject produce for the slightest flaw. From the growers' side, the biggest problem is quality control. Cooperative members have to be prepared to accept high grade out percentages on their crops and to harvest to strict size and quality specifications. Outright rejections of produce delivered are rare, but occasionally there are problems at the beginning and end of the growing season.

A close liaison between OGWW and OFF aims to keep members in touch with fluctations in crop prices and packaging and handling charges; a detailed monthly return is sent out giving a breakdown of all costs. The system is by no means without problems, but all aspects of the packhouse system are kept under constant review to try to bring down costs and improve returns.

With this in mind, and predicting a doubling of turnover in the next three years, OGWW is currently investing in new washing and handling equipment to be used in the packhouse. This investment and the employment of a coordinator was aided by grants from Food From Britain and advice from the Welsh Agricultural Organisation Society.

Research and development is seen as a priority. A series of one day seminars is held in the winter months to look at technical crop requirements and varieties. OGWW is also working with Elm Farm Research Centre and the Organic Advisory Service to provide advice to members.

Organic Farm Foods (Wales)

Organic Farm Foods (Wales) Ltd (OFF) is currently Britain's largest marketing operation for organically grown foods. The company is philosophically committed to the development of the organic market and to the Soil Association's standards. OFF occupies around 2,000 m² on an industrial estate on the outskirts of Lampeter in West Wales. The company's principle activity is prepacking for the major multiple retailers, but there is also a wholesaling operation supplying retail shops over a wide area of Britain and the company distributes cheeses produced by the neighbouring Welsh Organic Foods to the multiples. Annual turnover in 1988 amounted to more than £4 million.

In order to achieve this scale of trade in only two years, the company has had to invest heavily in modern equipment and facilities, including packing and grading lines, cold storage and processing equipment, with considerable assistance from the Development Board for Rural Wales. OFF's position in the market has enabled it to offer its services to other producers throughout the UK and to import considerable quantities of fresh produce from abroad.

The original staff of six has expanded to more than 100. The company runs two chilled distribution vehicles, including one 24 tonne capacity articulated lorry with a 12.5 m trailer. These vehicles travel daily to Bristol and the Midlands where the produce is delivered to national distribution depots. Return loads are usually taken, either of English, Scottish or foreign organic produce for packing in Lampeter. By 1988, some 120 tonnes of produce per week were being distributed from Lampeter, resulting in an annual 12 million overwrapped or bagged packages being offered to consumers on supermarket shelves, in addition to the organic produce supplied by other packers.

Organic Farm Foods has responded to the multiples' requirements for range, continuity, consistency of appearance and volume by asking its growers to accept rigorous production standards and by importing more

Plate 12.4   Pre-packing organic produce at Organic Farm Foods (Wales)   (A. Martin, courtesy of DBRW)

than 50% of the produce from abroad. Some of these lines, such as citrus fruits, grapes, out of season vegetables and exotics, will always be grown abroad, but British growers are expected gradually to develop their ability to supply staples such as potatoes, carrots and cabbage.

One of the problems associated with the insistence on high standards of external appearance by supermarket buyers is the economic utilisation of outgrades. OFF has made some inroads towards overcoming these difficulties by establishing a salad unit where fresh produce is washed and processed into ready-to-eat salad packs. This unit operates to extremely high hygienic requirements. It is able to use some of the carrot and cabbage outgrades, although overall quality specifications for produce must remain very high.

The company is expected to continue to evolve new ideas and products based on the needs of consumers and the interests of committed organic producers, thus influencing the development of the overall market for organic food while remaining competitive and increasing its overall market share.

Welsh Organic Foods

In 1984, four organic milk producers in West Wales, dissatisfied with selling their produce through conventional channels, came together to develop a cheesemaking and marketing initiative which would allow them to benefit from a potential, if at that stage non-existent, premium market.

Two of the group, Dougal Campbell and Gareth Rowlands, travelled to France to visit a range of cooperative cheesemaking enterprises, some of which were already producing organic cheese, and a decision was taken to concentrate on a soft Brie-type cheese.

A site was selected on the Lampeter industrial estate for reasons of existing three-phase electricity, adequate effluent disposal, grant aid and centralisation of buildings plus the added benefits of proximity to a local labour pool. Conversion of the premises to meet cheesemaking requirements proved problematic. The floors were levelled and surface drainage installed, and the two 140 m^2 units were brought up to good hygienic specifications. Plant and equipment, both new and second hand, were purchased from a variety of sources, both in Britain and abroad. Plant was installed capable of handling up to 2.5 million litres of milk per annum, despite an initial quota availability of only 800,000 litres. The total cost of the conversion and plant was around £100,000. This was financed by initial share capital, a loan from the Welsh Development Agency and an overdraft at the bank.

Initially, there were six farmer investors and two outside supporters. Three of the shareholders were supplier producers and a fourth, Dougal Campbell, was appointed managing director. The firm has since been taken over by Organic Farm Foods (Wales). Dougal and his wife Alex had extensive hard cheese making experience, having trained in Switzerland and developed a Cheddar-type cheesemaking enterprise on their own farm.

Production began in April 1986 with small quantities of the soft cheese, named Pencarreg. Milk was obtained from four established organic dairy farms in Ceredigion with a total of about 170 cows between them, in herds ranging from a small Jersey herd of around 12 cows to a 70 cow Guernsey herd. The 'mix' of dairy cows was around 50:50 Ayrshire/Channel Island, giving an average butterfat content above 4.5%.

The supplier farmers were asked to take particular care with milking hygiene, and regular bacteriological tests were taken. All the suppliers were required to hold the Soil Association symbol for milk production. Because all four of the initial suppliers had already been farming organically for some years, only minor management adjustments were necessary to obtain the symbol. These included moving away from routine antibiotic use for mastitis control and the introduction of organic cereal based concentrate rations.

The Milk Marketing Board (MMB) provides a 'clean' tanker to collect the milk on contract. The producers continue to be paid by the

MMB which then sells the milk on to the company. Suppliers receive an organic premium set at 1.5p/litre but, for cash flow reasons, this was withheld in the early stages. The premium was set as an incentive and was intended to cover any additional costs resulting from organic management.

Production and marketing difficulties including inconsistency due to the use of unpasteurised milk and the short shelf life of soft cheeses led to the introduction of milk pasteurisation and a second, harder cheese called Cardigan, which acted as a buffer for the variable demand for Pencarreg and made use of the surplus available milk supplies.

Plate 12.5   Turning Pencarreg cheeses at Welsh Organic Foods   (A.Martin, courtesy of DBRW)

It took a further two years before most of the marketing and production problems were sorted out. By mid 1988, sales and production were approximately in balance, with demand growing steadily and milk being imported from a farm on the Welsh border near Oswestry. The most serious difficulty facing Welsh Organic Foods is obtaining sufficient supplies of organic milk within the West Wales area. Although several local dairy farmers are considering conversion, it is unlikely that sufficient supplies will become available during the next two years.

The Welsh Organic Foods initiative has many parallels with cheese producing cooperatives in France, where producers in a particular region pool their milk to produce quality regional cheeses which reflect the particular characteristics of an area. It is an approach which could be adopted in other parts of Britain to produce other new regional organic cheeses of character and distinction.

Rachel's Dairy

Rachel's Dairy is the processing and marketing operation established by the Rowlands family at Brynllys farm near Borth in Dyfed. Initially, it grew from the processing of milk produced on the farm into butter and cream. Now the lines include clotted and untreated cream, natural and eight different fruit flavours of whole milk and low fat yogurt, as well as butter, cottage cheese and buttermilk.

Turnover in the first year for local sales of cream and some butter was £5,100. Growth has been rapid and by 1988, the dairy business turned over in excess of £120,000. In 1988/89, 100,000 litres of milk were processed, including 16,500 litres used for whole milk yogurt which was outside quota and allowed for an expansion of four cows in the size of the dairy herd.

A cautious approach to development was adopted from the beginning, care being taken not to overcapitalise in the early stages. Outlets were established and, crucially, these were followed through to discover what the problems were and to see if ways could be found to overcome them. People had to learn to handle the yogurts and cream differently, because they are produced without stabilisers and additives, making them not as consistent as consumers may be used to. At the same time, the demand has clearly been consumer led, with the consumer liking the product and asking for it specifically. Stores have then approached Rachel's Dairy for supplies as a direct result; the expansion in the Rachel's Dairy business has been solely through recommendation without the need for additional advertising.

Experience has shown the importance of staying in charge of the product as far as possible until it reaches the consumer. In the past other distributors were used, but this did not prove satisfactory. In order to ensure both quality and the right conditions for expansion, it has been necessary to take on distribution as well as processing, which involved the purchase of a 2 tonne refrigerated van for distribution and the establishment of a distribution round directly from the farm.

Several other problems have had to be overcome, including dealing with Trading Standards Officers and local Environmental Health Inspectors. One issue concerned the labelling of pots, with Trading Standards Officers requiring information to be printed directly onto the pots for each product. Not only does this require increased investment in stock, but it reduces the flexibility for moving in and out of individual products as the market dictates. Many of these issues are considered in more detail in Haines & Davies (1987).

The business cannot afford to stand still if it is to survive. Since its formation in 1984, Rachel's Dairy has been supplying a sector of the dairy products market which continues to expand. From the beginning, the products were test marketed in a 40 mile radius of Aberystwyth and, on proving themselves, have been supplied to the London and Midland markets. The operation continues to be run on the twin footings of local deliveries and distribution to a wider market in the North West of England and the West Midlands.

Another aspect of considerable importance is the design of packaging, which has already been shown to have a significant impact on sales. Rachel's Dairy packaging has recently been totally redesigned, from origination through to final art work, at a cost of £10,000 before a single pot was produced. The need to employ professional help where necessary is well recognised; the high costs are more than justified by the returns.

Further closely related developments on the farm are conservation projects, farm trails and a farm shop. Much work was undertaken in drawing up a suitable scheme under the Farm and Countryside Initiative, involving many agencies such as the Manpower Services Commission, MAFF, the Countryside Commission, the Forestry Commission and the Nature Conservancy Council. Work on the trails commenced in March 1987 and the trails and shop/tearoom were opened in 1988, the whole scheme costing the farm about £14,000. The farm trails are free, with a charge made for a conducted walk.

The farm trails idea is not purely altruistic. People will have the opportunity to see an organic farm in action, to sample the produce and then to go home and help to expand the markets further. In the future,

Plate 12.6   One of Rachel's Dairy's refrigerated vehicles showing the distinctive design which is used on packaging.

Plate 12.7   The Brynllys farm shop with a full display of Rachel's Dairy products
(A. Martin, courtesy of DBRW)

Gareth and Rachel Rowlands will concentrate exclusively on the farm shop and on conducted tours involving schools, colleges and other interested parties, representing an important public relations element. One of the sons, Johnny, has taken on the distribution side of the business, and the daughter, Sian, is responsible for the processing side. The other son, Mark, is taking over the running of the farm itself. In this way, the farm is able to support and provide livelihoods for the whole family at the same time as maintaining the organic ideal.

Further growth is likely to be limited by the ability of the farm to adjust milk production to seasonal variations in sales, although these are becoming less marked. Rachel's Dairy will have to face another significant problem as it expands, which is the fact that the Rowlands also have a stake in the Welsh Organic Foods cheesemaking venture, and some of their milk goes to the WOF plant in Lampeter. As Rachel's Dairy expands, less milk will be available to send from Brynllys to Lampeter, which could undermine the viability of the cheesemaking operation. In the long term, both Rachel's Dairy and Welsh Organic Foods need additional supplies of organically produced milk from local farms, but whether the current incentives and market potential are sufficient to persuade others to convert remains to be seen.

Frost Fresh – Tynyrhelyg, Llanrhystud, Dyfed

Tynyrhelyg, farmed by David and Barbara Rottner-Frost, is an 8 ha holding near Llanrhystud on the West Wales coast. When the property was bought in 1976, it comprised a house, out-buildings and 0.6 ha land.

Initially, goats were kept and vegetables produced for home consumption, very much in the self-sufficiency spirit of the time. However, largely in order to generate a cash income, a van round and market stall fruit and vegetable business were acquired as retail outlets in 1978. From the outset, organically grown produce was intended to be a defining characteristic of the business and fruit and vegetables were bought in from other local organic holdings.

At first, interest in organically grown produce among local customers was limited, but demand for the freshly harvested produce at the market stall steadily increased and customers appreciated both the availability of a wide variety of locally grown vegetables and also their freshness and flavour.

By 1981, although the van round was successful as a door to door retail outlet, it had become difficult to combine this four-day-a-week service with running both the market stall and the market garden as full time enterprises. The round was sold in February 1981, at which time it accounted for approximately 30% of retail sales. Although it had become a tie, it had provided both a steady cash income and, very importantly in a rural community, an introduction to customers who have remained loyal in subsequent years.

Between 1982 and 1984, 7.5 ha of adjacent land were purchased. This period also saw a greatly increased awareness of organic growing, with strong media interest and intense publicity efforts by individuals in the organic movement. Customers were coming for organically grown produce as such, not just for fresh, locally grown produce. Besides increasing production on a field scale and extending into polytunnel and protected cropping, Frost Fresh diversified and stocked Rachel's Dairy butter and cream, and the Murton's Aeron Vale yogurts. With the establishment of Organic Farm Foods, it was also possible to obtain imported organically grown lines of fruit and vegetables impossible to grow locally. With this development, the Soil Association symbol scheme became vital as a safeguard that both the homegrown and the imported organic produce were genuine. The Frosts applied for the symbol on their own holding and this was granted in 1984.

When the expansion of production in West Wales led to the formation of the Organic Growers West Wales cooperative, the Frosts became founder members, still believing, however, that there was a much greater potential for local sales than other growers believed.

There was, however, always some conflict between their being members of a cooperative to which growers should commit 100 per cent of their produce on the one hand, and being growers with a direct retail outlet of their own on the other. There were some initial benefits in supplying the cooperative, but the trend to producing for supermarkets involved too many penalties. It required greater specialisation and volume production at the same time as the trading business was expanding and requiring ever greater variety. As a result of this conflict, the Frosts eventually left the cooperative and concentrated on building up other sides of the business such as supplying local catering establishments which were increasingly demanding fresh seasonal produce and for whom the 'organically grown' label was also a selling point.

On leaving the cooperative, the Frosts returned to growing a very wide range of lines. Only partial success with autumn and winter salads in 1985, however, left them with a lot of small, immature crops. Joy Larkcom's book *The Salad Garden* provided an answer: to use these salads in a mixed pack marketed as saladini. Both through the market stall and also through the cooperative, these proved a great success. The success was repeated in 1986, and it was decided to open a retail store concentrating upon salads. The range included the saladini pack concept, extended to include packets of lamb's lettuce, radicchio, baby carrots etc., as well as prepared salads.

This seemed both a distinctive approach to selling organically grown produce, and also a means of adding value to product. The shop has a cool-counter selling prepared potato salad, leaf salads, coleslaw and salads made from seasonal vegetables such as beetroot, carrots and celeriac. Whenever possible, these are made with vegetables grown at Tynyrhelyg. A range of salad dressings is also available, made with their own fresh cut herbs, free-range eggs and Rachel's Dairy produce. Home baked rolls and local organic cheeses complete the range of takeaway produce.

1987 also saw the reintroduction of regular van deliveries, though not for retail sales. Deliveries to restaurants were expanded and the distribution of organic produce in the locality, that had been pioneered by Organic Farm Foods (Wales), was taken over. By combining the restaurant deliveries with this round, it has been possible to form a viable distribution system in a sparsely populated area with long distances between customers.

In 1989, Frost Fresh comprises the market stall and salad shop open six days a week and the distribution round. This gives great flexibility in planning crop production and great potential for mixing intensive growing of soft fruit, salads and herbs with more extensive cropping of root crops and brassicas. The Frosts see the interest generated by this system and the complex rotations, as well as being able to keep close relations with customers and consumers in the local area, as an essentially organic approach to the marketing as well as the production of organic food.

One potential conflict still remains. At Tynyrhelyg, the Soil Association's standards are strictly observed and all the crops are organically grown. In the trading enterprises, however, both organic and non-organic produce are sold. Although the Frosts prefer to sell only organically produced crops, a way has not yet been found to do this which is also commercially viable, but the organic produce is promoted wherever possible.

Overview

Without these pioneering and imaginative marketing initiatives, it is unlikely that producers in West Wales could have achieved the national prominence that they have over the last few years. The region now boasts not only Britain's largest organic vegetable cooperative, but also its largest supermarket supplier and cheese making operation, as well as

Plate 12.8   Frost Fresh's market stall in Aberystwyth   (Elm Farm Research Centre)

award winning individual farm projects. One thing is clear: their success is built on considerable hard work, patience and above all commitment. The results have not been achieved overnight, nor is there necessarily a ready market for new producers without significant investment in terms of time and capital. Given the investment and the commitment, however, the potential is significant and there is no reason why these approaches cannot be adopted in other regions.

REFERENCES AND FURTHER READING

Standards

Absolon, C. (1988) *Organic Standards—the key to credibility.* In: *Organic Farming—Growing Up.* Proceedings of RASE/ADAS Conference. Royal Agricultural Society of England; Stoneleigh

Holden, P. (1989) *UKROFS—the story so far.* New Farmer and Grower, 22 (Spring): 30–32

IFOAM (1989) *Standards for Organic Agriculture.* International Federation of Organic Agriculture Movements; Tholey-Theley, West Germany

Soil Association (1989) *Standards for Organic Agriculture.* Soil Association; Bristol

UKROFS (1989) *Standards for Organic Food Production.* United Kingdom Register of Organic Food Standards, Food From Britain; London

Consumer surveys

Allensbacher Institute (1985) *Biologischer Landbau—ein neuer Massenmarkt.* Allensbacher Berichte 20. Institut für Demoskopie; Allensbach

Altmann, M. & Alvensleben, R. v. (1986) *Alternative Nahrungsmittel—Verbrauchereinstellungen und Marktsegmente.* Forschungsberichte zur Ökonomie im Gartenbau, 59. Universität Hannover

Baade, E. (1988) *Analyse der Konsumverhaltens bei alternativ erzeugten Lebensmitteln. – Ergebnisse einer Kundenbefragung in München.* Agrarwirtschaft Sonderheft 119. Verlag Alfred Strothe; Frankfurt.

Böckenhoff, E. & Hamm, U. (1983) *Perspektiven des Marktes für alternativ erzeugte Nahrungsmittel.* Berichte über Landwirtschaft 61: 345–381

Dent, S. (1988) *Consumer Awareness and Attitudes to Organic Produce.* Convent Garden Marketing Authority Summer Scholarship Report. Edinburgh School of Agriculture

Dixon, P. & Holmes, J. (1987) *Organic Farming in Scotland.* Edinburgh School of Agriculture; Edinburgh

DMB&B (1986) *The Healthy Eating Study.* D'Arcy, Masius, Benton and Bowles; London

EFRC (1988) *Organic agriculture in the UK—the market for organic foods and the demands of consumers.* Elm Farm Research Centre; Newbury

Fallows, S. & Gosden, H. (1986) *Does the consumer really care?* Food Policy Studies Institute; University of Bradford

Henley Centre (1989) *The Market for Organic Food.* Report commissioned by The Development Board for Rural Wales; Newtown

Hill, C. (1986) *The Future for Organically Grown Produce.* Cooperative Development Division, Food From Britain; London

Jolly, D. A. *et al.* (1989a) *Marketing organic foods in California – opportunities and constraints.* University of California Sustainable Agriculture Research and Education Program; Davis CA

Jolly, D. A. *et al.* (1989b) *Organic foods – consumer attitudes and use.* Food Technology, November: 60–66

Lamb, C. G. (1989) *The Market for Organic Food Products.* Lincoln University, Christchurch, New Zealand

MAFF (1987) *Survey of consumer attitudes to food additives.* Vol 1. HMSO; London

MINTEL (1989) *The Green Consumer.* Special Report. MINTEL International; London

NOP (1987) *Public attitudes to the environment.* Digest of Environmental Protection and Water Statistics, 9

Presto (1986) *Eating what comes naturally.* Survey conducted by KMS Partnerships Ltd, July 1986. Presto (Argyll Stores Ltd); London

Stopes, C. (1988) *The demand for organic food.* New Farmer and Grower 19, (Summer): 14–17

Vogtmann, H. *et al.* (1984) *Was erwartet der Verbraucher von Nahrungsmitteln aus biologischem Anbau?* IFOAM Bulletin, 49: 4–6

Werner, J. & Alvensleben, R. v. (1984) *Consumer attitudes towards organic foods in Germany.* Acta Horticulturae.155: 221–227

Which? (1990) *Organic food.* Which? Way to Health, February: 13–17

Marketing initiatives

Anon (1989) *Unigate begin retailing organic milk.* New Farmer and Grower, 22 (Spring): 24–25

Clackworthy, L. (1988) *A buyer's requirements for the purchase of organically grown produce.* In: *Organic Farming – Growing Up.* Proceedings of RASE/ADAS Conference. Royal Agricultural Society of England; Stoneleigh

Farm shops. New Farmer and Grower Special Feature, 20 (Autumn 1988): 20–27

Goff, E. *et al.* (1989) *Marketing feature – milk, vegetables, cereals and meat.* New Farmer and Grower 25: 24–31.

Haines, M. & Davies, R. (1987) *Diversifying the Farm Business.* BSP Professional Books; Oxford. pp 211–217

Hunt, M. (1989) *Organic meat markets—a major food retailer's approach.* In: Chamberlain, A. T. *et al.* Organic Meat Production in the '90s. Chalcombe Publications.

Marketing Initiatives in Germany. Bioland Special Issue. 2/1989

Mason, L. (1989) *Organic milk—first step towards every doorstep.* Dairy Farmer 36(4): 18–24

MLC (1987) *Specialist Markets for Alternative Meat Production Systems.* Meat and Livestock Commission; Milton Keynes

Murdy, R. (1989) *What price greener foods.* Paper presented at RURAL Conference, Oxford, October 1989

Starling, B. (1988) *Making the most of your cereals.* New Farmer and Grower, 19 (Summer): 17–18

Weindlmaier, H. & Czempiel, A. (1989). *Vermarktung alternativ produzierter Milch.* Agrarwirtschaft 38: 273–280. (also Wendt, H. (1989) Agrarwirtschaft 38; 263–272)

Producer cooperatives

Bowers, P. (1988) *ECOP & Geest—a perfect partnership?* New Farmer & Grower, 18 (Spring): 34–36

Hunter, R. D. & Dart, R. (1988) *Setting up an organic group.* In: *Organic Farming—Growing Up.* Proceedings of RASE/ADAS Conference. Royal Agricultural Society of England; Stoneleigh

Stewart, C. (1988a) *Somerset Organic Producers—the first year.* New Farmer and Grower, 17 (Winter): 20–21

Stewart, C. (1988b) *Cooperative Marketing—the Welsh experience.* New Farmer and Grower, 19 (Summer): 20–21

General

AID (1987) *Vermarktung alternative erzeugter Agrarprodukte.* Auswertungs–und Informationsdienst; Bonn

Franco, J. (1989) *An analysis of the California market for organically grown produce.* American Journal of Alternative Agriculture 4: 22–27

Mabey, D., Gear, A. & Gear, J. (1990) *Thorsons Organic Consumer Guide.* Thorsons; Wellingborough

Recent publications

EC. (1991). Council Regulation (EEC) No 2092/91 of 24 June 1991 on organic production of agricultural products and indications referring thereto on agricultural products and foodstuffs. *Official Journal of the European Communities.* **91**(L198):1–15.

Mintel. (1993). *Vegetarian and organic food.* Mintel; London.

Geier, B., C. Haest and A. Pons (eds.) (1990) *Trade in organic foods.* IFOAM. Tholey-Theley, Germany.

Chapter 13

Physical and Financial Performance

One of the major factors which has limited the expansion of organic farming in the past is the belief that such systems are not financially viable. However, the cost-price squeeze in agriculture has forced farmers to rethink their farming operations and look again at low external input options. Declining real prices have become a permanent feature of agriculture, and the production of surpluses as a result of technological advance has led to the introduction of quotas for milk and further real price reductions for other products. For many farmers, the only option remaining is to look critically at cost levels. In addition, the premium prices available for organic products have tempted some farmers to think again about farming organically. Unfortunately, information on the current financial viability of organic systems is scarce and, apart from the occasional gross margin in the farming press, there has been little for an interested producer to get his teeth into. At the same time, a large number of producers who have taken the plunge and changed over to an organic system seem to be managing to stay afloat, some very successfully.

What are the factors which make this possible? On the one hand, it is claimed that the use of fertilisers and other agro-chemicals is economically justifiable so long as the returns from their use exceed their cost. In fact, authors such as G. W. Cooke have argued that the relationship between fertiliser use and output is such that the technical and economic optima are almost identical. Fertiliser use should be directed at achieving maximum yields. On the other hand, organic producers will claim that lower costs often compensate for reduced yields, even without the benefit of premium prices.

The answer to this paradox may lie in the failure of agricultural economists and scientists to adopt a holistic approach to the assessment and comparison of agricultural systems. Given the extreme complexity and high cost of adopting such an approach to agricultural research, this is fully understandable. A reductionist approach, where only one or two factors are varied at a time, is much better suited to statistical analysis and scientific 'proof', but the applicability to biological systems of results thus derived is severely limited.

A fair comparison of the economic performance of conventional and organic farming systems requires a) that the whole farm should be taken into account and b) that this should be done in such a way as to eliminate factors (climate, soil type, management ability etc.) which are not determined by the type of farming system being operated. Most comparative studies of organic and conventional farms have not taken these points sufficiently into account and the results obtained have therefore been inconclusive.

Even so, there is an increasing amount of information becoming available which at least gives an indication of the relative position of organic farms and, as more resources are put into research in this area, the quality of the data is also improving. The aim of this chapter is to highlight those areas affecting the farm business which are most likely to differ in an organic system and to give an indication of the size of difference which may be expected. It must, however, be emphasised that generalisations can be very misleading. Any serious assessment of the potential viability of an organic system for a particular farm must be done on an individual basis.

YIELDS

A reduction in yield levels, like paying tax, tends to bring many farmers out into a cold sweat, particularly in view of the pressure over the last few decades to produce ever greater quantities of food. In the current situation, where surpluses are the norm and the costs of maintaining surpluses are unacceptably high, less intensive methods of production with lower levels of output may be preferable to taking large areas of land out of agricultural production altogether. Whether this approach is financially acceptable will depend as much on government policy as on other factors.

Crop yields

In organic farming systems, crop yields are frequently higher than would be expected if the conventional farming text books were to be believed, but sound husbandry practices are needed to make the system work successfully. Comparing yields on farms is fraught with difficulties—varieties differ in their yield potential and the range of yields on organic farms varies just as much as on conventional farms.

The data in Table 13.1, from a study conducted in Switzerland from 1979 to 1981, show just how close organic and conventional yields can be. More recent work from the same source shows that the general trend

Table 13.1   Yields on organic and conventional farms in
Switzerland (1979–81).

Crop		Organic	Conventional partner[1]	Regional average
Wheat	(t/ha)	3.9	4.5	4.7
Rye	(t/ha)	4.4	–	4.5
Maize	(t/ha)	4.4	4.7	4.9
Oats	(t/ha)	4.2	5.0	4.9
Barley	(t/ha)	3.9	4.5	4.5
Potatoes	(t/ha)	31.1	31.4	36.3

[1]Paired conventional farms of similar type and size.
Source: Steinmann (1983).

for yields to increase from year to year applies to organic systems just as much as to conventional systems.

Most studies, including those conducted in Britain (e.g. Vine and Bateman, 1981), suggest that yields will on average be between 10 and 30% lower in an organic system. Stanhill (1990), in a review of comparative studies, found that the ratio of organic to conventional yields was normally distributed with a mean of 91%, and a modal value between 80 and 90%. A recent US study (National Research Council, 1989) concluded that reduced use of agrochemical inputs 'lowers production costs and lessens agriculture's potential for adverse environmental and health effects without necessarily decreasing—and in some cases increasing—per acre crop yields and the productivity of livestock management systems.' There is some evidence (Stanhill, 1990) that yield differences increase with an increase in intensity of the conventional farming system.

These figures hide considerable variations between different organic farms as they do between conventional farms. The Vine and Bateman study found a range of yields for winter wheat from 3.0 t/ha to 5.3 t/ha in 1978/79. This compared with a range of conventional average yields in different regions from 4.6 t/ha to 5.7 t/ha (not allowing for varietal differences). In 1985, some organic farmers were achieving yields of up to 7t/ha for winter wheat, and there would appear to be no particular reason why above average yields cannot be obtained regularly in certain situations. These results are supported by those from wheat variety trials in Britain (Table 13.2). It is very unlikely, however, that organic farmers will be able to match the top yields being achieved by conventional producers. It is interesting to note the performance of Maris Widgeon, which, although no longer on the NIAB recommended list, has been consistently high yielding in organic systems. In spite of this, both the Elm Farm Research Centre work and similar work in Germany confirm that the modern, higher yielding varieties also perform better in organic systems, although the ranking may not necessarily be the same (Figure 13.1). Although it cannot be claimed that wheat yields in organic systems have kept pace with the very rapid expansion in average wheat yields in the last few years, there would seem to be considerable potential for improving yields through the selection of appropriate varieties and improved management and husbandry.

Table 13.2 Yield (tonnes/ha) of winter wheat in trials carried out by Elm Farm Research Centre (Berkshire/Hampshire).

	1983/4	1984/5		1985/6
Variety	EF	EF	DF	DF
Avalon	5.70	4.92	4.93	5.56
Brimstone	–	5.83	7.69	8.48
Copain	4.54	4.91	5.37	–
Disponent	5.95	–	–	–
Fenman	–	5.64	6.25	–
Maris Widgeon	–	6.03	7.43	6.72
Mission	5.75	6.12	5.91	5.80
Moulin	–	–	–	7.48

(column header row: *Year and site*)

EF = Elm Farm, DF=Dove's Farm.
Source: Stopes (1987).

Figure 13.1   Yields of 23 varieties of winter wheat in organic agriculture, related to the yields in conventional agriculture (trials, 1982–84).

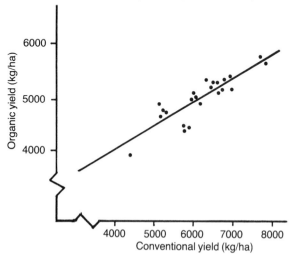

Source: Stöppler (1988).

The yield differences are less significant with other cereals and grain legumes may show no difference at all. Oil seed rape and sugar beet are not usually grown organically, so comparisons are not very helpful. More significant differences may occur with fodder crops such as forage maize and fodder beet, as well as horticultural crops and field vegetables. In some cases, this may be due to increased row spacings and other measures necessary to keep weeds under control, rather than a lack of nutrients as such. Improved technology for weed control can therefore play a major role in increasing crop densities and output. For example, while a ridge system for carrots can restrict yields to 20–25 t/ha, intensive organic bed systems in Germany have recorded yields as high as 80 t/ha. In these situations, comparisons of yields between conventional and organic systems are not very helpful without an understanding of all the factors which contribute to a particular yield being achieved. Regional differences also exist, but no attempts have yet been made to collate this information in Britain or elsewhere. Some data has been published for Scotland (Table 13.3), but the information does not allow any particular conclusions to be drawn.

Table 13.3   Yields of organically grown crops recorded in Scotland.

Crop	Average harvested yield (t/ha)	Range (t/ha)	Number of crops in sample
Wheat	3.1	–	1
Oats	4.3	3.7–4.7	4
Barley	4.3	3.0–5.7	2
Table swedes	22.5	15.0–30.0	2
Early potatoes	29.8	–	1
Main-crop potatoes	25.2	14.9–34.7	6
Seed potatoes	34.7	–	1
Carrots	27.3	10.0–44.6	2
White cabbage	24.8	–	1

Source: Dixon & Holmes (1987).

There have been several other studies in Europe, which although not directly reflecting British conditions, give an indication of what yields might be expected. Organic farms in the West German state of Baden-Württemberg have been the subject of particular attention, with research going back to the mid–1970s. Table 13.4 summarises the results from two studies in Baden-Württemberg, involving both organic and biodynamic farms. In the earlier study (MELU, 1977), the sample chosen is small, and the regional averages cover quite a wide area, but

Table 13.4 Yields in t/ha from organic and biodynamic farms in Baden–Württemberg.

| | Biodynamic farms only 1971–1974 average | | Organic and biodynamic farms 1983 survey of 200 holdings | |
	bio-dynamic	conventional	alternative[1]	conventional
Winter wheat	4.54	4.09	3.3 (1.0–5.3, 145)	4.7
Spring wheat	4.08	4.07	2.8 (1.0–4.6, 52)	3.9
Winter barley	4.05	4.22	3.5 (1.0–5.0, 28)	4.8
Spring barley	3.33	3.59	2.6 (0.7–4.2, 21)	3.7
Oats	3.90	3.66	3.2 (1.2–5.0, 36)	3.9
Rye	–	–	3.2 (0.8–5.2, 52)	3.8
Carrots	–	–	40.5 (6.0–80.0, N/A)	42.3
Early potatoes	–	–	13.8 (7.0–20.0, 20)	18.5
Main crop potatoes	22.80	22.70	16.5 (5.8–40.0,120)	22.6
Beetroot	–	–	29.9 (5.0–62.5, N/A)	32.6

[1]Yield range and number of farms in brackets.
Sources: MELU (1977) and Böckenhoff *et al.* (1986).

the more recent results are from a survey of 200 farms in the region (Böckenhoff *et al.*, 1986).

It should be noted that 1983 was a bad year for arable cropping generally in southern Germany and that both organic and conventional yields suffered as a result. Other work in Baden-Württemberg has shown a clear correlation between yields and soil type, with the differences between organic and conventional yields being significantly less on better soils. The range of yields indicated is very wide in many cases, but a further analysis showed that the top 25% of organic crops were equivalent to or above the conventional average. Individual cases approach the top yields which conventional producers can obtain.

During the conversion period, however, yields may well fall further as the new system becomes established. The biological processes which make the organic system work take time to become established. Nutrient deficiencies and weed problems can occur, although this is not always the case. The occurrence of low yields in the early stages of conversion, however, should not be taken as a guide to potential yield in an established organic system. There is no detailed evidence available to predict exactly what will happen during the conversion period, but the Baden-Württemberg study (Böckenhoff *et al.*, 1986) referred to above does indicate a link (Table 13.5). A follow-up study of 44 organic-biological farms in Baden-Württemberg (Dabbert, 1990) also established a link between yields and length of time under organic management (Figure 13.2). It should be noted that in both cases cross-sectional, not time series, data has been used so the results do not reflect what actually happened on individual farms during the conversion period. It

Table 13.5 Average annual yields by length of time under organic management in t/ha (or kg/cow), 1983.

| | Length of time under organic management (years) | | | | | |
	<2	3–5	6–10	>10	average	conventional
Cereals	2.9	2.7	3.2	3.3	3.1	4.3
Potatoes	21.0	11.4	16.8	16.1	16.2	22.2
Beetroot	9.2	25.0	28.0	33.6	29.9	32.6
Carrots	13.8	23.0	40.9	43.1	40.5	42.3
Milk (kg/cow)	3,912	3,685	3,895	4,126	4,013	4,231

Source: Böckenhoff *et al.* (1986).

is also important to consider relative rather than absolute yields when making any comparisons, given that the conventional yields are also lower than would be expected in Britain.

Livestock output

Livestock output is affected to some extent by changes in the management of the stock themselves, such as a shift to lower use of concentrates for ruminants and to the use of cereals and home produced grain legumes as opposed to fish meal and soya in livestock rations. The impact of this is often not very significant; the major impact arises from reduced forage output and reduced stocking rates.

The grass/white clover ley may be expected to yield 7 to 8 t dry matter (DM) per ha in an organic system given good conditions, with an associated stocking rate of 1.7 to 1.8 livestock units (LU) per forage ha.

Figure 13.2 Cereal yields in relation to period under organic management (farms in Baden-Württemberg, West Germany).

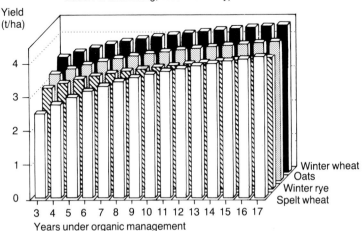

Source: Dabbert (1990).

This will obviously be reduced in more marginal areas or where factors such as soil type prove to be limiting. Short term red clover/Italian ryegrass leys might be expected to yield 10 to 12 tDM/ha and have been recorded yielding as much as 15 tDM/ha in organic systems. Lucerne, although more difficult to establish and less suited to the wetter western parts of the UK, can produce similar yields to red clover. Both are ideally suited to forage conservation in organic systems. The use of these crops can help to push stocking rates up towards 2 LU/forage ha.

There have been virtually no comparative studies of livestock output on conventional and organic farms. Vine & Bateman (1981) found that stocking rates on the organic farms they surveyed ranged from 32–132% of the conventional standards obtained from Farm Management Survey data. The average for 27 of the farms was about 80%. However, total livestock output per adjusted forage hectare was generally rather less than this at about 70% of the conventional standard. In trials at the North of Scotland College of Agriculture, Younie (1989) achieved similar overall liveweight gains for organic and intensive 18-month beef systems but at a 16% lower stocking rate (3·6 and 4·3 animals per hectare for the organic and intensive (conventional) systems respectively).

Steinmann (1983) in Switzerland found both lower yields per cow (associated with greater dependence on home grown forage) and lower yields per hectare (as a result of lower stocking rates) for dairy herds on the organic farms studied (Table 13.6). Schlüter (1987), in a study of biodynamic farms in southern Germany, found stocking rates 30–40% lower than the conventional average, except for one region where stocking rates were virtually identical.

Table 13.6 Milk yields on organic and conventional farms in Switzerland (1979–81).

Product		Organic	Conventional partner[1]	Regional average
Milk	(l/cow)	4,517	5,111	4,912
Milk	(l/forage ha)	8,609	10,669	11,254

[1]See Table 13.1.
Source: Steinmann (1983).

PRICES FOR ORGANIC PRODUCE

High prices for organic produce are frequently quoted in the farming press and have certainly played a role in the expansion of interest in organic production. In Britain, milling quality wheat can achieve a price of up to 100% above the conventional price, while in Germany organically produced wheat fetches as much as 200% more. In some cases, organically produced vegetables in Britain may qualify for premium prices more than 100% above their conventional counterparts. Such high prices do exist, but they are more the exception than the rule. In particular, they only apply to certain crops with well developed markets and marketing channels. The average premium for vegetables is more likely to be in the region of 25 to 50%.

Crop prices

Vegetables and cereals are the prime candidates for premium prices. Published data for ex-farm prices of organically produced crops are not readily available, but an indication is given in Tables 13.7 and 13.8. Even though some of these look attractive, they are very dependent on

Table 13.7 Ex-farm prices (£/t) for organically produced cereals (1989/90, previous year in brackets).

	Typical	Range	Premium (%)
Bread wheat	230 (180)	210–250	100+ (70)
Other milling wheat	180 (170)	–	70 (60)
Feed wheat	200 (150)	180–220	100 (50)
Malting barley	170 (–)	–	60 (–)
Feed barley	155 (135)	–	50 (35)
Oats	175 (160)	165–180	70 (55)
Rye*	200 (190)	200–220	70 (60)
Field beans	195 (200)	190–200	20 (20)

*in small, bagged lots.

Table 13.8 Ex-farm prices (£/t) for organically produced vegetables (1989/90*).

	Typical	Range	Premium (%)
Maincrop potatoes	200	160–250	100
Early potatoes	–	up to 700	100+
Carrots	330	200–530	50–150
Swedes	200	110–260	0–50
Onions	440	330–500	30–50
Leeks	550	260–770	25+
Beetroot	220	180–260	50–80
Parsnip	440	330–550	50–150
White cabbage	200	130–330	0–50
Little Gem (each)	10p	8p–15p	0–20
Courgettes	330–550	180–880	0–50

*prices as sold, after gradeouts. Prices in 1989/90 were generally better than the previous year but might not be maintained in 1990/91.

the market. Some organic farmers will argue that they are justified in charging more for their produce because it costs them more to produce, but they are only able to charge high prices where there is sufficient demand for the product. Demand and supply interact closely. The fact that premiums have continued to increase for some crops, especially milling wheat, is an indication that demand is still growing at a faster rate than supply.

Local markets, especially for vegetables, can very quickly become saturated and the premiums will begin to tail off. Even without the premiums, there are still some advantages to be had from selling direct to the retailer and benefiting from the increased margins. On a national scale, the market is nowhere near fully tapped. Many parts of the country still do not have access to organic produce, even though most multiples now stock it. For the foreseeable future at least, there seems to be plenty of scope for expanding demand by opening new markets at the same time as expanding supply, but a lot of work has to be put into building up these new markets. While it is very difficult to predict if and when the market for these products will become saturated, there seems no reason why the current prices should not remain stable for the next few years.

The premiums available for organically produced wheat and oats are closely linked to quality. These tend to range from 35% for poorer quality wheat to 80% to 100% for top quality wheats. Premiums for milling oats also vary between 30% and 60% depending on quality. Other arable crops, such as sugar beet and oil seed rape, have no ready organic market and therefore do not attract premium prices and generally speaking are not an economic proposition. This situation may change if herbicide costs increase and greater reliance on mechanical weed control once more becomes an economic proposition.

Livestock prices

Livestock products did not generally attract premium prices in the past because of the lack of an established market. The recent development of the market (Chapter 12) now means that premiums of 15 to 30% may be obtainable. At the same time, livestock account for the majority of the output on organic farms. This can represent a serious weak point in the profitability of an organic system, because the returns from livestock production, particularly beef and sheep, may not carry the fixed costs of farms in intensive arable areas.

Processing and direct marketing of dairy products can provide an opportunity to increase the margins for milk production. Producers taking part in the Unigate trials with fresh organic milk in south west

England are receiving a premium of 15% for their milk. Where markets do exist much of the potential premium may be absorbed by high distribution costs. One additional factor is that as the market for organically produced livestock products develops, the demand for organically produced feed cereals will also increase, with significant implications for feed costs.

VARIABLE INPUT COSTS

Variable input costs represent one of the chief differences between conventional and organic systems and can play a major role in compensating for reduced yields, especially where premium prices are not obtainable.

Crop variable costs

For most grain crops, a slight increase in expenditure on seed (higher seed rates are sometimes necessary for weed control and because of lower available nutrient levels) is compensated for by much lower costs for fertilisers and sprays. Differences in the costs of spreading manures and fertilisers will be minimal as not every course in an organic rotation will receive manure or slurry. In some cases, they may even be lower, because lower stocking rates will actually mean less manure or slurry is available to be spread.

Savings on sprays are likely to be significant. In cereals, weed levels frequently do not represent an economic threat anyway. Field vegetables such as carrots, however, will require active weed control measures, such as flame weeding (ca. £75/ha), hand weeding (up to £300/ha) and mechanical cultivation. Some organic farmers place great emphasis on extensive mechanical cultivations prior to seedbed preparation in order to obtain a clean seedbed, which can result in greatly increased cultivation costs.

Livestock variable costs

Concentrate and forage are the main costs which will be reduced as far as livestock are concerned, with greater dependence on home produced forage, fodder crops and concentrates. Where organically produced feed grains have to be purchased, however, costs are likely to be increased correspondingly. Another important aspect of a less intensive approach to livestock production may be reduced veterinary and medical bills, but there is no hard evidence to suggest what level of reduction can be expected and, in dairy systems, they are likely to be associated with increased replacement rearing costs.

GROSS MARGINS

The easiest way to illustrate the combined effect of lower yields, higher prices and lower variable costs is by looking at gross margins for individual enterprises. Tables 13.9 to 13.13 illustrate a range of different gross margins, including both actual figures from farms and estimated gross margins on the basis of standard data.

Crop gross margins

The comparison of organic and conventional winter wheat crops in Table 13.9 is derived from research by ADAS on Barry Wookey's farm in Wiltshire (Wookey, 1987). There are several points which should be noted. The difference in seed cost reflects the fact that varieties no longer in common use were used and therefore cost more to buy. In fact, the value is notional, because home grown, not purchased, seed was used. Direct drilling techniques were used for the conventional crops whereas extra passes were required for weed control in the organic system. The conventional average price has been calculated on the basis of the proportions of feed and milling wheat produced. Where a variety was suitable for milling, it is assumed to have been sold as such. The lower

Table 13.9 Average margins for twelve wheat crops (six organic, six conventional) in 1983 and 1984.

	Conventional		*Organic*	
Variable costs (£/ha)				
Seed	40.60		48.80	
Fertiliser	118.28		0.00	
Sprays	96.43		0.00	
Total variable costs	255.31		48.80	
Other allocatable costs (£/ha)				
Cultivations	65.38		75.72	
Harvesting	58.00		58.00	
Total costs (£/ha)	378.69		182.52	
Yield (t/ha)	7.4		4.4	
Price (£/t)*	108.32	115.00	160.00	200.00
Output (£/ha)*	801.60	506.00	704.00	880.00
Gross Margin (£/ha)*	546.29	457.20	655.20	697.48
Gross Margin less				
allocated costs (£/ha)*	422.91	323.48	521.50	697.48

*organic prices, output and gross margins calculated at conventional, lower and higher premium prices (see text).
Source: Wookey (1987).

premium price for the organic wheat is the price paid by another merchant, but the actual price obtained for wheat sold direct from the mill is the higher one. Finally, although 1983 and 1984 were good years for wheat yields, 1985 and 1986 were worse and the results may not always be as good as the ones shown here.

Vine and Bateman (1981) looked at a larger number of organic farms in Britain and calculated the gross margins for organically produced winter wheat on the basis of conventional prices as well as, where appropriate, the effect of organic premiums at the time (Table 13.10), although the data are now more than ten years old. These are compared with averages for the same size and type of conventional farm derived from Farm Business Survey data.

Table 13.10 Costs and returns for winter wheat (results for individual farms).

Yield (t/ha)		Ferts. (£/ha)		Sprays (£/ha)		Total* (£/ha)		Gross margin[†] (£/ha)			
org.	con.	org.	con.	org.	con.	org.	con.	org.	(org.)	con.	Year
3.0	5.6	0	38	0	17	28	83	242	(278)	421	1978
3.7	5.6	0	38	0	17	28	83	305		421	1978
4.3	5.7	9	40	0	44	35	110	362		412	1978
4.3	5.3	2.3	43	0	19	30	90	368	(416)	400	1978
4.7	4.6	26	38	3	17	54	83	369		331	1978
4.8	4.9	0	40	0	20	30	90	426		376	1978
4.9	5.3	28	52	34	35	87	115	378		398	1979
4.8	5.4	5.8	48	0	25	34	112	436	(535)	420	1979
5.3	5.1	15	52	9	35	51	115	456	(584)	375	1979
5.8	5.1	0	52	0	35	27.5	115	585	(668)	424	1979

* Total variable costs incl. fertilisers, sprays and seeds.
[†] Gross margins calculated at conventional prices; figures in brackets include organic premiums.
Source: Vine & Bateman (1981).

Unlike most other studies, Vine & Bateman were careful to treat each farm individually and not to average data from an unrepresentative sample. However, the data covered one year only, so that changes over time and under different weather conditions could not be assessed, a problem which also applies to many of the other comparisons which have been made. The wide variation between individual farms should be noted, emphasising the dangers of using gross margins alone as a standard for comparison.

The most important point is, however, that cash crops such as winter wheat cannot be grown continuously and are normally interspersed with grass leys which, unless utilised by a dairy enterprise, do not yield very high returns. Some producers have succeeded in establishing stockless systems, including grain legumes as cash crops, but they are few and far between. An example of this is given in Table 13.11, showing the gross margins for an Irish stockless organic system involving the following three course rotation:

- field beans;
- winter wheat (milling) followed by white clover green manure;
- winter oats followed by mustard green manure

Table 13.11 Typical gross margins for a stockless organic system.

	Beans	Wheat	Oats	
Yield (t)	3.50	4.00	4.00	
Price (£/t)	260.00	220.00	200.00	
Output (£/ha)	910.00	880.00	800.00	
Variable Costs (£/ha)				
Seed	125.00	65.00	65.00	
Green Manure		30.00	30.00	
Total	125.00	95.00	95.00	
Cultivations (£/ha)	182.50	182.50	202.50	
Gross Margin less cultivation costs (£/ha)	602.50	602.50	502.50	Average/year 569.17

Source: Stoney (1987).

In general, the data available are still very sparse. An attempt is made here to overcome this by illustrating a range of estimated gross margins for different arable and forage crops (Table 13.12) and vegetables (Table 13.13) at 1989 prices. Until more accurate and reliable data become available, these represent the best current estimates of organic crop gross margins.

Information from other countries
Although information on the economics of organic systems in other countries is not directly applicable to the British situation, the results confirm the main conclusions above. In West Germany, the 1977 study from Baden-Württemberg (MELU, 1977) found that, although yields were between 10% and 25% lower on organic farms compared with their conventional counterparts, lower variable costs resulted in similar gross margins, even before premium prices were taken into account (Table 13.14).

In Switzerland, Steinmann (1983) looked at more than 20 farms over a three year period and compared their results with both partner farms and regional averages and came up with similar findings. Both crop and

Table 13.12 Gross margins for cereals and forage crops – 1989/90 estimates*.

Crop	Winter wheat (milling)	Winter wheat (feed)	Malting barley	Milling oats	Field beans	Short term ley	Medium term ley
Yield (t/ha)†	4.5	5.5	4.0	4.5	3.5	10.0	7.0
Premium (%)	100	50	60	55	10	N/A	N/A
Convent. price (£/t)	110	100	105	110	180	N/A	N/A
Premium price (£/t)	220	150	170	170	200	N/A	N/A
By-products	80	80	70	70	0	0	0
Output (premium)	1070	905	750	835	700	N/A	N/A
Output (no premium)	575	630	490	565	630	0	0
Variable Costs							
Seed	60	55	40	45	60	40	25
Fertilisers	0	0	0	0	0	50	30
Other	5	5	5	5	5	0	0
Total	65	60	45	50	65	90	55
Gross Margin							
with premium	1005	845	705	785	635	N/A	N/A
without premium	510	570	445	515	565	−90	−55

*All £/ha unless otherwise stated.   †Dry matter in the case of forage crops.

Table 13.13 Gross margins for selected field vegetables – 1989 estimates*.

	Carrots	Parsnips	Cabbage	Maincrop onions	Maincrop potatoes	Swedes
Marketable yield (t/ha)	24	24	16	25	24	25
Premium (%)	100	100	40	30	100	50
Convent. price (£/t)	120	160	140	330	100	140
Premium price (£/t)	240	320	200	430	200	210
Output (premium)	5760	7680	3200	10750	4800	5250
Output (no premium)	2880	3840	2240	8250	2400	3500
Variable Costs						
Seed/Plants	130	40	470	1250	430	15
Fertilisers	25	25	25	25	25	25
Crop protection/drying	55	55	30	400	25	0
Casual labour						
planting/weeding	275	275	120	400	45	100
harvesting/grading	600	600	180	600	600	200
Packaging	210	210	650	220	150	210
Marketing (15% output)	865	1150	480	1615	720	790
Total	2160	2355	1955	4510	1995	1340
Gross Margin						
with premium	3600	5325	1245	6240	2805	3910
without premium	1152	2060	429	4115	735	2425

*All £/ha unless otherwise stated.

Note: Prices and costs for vegetable production can vary widely, so these estimates should be interpreted with caution. Allowances should be made for good and bad harvest years, including crop failures, and high gradeout percentages if selling to multiples.

livestock enterprises were analysed; the results for arable crops are shown in Table 13.15. Prices in Switzerland are generally very much higher than in the European Community. As a result, premium prices for organically produced food are not so great. In addition, the farms are smaller and much more labour and capital intensive than in Britain.

An alternative approach is to compare the results from situations where the systems are established specifically for the purposes of research and development. The Development of Farming Systems experiment at Nagele in the Netherlands has been going for more than ten years, and includes three systems, conventional, integrated (limited chemical use) and biodynamic on virtually identical soils and areas of similar size (Vereijken, 1986 & 1989; Zadoks, 1989). Again, because of lower costs and higher prices, the reduced yields in the biodynamic system still result in higher net returns (Table 13.16).

The best known study from the United States is the one by Lockeretz and his colleagues at Washington University, St. Louis, during the 1970s. Lockeretz's team compared a sample of 14 organic farms with 14 paired conventional farms over a period of three years from 1974–76. This was followed up using a larger sample of about 20 farms in 1977 and 1978 and comparing the results with regional averages. The results from this study are summarised in Table 13.17.

Given the small sample size, care should be taken with interpreting the results, but the consistency over the five year period is important. The study concentrated only on crop production, ignoring the livestock element and this is its major weakness. The size and proportion of the farm devoted to crop production was greater on the conventional than on the organic farms, and hence economies of scale may favour the conventional samples. In addition, although the returns to crop production were similar for the two groups, total farm income (which was not considered) would have been different because of the variation in enterprise structure between farms.

More recent studies from the United States include those of Helmers et al. (1986), Goldstein & Young (1987), Dobbs et al. (1988) and National Research Council (1989). The National Research Council report, which did not distinguish specifically between organic and conventional farms, concluded that although average production costs per unit of output varied markedly between regions, on the most efficient farms within a given region they were typically 25% and often more than 50% less than average costs on less efficient farms, indicating clearly that lower-imput systems could be economically efficient. Machinery, fertiliser, pesticide and interest charges were found to account disproportionately for differences in per unit production costs.

Table 13.14 Relative cereal yields and gross margins for biodynamic farms in Baden-Württemberg compared with similar conventional farms, 1971–74 (conv. = 100).

Region	1	2	3	4
Yield	90	74	86	91
Gross output	116	85	118	120
Variable costs	68	63	86	69
Gross margin (a)	137	95	133	141
Gross margin (b)	101	80	86	100

(a) at prices actually received.
(b) at conventional prices.
Source: MELU (1977).

Table 13.15 Crop yields and gross margins for organic and paired conventional (partner) farms in Switzerland, average 1979–81.

		Wheat (org.)	Wheat (con.)	Barley (org.)	Barley (con.)	Potatoes (org.)	Potatoes (con.)
No. of farms		10	10	7	7	8	8
Yields	(t/ha)	3.88	4.51	3.88	4.50	31.7	30.8
Price	(SFr/t)	1109	985	673	643	378	378
Output	(SFr/ha)	4898	4592	3753	3861	12404	11938
Variable costs	(SFr/ha)	631	890	362	709	2025	2982
Gross margin	(SFr/ha)	4267	3702	3391	3152	10379	8956

Source: Steinmann (1983).

Table 13.16 Average physical and financial yields of the three main marketable crops over 1982–1985, Nagele experiment.

	Conventional potatoes	sugar beet	winter wheat	Integrated potatoes	sugar beet	winter wheat	Biodynamic potatoes	sugar beet	winter wheat
Yield[1]	52.1	9.39	7.9	36.6	9.53	7.0	23.6	8.26	5.3
Output[2]	10.45	7.10	4.22	10.94	7.10	3.80	11.02	6.06	4.94
Costs[3]	5.51	1.82	1.28	4.49	1.61	1.08	2.43	1.11	0.62
Net return[4]	4.94	5.28	2.94	6.45	5.49	2.72	8.59	4.95	4.32

[1] t/ha (yield of sugar beet is expressed as t/ha sugar).
[2] price (DFl/t) × yield (t/ha). Special prices for biodynamic grain (0.85 DFl/kg, 70% higher than conventional) and biodynamic potatoes (0.50 DFl/kg, 100% higher than conventional).
[3] costs of pesticides, fertilisers, hired labour, seed, insurance and interest.
[4] output less allocated costs.
Source: Vereijken (1986) (updated).

Table 13.17 Mean economic performance, organic and conventional samples ($/ha) (interquartile range in brackets).

Year	Value of crops produced Organic	Value of crops produced Conv.	Operating expenses Organic	Operating expenses Conv.	Net returns Organic	Net returns Conv.
1974	393	426	69	113	324	314
	(348–469)	(346–501)	(49–84)	(96–131)	(264–380)	(249–383)
1975	417	478	84	133	333	346
	(326–479)	(422–479)	(72–101)	(104–148)	(269–385)	(294–353)
1976	427	482	91	150	336	333
	(371–534)	(407–524)	(70–109)	(126–164)	(282–427)	(270–377)
1977*	384	407	95	129	289	278
1978†	440	527	107	143	333	384

* 23 farms, not the same as 1974–76, compared with regional averages, not partner farms.
† 19 farms, also not the same as 1974–76.
Source: Lockeretz (1984).

Livestock gross margins

Comparative gross margins for livestock have only recently received detailed attention in Britain. The Vine and Bateman study, using data from 1978/79, found a very wide range of performance on the organic farms, with stocking rates ranging from 32–132% of the conventional standard. Expenditure on fertilisers and manures was mainly around 0–30% of the conventional standard, with one or two extreme examples. Expenditure on concentrates (both purchased and homegrown) varied from 0–130%, with the majority in the 30–70% range. The results make it very difficult to come to any generally applicable conclusion.

More recent results for beef production come from experimental systems (Lowman, 1989 and Younie, 1989) and from budgeting exercises (MLC, 1988 and Spedding, 1989). Results for 18-month finishing systems from three of these studies are summarised in Table 13.18. The comparisons suggest that premiums of between 10 and 25% (20–50 p/kg carcass weight) would be needed to achieve gross margins comparable with conventional systems.

The following points should, however, be noted. Lowman used forage of widely differing quality. The 'organic' land was essentially poor quality grassland which had received no fertilisers, rather than land managed organically with high clover levels. Comparisons on a per hectare basis are therefore very difficult. In addition, the policy of ensuring similar liveweight gains between the two groups meant that high levels of concentrates were fed to the organic animals, which along with the higher cost per tonne of organic feeds resulted in much higher concentrate costs per head. In contrast, Younie used similar land with a

Table 13.18 Comparative gross margins for 18-month beef systems from various sources

	Lowman (1989) Org Conv		Younie (1989) Org Conv		Spedding (1989) Org (high clover)	Beefplan average
Output						
Sales	550	564	543	540	522	522
less calf cost & mort.	140	140	175	175	241	206
plus surplus silage	0	0	12	32	0	0
Total output	410	424	380	397	281	316
Variable costs						
Calf rearing	56	53	33	12	0	0
Concentrates						
First winter			45	40	22	20
Grazing	180	39	11	4	7	6
Finishing winter			35	21	76	59
Forage	0	68	8	35	30	36
Vet & Med	9	31	2	9	12	12
Bedding	N/A		20	20	16	16
Other	N/A		0	0	14	14
Total variable costs	245	190	154	141	177	163
Gross margin per head	165	234	226	256	104	153
Gross margin per hectare	N/A		806	1108	320	589
Working capital per hectare	N/A		897	1063	1014	1107
Interest on capital	N/A		176	210	232	253
Gross margin per hectare after interest	N/A		630	898	88	336
Premium required to equal conventional GM/ha incl interest (p/kg carcass weight)	(48)		30		32	

high clover content for the organic system. As a result, although the cost per tonne of the organic concentrate was again higher, the cost of organic concentrates per head was only half that in Lowman's trial.

Spedding's comparisons (which also covered finishing store cattle and suckler beef) are based on hypothetical organic data and conventional MLC Beefplan averages. Lower output (due primarily to higher calf and mortality costs) and higher variable costs (especially concentrates and forage) result in gross margins per head and per hectare considerably lower than those found in either the Lowman or Younie studies, although a similar conclusion is reached about the level of premium required to achieve results comparable with conventional systems. The assumptions made by Spedding deserve more detailed consideration than is possible here; interested readers should consult the original paper to avoid drawing unwarranted conclusions. In particular, little account seems to have been taken of the potential for improved liveweight gains from forage with a high clover content as identified by Younie and other authors (see Chapters 8 and 10). Similar criticisms also apply to the earlier Meat and Livestock Commission costings (MLC, 1988) which concentrated more on 'semi-organic', low-input approaches such as Conservation Grade.

Data on milk production is even sparser. The only recent comparison published in Britain is that of Holden (1989). Holden took four organic farms and compared their results with the average of all Milk Marketing Board costed farms in 1988/89 (Table 13.19). As with beef production, reduced stocking rates and the higher cost of purchased organic feed were major factors in the poorer economic performance of the organic dairy herds. On the basis of these results, a premium of 4 to 5 pence per litre (20 to 25%) would be required to achieve the MMB average for conventional herds. Too much weight should not be attached to the results from the Channel Island herds which have their own financial characteristics.

Table 13.19 Comparison of conventional and organic milk production costings.

| | Farm | | | | MMB |
	A	B	C	D	average
Breed	Friesian	Guernsey	Jersey	Friesian	N/A
Average yield (l/cow)	5100	4500	3582	4950	5549
MMB milk price (p/l)	18.00	21.50	21.56	18.50	17.78
Concentrate use (t/cow)	1.0	1.35	0.8	1.1	1.4
Concentrate cost (£/t)	160	158	153	173	131
Margin over concentrates (£/cow)	758	754	652	725	794
Margin over feed and fertilisers (£/cow)	754	744	644	722	749
Stocking rate (LU/ha)	1.7	1.7	1.7	1.8	2.2
Gross margin (£/ha)	1281	1264	1107	1299	1644
Premium required to equal conventional gross margin (p/l)	4.2	4.9	8.8	3.9	N/A

Source: Holden (1989).

Comparative gross margins for milk production have been included in several studies abroad. Data from the Swiss study by Steinmann are presented in Table 13.20. The higher price per kg for organically produced milk reflects an element of direct marketing in the group of farms studied. Although the absolute figures are not particularly appro-

Table 13.20 Comparative yields and gross margins for milk production on organic farms in Switzerland, average 1979–81.

	Unit	Organic	Conventional partner[1]	Regional average
No. of farms		26	26	1465
Herd size	LU	22.3	24.9	26.5
% Dairy cows		74	69	69
Milk production	kg/cow	4517	5111	4912
Milk price	SFr/kg	0.87	0.80	N/A
Output:				
Milk	SFr/LU	2572	2461	2428
Other	SFr/LU	739	935	830
Total	SFr/LU	3311	3396	3258
Variable costs:				
Concentrates	SFr/LU	308	370	431
Vet & Med	SFr/LU	70	73	76
Other	SFr/LU	120	94	91
Total		498	537	598
Gross margin (excl. forage)	SFr/LU	2813	2859	2660
Stocking rate	LU/fha	1.87	2.08	2.44
Gross margin (incl. forage)	SFr/fha	5263	5826	5989

[1] See Table 13.1.
Source: Steinmann (1983).

priate to British conditions, the relative results are important. Gross margins per forage hectare (including forage costs) are very similar between the two systems, in spite of lower yields per cow and lower stocking rates.

General

Looking at gross margins for individual enterprises on their own, however, can be highly misleading. Organic systems are very dependent on interactions between enterprises, the most important aspect of which is well designed rotations. This means high gross margins for cereals and vegetables can only be achieved on an occasional basis as part of a balanced rotation, usually involving livestock and leys containing legumes, and are offset by these other enterprises. Conversely, reduced gross margins for livestock may be compensated for by higher gross margins for crops, so that it is not necessarily appropriate to look at beef or milk production simply in terms of the premium required to achieve comparability with conventional systems.

FIXED COSTS

Labour

It is frequently assumed that organic farms require considerably more labour than conventional farms. In fact, this is only the case for some enterprises, particularly those which require specific additional labour inputs, for example handweeding in field vegetables. Frequently, this type of labour is casual rather than regular labour. Another source of additional labour requirement lies often in the processing and marketing of produce. Although this is not limited to organic systems, marketing and processing activities tend to be more important to individual organic producers and their impact can be significant. The Vine and Bateman study found that paid labour tended to be lower on the organic farms than conventional farms, but when farmer and spouse labour was included, they were very similar.

For comparative data on labour requirements for different enterprises, it is necessary to look at studies from other European countries again. Table 13.21 shows estimates for labour use on organic and conventional farms in Switzerland. An analysis of the contribution of different enterprises to farm output in the Steinmann study suggests that vegetable production plays a more important role on organic farms compared with conventional farms, and this may explain to some extent the greater labour requirement per hectare of arable land. However, the Steinmann study was not able to analyse labour use in detail and other factors such as increased off-farm activities (advisory work, talks etc.) and a

Table 13.21 Labour requirements on organic farms compared with conventional partner and test farms in Switzerland, 1978–81.

No. of farms	Organic 21	Conventional partner[1] 21	Regional average 1030
SMD per farm	799	674	614
of which: family	469	430	446
paid labour	330	244	168
Family as % of total	59	64	73
Total labour units	2.7	2.3	2.1
Arable area/labour unit	7.1	8.1	8.0
SMD/ha utilisable area	46	40	41
SMD/ha arable area	43	37	37

SMD = standard man days.
[1] See Table 13.1.
Source: Steinmann (1983).

deliberate policy to support and train students on the part of some farmers may account for some of the differences in labour requirements.

The study by Schlüter (1986) of biodynamic farms in southern Germany is the first to collect data directly from farms and the results are summarised in Tables 13.22 and 13.23. The results show, as would be expected, great variability between farms and this affects any comparison which is made with data from conventional systems.

The comparison between biodynamic and conventional winter cereals (Table 13.23) shows similar labour requirements for seedbed prepara-

Table 13.22 Average labour requirements for different enterprises in hr/ha (bio-dynamic farms, Baden-Wüttemberg, average 1979/80).

Crop type	Number of farms	Soil culti- vation	Seedbed preparation, drilling	Manuring
Winter cereals	13	7.2	3.3	5.9
Range		2.9–12.3	0.9–6.7	2.7–11.7
Spring cereals	9	5.6	3.5	2.5
Range		2.6–9.8	1.5–5.0	0.0–5.0
Potatoes	11	9.5	39.0	13.6
Range		6.9–12.8	10.0–71.0	6.7–21.2
Field vegetables	11	8.3	19.8	6.8
Range		5.0–13.1	5.7–46.6	0.0–14.3
Fodder beet	6	12.0	11.0	19.87
Range		7.5–17.0	4.6–15.1	9.7–36.7
Maize	3	10.0	6.1	12.1

Crop type	Weed control, hand	Weed control, mechanical	Harvest	Total
Winter cereals	3.7	2.8	12.4	35.3
Range	0.0–8.3	0.0–7.0	8.0–20.8	21.6–55.2
Spring cereals	6.8	2.4	12.2	33.0
Range	0.0–11.9	0.0–6.1	1.0–19.0	22.5–44.2
Potatoes	61.8	15.2	225.8	363.9
Range	33.8–101.1	0.0–31.6	90.0–342.7	233.8–554.3
Field vegetables	280.9	9.8	277.3	602.9
Range	89.5–462.0	0.0–25.2	85.7–587.7	246.1–1239.0
Fodder beet	171.9	38.0	130.2	382.8
Range	25.0–269.0	6.5–48.3	75.9–203.9	238.6–564.0
Maize	54.8	6.2	17.0	106.8

Source: Schlüter (1986).

Table 13.23 Comparative labour requirements for different enterprises in hr/ha (biodynamic and conventional farms, Baden-Württemberg, 1979/80).

Area and crop	Number of farms	Seedbed preparation, ploughing	Weed control, manuring & biodynamic preparations	Harvest	Total
2–5 ha winter cereals					
Conventional average	4	15.1	3.1	9.6	27.8
Conventional range		6.0–29.3	2.1–4.7	6.6–15.4	17.7–41.3
Biodynamic average	6	12.7	17.2	14.1	44.0
Biodynamic range		7.2–24.0	11.1–27.5	9.2–22.0	27.5–59.0
1–4 ha potatoes					
Conventional average	3	13.0	21.8	69.1	103.8
Conventional range		7.4–20.7	8.4–46.8	51.0–95.1	78.7–124.2
Biodynamic average	4	39.2	81.9	163.2	284.3
Biodynamic range		7.9–64.0	66.6–121.1	116.3–270.5	233.8–375.1
Under 1 ha fodder beet					
Conventional average	4	14.0	88.8	119.5	222.3
Conventional range		7.5–26.7	54.8–116.2	62.5–240.0	158.1–321.5
Biodynamic average	4	25.5	261.0	152.2	438.7
Biodynamic range		10.0–53.3	158.6–389.0	96.1–209.0	238.6–608.8

Source: Schlüter (1986).

tion and drilling, but higher requirements for weed control and the biodynamic preparations. The biodynamic preparations require a considerable amount of time to prepare and these figures therefore overstate the additional labour use which might be expected in more straightforward organic systems. However, estimates by Dabbert (1990) for organic farms in Baden-Württemberg produced similar results.

These estimates are not directly transferable to the British farming situation for a number of reasons. Dabbert's estimate of about 30 labour hours per year for winter cereals is nearly three times that given in Nix's Farm Management Handbook for conventional winter wheat production, but is not much greater than the conventional average for that part of Germany (Table 13.23), reflecting differences in farm size, structure and level of mechanisation. In Britain, there is virtually no post emergence weed control in cereals (although some crops might well benefit from it), so it is unlikely that the labour requirement for cereals will be very different from conventional systems. The higher harvesting labour requirements for all three crops illustrated in Table 13.23 may be due to differences in mechanisation which the small samples used have not been able to eliminate.

Another survey of organic farms in Germany (Lösch & Meimberg, 1986), which looked at the whole membership of one of the major organic organisations, found that the additional labour use was much greater on small farms compared with larger farms (Table 13.24). This is largely due to the necessity for small farms to concentrate on high value crops such as vegetables which tend to be much more labour intensive.

Table 13.24 Available labour on organic farms in Germany compared with the national average, measured in full-time labour units per 100 ha.

| | *Utilisable Agricultural Area (ha)* | | | | | | | *All* |
	<1	*1–2*	*2–5*	*5–10*	*10–20*	*20–30*	*30–50*	*>50*	*farms*
Organic	285.4	100.0	48.4	23.7	12.0	7.5	5.8	4.2	9.3
National average	200.7	36.2	20.2	13.1	9.4	6.9	4.9	3.1	7.7

Source: Lösch & Meimberg (1986).

Apart from horticultural production, the other main area where labour requirements could be increased significantly by changing over to an organic system is when a specialised arable or livestock unit converts to a mixed farming system. The benefits of specialisation in terms of economies of size will be lost. There may also be potential gains in this situation, due to the fact that on many specialised farms, particularly arable farms, full time labour is underemployed for large parts of the year. A switch to more enterprises may allow for better seasonal labour utilisation and therefore lower labour costs per unit value of production.

Other fixed costs

Losing the benefits of specialisation also applies to other fixed cost items such as machinery servicing and depreciation, and to some extent to the servicing of capital borrowings. Requirements for working capital, however, are likely to be reduced because of the reduction in use of variable inputs such as fertilisers and sprays. Other fixed cost items are unlikely to differ between the systems.

The Vine and Bateman study found that, if anything, fixed costs were lower on the organic farms they surveyed, but some of this was attributable to greater thriftiness and lower machinery depreciation. The Swiss study by Steinmann also found reduced machinery costs on the organic farms, but these results are contradicted by the studies in Baden-Württemberg. The situation remains unclear. It seems safest to assume that, unless there are dramatic changes to the overall mix of enterprises, fixed costs are not likely to change significantly in an organic system.

Total Farm Income

There is also very little firm evidence on the level of incomes on organic farms in Britain, at least of the kind which we are used to from the Ministry of Agriculture's annual report 'Agriculture in the United Kingdom' or the more detailed Farm Business Survey reports on which it is based. The Vine and Bateman study found that, at the then current prices for conventionally produced farm products, the majority of organic farms had lower financial returns than they might have expected had they used conventional methods. Although variable costs per unit of output were generally lower, as were fixed costs to a lesser extent, the reductions in costs were not sufficient to offset the lower yields obtained with the result that net financial returns suffered (Table 13.25).

They added, however:

> Although the majority of organic farmers had lower financial
> returns, it should be noted: firstly, that the variety of systems
> and levels of management were such that some organic farmers
> outperformed the conventional average; secondly, that in the
> cases of some arable products, premiums for organic produce are
> available that can shift the organic farm from a position of poor
> returns (compared with conventional) to one of good returns;
> and thirdly, that the competitive position of organic farming
> could change if output prices changed relative to certain input
> costs.

Studies from other European countries conducted over a longer period than just one year show income levels comparable with similar conventional farms. The results from the Swiss study by Steinmann are summarised in Table 13.26. Agricultural incomes were similar between the organic and conventional groups. The higher labour requirements on the organic farms meant, however, that the agricultural income per person per day was about 20% lower than on the paired conventional farms.

Comparisons have also been included in the annual reports of the West German Federal Ministry of Agriculture for more than seven years. These have been severely criticised because, originally, a very small sample of 15–20 farms was compared with the average for all West German farms, whereas conventional farms were analysed by farm type, size and region. Most of the farms were situated in the agriculturally more disadvantaged, southern parts of Germany. The result was that the organic farms emerged very poorly from the exercise. The average farm size of the organic sample was around 35 ha, some 10 ha larger than the national average. The sample was subsequently compared with national averages for the 30–40 ha category, even though it is far from clear that the individual organic farms fell within this range.

The result of this somewhat casual approach has been a marked unwillingness on the part of organic farmers in West Germany to

Table 13.25   Net farm income per ha on organic farms in England and Wales as a % of conventional standard (1978/79).

Farm classification	Less than 0%	0 to 19%	20 to 39%	40 to 59%	60 to 79%	80 to 99%	100 to 119%	120 to 139%	Over 140%	Total
By farm size										
<50 ha	1	3	2	2	3	1	–	–	1	13
50.1–100 ha	1	–	3	3	–	–	–	1	2*	10*
100.1–200 ha	–	2*	1	–	–	1	1	–	1	6*
200.1–300 ha	–	–	–	–	–	–	–	–	1	1
Total	2	5*	6	5	3	2	1	1	5*	30*
By farm type										
Specialist dairy	1	2*	2	1	1	1	–	–	–	8*
Mainly dairy	–	–	2	–	–	–	–	–	1	3
Livestock, cattle & sheep	1	3	1	4	1	1	1	1	–	13
Mixed	–	–	–	–	–	–	–	–	4*	4*
Cropping, mostly cereals	–	–	1	–	–	–	–	–	–	1
General cropping	–	–	–	–	1	–	–	–	–	1
Total	2	5*	6	5	3	2	1	1	5*	30*

* two of the farms were 'semi-organic' and their inclusion is indicated by an asterisk.
Source: Vine & Bateman (1981).

Table 13.26   Income and returns to family labour on organic and conventional farms in Switzerland, average 1979–81.

	Organic	Conventional partner[1]	Regional average
No. of farms	20	20	1030
Utilisable agricultural area (ha)	17.2	16.8	15.0
Agricultural income (SFr)	54,417	55,819	52,937
Other income (SFr)	7,945	4,113	5,456
Total income (SFr)	62,362	59,932	58,393
Returns to labour:			
per family (SFr)	45,262	47,904	45,215
per person/day (SFr)	93.9	115.0	101.6

[1] See Table 13.1.
Source: Steinmann (1983).

participate further, and the sample remains small. The comparison of the most recent data available, for 1987/88, has been done with much more care, and includes 57 full-time farms (out of more than 2,000 organic farms in West Germany).

The results, summarised in Table 13.27, are very interesting, not least because they show that, as sample size increases and the method of selecting the comparison groups improves, the differences between the two groups decrease. Although the organic farms have only 82% of the output of the conventional farms, total costs are 77% of conventional, and the resulting farm incomes per farm are very similar. Because of the smaller average size of the organic farms in the sample, farm income per hectare and as a percentage of output is higher on the organic farms, but the income per family labour unit is about 15% lower. The differences in cropping areas and stocking rates, especially field vegetables and pulses, forage maize and pigs are also worth noting.

The smaller group of biodynamic farms studied in detail by Schlüter (1986) in Baden-Württemberg follow a similar pattern. Cropping enterprise output was considerably higher, but lower output for livestock enterprises and other activities resulted in similar farm output overall. Costs were much lower, with the result that both profit and farm income per ha were higher on the biodynamic farms than on the conventional farms. Profit per family labour unit and farm income per full-time labour unit were lower on the biodynamic farms in most cases in the first year of the study, but higher in the second year.

The whole farm results for the comparative systems at Nagele in the Netherlands are summarised in Table 13.28. These contrast sharply with the survey results discussed above. The poor results for the biodynamic system show up the main problem with the comparisons taking place at Nagele. The biodynamic unit was established as a labour intensive mixed dairying and arable system in an area which is almost exclusively arable.

The average hourly labour costs are taken as 27 DFl/hour, but the real returns per hour spent in the conventional, integrated and organic systems were only 15, 19 and 0 DFl respectively. The fact that farmers do not receive proper recompense for their labour is a fact of life. Most of the difference in the net returns is caused by the labour requirements of the biodynamic system. It is probable that a more appropriate organic system could have been developed which would not have been so labour intensive and hence would have been more viable in these particular circumstances.

Table 13.27 Comparison of conventional and organic farming data, West German Federal Ministry of Agriculture Annual Report 1989; 1987/8 results.

Parameter	Unit	Organic	Conventional comparison*	National average
Number of farms	Nos.	57	223	9,018
Farm size	ha UAA/farm**	27.65	28.36	30.71
Comparative value	DM/ha UAA	1,142	1,155	1,364
Full-time labour units	per farm	1.98	1.64	1.68
Full-time labour units	per ha UAA	0.071	0.058	0.055
Stocking rate	LU/ha UAA	1.09	1.43	1.84
Dairy cows	LU/ha UAA	0.50	0.48	0.49
Other cattle	LU/ha UAA	0.43	0.58	0.62
Pigs	LU/ha UAA	0.09	0.35	0.70
Poultry	LU/ha UAA	0.04	0.02	0.04
Cropping	% UAA	62.4	68.9	61.5
Cereals	% arable area	53.9	66.2	63.2
Potatoes	% arable area	4.1	2.1	2.3
Sugar beet	% arable area	0.0	2.9	5.6
Field vegetables/pulses	% arable area	11.6	6.9	15.6
Forage maize	% arable area	2.6	11.5	13.3
Other fodder crops	% arable area	27.9	10.3	6.4
Yields				
Wheat	t/ha	3.6	5.5	5.9
Rye	t/ha	2.9	4.0	4.2
Potatoes	t/ha	14.4	25.7	32.0
Milk yields	kg/cow	3,552	3,972	4,506
Prices				
Wheat	DM/t	1,011.50	373.80	389.90
Rye	DM/t	1,063.90	370.70	365.40
Potatoes	DM/t	544.50	204.30	145.10
Milk	DM/kg	0.76	0.63	0.63
Enterprise output	DM/ha UAA	4,753	4,385	5,725
Crops	DM/ha UAA	1,159	659	1,038
Livestock	DM/ha UAA	2,401	2,777	3,619
Costs	DM/ha UAA	3,611	3,350	4,276
Fertiliser	DM/ha UAA	27	252	256
Pesticides	DM/ha UAA	4	101	125
Feedstuffs	DM/ha UAA	307	451	924
Livestock	DM/ha UAA	168	338	482
Labour	DM/ha UAA	385	91	198
Profit	DM/farm	31,568	29,332	N/A
	DM/ha UAA	1,142	1,034	N/A
	DM/FLU ***	21,122	18,826	N/A
Profit as % of output	%	24.0	23.6	N/A

* average for farms of comparable type, business size and location.   ** utilisable agricultural area.   *** family labour unit.

**Table 13.28 Overall farm economic results of the three
farming systems over 1982–1985 (× 1000
DFl), Nagele experiment, the Netherlands.**

	Conventional	*Integrated*	*Biodynamic*
Farm size (ha)	17.0	17.0	22.0
Labour (person/year)	0.6	0.6	1.8
Total returns	118.1	111.1	168.3
Total costs	132.6	121.2	265.4
(of which farmer's own labour)	33.7	35.5	97.9
Net surplus	−14.5	−10.1	−97.1

Source: Vereijken (1986).

Models

An alternative to the use of direct comparisons is the use of hypothetical models which concentrate mainly on the technical differences between the two systems. This type of approach is very sensitive to the initial assumptions made, as is indicated in a study by Jaep *et al.* (1985) in West Germany. Results for four farm types have been calculated under several different sets of circumstances. Table 13.29 shows some possible solutions for one of the farm types, emphasising the role of enterprise combinations in determining farm income.

The importance of this kind of example is that it demonstrates that the relative economic performance will be different in each individual case, depending on the starting position. In addition, these results lend support to the argument that a change over to an organic system may help to ensure the continuation of smaller farms, particularly if high value vegetable crops can be introduced into existing extensive units. In view of the lack of first hand data in Britain, an attempt is made here to indicate the possibilities using hypothetical data (Tables 13.30 and 13.31).

In Table 13.30, the farm modelled is a 42 ha dairy farm on Grade II-III land. Org.1 represents an organic system without vegetables, consisting of four years grass/white clover ley, one year forage crops (rye and kale) and one year cereals for concentrates. The stocking rate assumed is about 1.7 LU per forage ha. Org.2a is the same farm with a five-year ley and one year (i.e. 7 ha) organically produced vegetables sold at current premium prices. Org.2b is the same as 2a, but without premium prices. The conventional example is a specialist dairy unit with a stocking rate of 2.25 LU/ha. Quotas have not been taken into consideration. The example is only intended to be illustrative—much of the data used are standard data from Nix's Farm Management Handbook and other sources.

Table 13.29 Summary of gross margins for crop and livestock production, fixed costs and net income for one farm type with different enterprise combinations (all values are DM).

Model 18 ha	Crops	Gross margins Livestock	Total	Fixed costs	Net farm income	Net farm income/FTLU*
Conv. 1	7,186	43,557	50,743	22,207	28,536	21,951
Conv. 2	41,795	40,550	82,345	27,033	55,312	55,312
Org. 1	41,732	39,322	81,054	31,403	49,651	29,206
Org. 2	39,768	27,179	66,947	23,135	43,812	33,702

* Full-time labour unit.
Note: The example farm is an 18 ha holding on marginal land in Hessia. Conv. 1 represents cereals, grassland and forage crops, with 12 dairy cows, 3 heifers and 12 breeding sows . Org. 1 is a direct comparison, including vegetables in the rotation, but with only 10 dairy cows, 2 heifers and 12 breeding sows. Conv. 2 is an arable and pig fattening unit, with sugar beet, cereals, 30 breeding sows and 440 fattening pigs. Org. 2 is based on cereals, vegetables, forage crops and grassland again, (an intensive pig unit not being possible in an organic system), with 16 fattening cattle, 1 breeding sow and 18 fattening pigs for livestock.
Source: Jaep et al. (1985).

Table 13.30 Gross margins, fixed costs and net farm income for a 42 ha specialised dairy farm—hypothetical example.

	Org 1 (£)	Org 2a (£)	Org 2b (£)	Conv (£)
Gross margin				
Grazing livestock	38,122	35,238	35,238	46,003
Less forage costs	−1,428	−1,050	−1,050	−7,114
Cash crops	3,455	24,970	11,920	0
Total gross margin	40,149	59,158	46,108	38,888
Fixed costs				
Labour: paid	3,000	6,000	6,000	3,000
unpaid	6,000	6,000	6,000	6,000
Machinery and power	11,760	11,760	11,760	11,760
Rent & rates, incl. notional	5,670	5,670	5,670	5,670
Other fixed costs	3,780	3,780	3,780	3,780
Total fixed costs	30,210	33,210	33,210	30,210
Management and investment income	9,939	25,949	12,899	8,679
Net farm income	15,939	31,949	18,899	14,679

In Table 13.31, the farm modelled is a 180 ha mixed dairy/arable farm, also on better land. The organic rotation consists of four years grass/ white clover ley, wheat, oats, field beans, wheat and barley. The stocking rate is assumed to be approx.1.7 LU per forage ha, with a herd size of 110 cows. Org. (a) is with and Org. (b) without premium prices. The conventional system consists of a dairy herd of 120 cows stocked at 2.25 LU per forage ha on 70 ha, with a further 70 ha of wheat and 40 ha

Table 13.31 Gross margins, fixed costs and net farm income for a 180 ha mixed cropping farm—hypothetical example.

	Org (a) (£)	Org (b) (£)	Conv. (£)
Gross margin			
Grazing livestock	64,498	64,498	74,602
Less forage costs	−2,400	−2,400	−11,858
Cash crops	66,198	55,298	69,212
Total gross margin	128,296	117,396	131,956
Fixed costs			
Labour: paid	14,400	14,400	14,400
unpaid	6,000	6,000	6,000
Machinery and power	29,700	29,700	29,700
Rent & rates, incl. notional	19,800	19,800	19,800
Other fixed costs	9,000	9,000	9,000
Total fixed costs	78,900	78,900	78,900
Management and investment income	49,396	38,496	53,057
Net farm income	55,396	44,496	59,057

of barley. Again, the example is only illustrative, as standard data have been used.

SOCIAL AND OTHER EXTERNAL COSTS

One should not lose sight of the fact that the profit and loss account does not tell the full story. Agricultural systems also entail costs for society which the farmer does not usually have to take account of. These not only include transfer payments in the form of the subsidies which farmers receive, either paid for by the taxpayer or by the consumer in the form of higher food prices. They also include the costs of putting right environmental problems caused by farming practices. The costs of ensuring clean water supplies without excessive levels of nitrates, the costs of monitoring foodstuffs for pesticide residues or cleaning up water courses after accidents at agrochemical plants, the costs of conserving threatened habitats and environmentally sensitive areas, to name but a few, all have to be taken into account when assessing the relative merits of different farming systems. If organic systems can contribute to reducing these 'external' costs, while at the same time enabling individual producers to earn a living, then the organic approach would appear to have much to recommend it from the point of view of society as a whole.

Another aspect in assessing the relative economic merits of organic and conventional systems is their impact on the local and national economy through employment in the agricultural supply industries and other areas. A reduction in the demand for agricultural inputs could well have adverse side effects in other industries. In an examination of this issue, Lockeretz (1989) concluded that although high-input systems had greater local economic benefits on a per hectare basis, the benefits per unit value of production were lower which meant that a greater proportion of the value of production left the local economy to pay for purchased inputs. The implication of this is that organic systems which are less reliant on purchased inputs may contribute more to local economic activity in the long term, particularly in rural areas which are highly dependent on agriculture, than less sustainable, high input systems.

OVERVIEW

Such information as is available on the economics of organic farming would seem to indicate that changing over to such a system certainly need not spell financial disaster. Although yields are likely to be lower, variable costs are also likely to be much lower, with little or no expenditure on synthetic fertilisers or sprays, and premium prices may be available for certain crops. Premium prices are only just becoming available for livestock enterprises which account for most of the output on organic farms, but processing and direct marketing can help to increase returns.

The combination is likely to result in similar or higher gross margins, except for livestock. Although labour costs may be higher, other fixed costs are likely to be similar or lower and the result is net farm incomes which are comparable to or somewhat lower than conventional systems.

For the farmer interested in changing over to an organic system, the question of what effect the change-over will have on his or her income can only be answered with reference to the current position on the farm. Mixed and livestock farms without intensive pig or poultry units are likely to be in the most favourable position, while all-arable farms will have to introduce livestock enterprises, at least until stockless organic systems have proved themselves a viable alternative. Farms already struggling under a heavy debt burden are unlikely to solve their problems by adopting an organic approach, even if the premium prices look tempting.

REFERENCES AND FURTHER READING

Böckenhoff, E. *et al.* (1986) *Analyse der Betriebs- und Produktionsstrukturen sowie der Naturalerträge im alternativen Landbau.* Berichte über Landwirtschaft, 64 (1), 1–39

Dabbert, S. (1990) *Zur optimalen Organisation alternativer landwirtschaftlicher Betriebe.* Agrarwirtschaft Sonderheft 124. Verlag Alfred Strothe; Frankfurt

Dixon, P. & Holmes, J. , (1987) *Organic Farming in Scotland.* Edinburgh School of Agriculture

Dobbs, T. L. *et al.* (1988) *Factors influencing the economic potential of alternative farming systems.* American Journal of Alternative Agriculture 3: 26–34

Goldstein, W. A. & Young, D. L. (1987) *An agronomic and economic comparison of a conventional and a low-input cropping system in the Palouse.* American Journal of Alternative Agriculture 2: 51–56

Helmers, G. A. ; Langemeier, M. R. & Atwood, J. (1986) *An economic analysis of alternative cropping systems for east-central Nebraska.* American Journal of Alternative Agriculture 1: 153–158

Holden, P. (1989) *What price organic milk?* New Farmer and Grower (Autumn): 28–29

Jaep, A. *et al.* (1985) *Konventioneller und alternativer Landbau—ein betriebswirtschaftlicher Vergleich.* IfB Landwirtschaftliche Fachinformationen 143/85. Hessisches Landesamt für Ernährung, Landwirtschaft und Landentwicklung, Kassel

Lockeretz, W. (1989) *Comparative local economic benefits of conventional and alternative cropping systems.* American Journal of Alternative Agriculture 4: 75–83

Lockeretz, W. *et al.* (1984) *Comparison of conventional and organic farming in the Corn Belt.* In: Bezdicek, D. *et al.* (eds.) *Organic Farming: Current Technology and its Role in a Sustainable Agriculture.* American Society of Agronomy, Special Publication No. 46; Madison, WI

Lösch, R. & Meimberg, R. (1986) *Der "alternative" Landbau in der Bundesrepublik Deutschland.* Ifo Studien zur Agrarwirtschaft, 24

Lowman, B. (1989) *Organic beef production.* In: Chamberlain, A. T. *et al. Organic Meat Production in the '90s.* Chalcombe Publications

MELU (1977) *Auswertung drei-jähriger Erhebungen in neun biologisch-dynamisch bewirtschafteten Betrieben.* Baden-Württemberg. Ministerium für Ernährung, Landwirtschaft und Umwelt; Stuttgart

MLC (1988) *Beef from alternative production systems.* In: *Beef Yearbook 1988.* Meat and Livestock Commission; Milton Keynes

National Research Council (1989) *Alternative Agriculture.* National Academy Press; Washington DC

Schlüter, C., (1986) *Arbeits- und betriebswirtschaftliche Verhältnisse in Betrieben des alternativen Landbaues.* Agrar-und Umweltforschung in Baden-Württemberg, 10. Ulmer Verlag; Stuttgart

Spedding, A. W. (1989) *Organic beef production—the production economics of alternative systems.* In: Chamberlain, A. T. *et al. Organic Meat Production in the '90s.* Chalcombe Publications.

Stanhill, G. (1990) *The comparative productivity of organic agriculture.* Agriculture, Ecosystems and Environment, 30: 1–26

Steinmann, R. (1983) *Der biologischer Landbau—ein betriebswirtschaftlicher Vergleich.* Schriftenreihe der FAT, 19. Forschungsanstalt für Betriebswirtschaft und Landtechnik FAT; Tänikon, Switzerland

Stoney, R. (1987) *Organic agriculture in Ireland—an appraisal.* IFOAM Bulletin, 1: 8–11

Stopes, C. (1987) *Myth of old varieties must be finally buried.* New Farmer and Grower, 15 (Summer): 15–16

Stöppler, H.; Kölsch, E. & Vogtmann, H. (1988) *Suitability of winter wheat varieties for ecological agriculture.* In: Allen, P. & Dusen, D. van (eds.) *Global Perspectives on Agroecology and Sustainable Agricultural Systems.* Proceedings of 6th IFOAM International Conference, University of California, Santa Cruz

Vereijken, P. (1986) *From conventional to integrated agriculture*. Netherlands Journal of Agricultural Science, 34: 387–393

Vereijken, P. (1989) *Experimental systems of integrated and organic wheat production*. Agricultural Systems 30: 187–197

Vine, A. & Bateman, D. I. (1981) *Organic Farming Systems in England and Wales—practice, performance and implications*. Department of Agricultural Economics, UCW Aberyswyth

Wookey, B. (1987) *Rushall—the story of an organic farm*. Blackwell; Oxford

Younie, D. (1989) *Eighteen-month beef production—organic and intensive systems compared*. In: Chamberlain, A. T. *et al*. *Organic Meat Production in the '90s*. Chalcombe Publications

Zadoks, J. C. (1989) *Development of Farming Systems—evaluation of the five year period 1980–1984*. Pudoc; Wageningen

Recent publications

Bateman, D. I. (1993). The financial and economic performance of organic farming. *Journal of the Agricultural Society, University of Wales, Aberystwyth*.

Bulson, H. (1993). Organic cereals—what to sow. (A review of variety trials and yield levels) *Elm Farm Research Centre Bulletin*, 9:2–4.

Lampkin, N. H. and D. I. Bateman. (1993). *The economics of organic farming in Wales, 1989*. Discussion Paper Series 93/3. Centre for Organic Husbandry and Agroecology, University of Wales; Aberystwyth.

Lampkin, N. H. and S. Padel (eds) (1994). *The economics of organic farming—an international perspective*. CAB International, Wallingford.

Murphy, M. (1992). *Organic farming as a business in Great Britain*. Agricultural Economics Unit, University of Cambridge; Cambridge.

UWA/EFRC (1994). *1994 Organic farm management handbook*. Elm Farm Research Centre/Department of Agricultural Sciences, University of Wales, Aberystwyth.

Chapter 14

Converting to Organic Farming

It should be clear by now that farming organically is much more than just abandoning chemicals or using muck on potatoes. It entails, in most cases, quite radical and dramatic changes to the farming system as a whole and requires a fundamentally different approach to the management of a farm.

These changes cannot be entered into lightly; the decision to convert all or part of a farm requires a high level of commitment on the part of the farmer or manager if it is to succeed. In particular, the decision to convert carries with it a high element of risk and uncertainty as far as the financial viability of the farm is concerned, and this is compounded by the current lack of detailed information and advice. In the initial stages of conversion, a wide range of problems can arise, and these can seriously undermine the confidence of the person undertaking the conversion. It is at times like these, when the going seems difficult, that the support of others becomes vital. One's immediate family, close friends, farm workers, neighbours and others who have an interest in the financial wellbeing of the farm can all play a crucial role. Without their support, the conversion process can become very difficult indeed.

This element of personal commitment and support for the organic farming concept is often overlooked. It has proved on many occasions, however, to be a major stumbling block, often outweighing the problems caused by changes in production techniques. Conversion of the farming system has to begin with a personal conversion, in terms of attitude and approach, of all the people who have a significant influence on the running of the farm. This may require a long lead-in period, during which time as much information as possible is assembled.

Information can come from a wide range of sources, including both conventional and organic advisory services and a wide range of organic farming publications and journals. Perhaps more importantly, personal contact with other organic producers at farm walks, seminars and conferences, allows experiences to be exchanged and first-hand assessments of working organic systems to be made. Direct personal contact is often far more convincing than any quantity of written material.

Part of the initial preliminary conversion period may also involve experimentation on one's own farm, in the form of a field which is set aside for the "organic treatment" while the rest of the farm continues to be managed conventionally. This approach has many attractions, but there are also disadvantages. The most important one is that the use of just one field often does not allow for the development of a suitable rotation for the system, or for the necessary adjustments of techniques and mechanisation, especially if emphasis is placed on high-value cash crops such as field vegetables or milling wheat. Under these circumstances, many of the biological processes and interactions which are necessary for the organic system to function may fail to become established. Rotations can take several years to become properly established and for their full effects to show through. It is therefore unwise to place too much reliance on the results from one field in isolation, even if it proves to be successful. If this approach is taken, further information should be sought from advisors and/or experienced producers to assess the outcome.

The approach taken in this book has been to emphasise the need to consider organic farming in terms of a "whole-farm" system, integrating both crop and livestock enterprises, and not to consider individual crops in isolation. The principles and practices which form the basis of successful organic farming systems have been described in the preceding chapters. The physical conversion of the farm is based on the application of these principles in a way which is appropriate to the individual farm. There is no general prototype or recipe which can be copied. Every farm is unique in terms of the environmental and climatic constraints under which it has to operate, as well as the resources (land, labour, capital etc.) which are available to the farmer. Farms are also very different as far as their previous development is concerned. The greater the degree of specialisation and intensification which has taken place, the greater the change which will be required to re-introduce diversity and scale down the intensity of individual enterprises. All these factors play a role in determining the suitability of a farm for conversion to an organic system. In theory at least, there is no reason why suitable systems cannot be developed to meet most farming situations, although some are more difficult than others. The greatest problem usually arises where economic pressures resulting from previous activities, such as capital investment in advanced dairying facilities or an intensive poultry unit, prevent the switch to an organic system. The issue of whether the farm itself is suitable for conversion to an organic system therefore has to be given very careful consideration before the actual conversion process is embarked upon.

In the past, the conversion has usually been done in a very haphazard manner. There are many examples where farmers have simply stopped applying nitrogen and then found that crops fail. Crops have also succumbed to massive weed infestations because the conversion started at an inappropriate point in the rotation. This trial-and-error approach characterised the development of organic farming in Britain. It was associated with a severe lack of sound information and advice as well as the absence of appropriate research and development work. It was also influenced by the large number of newcomers to organic farming who were also newcomers to farming itself and who experienced many of the problems which anyone new to farming would have done.

The situation has now changed markedly. The crisis in agriculture during the 1980s has resulted in many more farmers becoming involved who already have experience, either through farming conventionally on their own account, or having been brought up on conventionally managed farms and being in a position to take over the farm from parents. Information is more readily available, particularly as a result of the formation of producer organisations like British Organic Farmers and the Organic Growers Association. Advice can be obtained through the Organic Advisory Service, a service run by Elm Farm Research Centre in co-operation with British Organic Farmers, the Organic Growers Association and the Soil Association. Research work specific to organic agriculture is being conducted by Elm Farm Research Centre and at several publicly funded institutions in Britain and elsewhere in Europe.

These developments mean that many of the mistakes which others have made when converting to an organic system can be avoided. In particular, much greater emphasis is being placed on detailed planning of the conversion before the process actually starts. Conversion planning allows many of the implications to be considered fully, including both changes in production methods and the financial consequences. Experience with formal conversion planning is still limited, but where it has been carried out, many of the more serious problems have been avoided and the results are very encouraging.

Conversion Planning

An analysis of the experiences of farmers who have converted their farms to organic systems indicates that the main problems which are encountered during the conversion process include shortage of forage (due to a reduction in yields and increased reliance on home-grown forage), excess protein in the rations leading in some instances to health problems with livestock, weed control (notably docks, couch and thistles), unexpectedly high workloads in peak periods and financial difficulties. Planning the conversion entails being aware of the problems which can occur and attempting to avoid them by preparing for them in advance.

The preparation of a conversion plan is normally done by an adviser in conjunction with the farmer. The first stage is an assessment of the motives behind the conversion and whether they form a suitable basis for a successful conversion, together with an analysis of the current situation on the farm, including general farm details (size, layout etc.), soils (texture, nutrient status etc.), climate (rainfall, growing season length), current enterprises on the farm, manure handling systems, livestock housing and potentially limiting factors (e.g. labour/capital availability). The second part involves the development of a target organic system and a plan for the transition between the current, conventional system and the organic endpoint.

The procedure is often easier on mainly grassland farms, where livestock determine the farming system, than on mainly or all arable units where the range of options is wider and more complex. In the former situation, especially on dairy farms, the first consideration has to be the number and type of stock to be kept, taking into account adjustments in stocking rates which may be necessary, as well as personal preferences for particular types of stock or specific enterprises.

On lowland farms, a target of around 1.2 LU/ha taken over the whole farm, or 1.7 LU per forage ha, provides a useful basis for planning. This allows for a proper balance between forage legumes and cash crops to be achieved, as well as allowing for most of the concentrate requirements to be met from home-grown produce. The next stage of the plan must therefore be aimed at meeting the forage and fodder requirements for the proposed stocking levels. Suitable rations need to be developed, allowing for the fact that the legume content of the grassland will tend to increase both voluntary intakes and protein levels in the ration. The area of land needed to support the livestock enterprises can then be determined, but will also be affected by the amount of bought-in conventional and organic feeds likely to be used (conventional feedstuffs are limited to 20% of total dry matter fed for dairy cattle and 10% for other ruminants under the Soil Association Standards (see Appendix 1).

The remaining land can be considered to be available for the production of cash crops, which need to be selected according to the preferences of the farmer as well as the physical and financial constraints imposed by the individual circumstances of the farm. At this point, sufficient information is now available to allow an appropriate rotation to be developed.

The design of the rotation is probably the most important element of the conversion plan. In addition to producing sufficient feed for livestock and maintaining the output of livestock and cash crops so that the farmer can obtain a satisfactory income, the rotation will contribute to the minimisation of weed, disease and pest problems, the maintenance of soil organic matter levels and soil structure, the provision of sufficient nitrogen and the minimisation of nutrient losses. These issues and the subject of rotation design have already been discussed in detail in Chapter 5. It is, however, very important to maintain flexibility within the rotation, both in terms of the length of the grass/clover ley and the specific crop to be grown. It may well be more appropriate to plan the rotation in general terms (e.g. forage legumes followed by a winter cereal cash crop) rather than taking specific crops (e.g. 4 years lucerne followed by winter wheat), although both approaches have their uses.

The feasibility of the rotation should be checked, in terms of nutrient removals (preparation of a nutrient budget), weed, disease and pest control and labour/machinery requirements. The changes in total labour requirements and availability, both regular and casual, and its seasonal distribution need to be considered. In addition, problems caused by unequal field sizes, different soil types, previous cropping of specific fields and so on may entail the development of more than one rotation for use on the farm.

Once the rotation and stocking rates have been determined, an estimate of the quantity of manure available can be produced, and a manuring plan prepared where the manure is allocated to the most appropriate points in the rotation.

Technical questions, such as measures for soil improvement, direct weed, pest and disease control techniques, appropriate cultivations and suitable manure management systems are less important at this stage of the planning process, although they do need to be considered at some stage before the conversion process gets underway. However, mechanisation needs to be assessed in terms of changes to tillage practices, requirements for weed control (harrows, flame weeders etc.) and for harvesting. Capital investment in improved livestock housing, feeding and manure handling systems as well as facilities for processing, packaging and marketing of produce may also be necessary.

Only when the target system has been developed within an agri-cultural/ecological context is it really appropriate to consider the financial implications in detail. Obviously, no system could be expected to be adopted which is not also financially viable, but the primary consideration should be a system which works agriculturally and is ecologically sustainable, with as few compromises imposed by economic constraints as possible. In this sense, a satisfactory level of income to ensure the financial survival of the farm business and meet the needs of the farm family is a more appropriate goal than profit maximisation which may, and often does, conflict with environmental considerations and sound agricultural practice.

The financial implications can be dealt with at several levels. The simplest is to calculate gross margins for the different enterprises on the basis of estimated yields and current prices and premiums as an attempt to represent the situation when the farm is finally converted. A more detailed financial assessment would involve considering fixed costs as well, and may also involve capital investment appraisal of specific items such as machinery purchases or additional housing for livestock. A further stage could involve the preparation of annual budgets during the conversion period. This requires a great deal of co-operation between the farmer and the adviser, particularly in terms of analysing the current financial circumstances of the farm and using these details for future planning.

While premium prices can make organic farming economically as well as environmentally attractive, they can only be achieved where a viable marketing strategy has been established. This means that marketing options have to be investigated, both in the locality and further afield, before proceeding with a particular plan, especially if the financial viability of a crop or enterprise is highly dependent on achieving good premium prices.

Much of the procedure for planning the conversion described here would apply also to mainly arable or stockless farms, except that the rotation itself would provide the starting point for the planning process, and the extent of any livestock enterprise is less clearly defined. Specific problems occur where livestock have to be reintroduced onto a farm and a stockless organic system may be worth considering. In some cases, the farm will be able to support more livestock than can be housed and further investment, or a livestock enterprise involving out-wintering of livestock, should be considered. The parameters within which the planning process takes place are not usually as obvious as with livestock farms. This can complicate the planning process, although it can also allow for greater flexibility .

The planning process does not only involve the design of a feasible, "target" organic system. It must also set out the steps that need to be taken during the conversion process and the timescale over which the conversion takes place. Probably the most important consideration when developing a timescale for conversion is to start slowly. Experience needs to be gained with new crops, new techniques and the potential output from the system. Ideally, the conversion should start with only a couple of fields entering the rotation, using this opportunity to see what happens over at least one, preferably two, seasons. Once some experience has been obtained, more fields can be converted, and the original fields will progress into the later stages of the rotation. The original fields are therefore always a couple of years ahead of the rest of the farm and any mistakes which are made as part of the learning process should not have a disastrous impact on income. If things go wrong, a (large) element of flexibility may be required to get round the problem, but once the skills have been mastered, the conversion can proceed at a much faster rate. It is of course possible to attempt a conversion of the whole holding at once, but the risks inherent in this approach are great and it is not recommended unless there are specific circumstances which warrant it.

Part of the process of developing a conversion plan, therefore, is to prepare a programme for the changeover of the farm, field by field, over a period of years. As with the design of the rotation, this programme has to incorporate a great degree of flexibility. Ideally, the whole conversion plan should be redesigned each year to take account of experience and of actual developments on individual fields. In many cases, the target system will also change, as certain crops prove to be less than suitable for the farm or other options look more attractive. Unfortunately, the work involved in annual replanning is often considerable and seldom achieved, but continuing input from outside sources will help overcome some of the problems during the conversion period. Even in a situation where the organic system which finally evolves looks nothing like the original target system, the planning process still plays a crucial role in ensuring that the various implications of embarking on a conversion have been thought through before any action is taken.

Converting livestock enterprises

The livestock enterprises on the farm will probably be the most difficult to change to an organic management system and their conversion will therefore need to be done carefully and gradually. The following steps indicate an approach which can be taken so as to avoid many of the pitfalls.

The first step is the reduction of animal stress by:

- introducing more natural rearing systems (using nurse cows or artificial suckling systems; feeding whole milk; increasing the pre-weaning period; reducing the use of concentrate feed where possible; optimum timing of concentrate feeding);
- introducing family or other appropriate small groups;
- modifying the housing system (increasing the use of appropriate bedding (i.e. straw); increasing the quantity and quality of available space; allowing flexibility in housing dates).

Whereas the first step involves making adjustments to the farm buildings and work routines, the next step requires the confidence on the part of the farmer to eliminate practices conventionally regarded as essential but which can be reduced, i.e. the reduction of routine veterinary inputs by:

a) ending practices such as antibiotic dry cow therapy for dairy cows;
b) reducing the use of insurance vaccines;
c) reducing routine worming;
d) introducing the use of alternative, system enhancing treatments like homoeopathy.

Point a) requires an increase in observation and management expertise, point b) is a matter of confidence and point c) requires an adjustment to grazing systems. Point d) can be slowly introduced on simple cases.

The third step involves the integration of the livestock enterprises into the whole farming system, a step which is essential for the ultimate success of an organic farm, i.e. the introduction of balance. This can be achieved by breeding for the system; adjusting the farm rotation and cropping; emphasising organically produced, home-grown food; modifying stocking rates and developing effective marketing. These points require long-term development and may mean large scale adjustments to the whole farming system. It is important that the animal is suited to and can take advantage of the system, for example a cow that can take in large amounts of forage rather than rely on concentrates for production. The cropping on the farm may need to be adjusted to provide adequate food of the right quality.

FINANCIAL PROBLEMS DURING THE CONVERSION PERIOD

The husbandry difficulties associated with establishing a new organic system and reduced yields without the benefit of premium prices can lead to financial problems during the conversion period which might not be foreseen if simple "before" and "after" calculations are done. First-hand information on what actually happens during the conversion period is currently very limited, although there are now several research programmes underway to investigate the issues more closely. In order to provide an indication of the sort of financial implications which may be involved on different farms, hypothetical budgets have been prepared for four different farm types, starting with a conventional system and working through a five-year conversion period to an organic endpoint.

Model 1—a 32 ha specialist dairy farm

The first example is a 32 ha specialist dairy farm, similar to many found in West Wales. The land is relatively low-lying, with moderate rainfall and the whole area can be cultivated. The quota assumed for the farm is 225,000 litres, which, because of further cuts in quota, is assumed to fall to 200,000 litres by the end of the conversion period. It is a typical family farm, with the farmer working full time and the spouse working part time on the holding. The buildings are adequate for the size of the farm, but machinery is limited to basic essentials such as fertiliser spreaders and forage conservation implements.

The current (conventional) cropping consists of 30 ha of permanent grass which is fertilised intensively to maintain a stocking rate of 2.25 LU/ha. The livestock enterprises include a 50 cow, autumn calving dairy herd yielding an average of 4,500 l/cow and year. Replacements are also reared on the farm and tack sheep are kept over the winter. The cows are fed on self-feed clamp silage over the winter and about 0.8 t/cow of concentrates. The replacements are fed on a straw diet, supplemented with barley and a protein supplement.

The conversion is based on an eight year rotation aimed at maintaining milk output up to quota if possible, consisting of:

- 2 years red clover/Italian ryegrass short-term ley;
- 1 year fodder crop (e.g. kale, fodder beet etc.);
- 5 years longer-term white clover/perennial ryegrass ley.

The higher nutritive value of the clover swards, combined with the use of high energy fodder crops and an increased quantity of (organically produced, home mixed) concentrate feed (1 t/cow) should enable individual cow yields to be increased while reducing cow numbers, so that total milk production remains within quota.

Labour and financial budgets have been estimated for the period from the start of conversion (Year 0 = conventional) through to a fully organic system (Year 7) where all the land and the livestock would qualify for the Soil Association's symbol. No account has been taken of possible processing activities. It is assumed that all the milk either continues to be sold to the Milk Marketing Board at conventional prices, or is sold for processing for a small premium. Figure 14.1 illustrates the financial results during the conversion period, additional details are given in Table 14 .1.

It should be noted that the absolute value of the net farm income (NFI) estimated is higher than would be expected for this size of farm. One reason is the assumption concerning milk price which was estimated at 18 pence per litre (ppl), rather than the recently prevailing

Figure 14.1 Changes in income during the conversion period on a hypothetical 32 ha specialist dairy farm.

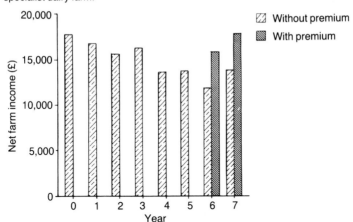

average of 16 ppl. For this reason, a sensitivity analysis has been included to show the impact on NFI of different prices. The sensitivity analysis can also be used to estimate the effect of a 1 or 2 ppl premium on the organically produced milk. Even taking the sensitivity analysis into account, the NFI is still high, but it is the relative changes which are important, rather than absolute figures .

From Table 14.1, it can be seen that, over the five year conversion period and assuming that the milk price remains at 18 ppl, the total reduction in NFI will be £14,115 or about £90 per ha per annum. Under the assumptions made here, the farm does not succeed in restoring income to pre-conversion levels. This is largely associated with the intensity of production assumed; the impact is not likely to be so significant on less intensive holdings.

There are, however, several qualifications to be made to these conclusions. Firstly, the reduction in NFI would have been approximately £3,250 anyway, as a result of the 10% reduction in quota which it has been assumed will take place over the next five years. Secondly, it may be possible for holdings to obtain a premium for organically produced milk. As an example, if the milk price is 16 ppl for conventionally produced milk, and a premium of 2 ppl can be obtained for organically produced milk, the NFI for the conventional farm would be £17,712 and for the fully converted organic holding (Year 7) the NFI would be £17,784. Taking the impact of quota reduction into account as well, it can be seen that there may be situations where the conversion, even at this intensity of production, results in an improvement of the income situation. This only emphasises the need to consider each

Table 14.1 Summary of results from a hypothetical 32 ha specialist dairy farm.

	Conventional	Year 1	Year 2	Year 3	Year 4	Year 5	Year 6	Fully organic (Year 7)
Area converted (%)	0	10	25	50	75	100	100	100
Gross Margins (£)								
Dairy	32,205	31,561	30,273	31,394	29,934	29,204	29,204	28,474
Dairy replacements	3,953	3,624	4,283	3,953	2,006	2,006	0	2,808
Sheep	1,300	1,300	1,300	650	650	650	650	650
Forage	−3,996	−4,087	−4,178	−3,831	−3,118	−2,378	−2,278	−2,398
Total	33,462	32,398	31,677	32,167	29,471	29,482	27,577	29,534
Fixed Costs (£)								
Machinery	3,500	3,500	3,500	3,500	3,500	3,500	3,500	3,500
Property	1,100	1,100	1,100	1,100	1,100	1,100	1,100	1,100
General	2,200	2,200	2,200	2,200	2,200	2,200	2,200	2,200
Total	6,800	6,800	6,800	6,800	6,800	6,800	6,800	6,800
Profit before Depreciation & interest (£)	26,662	25,598	24,877	25,367	22,671	22,682	20,777	22,734
Management & Investment income (£)	12,212	11,148	9,927	10,417	7,721	7,732	5,827	7,784
Net farm income (£)	22,212	21,148	19,927	20,417	17,721	17,732	15,827	17,784
Quota (litres)	225,000	220,000	215,000	210,000	205,000	200,000	200,000	200,000
Net farm income values (£) if:								
Milk price = 16p/litre	17,712	16,748	15,627	16,217	13,621	13,732	11,827	13,784
Milk price = 17p/litre	19,962	18,948	17,777	18,317	15,671	15,732	13,827	15,784
Milk price = 19p/litre	24,462	23,348	22,077	22,517	19,771	19,732	17,827	19,784
Milk price = 20p/litre	26,712	25,548	24,227	24,617	21,821	21,732	19,827	21,784
Labour Requirements								
Total calculated hours	2,798	2,758	2,753	2,575	2,300	2,215	2,032	2,210
Total standard man days	510	506	496	437	413	400	383	390

individual sitation separately: many dairy farmers interested in conversion are farming less intensively, and the impact of conversion may
well be less severe than this model seems to predict.

The labour budget shows a fall in the total labour requirements as a
result of conversion to organic production, but this is largely due to the
decline in dairy cow numbers; the marginal changes may well be less
than the average data used suggest.

Model 2 – a 67 ha lowland cattle and sheep farm

The second hypothetical example is of a 67 ha lowland cattle and sheep
farm with some arable land used for growing cereals and fodder crops for
stock feed. Of the total utilisable area of 65 ha, only 45 ha are assumed to
be suitable for cropping. The farm is managed by the farmer and spouse,
with contractors used for silage making, ploughing and harvesting.
Adequate housing is available for both the cattle and the sheep
enterprises.

The cropping in the conventional system consists of 20 ha permanent
grass, 38 ha temporary grass, and 4 ha each of spring barley and stubble
turnips. The livestock enterprises include 280 ewes with a lambing
percentage of 160, 40 suckler cows and offspring, which are finished on
the farm, together with about 30 purchased stores. The overall stocking
rate is about 2.0 LU per forage ha.

The aim of the conversion is to maintain the current balance of
livestock as far as possible, converting the forage to legume-based leys,
continuing to produce some cereals for stockfeed and eventually introducing a small area of field-scale vegetables. The proposed organic
rotation for the cultivable land is given below:

Year   Crop
  1–3 Medium-term grass/white clover ley
   4   Winter oats (cash crop)
  5–6 Short-term Italian ryegrass/red clover mix
   7   Root vegetables
   8   Spring cereal undersown with ley mix (for livestock feed)

The 21 ha of permanent grass on land which is not cultivable will
remain, but fertilisers are to be withdrawn and clover introduced. The
new rotation on the cultivable land can be established by ploughing up
the old ley and following it with a spring cereal crop undersown with the
organic ley mix, using fertilisers to ensure establishment, but no sprays.
No prohibited fertilisers should be used subsequently. The stocking rate
is estimated to be 1.7 LU/forage hectare (fha) on the organically
managed section after falling to 1.5 LU/fha during the conversion

period. The intention is that the sheep should be folded on the red clover aftermath, rather than stubble turnips, once the conversion is completed.

The results for this hypothetical farm example are illustrated in Figure 14.2, with further details in Table 14.2. The basic solution assumes premium prices for vegetables of up to 100% and for cereals of 35–50%. The situation in Year 5 and Year 9 (when the organic system has become fully established) has been calculated with and without premiums.

Figure 14.2   Changes in income during the conversion period on a hypothetical 67 ha lowland cattle and sheep farm.

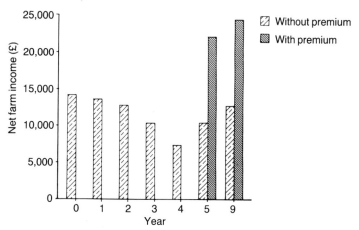

The production of arable and horticultural cash crops provides an opportunity to increase farm income substantially, if premium prices can be obtained without undue marketing costs. At the same time, the output of beef and sheep will be reduced. Even if no premium prices are obtained, the final result is not much worse than the original conventional situation. During the conversion period, however, the model predicts that farm income will fall substantially: in the first five years of conversion the total reduction in NFI is £4,734 if premium prices are obtained, £16,435 if they are not. The latter works out at an average income reduction of £51 per ha per annum during the first five years of the conversion period.

If no premium prices are obtained, NFI will continue to be reduced for a further period until the organic system is fully established. It has been assumed that no premium prices are obtained for livestock products. It may be possible, however, that the market will have developed to the extent that premium prices are obtainable. A 10% increase in the price of lamb will result in a £1,200 increase in NFI, a

Table 14.2 Summary of results from a hypothetical 67 ha lowland cattle and sheep farm.

	Conventional	Year 1	Year 2	Year 3	Year 4	Premium Year 5	No Premium Year 5	Premium Fully organic (Year 9)	No Premium Fully organic (Year 9)
Area converted (%)	0	10	25	50	75	100	100	100	100
Gross Margins (£)									
Beef	21,987	20,962	20,174	18,035	15,491	11,731	11,731	13,368	13,368
Sheep	14,902	14,902	13,837	12,773	12,773	11,094	11,094	11,094	11,094
Forage	−9,155	−8,887	−7,874	−7,125	−6,549	−4,114	−4,114	−4,385	−4,385
Cereals	972	1,164	1,164	1,164	1,164	3,984	3,104	5,092	4,212
Vegetables						14,923	4,102	14,923	4,102
Total	28,706	28,141	27,301	24,847	22,879	37,618	25,917	40,093	28,392
Fixed Costs (£)									
Machinery	3,500	3,500	3,500	3,500	3,500	4,000	4,000	4,000	4,000
Property	1,500	1,500	1,500	1,500	1,500	1,500	1,500	1,500	1,500
General	2,300	2,300	2,300	2,300	2,300	2,300	2,300	2,300	2,300
Total	7,300	7,300	7,300	7,300	7,300	7,800	7,800	7,800	7,800
Profit before depreciation & interest (£)	21,406	20,841	20,001	17,547	15,579	29,818	18,117	32,293	20,592
Management & investment income (£)	4,181	3,616	2,776	322	−1,646	11,093	−608	13,411	1,710
Net farm income (£)	14,181	13,616	12,776	10,332	7,354	22,093	10,392	24,411	12,710
Labour requirements									
Total calculated hours	2,725	2,703	2,566	2,503	2,417	3,054			
Total standard man days	412	409	392	393	381	610			

25% increase will add £3,000. For beef, a 10% price premium would add £1,400 to NFI, 25% would increase NFI by £3,500.

The introduction of root vegetables into the rotation has a marked effect on labour requirements. To a large extent, this may be met by casual labour, so that the family labour input is not necessarily significantly increased. It has been assumed that contractors would still be used for silage making and for harvesting, but the cultivations would be taken on by the farmer. Investment in cultivation and weed control implements, as well as vegetable storage and improvement of the manure handling facilities, would be required.

Model 3 – a 140 ha arable farm

The third example is for a moderate-sized arable farm of 140 ha, of which 136 ha is utilisable and can be cultivated. Labour, machinery and buildings are appropriate to the type and size of farm. The current conventional rotation consists of wheat, barley, barley, oilseed rape. No livestock are kept. This is where one of the main problems can arise with the conversion of arable farms, but in this case it has been assumed that the farmer is prepared to accept the introduction of a lowland sheep enterprise. The proposed rotation for the organic system consists of:

> Red clover/ryegrass (two years)
> Winter wheat (milling)
> Oats and peas (feed)
> Field beans
> Winter wheat (feed)
> Winter barley undersown

During the conversion, the rotation would be entered at two points: in year one with spring barley, fertilised and undersown with a short-term ley and in year four with field beans, followed by winter wheat (which may gain a small premium as in-conversion).

The results are summarised in Figure 14.3 and Table 14.3. If currently obtainable premiums on cash crops are included, the budgets predict a substantial increase in NFI as a result of conversion, notwithstanding the need to employ additional labour to cope with the new livestock enterprise. The sensitivity analysis, however, shows how important the premium prices are. If they are taken away, the result is a significant reduction in NFI compared with the initial, conventional situation. Premium prices for livestock have not been included, but the sensitivity analysis provides an indication of the potential impact which they, too, could have on the overall results. Investment in sheep housing and the purchase of breeding livestock, as well as the development of manure handling and storage facilities, will be required.

Figure 14.3 Changes in income during the conversion period on a hypothetical 140 ha arable farm.

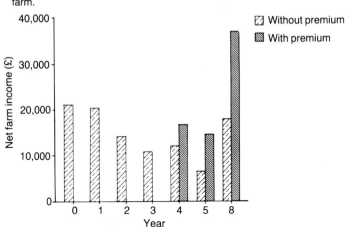

Model 4 – a 98 ha upland sheep farm

As a final example, a 98 ha upland sheep farm is included here to illustrate the possibilities in more marginal circumstances. Many of these farms, which traditionally used to carry both cattle and sheep, have concentrated their efforts solely on sheep in response to the poor economic returns to beef production. Intensification has been encouraged by Hill Livestock Compensatory Allowances on a headage basis, and better techniques for hill land improvement and grazing management.

This example assumes 20 ha of rough grazing and 75 ha of improved grazing, of which 25 ha are cultivable. The flock of 1,100 ewes is managed by the farmer and a full-time paid worker, with additional labour input from the spouse. Contractors are used for silage making, and additional casual labour is employed at lambing. Fertiliser use on the improved grassland is fairly intensive, enabling a stocking rate of 1.5 LU/ha to be maintained. The lambing percentage assumed is 120.

The conversion would involve the strip seeding of clover into the permanent grassland and the development of a clean grazing system, possibly combined with strategic dosing to minimise the use of anthelmintics. To assist grazing management and the control of parasites, a beef enterprise will need to be re-introduced into the system. Winter housing of the stock would be desirable, to improve manure management and to enable earlier lambing, so that a higher proportion of the lambs can be sold finished before dipping. The rough grazing would remain unimproved. The stocking rate assumptions are 1.0 LU/ha on land in conversion, 1.4 LU/ha after conversion.

Table 14.3 Summary of results from a hypothetical 140 ha arable farm.

	Conventional	Year 1	Year 2	Year 3	Year 4	Year 5	Fully organic (Year 8)
Area converted (%)	0	14.3	28.6	42.9	71.4	100	100
Gross Margins (£)							
Sheep			9,180	18,360	18,360	18,360	22,032
Forage			-3,622	-7,244	-7,244	-7,244	-7,613
Cereals	58,829	58,137	46,864	37,946	47,846	45,726	64,775
Total	58,829	58,137	52,422	49,062	58,962	56,842	79,194
Fixed Costs (£)							
Labour	0	0	0	0	4,000	4,000	4,000
Machinery	9,700	9,700	9,700	9,700	9,700	9,700	9,700
Property	3,400	3,400	3,900	3,900	3,900	3,900	3,900
General	5,000	5,000	5,000	5,000	5,000	5,000	5,000
Total	18,100	18,100	18,600	18,600	22,600	22,600	22,600
Profit before depreciation & interest (£)	40,729	40,037	33,822	30,462	36,362	34,242	56,594
Management & investment income (£)	11,109	10,417	4,201	842	6,742	4,621	26,973
Net farm income (£)	21,109	20,417	14,201	10,842	16,742	14,621	36,973
Impact on NFI (£) of:							
No premiums for organic cash crops (£)					-4,712	-8,099	-18,978
25% organic premium for lamb (£)				6,080	6,080	6,080	7,296
Labour requirements							
Total calculated hours	1,561	1,668	2,433	3,217	3,171	3,111	3,423
Total standard man days	204	219	330	440	440	440	486

The results are illustrated in Figure 14.4 with additional details in Table 14.4. No premium prices have been assumed, but the sensitivity analysis shows the potential impact of 10% and 25% price premiums for beef and sheep. Even without the benefit of premium prices, the reduced forage costs appear to compensate for the lower stocking levels so that Net Farm Income at the end of the conversion period is marginally higher than at the beginning, but the reduction during the conversion amounts to nearly £15,000.

Overview

The models indicate that it is possible for holdings to convert without a substantial reduction in income in the long term, and with the possibility of increasing income by producing for a market which is prepared to pay higher prices. The conclusions of the models are supported by current research on actual farms which is being conducted at the University College of Wales, Aberystwyth, but no published results are currently available.

The other major conclusion from the models is that there is unlikely to be a significant difference in labour use on farms once converted, unless labour-intensive enterprises such as vegetables are introduced into the farm system, or livestock are introduced into an arable system. Where an increase in labour use is likely is in the processing and marketing of the produce, but this has not been examined here.

Figure 14.4   Changes in income during the conversion period on a hypothetical 98 ha upland sheep farm.

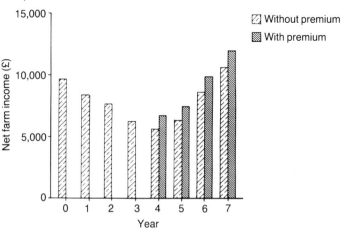

Table 14.4 Summary of results from a hypothetical 98 ha upland sheep farm.

	Conventional	Year 1	Year 2	Year 3	Year 4	Year 5	Year 6	Fully organic (Year 7)
Area converted (%)	0	12.5	25	50	75	100	100	100
Gross Margins (£)								
Beef					1,264	2,527	3,791	5,055
Sheep	35,586	34,004	32,581	30,022	26,943	25,095	26,173	26,943
Forage	−8,363	−8,003	−7,343	−6,248	−5,018	−3,698	−3,758	−3,788
Total	27,224	26,002	25,238	23,775	23,189	23,925	26,207	28,210
Fixed Costs (£)								
Labour	7,000	7,000	7,000	7,000	7,000	7,000	7,000	7,000
Machinery	2,000	2,000	2,000	2,000	2,000	2,000	2,000	2,000
Property	1,100	1,100	1,100	1,100	1,100	1,100	1,100	1,100
General	1,500	1,500	1,500	1,500	1,500	1,500	1,500	1,500
Total	11,600	11,600	11,600	11,600	11,600	11,600	11,600	11,600
Profit before Depreciation & interest (£)	15,624	14,402	13,638	12,175	11,589	12,325	14,607	16,610
Management & investment income (£)	2,623	1,402	638	−825	−1,411	−675	1,607	3,610
Net farm income (£)	9,623	8,402	7,638	6,175	5,589	6,325	8,607	10,610
Impact on NFI (£) of:								
10% premium on Symbol Standard lamb (40%)					1,010	940	980	1,010
25% premium on Symbol Standard lamb (40%)					2,520	2,350	2,450	2,520
10% premium on Symbol Standard beef stores					100	200	280	380
25% premium on Symbol Standard beef stores					250	500	700	950
Labour requirements								
Total calculated hours	4,990	4,780	4,600	4,699	4,328	4,127	4,367	4,555
Total standard man days	695	668	645	628	585	563	596	622

THE CONVERSION PROCESS IN PRACTICE

The fact that the conversion to organic farming can be achieved on a wide range of farm types and sizes can be illustrated by looking at four actual examples, each of which has approached the conversion in different ways.

A 45 ha dairy farm in Shropshire

In 1975, Edward Goff returned from Zambia, where he had been teaching agricultural science, and took on the tenancy of his grandfather's farm. He established a dairy herd and, with the help of an FHDS grant, was able to put up a milking parlour, slurry store and cubicle housing. Inspired by the work of the Mayalls at the nearby Lea Hall Farm and the writings of Newman Turner, Edward Goff tried early on to manage a 3 ha field organically, but without much success. Simply abandoning the field to its own devices without the help of fertilisers proved fruitless. It was only when a conference was held at Stoneleigh, following the formation of British Organic Farmers, that he really had the opportunity to meet other organic dairy farmers. A tour of German and Swiss organic dairy farms, supported by a grant from Walford College of Agriculture, finally gave him the confidence to take the organic option seriously and to proceed with the conversion of the whole farm. In 1984, Wales Gas placed a pipeline right across the farm and the resulting reseeding work acted as the starting point for the conversion.

Edward Goff was the first farmer in Britain to start a conversion with the help of a conversion plan drawn up by advisers from Elm Farm Research Centre. The use of conversion plans was in its infancy, not only in Britain but also elsewhere in Europe, and the plan may well have been of more benefit to the advisors than to Edward himself. However, because many of the issues were thought out in advance, the conversion of the farm was able to proceed apace, with the majority of the farm converted within four years and the Soil Association symbol being awarded in 1988, much faster than originally foreseen.

The farm consists of 45 ha of mainly level land at an altitude of about 80 metres. The soil is a light sandy loam and rainfall is about 750–800 mm per annum. The objective of the conversion plan was to maintain stocking as high as possible, although quotas caused a reduction anyway. At the same time, the aim was to meet as much of the fodder requirement as possible from the farm itself, rather than buying in large quantities of conventionally produced concentrates.

Reseeding the old and largely unproductive conventional leys with short and long-term grass/clover leys, including Clifton Park mixtures, formed the basis of the conversion. Grazing rye or oats and vetches have been used preceding long-term leys as an entry to the organic rotation, as have short-term Italian ryegrass/red clover leys which proved to be the most productive organic pioneer crops. The production of fodder beet as a high energy feed was tried and proved to be very successful (see Chapter 10).

The total number of cows in the herd has been forced down, as a result of quotas and the activities of Wales Gas, from over 70 in 1981/2 to 63 in 1988/9, but this has been accompanied by an increase in the number of replacements reared on the farm so that the overall stocking rate is now nearly 1.9 LU per forage hectare. Over the same period, 6 ha of the farm have been transferred from grassland to the production of fodder beet and cereals. Reliance on purchased organic feeds (oats, wheat, spring beans and seaweed meal) has increased and the 20% allowance of conventional feed is taken up by brewers' grain and linseed.

At present, the cows are treated largely conventionally, but alternative approaches such as homoeopathy are being investigated. Regular reseeding and the use of silage aftermaths have provided sufficient clean grazing to dispense with wormers. Dry cow therapy has also been abandoned with relative ease, although two severe cases of summer mastitis led to a change in the calving pattern to avoid having dry cows

Plate 14.1   Edward Goff standing in a recently converted grass/ clover ley

in August. The very few cases of mastitis which do not respond to Uddermint are reluctantly treated with antibiotics, incurring the long withdrawal period of one month for Soil Association symbol standard milk. The most intractable problems are breeding irregularities, which can no longer be sorted out with Estrumate.

Now that the Soil Association Symbol has been obtained, the farm is able to supply milk to the Welsh Organic Foods dairy in Lampeter. However, the establishment of a milk processing activity on the farm is seen as a priority.

In spite of the fact that the milk was sold to the Milk Marketing Board without the benefit of premium prices during the conversion period, and that the premium currently obtained is very low, the profitability and output of the farm has held up well. An increase in dairy enterprise output combined with significantly reduced concentrate and forage costs resulted in a dairy gross margin including forage costs of £662/cow in 1986/7 compared with £505/cow in 1982/3. In 1987/88, fertiliser costs fell to only £4/ha overall and total forage variable costs were less than £2,000. Although fixed costs increased due to greater reliance on paid labour and contractors for cultivations and harvesting, profits in 1987/88 were 30% higher than under conventional management in the 'good' year of 1982/83.

A 105 ha mixed dairy/arable farm in Dorset

Manor Farm has been in the Best family for over 50 years and Will and Pam Best have farmed it for the last 20. Originally 60 ha, it has been increased by purchases over the years to 105 ha, all in a ring fence. The farm runs to 200 metres up a valley in central Dorset, one of several which run roughly parallel, separated by chalk downs. It is very attractive, well hedged, with a 4 ha wood in the middle, a hardwood shelter belt along the northern boundary, two small copses and some patches of shrub and gorse. It is well sheltered from the west and north, but exposed to the east. The soil is mainly light and stony, free draining and easy working, but about 24 ha are clay and require more careful management. There is not one level field, but little is unploughable. Rainfall is around 1,000 mm per annum.

The policy on taking over the farm was to buy some Friesian heifers and start milking them while putting most of the land into cereals. The herd was built up to 80 cows plus followers. The profits, which were quite good in the '70s and much helped by 40% grants, were put into buildings and machinery until grain drying and storage for 200 tons of cereals, a basic kennels/open silo/herring-bone set-up for 80 cows and,

Plate 14.2   Light, flinty soils are characteristic of most of Manor Farm   (W. Best)

more recently, a silo and covered yard for 50 young cattle had been established.

During the dairy expansion, a school leaver was taken on as an ATB apprentice; 15 years later, Philip Hansford is still at the farm as a very good dairyman and a convinced user of homoeopathy. A silage making syndicate was formed with two neighbours which can handle a 2,000 ton first cut. The Bests run their own combine, doing some contracting with it.

At first, the conventional wisdom was accepted and the Bests un-questioningly used large quantities of nitrogen, whatever sprays were deemed necessary, and plenty of dairy concentrates and antibiotics. Disenchantment was gradual; the writings of Schumacher, the influence of the self-sufficiency movement, the results of homoeopathy (particularly on their children's health) all increased their environmental awareness, while around them the air was being polluted with more and nastier pesticides and the pressure to use them was continually increasing.

The cows were moved onto a low-concentrate, flat-rate system and kept a little more naturally. Mastitis and lameness decreased markedly, while milk yields dropped only a little. Some lucerne was put in as a nitrogen fixing forage, a small sheep flock and a couple of sows in huts were introduced. The Bests began to study homoeopathy and gave up the more potent sprays.

At the same time, however, the Bests felt they were only chipping away at the problem, making life considerably more difficult for themselves and being unsure where it was all leading. Then they went to a Soil Association regional meeting, were much inspired and soon discovered British Organic Farmers and Elm Farm Research Centre. By the time Will had attended the 1985 Cirencester conference and a BOF/Elm Farm seminar on conversion, the "conversion" of the farming couple had effectively been achieved. All that remained was to convert the farm!

With the assistance of the Organic Advisory Service based at Elm Farm, a detailed conversion plan for the farm was drawn up, with nutrient budgets, rotations for each field and so on. This formed the framework for the conversion, but the details changed as the system evolved. The plan allowed for a conversion period of five or more years, but it was decided to accelerate it considerably and within a year of receiving the plan chemicals were used for the last time. By 1988, full Soil Association symbol status had been achieved after a conversion period of only four years.

Any conversion plan has to take into account a wide range of factors and, if possible, should be a development of the existing system, rather than a major re-organisation. As much use as possible should be made of the existing stock, buildings, machinery and especially the aptitudes and experience of the people involved in the farm. This is reflected in the basics of the Best's plan, which are:

1. A slight reduction in cow numbers, necessary anyway to get down to quota, with a concomitant decrease in heifer numbers. This gives 70 milkers, with around 15 heifers reared each year for calving at 2.5–3 years old. The rest of the calves are sold after a week or two suckling, some at least to organic rearers.
2. A reduction in the cereal area to 30 ha, at least 15 ha of which is milling wheat for sale and the rest oats and barley for feeding on the farm. Half of the wheat is Maris Widgeon, which is cut with a binder and the straw sold for thatching reed.
3. An increase in sheep numbers from 30 to 100 ewes. It is felt that it must be right to have as many complementary enterprises as possible and hoped that, as well as helping the fertility and the weed control, the sheep will make a useful financial contribution.
4. Replacement of the conventional leys with lucerne and short-term red clover/Italian ryegrass leys for silage and white clover based swards for grazing. So far the leguminous forages have yielded consistently and well, and make excellent silage provided they get enough wilting. They need 12 good hours more wilting than grass. The white clover leys are going well with rotational grazing and Autumn manuring.

Plate 14.3   Harvesting Maris Widgeon for thatching straw   (W. Best)

5. A basic rotation over much of the farm of:

   - four years grass/white clover ley or lucerne
   - wheat
   - barley or oats
   - two years ryegrass/red clover
   - wheat
   - barley or oats

   For the fields close to the dairy a rotation more suited to fodder production is followed, involving 4–5 years grass/ white clover ley; grazing rye; kale or other forage; undersown barley.
6. Management of the livestock to Soil Association symbol standard. Living without the use of vaccines, antibiotics, hormones, organo-phosphorus dips and so on is a learning process which is continuing all the time, but with everyone on the farm keen on homoeopathy, good progress has been made.
7. Increase in the pig herd from one sow to five. A useful sideline, as demand for pork and sausages is strong.
8. Replacement of the cow kennels with a covered yard, and the installation of a low-rate irrigation system for the dirty water run-off from the silo, feed and collecting yards. This is a considerable investment costing about £20,000, but in the long run should not be

as costly as losing nutrients and possibly polluting the ground water with a messy and inefficient slurry system. Straw can be obtained cheaply from neighbouring farms and makes good manure.

9. Development of marketing. It is necessary to explore all avenues, but a farm shop looks the best bet, and that is what is being worked towards, combined with an increase in processing, e.g. wheat into flour, milk into yogurt or cheese.

Since the conversion started, another farming couple, Hugh and Patsy Chapman, have come to the farm to join forces with Will and Pam. Their sheep helped to provide the necessary flock expansion, but their major role has been in the establishment of a vegetable enterprise on some additional, low-lying land. The importance of this development lies in the fact that vegetable growing is often crucially dependent on the import of manures from off the holding. In the long term, the only way in which a sustainable system can be developed is through the integration of vegetable production with the other activities on a farm, or for close links to be made between organic farmers and growers. The arrangement which the Bests and the Chapmans have made will mean that they can now work together in the development of the whole farm as a well balanced organic system, eliminating the need for off-farm (often conventional) sources of manure.

All this makes big demands on labour and management, added to the already increased complexity of the actual farming system. More work means more people, more people means more communication, more paperwork, more management time. Will Best feels that the years of conversion are hard and show little financial return, but argues that neither the reasons for doing it nor the vision of how it should all work out should be lost sight of. Recognition also needs to be given to the contribution of all the people who work on the farm, including the dairyman, the general farm worker Andrew Daw (now lecturer with responsibility for organic farming at Worcestershire College of Agriculture), and the various students and part-timers, because without their support the conversion would have been even harder.

A 36 ha mixed farm in West Wales

This farm near Cardigan, which was bought in 1984, had previously been part of a much larger unit, now split between Jeremy and Valerie Harding and neighbouring farmers. It was effectively derelict, having been farmed in a very haphazard manner, much of it neglected completely. Faced with this, and the conviction that farming conventionally was wrong, they decided to convert the whole farm immediately. For many farmers, an immediate changeover would be disastrous. In this case there was no existing system to change from, the bank was willing to give them some flexibility in the early years to set up their new system, and so it seemed that this was the most appropriate way forward. Re-introducing chemicals would only have made the situation worse.

Jeremy Harding was brought up on a farm; his father had farmed organically since the late '40s, most recently on a smallholding only a few miles away, and is a long-standing member of the Soil Association, firmly committed to the organic philosophy. In spite of this, Jeremy chose not to go into farming immediately and instead ended up working for the food processing company Findus and its parent multinational, Nestlé. Over the years, he became very disillusioned with the whole food processing industry. The work he did on TVP (texturised vegetable protein) and other products strengthened this disenchantment. Then, in 1982 he was made redundant. After this, he wanted to determine his own future and not to face the threat of redundancy again. Here was the opportunity to get back into farming, with the advantage of his father being there and able to provide valuable experience and active help.

The farm consists of about 36 ha on mostly level land, the majority of which is cropable. As it is close to the coast, the rainfall is not too high and arable crops, including cereals, are possible. The current rotation consists of:

- 4–5 year Clifton Park type ley;
- cereals (a good market exists in the area for organically produced feed grain);
- brassicas (mucked in the previous autumn);
- grain legumes;
- potatoes (the straw from the peas and beans is ploughed in and the land is mucked the previous autumn);
- carrots, or other vegetables, then back to the grass ley.

Catch crops, including mustard and forage rye, are also included from time to time. The reasons for this particular rotation are practical; vegetables are important because they provide the sort of financial returns which will enable the farm to survive with heavy borrowings, but the stalks from the brassicas create problems with other vegetables and the potatoes may suffer from wireworm attack if they follow the ley too closely. Weed problems such as couch and pests like leatherjackets have arisen but the rotation appears to be getting on top of these. The silty soils common to this area of Wales can be a serious problem when wet, and are often low in important nutrients such as potash and phosphate. Various measures, including the use of rock phosphate, slag and seaweed meal and sprays have been undertaken to try to correct the deficiencies. One of the biggest problems during the conversion period was the appalling weather and bad growing seasons which seriously affected timing and led to poor early results.

On the livestock side, a flock of sheep was introduced, but these have been sold again to give a chance for other aspects of the system to become better established. The main livestock enterprise is beef. Calves are purchased from September onwards, not necessarily organic, but are reared on whole milk and fed on organically produced feed and forage, with the aim of finishing them by mid-summer of their second year. Problems exist with both the type of calves purchased, in terms of meeting consumer requirements, and the fact that the financial returns do not fully justify the use of land carrying high overheads for livestock. In winter, the cattle are housed in straw yards. The straw and most of the feed is produced on the farm, with the manure returned to the land.

Some of the land qualified immediately for the Soil Association symbol, and the whole farm achieved organic status within three years. The vegetables are sold through Organic Growers West Wales (see Chapter 12), a producer cooperative set up to coordinate the production and marketing of horticultural crops. Many of the vegetables are packed and distributed to major supermarkets such as Safeway and Sainsbury's by Organic Farm Foods (Wales), based in Lampeter. The cereals are sold to local organic dairy farmers who supply the Welsh Organic Foods cheese factory, also based in Lampeter, and who need organically produced feed grain to meet the Soil Association symbol requirements.

From the organic point of view, there are definite signs that things are moving in the right direction, but meeting the requirements of the vegetable packhouse in terms of size and quality has created problems.

Plate 14.4   Jeremy and Valerie Harding with some of their beef cattle   (A. Martin, courtesy of DBRW)

Understanding techniques and accurately predicting soil fertility status are essential, but other problems have also arisen because of the run-down state of the farm. Jeremy Harding admits that the conversion is a gamble, but he is quietly optimistic that the system will work financially in the long term.

A 650 ha mixed arable farm in Wiltshire

As testimony to the fact that it is not only small farms which can be run organically successfully, Barry Wookey's farm in Wiltshire has often featured prominently. The full story has been told in his book *Rushall -the story of an organic farm*. The Wookey family took over the farm in 1945 and it was run by Barry Wookey's father until his death in 1964. Barry Wookey started to convert the farm in 1970, influenced jointly by the noticeable decline in partridge numbers which was taking place on the farm (the partridge chicks rely for their survival on the insects associated with wild flowers and weeds, many of which were eliminated by the use of effective herbicides), by Rachel Carson's *Silent Spring* and by the thalidomide tragedy.

The conversion process started, as has so often been the case on organic farms, in a haphazard manner with the idea that all that was necessary was to stop using fertilisers and sprays. This was compounded by an attempt to grow a second winter wheat and the result was disastrous. But the lessons were learnt and experience showed that it takes some five to six years for the soil to regain its full potential. The conversion of the remainder of the farm was more carefully planned. Initially, it was envisaged that the conversion of the whole farm would take until 1990 to complete, but in fact the process was finished by 1985.

The original rotation was a simple one: a three-year ley followed by two years winter wheat, the second undersown to another ley. This, however, has the disadvantage that only about 40% of the land is down to arable crops and an alternative approach was needed which was better suited to arable farming conditions. A new rotation recently introduced consists of a three-year ley, grazed by cattle and sheep in the first two years and cut for hay followed by a bastard fallow in the third year. This is followed by two years winter wheat, undersown in the second year with ryegrass/red clover which is cut for hay in the following year. The aftermath is grazed by sheep and the ley is then ploughed in to be followed by winter wheat and winter oats. The one-year red clover ley does not really provide sufficient nitrogen for two winter corn crops, so a possible development involves undersowing the penultimate winter corn crop with pure red clover, allowing sheep to graze it in the autumn and ploughing it in in the spring before sowing a spring cereal, either wheat or oats. It is obvious that even now, a few years after the last field has been converted to an organic system, the development work is continuing. The organic system is never static or fixed, but continuously evolving in the light of experience and individual circumstances, as well as new information which becomes available from other sources.

During the conversion process, some important lessons were learnt from the mistakes which were made. They included the problems caused by leaving the land bare over winter and spring, and summer cultivations resulting in nutrient losses from which the following crops never really recovered. In the end, the use of undersowing for the establishment of leys proved to work best, and the use of undersowing and a bastard fallow after the three-year ley have become an established part of the system. The intensive cultivation associated with the bastard fallow after the ley is controversial, because much of the nitrogen fixed by the ley will be mineralised and lost through leaching, itself a matter of environmental concern, and will no longer be available to the subsequent wheat crop. Barry Wookey, however, considers this a price which has to be paid for keeping a good, clean farm and preparing a clean seedbed for the wheat.

This farmer's approach also illustrates the fact that there are different ways of starting the conversion of a field, depending on individual circumstances. In his case, he emphasises the need to start with a clean field, using chemicals after harvest and following this with conventionally produced spring barley. This crop should be undersown with a grass/clover ley mixture, which will start growing away after the barley is harvested, and from then on, no further chemicals, fertiliser or spray, may be used. If possible, he recommends a top-dressing of well rotted manure during September or early October in the first autumn, so as to get the process of building up soil fertility and biological activity going as soon as possible.

Barry Wookey accepts that the conversion period will be a difficult one for any farmer, and that there will be many times when confidence is sapped and the farmer questions the point of continuing. The farmer needs at this stage to have the courage of his or her convictions and to persevere. Once again, the crucial point about the "conversion" of farmer and family before starting the conversion of the farm itself is clearly illustrated. At the end of the conversion period, however, the rewards of a working organic system will, he argues, be worthwhile and result in a farm which it is a pleasure to work and a joy to behold.

REFERENCES AND FURTHER READING

Benecke, J.; Kiesewetter, B. & Urbauer, H. (1988) *Bauern stellen um—Praxisberichte aus dem ökologischen Landbau.* Alternative Konzepte 62. C F Müller; Karlsruhe

Culik, M. & Liebhardt, W. C. (1984) *Conversion project progress report 1981–1983.* Rodale Research Centre; Pennsylvania

Dabbert, S. & Madden, P. (1986) *The transition to organic agriculture—a multi-year simulation model of a Pennsylvania farm.* American Journal of Alternative Agriculture 1: 99–107

Rantzau, R. ; Freyer, B. & Vogtmann, H. (1986) *Betriebliche Erfordernisse und Konsequenzen bei der Durchführung des biologischen Landbaus.* Gesamthochschule Kassel, Witzenhausen

Wookey, B. (1987) *Rushall—the story of an organic farm.* Blackwell; Oxford

Recent publications

Lampkin, N. H. (1993a) *The economic implications of conversion from conventional to organic farming systems.* PhD Thesis. Department of Agricultural Sciences, University of Wales; Aberystwyth.

Lampkin, N. H. (1993b). *The impact of conversion to organic farming on eligibility for arable aid, beef special premium, suckler cow premium, sheep annual premium and hill livestock compensatory allowances.* Discussion Paper Series 93/5. Centre for Organic Husbandry and Agroecology, University of Wales; Aberystwyth.

Chapter 15

The Wider Issues

The main purpose of this book has been to concentrate on the principles and practices of organic farming, to provide an introduction to organic food production techniques for the benefit of people who are actively involved with the agricultural industry. In Britain, the agricultural industry is less significant than in other countries, but it is still of economic importance and in other ways it has a significant impact on society as a whole. Agriculture still makes use of most of the land and the effects of agricultural production techniques on the environment are therefore particularly visible. More importantly, every person in Britain is dependent on the food produced by the agricultural industry, and whether we acknowledge it or not, each one of us has an interest in the way our food is produced.

Much of the recent expansion of interest in organic farming has come from consumers who for a variety of reasons have decided that organic food is worth buying and who are prepared to pay higher prices for it. The motives which lead consumers to demand organic produce range from a personal desire for 'healthy' food to a more altruistic belief in the ability of organic production methods to contribute to wider social goals such as protecting the environment or benefiting developing countries. How far any of these particular beliefs are valid is a matter for debate and the claims which are made are often seen as being controversial.

Organic food production, however, cannot take place outside this context. It is important, therefore, to include within a book such as this, which is primarily production-oriented, a discussion of some of the most important issues affecting the way organic agriculture is perceived by the general public. Producers themselves need to understand these issues if they are to enjoy the benefits which the premium market for organic produce brings and not risk destroying the confidence of consumers by perpetuating claims and myths which cannot be substantiated.

The Quality of Organically Produced Foods

For a long time, the image of organic produce in the consumer's mind has been one of food which is in some way 'better' and more 'healthy'

557

than that produced under conventional agricultural systems. The primary market for organic produce has thus been in the health-food sector. Actually proving or substantiating the supposed health benefits has been more difficult, and the agro-chemical industry has been quick to defend itself by pointing to studies comparing 'organic' and 'conventionally' produced food which claim to show that there are no differences in quality. Unfortunately, the evidence and arguments which originally formed the basis for these claims have hardly been allowed to re-surface as statistics fly backwards and forwards overhead. The work of scientists such as Sir Robert McCarrison and well known experiments including the Peckham Experiment (Pearse & Crocker, 1943) and Pottinger's Cats (Pottinger, 1946), which showed adverse effects on cats fed cooked food and pasteurised milk, provided much of the evidence which was later used by people like Lady Eve Balfour to promote organic farming within a much wider context which included human health and nutrition.

Since the time that Eve Balfour was writing and McCarrison was carrying on his work in the field of human nutrition in the first half of this century, the question of diet and health has, if anything, become more significant rather than less. Although much more is now known about the role of individual nutrients in human nutrition, there is still no real consensus about the ideal diet: the mass of contradictory literature on the subject bears testimony to this. In addition, the consumer has become increasingly concerned about the possible dangers from small quantities of food additives and pesticide residues in food. The mounting medical evidence linking additives and pesticide residues to food allergies and increased cancer risks has been the primary stimulant; the market has not been slow to respond to the clamour for 'additive-free' and 'chemical-free' food. Unfortunately, perhaps, the concerns about the wholesomeness of individual food products has tended to detract from serious consideration of the overall composition of the diet.

The issues are clearly complex and this is not the place to try to resolve all of them. At the same time, the question of whether food produced organically is 'healthier' or 'better' than conventionally produced food is an important one to many people. The problem is, what is meant by 'better'—how is quality defined?

The definition of food quality

The quality of foodstuffs cannot be defined in terms of any one single, measurable characteristic. In fact, it is usually assessed by three principal criteria:

1. Appearance (size, shape, colour, freedom from blemishes and a taste specifically associated with the individual product).

2. Technological suitability (specific attributes which determine a food-stuff's suitability for processing or storage — e.g. sugar content of sugar beet, nitrogen content of malting barley, suitability of milling wheat for bread making).
3. Nutritional value (content of beneficial nutrients, such as protein, vitamins etc. and content of harmful substances such as nitrates, natural toxins, pesticide residues and heavy metals).

In many cases, these quality characteristics can be measured quantitatively and thus provide a basis for comparison. There is, however, a further aspect to quality which is much more difficult to pin down. This is the subjective, ideological values which each individual consumer has and which form a crucial part of the overall image of the product held by the consumer. Whether this is a resistance to buying produce from South Africa, or deliberately buying Nicaraguan coffee (both usually highly motivated actions), subjective values will play a role in the consumer's perception of quality. Increasingly, the consumer is associating with organic food the concept that the production methods used are more sympathetic to the environment and this plays an important part, therefore, in the comparison of conventional and organically produced food.

Appearance

External appearance
External appearance is the most obvious aspect of quality as far as the majority of consumers is concerned. In this respect, organically produced food sometimes fails to match the perfection achieved by the use of agrochemical controls, especially with fruit and vegetables. In many cases, however, this need not be so and there is often no excuse for attempting to sell produce of substandard appearance simply on the grounds that it has been produced organically. Where problems which are less easy to deal with do occur, such as scab on apples, then effort has to be put into educating consumers to accept these types of blemish, so long as the product itself is sound.

Taste
Many consumers are prepared to make allowances for external appearance if they feel that other aspects are more important. One of these is taste. It is a widely held belief that organically produced food tastes better, but conclusive scientific evidence to prove that this is the case is hard to come by.

Part of the problem lies in the fact that what tastes good is often determined by the tastes which people are already used to. A person raised on UHT (ultra heat treated) milk may have a completely different perception of the taste difference between UHT and non-UHT milk than someone raised on fresh, unpasteurised milk. A trained tasting-panel might well come up with results which contradict the preferences of the 'average' consumer.

The second major problem is that there is a wide range of factors which affect the taste of a product, some of which can be considerably more important than the production system. The prime example is variety—the difference in taste between Cox's Orange Pippin and Golden Delicious apples, or between Channel Island and non-Channel Island milk, is very distinctive. Regional variations complement varietal influences, as any wine buff knows, and these regional differences are determined by soil types as much as by climatic differences.

The effect of the production system on taste cannot, however, be ignored. The way food is produced can influence the nutrient composition of foodstuffs, as well as other aspects such as dry matter content (see, for example, Maga, 1983). These must have an impact on taste and texture, but whether this is for better or for worse often depends on individual inclinations.

Detailed research on quality differences in West Germany (Schuphan, 1976; Wistinghausen, 1979; Abele, 1987) has pointed to growth rates and physiological maturity of crops when harvested as having a significant effect on certain nutrients, among them sugar levels. If sweetness is an important determinant of 'good' taste, then the evidence from the West German work would support the argument that food produced organically can taste better. One recent study, by Lindner (1985) using a panel of 30–50 consumers who were deliberately not informed about the basis of the comparison, found that vegetables produced organically under carefully controlled experimental conditions did taste better. However, in the same study, a panel of trained tasters found no significant differences (Table 15.1). Duden (1987) and Bulling et al. (1987) have also found taste differences in favour of organically or biodynamically produced tomatoes and potatoes respectively.

In Britain, research at the Edinburgh School of Agriculture (Lowman, 1989) found that a panel of nearly 200 consumers showed a significant preference for organically produced steaks in terms of overall appearance, and in particular the organic steaks were classed as being leaner. Overall eating quality was also significantly higher, although none of the individual components of eating quality was significantly higher in statistical terms. However, these results are contradicted by a study from the North of Scotland College of Agriculture (Younie, 1990) which found no differences in flavour or improved eating quality in organic beef from 18-month-old Hereford × Friesian steers. Trials by the Potato Marketing Board (PMB, 1989) found that organically produced potatoes had fewer 'off-flavours', but that the potatoes tended to disintegrate more on cooking.

In general, though, the evidence with regard to taste is indicative rather than conclusive, with many of the studies suffering from significant weaknesses; extravagant claims for the supposed advantage of organic produce in this respect need to be treated with caution. At the same time, the argument that there is no difference in taste between organic and conventionally produced food is both empirically and theoretically unsound. Further work is needed under much more carefully controlled conditions (Meier-Ploeger, in Meier-Ploeger & Vogtmann, 1988) for the taste issue to be properly and objectively evaluated.

Technological suitability

Storage quality and post-harvest behaviour
Various West German studies have considered some of the factors relating to storage ability and post-harvest behaviour of produce from different production systems. As with the taste issue, growth rates and physiological maturity at harvest seem to play an important role. Crops grown organically, with slower growth rates and greater physiological maturity at harvest, have been shown in carefully controlled trials to have a longer storage life (e.g. Samaras, 1977; El-Saidy, 1982 and Abele, 1987). Respiration rates and enzyme activity have also been shown to be lower in organically produced vegetables, leading to reduced storage losses (Table 15.2).

El-Saidy (1982) concluded that the better storage life of spinach grown with organic manures may be associated with lower free amino acid content, a nutrient attractive to bacteria and related closely to levels of nitrogen fertiliser use. The organically manured spinach had the

Table 15.1 Comparative taste tests on vegetables produced organically and conventionally (10 yr av).

	% better score than conventional	
	Fresh	*Stored*
Celery	11	29
Carrots	−4	−8
Beetroot	19	15
Cabbage	17	—

— not statistically significant.
Source: Lindner (1990, pers comm.)

Table 15.2 Storage losses (%) for vegetables grown with different fertilisers.

	Fertiliser type	
	Mineral	*Organic*
Carrots	45.5	34.5
Kohlrabi	50.5	34.8
Beetroot	59.8	30.4
Various vegetables (average)	46.2	30.0

Source: Samaras (1977).

lowest level of biochemical changes and numbers of total bacteria. Abele (1987), on the other hand, found no significant storage differences related to manuring under optimal storage conditions, but under stress conditions (e.g. temperature, humidity, injury, grating or chopping), differences emerged in favour of lower fertiliser levels, organic manuring and biodynamic preparations.

Ahrens (in Meier-Ploeger & Vogtmann, 1988), summarising all the research carried out by his department at the University of Giessen, which includes the work by El-Saidy and Samaras, reports that in more than 75% of the cases, the post-harvest behaviour of crops which had received only organic manure was superior, and that increasing levels of nitrogen use were particularly detrimental. The greatest differences occurred with respect to perishability, shrivelling, colonisation by epiphytic microorganisms, peroxydase activity, nitrite formation and vitamin C breakdown.

Other aspects

Much less work has been done on other aspects of the suitability of organically versus conventionally grown foodstuffs for processing. Protein levels in organically grown cereals may be lower and this will mitigate against their use for bread-making flour and other purposes. This whole area is currently under scrutiny because many of the tests for bread-making quality were developed for refined, white flour, whereas most consumers of organically produced food are interested in wholemeal flour and bread. Higher dry matter levels in organically produced vegetables can be advantageous as far as some specific processing operations such as pickling are concerned.

Nutritional quality

The domestic consumer is usually more interested in the nutritional value of food than in its suitability or otherwise for processing and storage. More specifically, concern is oriented towards 'negative' aspects such as pesticide residues, food additives, fats and, to a lesser extent, nitrates, than towards positive factors such as protein, vitamins and trace elements.

Pesticide residues

A reflection of this concern with the presence of harmful constituents in food has been the tendency to refer to organically produced food as chemical or pesticide free, and to make this a major selling point. 'Chemical free' is a catchy, but meaningless, term. Looking at the issue simplistically, all foods are made up of chemicals of various types. They would not exist if they were really chemical free. Obviously, this is not what is intended by the term, but even freedom from harmful chemicals, pesticide residues and heavy metals cannot be guaranteed. The state of the environment is such that pesticide residues in the soil, air pollution from spray drift and industrial sources and heavy metals from sewage sludge etc. will inevitably contaminate crops to some extent, even if the production methods used do not involve pesticides or contaminated manures. Further contamination can occur during transport, processing and marketing of the produce so that, with the best will in the world, food produced organically cannot be guaranteed to be completely chemical or pesticide free. Claims to this effect are misleading and detract from the real contribution which organic food production can make.

At the same time, the risk of produce grown organically being contaminated with pesticide residues is much smaller than with conventionally produced crops—because pesticides are not used. Some studies have been published which claim that there is in fact no difference between organic and conventional produce as far as pesticide residues are concerned. On closer examination, however, these studies can be criticised on the basis of misleading statistical manipulation, and more importantly that the food tested could not be guaranteed to come from genuine organic production. Where food genuinely produced using organic methods has been tested, the results have been much more clear-cut (Tables 15.3–15.5). In the cases where residues were found, they were often of the long-lived DDT-type pesticides which still contaminate the environment many years after having (officially at least) been withdrawn from general use.

The possible effects of even very low levels of pesticide residue in the human diet and in that of animals are only just being discovered. Even if not directly toxic, pesticide residues may affect the fertility of animals and the health of their offspring, as well as disrupting the chemical communications systems on which many lower lifeforms depend. The concentration of chlorinated hydrocarbons such as DDT in mothers' milk has given sufficient cause for concern that, in some European countries, the official recommendation is that the risks of breast-feeding a human baby after six months outweigh the benefits.

A diet based on organically produced food, with a lower risk of contamination, can help considerably to alleviate the pesticide residue problem. A recent French report (Aubert, 1987) showed that the contamination of mothers' milk in France by DDT, HCH and PCBs (an industrial pollutant) was still above the World Health Organisation's maximum recommended limits in 1986, but at lower levels than in 1972 (Figure 15.1). These pollutants are deposited and concentrated in

**Table 15.3 Pesticide residues in fresh fruit and
vegetables sampled in Basel,
Switzerland, 1980–83.**

	Conventional	Organic
Number of samples	856	173
No residues detected (%)	60.9	97.1
Tolerable residue levels (%)	32.9	2.9
Excessive residue levels (%)	6.2	0

Source: Schüpbach (1986)

**Table 15.4 Pesticide residues in fruit and vegetables measured by the
Chemischen Landesuntersuchungsanstalt Sigmaringen (% in
brackets).**

	Organic			Conventional			
Year	Sample size	Residue free	Below 0.01 mg/kg*	Above	Sample size	Residue free	Below max. perm. lev.	Above
1983	43	42 (98)	1 (2)	0 (0)	484	222 (46)	249 (51)	13 (3)
1984	108	100 (93)	7 (6)	1 (1)	383	180 (47)	191 (50)	12 (3)
1985	43	37 (86)	6 (14)	0 (0)	456	244 (53)	200 (44)	12 (3)

*less than 0.01 mg/kg represents presence in trace amounts only.

Source: Reinhard & Wolff (1986).

**Table 15.5 Pesticide residues in vegetables by type measured by the Chemischen
Landesuntersuchungsanstalt Sigmaringen, 1985 (% in brackets).**

	Organic			Conventional			
Vegetable	Sample size	Residue free	Below 0.01 mg/kg*	Above	Sample size	Residue free	Below max. perm. lev.	Above
Red cabbage	1	1	–	–	2	1	1	–
White cabbage	1	1	–	–	5	4	1	–
Mangold	1	1	–	–	1	1	–	–
Lettuce	5	5	–	–	48	9	36	3
Courgette	1	1	–	–	1	1	–	–
Green pepper	1	1	–	–	32	9	18	5
Tomatoes	4	3	1	–	18	13	5	–
Kohlrabi	3	3	–	–	16	13	3	–
Carrots	6	2	4	–	7	6	1	–
Beetroot	2	2	–	–	2	2	–	–
Total	25	20	5	–	132	59	65	8
Per cent	100	80	20	–	100	45	49	6

*see Table 15.4.

Source: Reinhard and Wolff (1986)

human fat, to be released later in breast milk, where the concentration of DDT can be as much as 127 times as high as in cows' milk. One option to reduce DDT levels is to get rid of surplus fat before pregnancy. The other is to eat organically produced food. Aubert showed that women on a diet containing 80% organically produced food had only 30% of the concentration of DDT in their milk (Figure 15.2).

Nitrates in vegetables
Pesticide residues are not the only problem arising from modern agricultural techniques. Increasingly, nitrate levels in vegetables are causing concern, although most attention so far has been focused on nitrates in water supplies (Dudley, 1986; Vogtmann & Biedermann, 1985). About 70% of average daily nitrate intake comes from vegetables, compared with only 20% from drinking water. Nitrates are taken up very readily by crops, and if they are not utilised immediately in the formation of protein, they are stored in the cells in their original form. There is then the risk that when nitrates are ingested or cooked, they convert to nitrites which can potentially combine with amines to form carcinogenic nitrosamines. Nitrites can also form carcinogenic compounds with certain pesticide residues, such as the dithiocarbamates (used as fungicides). However comprehensive the safety tests for pesticides are, it is almost impossible to test all the different 'cocktails' (chemical combinations of different pesticides and their breakdown products) which can arise. The nitrite/dithiocarbamate cocktail is only one example, but it is a good illustration of how little account has been taken of the interactions which can occur and the problems which arise from introducing alien substances into the environment.

The uptake and utilisation of nitrates by plants is influenced by a variety of factors, including soil type, climate, light intensity, variety and fertiliser use. Leafy vegetables, such as lettuce and spinach, are particularly susceptible. Work in Switzerland (Temperli *et al.*, 1982; Vogtmann *et al.*, 1984) has compared nitrate levels in vegetables from organic and conventional production systems and shows clear differences between the two (Figure 15.3). Not only is nitrate accumulation lower in organically produced vegetables (except in winter where light intensity is low), but the ratio of protein-nitrogen to nitrate-nitrogen is much higher. This is significant, because protein is often measured simply on the basis of the nitrogen content of a foodstuff and little or no account is taken of the quality of protein present. In Britain, Elm Farm Research Centre has also conducted research into the nitrate issue (EFRC, 1988; Stopes *et al.*, 1988), finding high nitrate levels in produce labelled as from both organic and conventional production systems, which emphasises the risk associated with vegetable production under sub-optimal light conditions.

Figure 15.1 Contamination of human milk with pesticides and other pollutants in France, 1972 and 1986.

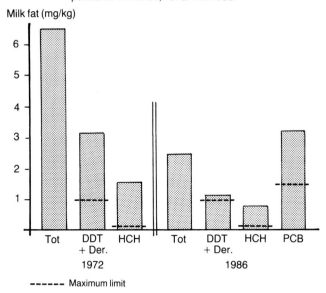

Milk fat (mg/kg)

------ Maximum limit

Figure 15.2 Contamination of human milk with chlorinated hydrocarbon pesticides in relation to the proportion of organically produced food in the diet.

Total content
of chlorinated
hydrocarbons
(mg/kg milk fat)

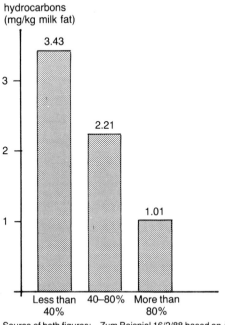

Source of both figures: Zum Beispiel 16/2/88 based on Aubert (1987).

Figure 15.3   Nitrate content and ratio of protein- to nitrate-nitrogen in lettuces grown organically and conventionally. Nitrate content is significantly lower in the organically produced lettuces except in situations of low light intensity.

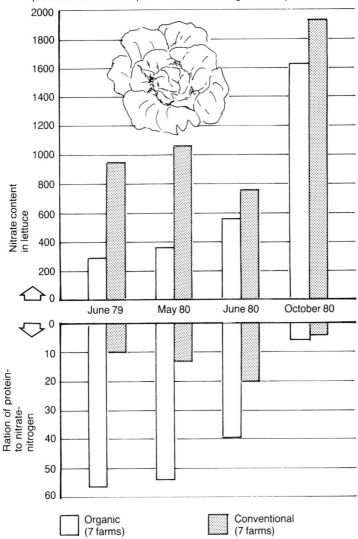

Source: Temperli *et al.* (1982), as illustrated in Vogtmann, H. (1985) *Ökologischer Landbau*. Pro Natur Verlag; Stuttgart

Other toxins and contaminants

Eating organically produced food is an important way in which the risks associated with pesticide residues and nitrates can be reduced. It is also true, however, that many 'natural' substances are also poisonous: some bacteria and fungi produce toxins which can be harmful to human health; ergot and aflatoxin (found in peanuts) are prime examples. There is no evidence that the occurrence of natural toxins is more prevalent in organic systems than conventional systems, but careful management is obviously necessary to avoid these sorts of problems arising. Where chemicals are used to control the bacteria or fungi which produce the toxins, they kill only the organism, while the toxin still remains. This is a major cause for concern with food irradiation as well, as the product may appear to be wholesome when it has in fact been contaminated (Webb & Lang, 1989).

Mineral and vitamin composition
The presence or absence of harmful substances in food is still only one side of the nutritional value question. It is well known that the use of significant quantities of mineral fertilisers affects the nutritional composition of crops. In an agricultural context, hypomagnesaemia (milk fever) in dairy cows is quite clearly linked to the luxury uptake of one element (potassium) at the expense of another (magnesium). It is the interaction of the different elements and fertilisers with each other which is important here.

Various studies (e.g. Schuphan, 1975 & 1976; El-Saidy, 1982; Fischer & Richter, 1986; Lairon *et al.*, 1986; Abele, 1987; Bulling *et al.*, 1987; Kerpen, 1988) have shown increased nitrogen fertiliser use resulting in not only higher nitrate levels, but also higher levels of free amino acids, oxalates and other undesirable compounds, as well as lower levels of vitamin C in particular. Calcium, phosphorus, magnesium and sodium contents are also affected by levels of fertiliser use, as are trace elements (Table 15.6 and Figure 15.4). The use of organic manures and appropriate soil management practices in organic agriculture means that a much wider and more balanced range of nutrients is available to crops than is the case when readily soluble NPK mineral fertiliser is applied and taken up directly by the plant. Comparisons of organically and conventionally produced food have in many cases confirmed these differences, but where soils are already well supplied with nutrients, the differences tend to be less marked.

The effect on health

What effect do all these differences have on human health? This question is much more difficult to answer. While the role of individual nutrients is fairly well understood, the interactions between nutrients and other substances contained in food is much more complex. The ideal situation would be one where groups of people could be fed on identical diets using either organically or conventionally produced food. The problems of setting up such an experiment to yield meaningful results are

Table 15.6 Relative yield and composition of vegetables grown with composted manures compared with mineral fertilisers; results of a 12 year experiment.

Yield: 24% lower

Desirable components:
    23% higher dry matter
    18% more protein
    28% more vitamin C
    19% more total sugar
    13% more methionine (an important amino acid)
    77% more iron
    18% more potassium
    10% more calcium
    13% more phosphorus

Undesirable components:
    12% less sodium
    93% less nitrate
    42% less free amino acids

Source: Schuphan (1975).

Figure 15.4   The interaction between nutritional components of spinach as applied nitrogen levels are increased (average over 3 years, 1964–1966, application rates 0, 80, 160, 240 and 320 kg N/ha).

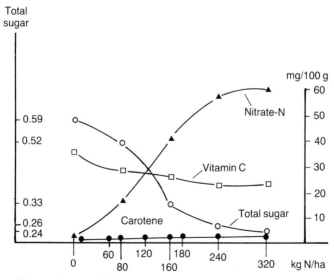

Source: Schuphan (1976) adapted.

phenomenal. Differences between individuals, their lifestyles and exposure to other environmental factors all serve to confuse the picture.

A compromise approach is to test the effects on animals instead, and several studies of this type exist. In a series of experiments (Aehnelt & Hahn, 1973), semen quality of bulls and rabbits fed on herbage fertilised with either mineral fertilisers or compost was found to be adversely affected by the use of mineral fertilisers. Various factors could be involved here, including concentrations of specific organic compounds, the mineral composition of the individual plants, and the effects on species composition of the herbage resulting from different fertiliser regimes. Comparing herbage of different species composition does not necessarily invalidate the comparison, as the production system may well be the determining factor anyway.

Other studies, including those by McCarrison in 1926 (Figure 15.5) and Gottschewski (1975), have indicated differences in growth rates and mortality, but some (e.g. McScheehy, 1977) have found no differences at all. More recent work by Edelmüller (1984) found that all fertility parameters including total birth and finishing weights were better where rabbits were fed on organically produced food, although average weights were no different (the organically fed rabbits had more offspring). In a parallel study by Plochberger (1984), chickens fed conventionally produced feed had higher growth rates in the first generation after 4 and 8 weeks, but after 32 weeks in the second generation the organically fed chickens had reached a greater weight. Egg weights were greater in both years for the organically fed chickens. The weight of egg yolk was also higher, but the egg white weight was greater for the conventional group. In a food choice test, the conventionally fed hens were more selective and clearly preferred the organically produced feed.

So many factors play a role in determining the overall physiological value of food that it is often difficult to isolate those which result directly from the production system. One of the criticisms of the earlier studies with animals is that the composition of the diet differed between the two groups. Staiger (1986) attempted to overcome this by ensuring that the chemical analysis of the diets was similar for two groups of rabbits fed organically and conventionally produced feed. Even then differences occurred, with the organically fed rabbits achieving greater pregnancy rates, more embryos per doe and more births per doe. There were, however, no differences in proportion of dead to live births, average birth and finishing weights or mortality. The conventionally fed rabbits were more susceptible to disease. One interesting observation, noticed also by Edelmüller, was that after one week, the faeces from the organically fed rabbits were covered in fungal growth, whereas faeces from the other groups showed no growth.

Figure 15.5   Influence on the body weight of doves of a supplement of millet grown using different types of fertiliser added to a base diet of rice.

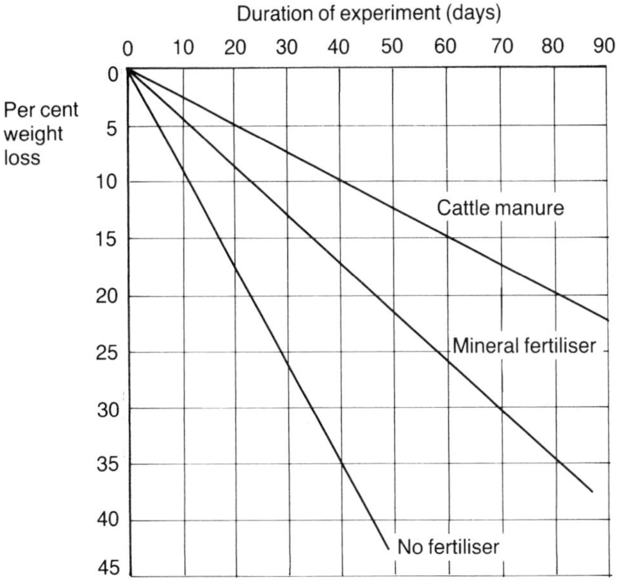

Source: McCarrison (1926), illustrated in Vogtmann (1985) (see Fig. 15.3).

Novel methods for assessing quality holistically

There have been attempts to assess quality differences between organic and conventional production systems in other, more novel ways, including:

- image-forming techniques such as copper-chloride crystallisation and chromatography (Knorr, 1982 & 1984; Knorr & Vogtmann, 1983; Balzer-Graf & Balzer in Meier-Ploeger & Vogtmann, 1988) and water-droplet patterns (Schwenk in Meier-Ploeger & Vogtmann, 1988);
- physical/chemical techniques such as counting photon emissions from samples of food (Popp in Meier-Ploeger & Vogtmann, 1988; Teubner, 1983) and measuring electrical conductivity and other electro-chemical properties of food (Hoffmann, 1988 and Hoffmann in Meier-Ploeger & Vogtmann, 1988);
- microbiological and biochemical techniques (Huber et al. in Meier-Ploeger & Vogtmann, 1988; Kerpen, 1988).

Of particular interest is the technique of counting photon emissions. Every living organism emits biophotons or low-level luminescence (light with a wavelength between 200 and 800 nanometres). This light energy is thought to be stored in the DNA during photosynthesis and is transmitted continuously by the cell. It is thought that the higher the level of light energy a cell emits, the greater its vitality and the potential for the transfer of that energy to the individual which consumes it. Significant differences have been found in favour of organically produced food (Figures 15.6 and 15.7), but differences also occur with respect to location, freshness and stage of maturity (ripeness).

Although these novel techniques are able to show up sometimes quite dramatic differences, there is often no real way of determining whether they represent improvements or otherwise in nutritional quality, particularly with the image-forming techniques. Further developments with these techniques are, however, taking place (EFRC, 1989) with the possibility that they could lead to a new insight into the meaning of quality.

Overview

Much of the evidence currently available clearly indicates that quality differences do exist, for nutritional value as for taste and other quality criteria, between organic and conventionally produced food. The evidence is not yet sufficient, however, to provide conclusive proof that organically grown food is always better and healthier for the consumer: further detailed research is needed before such claims can be fully justified.

In a situation where there is insufficient conclusive evidence to guarantee that organically grown food is of different or higher quality as far as external, technological or nutritional characteristics are concerned, any definition of organically grown food in terms of the end product is meaningless. Food free of pesticide residues has not necessarily been produced organically and food contaminated with residues could have been produced organically, but contaminated from external sources. Organically produced food can only be defined in terms of the production system and this is the basis of the Soil Association's standards and other similar schemes operated in many other countries around the world.

It is not only organically produced food which demonstrates the consumer's interest in and concern about the way food is produced. Animal welfare issues have often been prominent in recent years and have found expression in the demand for free-range eggs and the swing towards vegetarianism. A large section of the public is not prepared to accept the inhumane methods of factory-farming, whether for poultry, pigs or veal, which deny many of the natural behavioural needs of

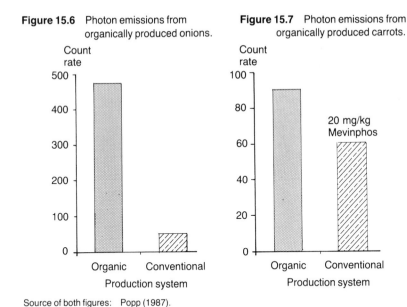

Figure 15.6 Photon emissions from organically produced onions.

Figure 15.7 Photon emissions from organically produced carrots.

Source of both figures: Popp (1987).

livestock and often create appalling conditions for the workers who have to look after them. The organic approach emphasises the need to meet the behavioural and ethological requirements of livestock, not only for philosophical reasons, but also for the management and husbandry goals of reducing stress and allowing the animals' own, natural processes to cope with disease without resorting to drug therapy unless absolutely necessary.

Many consumers are also seriously concerned about the impact of agricultural practices on the environment. It is the production system which has the greatest impact on the environment and, increasingly, the consumer is recognising this as an additional quality factor, linked to the subjective or ideological beliefs which are central to any comprehensive quality concept. Consumers recognise that an agricultural system which consciously avoids the use of environmental poisons, encourages soil life, enriches the landscape through plant diversity, rotations, etc. and reduces the risk of erosion through continuous soil cover is worth encouraging. They are prepared to say so as well, either by applying political pressure through groups such as the Soil Association, the Council for the Protection of Rural England and Friends of the Earth, or by consciously buying (and paying more for) organically produced food.

THE ENVIRONMENTAL IMPACT OF ORGANIC FARMING

A major case for organic farming rests on its ability to reduce or eliminate many of the worst environmental consequences of modern production systems, including loss of wildlife and wildlife habitats, pollution of the environment and the excessive usage of non-renewable resources. There have been few published in-depth assessments of the impact of organic farming systems on wildlife and its habitats in Britain and virtually no field studies have been conducted on functioning organic farms in the UK, although some studies have been conducted in West Germany, Denmark and the United States. Consequently, any assessment must be based primarily on an examination of the practices and inputs which characterise organic farming systems.

Farming systems and ecological diversity

Organic farming is clearly not simply conventional farming without the use of inorganic fertilisers and pesticides. Whereas conventional farming methods attempt to substitute for natural production processes, organic farming attempts to enhance them, using a system which, to a large extent, mimics natural ecosystems in terms of species and trophic level diversity.

The most fundamental element of this mimicry, common to most organic farming systems, is the mix of arable and livestock enterprises which enables these farms to maintain many of the cyclical processes characteristic of natural ecosystems. In this respect at least, organic farming is similar to the traditional, rotational farming practised before the advent of inorganic fertilisers and pesticides. It was precisely this form of agriculture which established the structural and biological diversity of the British countryside, which so many conservation bodies now wish to preserve. Modifying conventional systems will not achieve this, as crop diversity is antithetical to their design, specialisation being of key importance.

The large monocultures of conventional farming systems not only limit the variety of in-crop habitats available to wild flora and fauna, they also ensure that there is an uneven temporal distribution of the resources on which many wild animals are dependent. The trend to autumn sowing of cereal monocultures over large parts of the country is the most extreme example of this. For many animals, these crops provide significant resources for only a short period. For the remainder of the year, the animals will be dependent on the ever-diminishing area of non-crop habitat.

The emphasis on specialisation in conventional systems also limits the range of non-crop habitats available to wild flora and fauna. This effect is particularly evident in areas dominated by intensive cereal growing, such as East Anglia and the East Midlands of England. The decline of some farmland bird and mammal species (e.g. the grey partridge and the hare) has been directly linked to this trend toward agricultural specialisation. Indeed, the reduction in habitat and landscape diversity has been cited by some ecologists as the single most important agricultural impact on wildlife (e.g. O'Connor & Shrubb, 1986).

The shift away from mixed farming has made field boundaries (necessary for stock management) redundant in the intensively farmed eastern and southern parts of England. Hedgerow removal in those areas is still proceeding at a rate of 2–3,000 miles per annum. The mixed design of organic systems preserves the function of these boundaries, and hence the boundaries themselves. 'Set-aside' policies to deal with the current CAP surplus problems may lead to a further erosion of these and other non-crop habitats as farmers extend their cropping base prior to its censusing. This effect has already been observed in the Penwyth Environmentally Sensitive Area in Cornwall, and in Lincolnshire, Northamptonshire and Wiltshire. There is, however, also a problem in other parts of Britain, notably Wales and the South West, where modern hedgerow maintenance techniques are leading to the degeneration of hedgerows and a considerable reduction in their value as habitats for wildlife. Organic techniques on their own may be insufficient to reverse this decline, unless traditional approaches to hedge management are revived. Hedges are of great value, not only as containment for livestock and habitats for wildlife, but also because of their influence on field microclimate (Figure 15.8). They serve as windbreaks, and the increased yield of crops within a hedged field more than compensates for the loss of yield in the immediate vicinity of the hedge (Figure 15.9). An area devoid of hedges can have a mean annual air temperature several degrees lower than the same area when it is properly hedged, and water evaporation is greatly increased. Recent research has also indicated the value of hedges in controlling soil erosion and they can provide habitat for some of the predators of insect pests. Hedges of various types must therefore be seen as an important component of organic farming systems.

Diversity within the crop in organic farming systems is also substantially greater than in conventional systems. This is both by deliberate design and by virtue of the less rigorous control of weeds achieved without herbicides. Undersowing of crops (e.g. of barley with the grass/clover ley that will take its place in the rotation in the succeeding year, or of virtually any crop with a leguminous green manure), retains a key position in organic systems. This practice, which has all but completely died out in conventional systems, has been shown to support a higher diversity and abundance of insect species which are food for farmland

Plate 15.1   Traditional hedgelaying techniques make hedges more stockproof and are beneficial for wildlife. Hedges should be trimmed only every two to three years, and then only during the period from September to March to avoid harm to nesting birds and other wildlife. (Elm Farm Research Centre)

birds and mammals. The effect of undersowing can spread well beyond the edges of the field in question (Vickermann, 1978), causing significant increases in insect abundance over the whole farm.

Herbicidal weed control has virtually eliminated broad-leaved weeds from cereals and other crops. While not being in danger of extinction at a national level, these floral species are subject to local eradication. The retention of at least some specimens is desirable on aesthetic grounds, and the seeds of a few of these species are important food resources for some diminishing farmland bird species. The control of these weeds in rotational, herbicide-free organic systems does not approach that in conventional systems, a situation which may also be influenced by lower crop plant densities. (See Colour plate 27.)

A significant number of recent studies have shown that the abundance and diversity of wild flowers and insects are higher in organically managed arable crops than conventionally managed ones, both in the field centres and in field margins (Dritschilo & Wanner, 1981; Braunewell *et al.*, 1986; Ingrisch & Glück, 1987; Wolff-Straub, 1987; Zwingel, 1987; Frieben, 1988; Ries, 1988; Plakolm, 1989 and Elsen, 1989). Elsen compared two biodynamic row crop fields with two immediately neighbouring conventional row crop fields and found that the median number of wild plant species in the biodynamic field margins

Figure 15.8 Effect of 2.5 m high hawthorn hedges on windspeed and evaporation (wind from left).

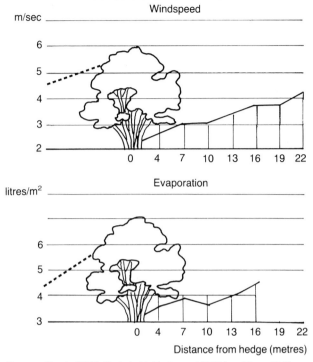

Windspeed

m/sec

Distance from hedge (metres)

litres/m²

Evaporation

Source: Kreutz (1956), illustrated in Vogtmann (1985) (see Figure 15.3).

Figure 15.9 Cereal yield losses and gains as a function of distance from windbreaks.

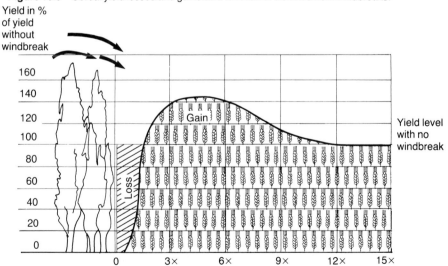

Yield in %
of yield
without
windbreak

Yield level
with no
windbreak

Distance from windbreak as a factor of windbreak height

Source: Kreutz (1956), illustrated in Vogtmann (1985) (see Figure 15.3).

was 25, compared with 16 in the conventionally managed field margins, while in the field centres, the median number of species in the bio-dynamic fields was 18 compared with only two in the conventional fields. Some endangered 'red list' species were occasionally found on the biodynamic fields, but were not found on the conventional fields. Plakolm found an average of 29 species in the organically managed cereal fields, compared with only 15 in the conventional fields. Weed grasses, however, were more common in the conventionally managed fields due to the nature of the weed control practices employed.

Increased populations of arable weed species and insects, as well as more varied habitats resulting from a mixed rotation, are likely to influence bird populations on organic farms. Two United States studies comparing organic and conventional farms (Ducey et al., 1980, and Gremaud & Dahlgren, 1982, reported by Youngberg et al., 1984) both concluded that the higher bird population densities found on the organic farms were linked to the greater diversity of crops and presence of grassland on those farms. Ducey et al., working on two Nebraska farms, found six times the number of bird territories on an organic farm compared to an adjacent conventional farm. Gremaud & Dahlgren found bird population densities in organically managed fields between six and eight times those in conventionally managed fields. In Britain, research by the British Trust for Ornithology (Anon., 1989) found significantly higher levels of soil invertebrate feeding winter birds on organic farms than on conventionally farmed areas. The BTO study looked at eight species including rooks, fieldfares, jackdaws and redwing on winter grassland and temporary cereals. The fieldfare appears to have done particularly badly under conventional systems, producing populations 20 times greater on the organic farms. In Denmark, a comparison of conventional and organic farms by the Danish Ministry of the Environment (Braae et al., 1988) also revealed significantly higher bird populations, notably of skylarks, grey buntings, swallows, lapwings and linnets, on organic farms.

Differences in the methods of soil fertility maintenance and the approach to the provision of crop nutrients in the two farming systems have far-reaching conservation implications. On intensive conventional farms, little attempt is made to cycle crop nutrients through the system. They are applied in surplus in a highly soluble form, immediately available to the crop and consequently prone to leaching and other forms of loss. High rates of inorganic nitrogen fertilisation reduce the botanical diversity of permanent pastures and can reduce soil faunal populations (e.g. of earthworms) on which some farmland birds and mammals are heavily dependent. The manures used on organic farms contain nutrients in less readily available (and therefore less leachable) form, and do not reduce the botanical diversity of grass swards at the rates used in organic systems. A herbal component is often deliberately incorporated in organic swards (as a means of tapping soil nutrient reserves), and some other invasive herb species sprayed with selective herbicides on conventional farms are also tolerated. Species composition is also influenced by livestock grazing management, where the appropriate, extensive use of livestock can prove beneficial.

Although, on balance, organic farming systems probably help to preserve species-rich grassland, organic practices, especially the use of short- and medium-term leys, can be quite intensive and may also involve the ploughing up of old grassland. Hill land improvement and drainage are also important to organic producers under economic pressure. There is, however, often a willingness on the part of organic producers to consider conservation issues positively, and this willingness is reflected in the Soil Association's standards (Appendix 1) and conservation guidelines. The contribution which individual organic producers can make to nature conservation and the environment is being recognised increasingly by awards in a wide range of farming and conservation competitions, including the *Country Life* Conservation Competition. (See Colour plate 28.)

Despite this, there is still a risk that old pastures will be ploughed up to produce 'organic' vegetables, or that land containing valuable species will be drained without thinking of the consequences to wild flora and fauna. Marshes, bogs, and lake and pond margins all support a wide range of plants and animals that are not found elsewhere. Five of the 19 plant species which have become extinct in Britain in the last three centuries were 'wetland' species, and a further seven are endangered. A large number of insects are also threatened. The breeding numbers of many wetland bird species, such as snipe and redshank, have been greatly reduced, and some species have ceased to breed in this country at all.

A further threat to wetland habitats comes from the excessive use of peat in horticulture. Although there is no ideal substitute with the same structural properties as peat, the proportion of peat in the growing medium can be reduced by reliance on organic matter from other sources and the use of bare-root transplants, although bare-root transplants from conventional sources are no longer allowed under the Soil Association's standards. Great care therefore needs to be taken when converting a farm, or planning farm improvements, that these issues are not neglected.

The impact of farming practices on aquatic resources

The imbalance of intensive conventional arable and livestock systems results in a linear throughput of the major plant nutrients (nitrogen,

phosphorus and potassium) and concommitantly high losses of some of
these nutrients to surface and groundwaters. In groundwaters they pose
a so-far inadequately defined threat to the human population's water
supply, in surface waters they can dramatically reduce the abundance
and diversity of aquatic flora and fauna. This effect, eutrophication, is
generally limited by the supply of phosphorus in the aquatic system, a
nutrient which is far less prone to leaching from the soil than nitrogen in
inorganic forms. However, intensive livestock units with poor waste
handling facilities and insufficient land on which to apply these manures
are important point sources of phosphorus. On at least one river in
eastern England, the Waveney, such livestock units are the major source
of water pollution; phosphorus in the waste causes eutrophication
problems, ammonia causes fish and invertebrate kills and the organic
content deoxygenates the river water by virtue of its high biological
oxygen demand (BOD).

Phosphorus can also enter surface waters from agricultural land,
bound to eroding soil particles. Intensive arable farming has recently
begun to accelerate erosion rates in some parts of the UK, and this
source of phosphorus may become significant in areas like East Anglia.
The eroded soil particles themselves also contribute to the degradation
of surface waters, as a sedimentary pollutant. Sediment pollution of
lowland rivers with gravel beds (e.g. the Hampshire Avon) is currently
threatening stocks of some fish species important in commercial and/or
recreational terms. As has been explained in chapter 2, it is now well
established that organic farming methods conserve soil structure and
prevent erosion (USDA, 1980; Arden-Clarke & Hodges, 1987; Reganold
et al., 1987).

Nitrates in the absence of phosphorus are not considered to be a major
problem as far as eutrophication is concerned, but there is concern that
nitrates in drinking water supplies represent a threat to human health.
The primary sources of nitrates are the increased levels of artificial
nitrogen fertiliser use, bad timing and over-application of organic
manures and the ploughing-up of grassland. Organic producers particularly, because they are concerned with conserving nitrogen within
the farm system, need to pay specific attention to the timing and rates of
manure applications. The emphasis should be on spring applications,
when crops are growing and able to benefit, and on limiting the
quantities of manure applied to below 50 t/ha or, as laid down in some
European standards, the equivalent of 2.5 grazing livestock units per
hectare.

Leaching of nitrates as a result of ploughing up a grass/legume ley can
be a significant problem on organic farms as much as conventional
farms. However, emphasis is placed on minimising the nitrogen leached
by following the ley with a fast growing green manure or other suitable

crop which will take up most of the nitrogen mineralised in the crucial autumn period. Where possible, spring cultivations are preferred, but timeliness requirements with respect to weed control mean that some compromises may need to be made.

Nutrients are seldom available in surplus quantities on organic farms, and the consequent close attention to waste handling systems, the rational use of manure and good rotation design result, on balance, in reduced nutrient losses from organic farms compared with conventional systems. Furthermore, the nutrient sources employed on organic farms are often less prone to leaching than those employed on conventional farms. For example, there is now evidence to suggest that nitrogen leaching rates from cut or grazed clover/grass swards are approximately one order of magnitude lower than those from pure grass swards receiving large applications of inorganic nitrogen fertiliser (Garwood & Ryden, 1986).

Silage effluent represents, in fact, a far greater water pollution risk than organic manures. Again, the problem is likely to exist on both organic and conventional holdings, although the advantages of big-bale silage to organic producers, and the prohibition of silage additives, may mean that a greater proportion of silage is pre-wilted and stored safely on organic farms.

Pesticides and the environment

The use of a wide range of synthetic pesticides in conventional farming systems remains one of the most significant, but least quantified, impacts on wild fauna and flora. The majority of these biocides have a wide spectrum of activity and their broadcast in sprays means that they are applied against ecosystems, rather than directly to pests. The impacts of the first generation of pesticides, the persistent organo-chlorines, remain with us today (e.g. in the depressed sparrowhawk population in the south east of England) and some of them are still in regular use. There appears to be continuing illegal use of some of the most damaging of these compounds, especially in the more intensively farmed areas of Britain. Increases in the levels of dieldrin in kingfishers were recorded by the Institute for Terrestrial Ecology in 1981, and between 1983 and 1986 in sparrowhawks and herons. Dead otters with potentially lethal levels of dieldrin residue have been found in East Anglia between 1982 and 1985. Carbamates and some broad-spectrum fungicides may also pose a threat to aquatic vertebrates and inverte-brates at very low concentrations. Herbicides used extensively in British agriculture currently occur at detectable, but apparently not harmful, concentrations in surface waters in some parts of the country (e.g. Croll, 1986).

The organophosphorus insecticides which replaced the organochlorines are less persistent, but generally more directly toxic to vertebrates. According to the Royal Commission on Environmental Pollution, this is likely to exacerbate the problems associated with spray drift. Many of these compounds are no less broad spectrum with respect to insect species than were the organochlorines, and their application can decimate populations of all the insect species living in or close to farm crops. The same is true of at least some of the new pyrethroids and many of the carbamates. It has also recently been shown that some of the new foliar fungicides, widely applied to conventionally grown cereal crops, have a very powerful insecticidal action.

The sub-lethal and indirect effects of these compounds are particularly difficult to detect and may only be revealed by years of careful research. It has been estimated that the intensive use of pesticides in cereal ecosystems has seriously depleted or currently threatens several species of mammals, 14 species of birds, 90 species of flowering plants and 800 species of insects (Potts, 1988). Given our awareness of the subtle yet far-reaching nature of the effects of many of the less selective compounds, their continuing intensive and widespread use must remain a major concern for conservation bodies.

The environmental impact of pesticide use stretches beyond the impact on wildlife to the impact on human health, including not only the people (most of us) eating food contaminated with pesticide residues, but also the workers involved in their manufacture and application. The risks are increased with conventionally produced wholemeal flour and bread, and jacket potatoes, where the parts of the plant where pesticide residues are concentrated are retained. Yet even now maximum limits on residue levels are only just being introduced and limits on lettuce and potatoes will not be enforced during a five-year phase-in period. In West Germany, where maximum limits on pesticide residues have been enforced for several years, residue contamination often exceeds these limits, not only in home-grown crops, but also where livestock have been fed concentrates containing imported ingredients.

Poisonings of agricultural workers, particularly in developing countries, are commonplace, and there is serious concern about increased incidences of cancer among agricultural workers generally. The manufacture of pesticides leads to accidental and deliberate environmental pollution, as toxic wastes are disposed of, or when disasters such as that at Bhopal in India and the Sandoz leak into the Rhine occur. Deliberate flouting of safety codes by pesticide firms exacerbate the problems. The pressures have become so intense that no new pesticide ingredient was registered in Britain between 1987 and 1989. Yet, at the same time, the resistance of pests and weeds to biocides is increasing rapidly, making many existing pesticides ineffectual. Any system, such as organic farming, which offers a possibility of a reduction in the use of these pesticides must therefore be given serious consideration.

Other forms of pollution

Apart from the problems of pesticides and water pollution by animal wastes and silage effluent, there are two other areas of concern. One is the odour resulting from intensive livestock systems, which attracts many complaints. Much of this problem is due to poor storage of slurry and livestock manures. Composting and slurry aeration, which promote aerobic as opposed to anaerobic decomposition of livestock wastes, overcome many of the worst odour problems and are recommended practices in organic systems. The other main issue which attracts public concern is straw burning. The air pollution problems caused by this practice, as well as the damage to wildlife and hedges, are well known. The loss of organic matter for recycling is also a consideration in organic systems and, for these reasons, straw burning is not considered to be an acceptable organic farming practice. There is, however, a case for managing heather and other upland vegetation through the use of burning. Much of the damage to upland ecosystems which has taken place has been as a result of overgrazing by sheep and the decline in traditional burning; on the other hand, too frequent burning will destroy the heather, allowing it to be gradually replaced by grassland.

Sustainability

In the long term, the most telling argument for a switch of political, scientific and technical support away from conventional farming methods towards more organic ones, is that only the latter are sustainable. Sustainability can be defined in a variety of ways, but it is closely linked to non-renewable resource use in the widest sense, including inherent soil fertility decline, soil erosion, the declining effectiveness of existing pesticide inputs, the breakdown of resistance in new cereal varieties, the drift of labour, disillusioned with the working conditions and economic returns to agricultural production, away from the land, and particularly the use of fossil energy and mined inputs such as potash and phosphate.

Non-renewable resources

As far as the mined inputs are concerned, greater emphasis will have to be placed in future on total recycling of 'waste' material and the reduction of use in less essential areas. Phosphate reserves are estimated to be sufficient for 400 years, but a considerable proportion of phosphate use is in detergents and washing powders which make a significant contribution to the eutrophication problem. Changing to non-phosphate based detergents would be a big step forward. Potash reserves are likely to be a greater problem, but for both essential nutrients, all avenues to avoid losses and wastage, including the more efficient collection and recycling of household organic wastes and sewage, will need to be explored. Organic farming, by reducing the use of these fertilisers, makes a significant step in this direction.

Fossil energy

The use of fossil energy for fertiliser and agrochemical manufacture, and for tractor power on farms, is excessive. Agriculture is ideally placed to be a net producer of energy, yet we are now in a situation where more energy is used to produce crops than is provided by those crops (Figure 15.10). Several studies (Table 15.7) have shown conclusively that organic systems require up to 60% less fossil energy per unit of food produced, even when the additional fuel used for weed control cultivations is taken into account. The Vine and Bateman (1981) study of organic farming in Britain came to similar conclusions, although the results were not as clear-cut as some of the other studies. A study by Pimentel *et al.* (1983) using standard data, however, indicated that organic vegetable and fruit production might be more energy intensive than conventional systems, but this has not been the subject of other detailed investigations. Livestock production, which is usually extensive and less dependent on purchased feedstuffs, has also not been studied in detail, but is likely to be more energy efficient in organic compared with conventional systems.

Table 15.7 Relative fossil energy requirements of organic and conventional farming systems (organic as % of conventional).

Author	Vine & Bateman	Kaffka	Mercier & Crouau	USDA	Klepper *et al.*	Harwood
Date	1981	1984	1976/78	1980	1977	1985
Country	Britain	Germany	France	US	US	US
Output	Whole farm	Wheat	Wheat	Cereals	All crops	All crops
Energy use/ha	25–100	20	50	42–85	–	50–90
Energy use per unit of output	50–100	26	55–60	50–87	40	50–80

Conventional systems are not sustainable given their physical, chemical and biological impacts on the soil, their excessive consumption of non-renewable resources and their far-reaching side effects on the global ecosystem. The only farming systems developed today, which offer a significant degree of sustainability in respect of all these resources, all have a basis in the organic model of carefully designed crop rotations, maximal internal cycling of nutrients, and enhancement of, rather than substitution for, natural biological processes.

Figure 15.10 Energy ratios for food production.

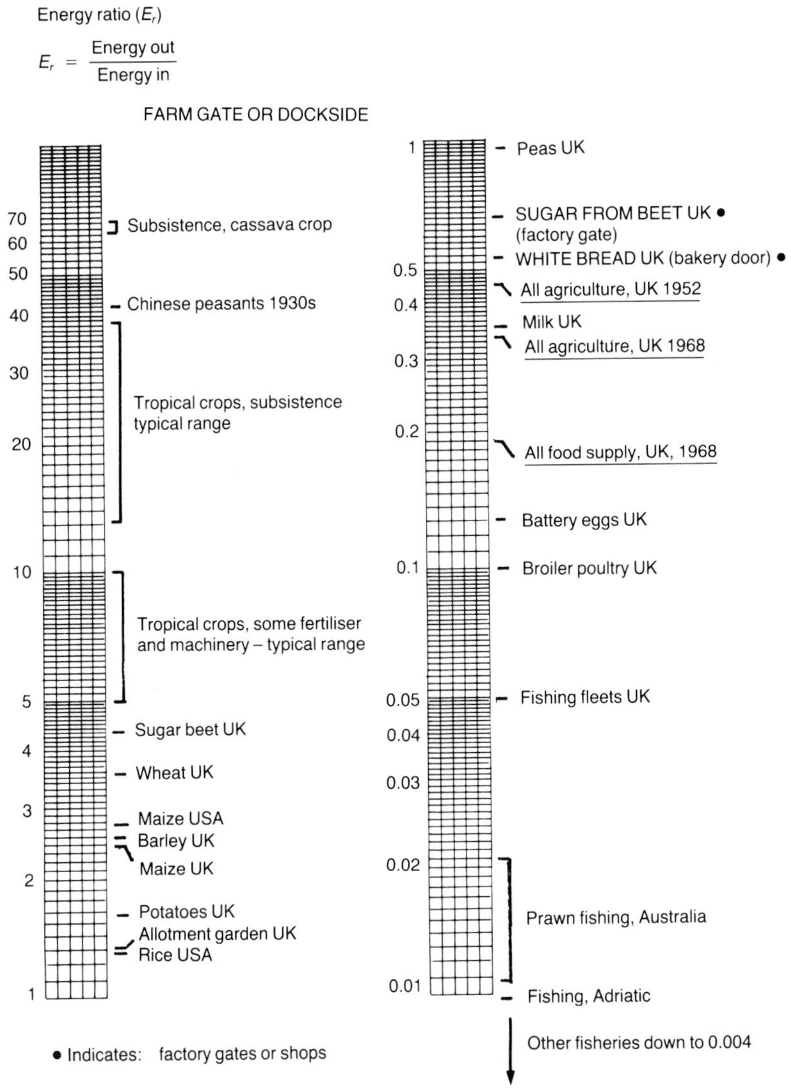
Energy ratio (E_r)
$E_r = \frac{\text{Energy out}}{\text{Energy in}}$
FARM GATE OR DOCKSIDE
70
60
50
40
30
20
10
5
4
3
2
1
Subsistence, cassava crop
Chinese peasants 1930s
Tropical crops, subsistence typical range
Tropical crops, some fertiliser and machinery – typical range
Sugar beet UK
Wheat UK
Maize USA
Barley UK
Maize UK
Potatoes UK
Allotment garden UK
Rice USA
1
0.5
0.4
0.3
0.2
0.1
0.05
0.04
0.03
0.02
0.01
Peas UK
SUGAR FROM BEET UK •
(factory gate)
WHITE BREAD UK (bakery door) •
All agriculture, UK 1952
Milk UK
All agriculture, UK 1968
All food supply, UK, 1968
Battery eggs UK
Broiler poultry UK
Fishing fleets UK
Prawn fishing, Australia
Fishing, Adriatic
Other fisheries down to 0.004
• Indicates: factory gates or shops

Source: Leach, G. (1976) Energy and Food Production. IPC Science and Technology Press.

Organic Farming and Developing Countries

The events of recent years in the Sahel, Ethiopia and elsewhere cannot fail to make us aware that, in spite of great technological advances in agriculture, we are a long way from the cherished dream of the '60s and '70s—that of 'feeding the world'. Rapid population growth, combined

with a declining agricultural base as soil is destroyed through erosion
and desertification, means that the problem has, if anything, got worse
in many parts of the world. It is therefore an important question whether
we should be encouraging organic farming at all, associated as it is with
lower yields, when there are still so many mouths to feed.

As recently as 1987, in a foreword to a report on organic farming in
Scotland, Professor J. C. Holmes wrote:

> While it is true that some aspects of organic production are very
> relevant to lower input farming, it is difficult to see how India,
> without the use of synthetic fertilisers, could maintain its
> currently achieved self-sufficiency or how Africa, already short of
> food and the gap between population growth and food
> production widening, can cope except by plant nutrient
> deficiencies being made good by synthetic fertilisers.

The case against organic farming with respect to agricultural development has been made even more forcefully by Samuel Aldrich who,
writing in the *Rhodesian Farmer* in 1978, concluded that:

> organic farming is not a viable system with which to meet food
> needs in today's world.

Superficially, these arguments are persuasive. Close examination, however, of the record of conventional approaches to agricultural development and the so-called Green Revolution presents a different picture.
Francis *et al.* (1986) argue:

> Increased food production and greater income for farm families
> are primary goals of agricultural development in the Third
> World. Most strategies to achieve these goals are unrealistic in
> assuming that limited resource farmers can move out of basic
> food production in multiple cropping systems to high-technology
> monocropping for export. These strategies are based on
> petroleum based inputs that demand scarce foreign exchange.
> They may include excessive use of chemical fertilisers and
> pesticides, which adds unnecessary production costs, endangers
> the farm family, and degrades the rural environment.
> Dependence on export crops and world markets is economically
> tenuous, especially for the small farmer.

The benefits of the Green Revolution have been obtained by the larger
farmers with sufficient capital to buy the seeds and chemical inputs
necessary to maintain the system. Their greater economic power has
enabled them to concentrate land resources, resulting in increased social
divisiveness and a rapid increase in the numbers of 'landless poor' —

farmers unable to escape the trap of poverty because they have no access to land of their own and must resort to working for other farmers under severely exploitative conditions.

Increased food production in the so-called 'more developed countries' has not helped either. Surplus food is disposed of at prices below the cost of production, with the result that most developing countries simply cannot afford to produce the crops themselves, and certainly cannot afford the fertiliser and other chemical inputs which are currently regarded as essential. Food aid is a short term solution to the problem of food shortages—in the long term, the agricultural resources of developing countries need to be managed in such a way that they are able to produce a greater proportion of their own food requirements.

At the same time, agriculture in industrialised countries is highly dependent on imports of cash crops for feeding to livestock and therefore competes directly with developing countries for essential foodstuffs. A major example of this is the export of soya from countries like Brazil. Leaving aside the issue of irreparable damage to the rainforests as land is cleared for agricultural use only to be rendered useless after a few years, Brazil exported soya beans produced on 8.2 million hectares in 1982, sufficient to feed 40 million pigs in Europe. The same area of land could have produced sufficient protein, in the form of black beans, to feed 35 million people, or energy, in the form of maize, to feed 59 million people. In fact, more protein is imported from developing countries to feed European livestock than is exported back in the form of food aid.

Developing countries are caught in a Catch 22 situation. They have to export cash crops so as to obtain the currency necessary to pay the interest on loans provided by the industrialised countries for over-ambitious, capital-intensive development projects, including high-technology agriculture. At the same time, the pressures caused by this situation are resulting in severe environmental damage and a declining agricultural base, making it even more difficult for real development to take place. The gap between rich and poor countries, and between rich and poor people in developing countries, can only continue to widen.

Problems also occur with the use of pesticides. Imported livestock feeds carry residues of pesticides banned in many developed countries, but still marketed avidly in large parts of the Third World. More serious still is the large number of poisonings among Third World agricultural workers through the mis-use of pesticides and accidents such as that at Bhopal.

Under these circumstances, the argument is not 'can organic farming feed the world?' but whether continuing with current approaches will ever solve the problems of developing countries.

An ecological approach to agricultural development

Many people working in the field of agricultural development now recognise the need for a change in direction, a change in perspective. The argument is certainly not, or should not be, one of taking organic agriculture as currently practised in many industrialised countries and transplanting the system directly into a developing country situation. It is instead one of concentrating on the need to apply the perspective of an ecological approach to agriculture; this means relying primarily on existing indigenous resources, biological processes and ecological inter-actions to maintain soil fertility, meet the nutrient requirements of crops and livestock, control pests and diseases and provide sufficient food and economic security for rural communities in developing countries.

Francis *et al.* (1986) argue that:

> Future agricultural production systems can be designed to take better advantage of production resources found on the farm. Enhanced nitrogen fixation, greater total organic matter production, integrated pest management, genetic tolerance to pests and to stress conditions, and higher biological activity all contribute to resource-use efficiency. Appropriate information and management skills substituted for expensive inputs can further improve resource-use efficiency. On the whole farm level, appropriate cropping on each field can be integrated with animal enterprises, leading to a highly structured and efficient system. Such systems can serve the need of national agricultural sector planners, who in many countries are concerned with increased self-reliance in farming inputs and in production of basic food commodities.

Altieri & Andersson (1986) take these points further:

> the major technological problem that development projects constantly face is that global recommendations prove to be seriously unfit for the highly localised heterogeneity of peasant farms. The many forms of agriculture found in Third World countries result from variations in local climate, soils, crop types, demographic factors and social organisation, as well as from more direct economic factors such as prices, marketing, availability of capital and credit. What is required is an integrated approach that considers this complexity since many different causal factors interact. Cropping systems and techniques specifically tailored to the needs of specific agroecosystems would result in a more fine-grained agriculture, based on a mosaic of appropriate traditional and improved

genetic varieties, local inputs and techniques with each
combination fitting a particular ecological, social and economic
niche . . . (encompassing) all interactions between humans and
the food-producing resources within both small (field level) or
large (regional level) geographic units. A study of 'ecology' of
such resources must necessarily emphasize the environmental
relationships of agriculture, but always within the social,
economic and political context.

Conway (1985) lists four important factors underpinning such systems:

- sustainability, or the ability of an agroecosystem to maintain production through time, in the face of long-term ecological constraints and
  socio-economic pressures;
- equity, or the evenness with which the products of agroecosystems are
  distributed among the local producers and consumers;
- stability, or the constancy of production under a given set of
  environmental, economic and management conditions, and
- productivity, a quantitative measure of the rate of and amount of
  production per unit of land or other input.

Also fundamental to this ecological approach is the recognition
(Altieri & Anderson, 1986) that

throughout developing countries, small farmers have developed
or inherited complex farming systems well adapted to the local
farming conditions; this has allowed families to meet their
subsistence needs for centuries, even under such adverse
environmental conditions as marginal soils, drought or floods.
Generally, these systems are highly diversified and are managed
with low levels of industrial technology. They rely mainly on
local resources, such as hand or animal labour and natural soil
fertility, in a system usually maintained through bush fallow and
the use of legumes or manure.

There is a considerable amount which can be learnt from traditional
systems, such as paddy rice culture in South-East Asia, shifting cultivation or swidden agriculture widespread in Africa and part of the
Amazon, and raised field agriculture extensively used by the Aztecs in
the Valley of Mexico (the 'chinampas'), but also found in China,
Thailand and other areas to exploit the swamp lands bordering the local
lakes. One of the major failures of development programmes has been
the lack of recognition of the inherent value of traditional systems and a
prejudice in favour of a modern, high-technology approach.

There are now, throughout the world, many examples of development programmes where an ecological approach to improving traditional systems has been given priority. Altieri and Anderson (1986) review such programmes in Latin America, where the

> general approach is to take existing peasant production systems and to use modern agricultural science to improve their productivity, progressively and carefully. The programmes have a definite ecological orientation and rely on resource-conserving and yield-sustaining production technologies. They emphasise an ecological engineering approach in which the various components of agroecosystems including crops, trees, soils and animals interact in a way that enhances use of internal resources, cycling of nutrients and organic matter, and trophic relationships among plants, insects, pathogens and weeds that foster biological control.

This approach is illustrated by:

- an attempt to achieve the diversity and stability originally characteristic of the productive traditional agroecosystems of Tabasco, Mexico, where researchers from the local agricultural colleges in collaboration with local peasants installed production units based on the prehistoric traditional chinampas and multi-layered, species rich, kitchen gardens (*huertos familares*) (Gliessman, *et al.*, 1981);
- a parallel Mexican project, where integrated farms were established in the state of Veracruz to help farmers make better use of their local resources (Morales, 1984), using unique designs based on the chinampas and Asiatic aquaculture systems and ensuring that all animal wastes were returned as fertiliser for the fields and fish ponds;
- a project in the highlands of Bolivia, where the agro-pastoral economy has been radically modified and peasants have become more dependent on agricultural chemicals. Members of the Projecto de Agrobiologica de Cochachamba are helping peasants to recover their production autonomy. To replace the use of fertilisers and meet the nitrogen requirements of potatoes and cereals, intercropping and rotational systems have been developed which use the native species *Lupinus mutabilis* (Augstburger, 1983). The systems are higher yielding than corresponding potato monocultures and virus incidence is greatly reduced.
- in Chile, 0.5 ha model farms have been established where the aim is to provide most of the food requirements for a family with scarce capital and land. The critical factor is diversity. Crops, animals and other farm resources are assembled in mixed and rotational designs to

optimise production efficiency, nutrient cycling and crop protection (CET, 1983).

Many of these traditional Latin American systems are also described in Allen & van Dusen (1988). However, it is not only in Latin America where such initiatives are taking place. In India, the Vigyan Shiksa Kendra or Science Education Centre in Uttar Pradesh is working to develop appropriate ecological farming techniques in the region, while recognising that it is only when several of the most urgent needs of a disadvantaged population are addressed that one can hope to develop a viable ecological agricultural system (Prakash, 1988). The Amarpurkashi village project (Sweeney, 1984 & 1986) envisages more direct application of organic methods adopted from industrialised countries than many of the projects, but still within the context of a wider rural development programme based around the concept of a Rural Polytechnic. Another example is that of the Vikas Maitri developmental association in the Chotanagpur plateau area of India, which recognised many of the inequitable consequences of the Green Revolution and is seeking to put them right using appropriate technology and ecological practices (Soil Association, 1984).

In Africa, and in particular Tanzania, Rwanda, Kenya and more recently the Sahel countries, especially Burkina Faso, considerable interest is being shown in ecological approaches, not just at small project level, but also at government level. Interest from Tanzania and contact with researchers in the field of ecological agriculture resulted in a major policy initiative to encourage ecological methods, at least on the subsistence farms (Semoka, 1983). In Kenya, the Kenya Institute of Organic Farming was established in 1986 in order to help Kenyan farmers revive and update their traditional methods of farming (Guepin, 1987).

Detailed practical experience, however, comes from projects such as those in the mountain areas of Rwanda and Tanzania based on the 'ecofarming' concept (Egger & Martens, 1987; Egger, 1988; Kotschi et al., 1989). This involves working with

> a range of possible solutions, ranging from a kind of imitation of the original forest using a mixture of trees (*Grevillea*, mango, avocado and others), bananas, coffee and cocoyams as in the Kilimanjaro region, via intermediate form rich in trees, but more scattered in a savanna-like pattern, containing now more field crops like maize, beans, sorghum, millet, sweet potatoes, cassava and others, to forms similar to industrialised agriculture characterised by a preponderance of open field cultures without trees.

Figure 15.11   The ecofarming concept: integration of tree lines and fields in the contour-parallel terraces. Small forests and hedges are an integrated part of the farm's total biotope mosaic.

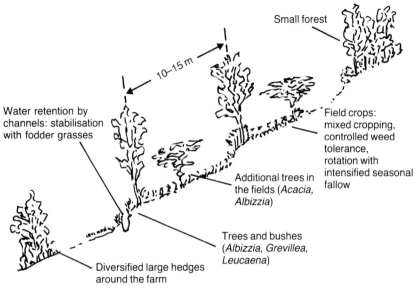

Small forest

10–15 m

Water retention by channels: stabilisation with fodder grasses

Field crops: mixed cropping, controlled weed tolerance, rotation with intensified seasonal fallow

Additional trees in the fields (*Acacia, Albizzia*)

Trees and bushes (*Albizzia, Grevillea, Leucaena*)

Diversified large hedges around the farm

Source: Egger & Martens (1987).

In 1989, the movement for an ecological approach to agricultural development took a major step forward with the holding of the 7th IFOAM International Scientific Conference in Burkina Faso, West Africa (IFOAM, 1990; Milis, 1989; Smith & Finney, 1989). For the first time, the worldwide organic movement's biennial conference was organised and sponsored by a national government with a firm policy commitment to ecological agriculture. The conference, on the theme of agricultural alternatives and nutritional self-sufficiency, was attended by more than 600 people from 48 countries and received widespread attention throughout Africa.

Participants had the opportunity to see and hear of many of the initiatives taking place in Burkina Faso, including the rehabilitation of the traditional Zai system, which involves planting seeds into individual compost-filled holes, as opposed to the more conventional rows, and a campaign to provide every household with a composting area. Other schemes included animal enclosures to prevent over-grazing, agroforestry projects, mini water barrages, and women-only cooperatives to encourage fresh vegetable production and enhance the economic independence of women in Burkinabe society.

Figure 15.12   Within-farm biomass recycling pathways of two main elements in ecofarming:
(a) integration of trees; (b) intensive green manuring by bush fallow.

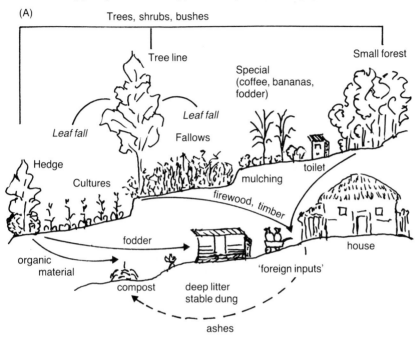

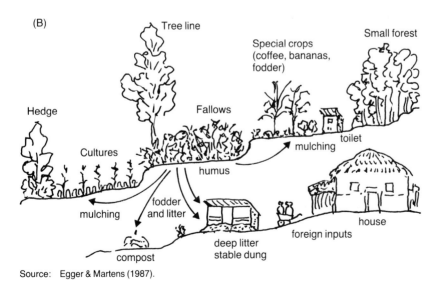

Source: Egger & Martens (1987).

Plate 15.2   The traditional Zai technique from Burkina Faso involves sowing seeds into holes filled with compost. Although more labour intensive, yield advantages are significant.

It was clear, however, that significant problems still remained, not least being the inadequate controls on trade in pesticides in the village markets and the problem that many Burkinabe extension workers trained abroad have received a conventional agricultural education and are not well placed to implement their government's policy. These and other problems were discussed at length during the conference, and gave rise to the 'Ouagadougou Appeal' at the end of the conference:

> To all States and Governments, international organisations, non-governmental organisations and all involved people and institutions
> Considering:
> – the serious food deficit in many developing countries;
> – the disturbing nutritional situation for large numbers of the planet's inhabitants;
> – the rapidity and extent of the erosion of soils, desertification and pollution of the environment;
> the organisers and the 600 participants from 48 countries declare:
> 1. that for developing countries, ecological agriculture is not an alternative, but a necessity imposed by local conditions;
> 2. that there is a need to take into account agro-ecological practices in attaining food self-sufficiency;
> 3. that there is an urgent need to unlock funds to permit massive advances in our knowledge of these systems and their application in the field;
> 4. that it is essential to develop local and international information networks on these methods.
>
> The conference.

Plates 15.3 and 15.4   Composting is a major feature of agricultural development in Burkina Faso. Many households have purpose built composting sites (above), *while banners exhort the benefits of composting for nutritional self-sufficiency* (below).

Far from threatening the populations of developing countries with mass starvation, ecological methods such as organic farming allow farmers to take control again and escape dependence on purchased inputs, improving output while at the same time reducing the risk and inequities which have been associated with the high technology approach to agriculture. It is, however, impossible in the space available to do justice to the range of projects which are now working from this ecological perspective. Details of projects in Sri Lanka (Adelhelm &

Plate 15.5   The enclosure of livestock is a critical first step in creating the ecological conditions for tree survival and agro-forestry programmes in the Sahel.

Kotschi, 1988), China (Zhengfang & Gensheng, 1989; Jolly, 1988), Thailand (MacKay, 1988), Mexico (Neugebauer, 1988), Brazil (Nickel, 1988), Nigeria (Olatunji, 1989) can be found in the literature and several other projects have been described in the proceedings of the 1986 IFOAM conference at the University of California, Santa Cruz (Allen & van Dusen, 1988) and the 1989 IFOAM conference in Burkina Faso (IFOAM, 1990) . Major international conferences on ecological approaches to agricultural developments have also taken place in China (1987), Bolivia (1989)—which saw the formation of the autonomous Latin American regional group of IFOAM—and New Delhi, India (1990).

Further information can be obtained from the International Federation of Organic Agriculture Movements (IFOAM) and AGRECOL, a Switzerland-based information network for ecologically sustainable development projects, as well as other organisations and individuals listed in Appendix 4. AGRECOL publishes a comprehensive source book and bibliography (AGRECOL, 1988) on ecological approaches, including abstracts of major books, details of periodicals and addresses of organisations as well as keywords and references for more than 250 other books.

ORGANIC FARMING AND AGRICULTURAL POLICY

Agricultural policy in the European Community has seen some dramatic changes during the 1980s, with greater emphasis being placed on the reduction of surplus output and protection of the environment. These changes have arisen from a growing recognition of the problems and inconsistencies of existing agricultural policies, which include:

- the waste of resources in producing and disposing of surplus food;
- the inequitable distribution of income between consumers, especially those on low incomes, and farmers and between small and large farmers, as well as the significant likelihood that the main beneficiaries of agricultural price support are in fact landowners, through higher land prices, and the agricultural supply industry;
- increasing farm size and reduced opportunities for new entrants to agriculture, due to rising land prices and the falling number of tenancies;
- a decline in the rural population, particularly those earning a living from rural resources rather than just living in rural areas;
- damage to the environment from agricultural pollution (silage effluent, slurry, nitrate leaching, pesticides) and the destruction of sensitive ecological habitats;
- increasing concern about the quality and safety of food and water supplies (pesticide residues, nitrates, salmonella and listeria).

These problems have arisen as a result of the failure to reconcile the conflicting objectives on which current agricultural policy (the Common Agricultural Policy of the European Community) is based, and un-foreseen consequences arising from the implementation of individual policies. It is clear that these problems are now being recognised, but this should not be surprising given their magnitude.

Even so, the solutions remain distant. Policies are introduced or discussed which effectively only touch on the issues. There is no longer a consensus on the future role of agriculture in the rural economy. In particular, we are faced with a choice between continued intensification of agriculture, but on a smaller area of land with the surplus land being diverted to other uses such as forestry, recreation (golf courses?) and tourism, and extensification, where farmers continue to play a prominent role in the rural economy, managing the countryside productively, but at lower levels of output. Each strategy has implications for rural employment and the character and survival of rural communities. The choice is therefore more than just a matter of economics, it is also political.

The potential contribution of organic farming

Organic farming, producing lower levels of output but supplying a market which is hungry for its produce, may be in a position to alleviate some of these problems. Sufficient evidence exists, as has been indicated earlier in this chapter, to support the view that organic farming can contribute to environmental protection objectives and to the enhancement of food quality, while at the same time helping to reduce the output of surplus commodities. The size and the immediacy of the surplus problem, however, is such that any conceivable short-term expansion of organic production could be no more than a partial solution. It has been estimated (Lampkin, 1989) that if 10% of Britain were to turn organic, total cereal output would only fall by 3% (Table 15.8), which is about 10–20% of the current cereals surplus.

Table 15.8 Relative changes in output (%) resulting from a 10% conversion of UK agriculture to organic production.

Cereals	− 3.0
Potatoes	− 1.3
Sugar beet	− 5.8
Grain legumes	+17.4
Oil seed rape	− 6.1
Milk	− 1.9
Beef	− 1.3
Sheep	− 1.9

Source: Lampkin (1989).

There is another side to the surplus question which has formed a key objective of post–1945 agricultural policy: the need to ensure adequate food supplies in times of crisis and to ease balance of payments problems. Farmers in Britain have achieved an unparalleled degree of self-sufficiency in recent years, which on the face of it indicates that this key objective of agricultural policy has been achieved with great success. In fact, the current level of self-sufficiency has only been achieved on the basis of greatly increased imports of oil and livestock feeds, neither of which would necessarily be available in a future crisis situation. It can be argued that organic farming, which is much less reliant on these inputs, is better placed than conventional agriculture to maintain production in a crisis situation. Indeed, the levels of output reduction as a result of large scale conversion to organic farming indicated in Table 15.8 would, in the case of 100% conversion to organic production, only result in a 30% reduction in cereals output, just sufficient to remove current surpluses. Mass starvation, as has been predicted by some critics of organic farming, is extremely unlikely!

Inequity of income distribution
Organic farming will not, however, help directly to reduce the inequity of income distribution due to agricultural policies which are based on

high food prices funded by the consumer rather than by the taxpayer. It is clear that even higher prices for organically produced food will be disadvantageous for lower income groups. However, committed purchasers of organically produced food tend also to emphasise unprocessed foods, with the result that the total cost of the weekly shopping bill need not be very different and has been shown in one German study to be 7% less (Brombacher & Hamm, 1990).

At the same time, many of the social and environmental costs arising from current agricultural practices are not borne directly by farmers, but are carried by society as a whole. Organic producers, by avoiding certain practices, are taking some of these external costs on themselves. Unless all farmers are required to abandon potentially harmful practices, this process of 'internalising' social and environmental costs can only be sustained in a competitive environment either through premium prices or through agricultural policy support.

Labour use and rural employment
Organic farming is often associated with increased labour use and in theory could help to maintain or even increase rural employment. In practice, labour requirements are not necessarily higher and, where they are, the labour required is often casual and seasonal. Only in cases where substantial diversification takes place, such as the re-introduction of livestock onto an arable farm, is there likely to be any impact on the requirements for regular labour.

Of greater importance is the potential for maintaining the viability of smaller holdings. The reduced use of off-farm inputs can mean that requirements for working capital are considerably reduced, easing the financial situation to some extent. Producing an output for which there is a clear demand and possibly premium prices can also help to improve returns.

There is also the possibility that organic farming is more likely to provide farm-related employment, through on-farm processing, direct marketing and other activities which are able to benefit from the organic "image". Perhaps the most significant impact is at the marketing and processing level, where larger scale operations employing many people, but sited in rural communities close to the point of production, can reverse the trend of exporting these types of jobs to urban areas.

The problem of people leaving the land, leaving rural areas, goes deeper still than simply the availability of employment. Particularly at a time when prospects for employment in urban areas are no better, why should the trend be continuing? The lack of services such as schools, public transport, shops etc., and recreation and entertainment facilities, may be a significant factor causing young people to contribute to the vicious cycle of declining population and reduced services. There is, however, another factor which deserves more attention: farming no longer seems attractive to young people as a way of life. Is this because the pressures of specialisation and intensification, associated with increased indebtedness and estrangement from both work (monotony) and nature, have created a situation where the only real choice left is to get out? Many farmers have found that by converting to organic production, farming once again becomes enjoyable. The need no longer to have to handle pesticides which may be harmful to the farmer's health, the more diverse range of farming activities, the contribution to the environment, the closer contact with the consumer and support rather than criticism from the general public all help to provide an added level of job satisfaction. If this is the case, perhaps the whole approach to rural and agricultural policy requires a much more fundamental reassessment than is currently taking place.

Policies to support organic farming

Given that organic farming would appear to have a contribution to make to current output reduction and environmental protection objectives, while also being market orientated, there would seem to be a case for policies to support organic farmers and producers who wish to convert. Such an approach would also be politically popular, with 55% of the British population favouring support for organic farming against only 20% for set-aside in a Gallup/Daily Telegraph poll in late 1988 and, in a more recent study, 80% preferring support for organic farming to set-aside (Henley Centre, 1989). Most of the major political parties have recognised this and have committed themselves to further support for organic farming.

Financial support
The period of conversion from conventional to organic production, during which producers experience lower yields and increased investment expenditure without the benefit of premium prices, is one which is recognised as being difficult financially. The information presented in Chapter 14 and other models prepared by the author (Lampkin *et al*, 1987; Lampkin & Midmore, 1988) indicate that the reduction in income during the typical five year conversion period for a whole farm may amount to as much as £200 per farm hectare per year on an intensive arable farm in the absence of any organic premium, but this can be reduced to less than £100 per hectare per year where livestock are involved and less than £50 per hectare per year on extensive hill and upland farms.

It is difficult to see this as a form of investment by the farmer in future higher returns because there is no guarantee of improved incomes in the long run. There is therefore a strong case for financial support for a period of up to five years for producers converting their farms. Such schemes already exist in Denmark, Sweden, Norway, Switzerland and West Germany.

The Danish scheme, involving more than £1 million annually, is currently the largest. It covers the development of information and marketing services as well as financial assistance during the conversion period and involves payments of about £300/ha per year over a three year period. For livestock farmers, the payment has recently been increased from £220/ha to £330/ha. There is also an organic extension service run by five full-time agronomists who are employed by the farmers' union and one of Denmark's twenty-five agricultural colleges is devoted entirely to organic farming. The major marketing initiative has been with milk, which is sold though the major supermarkets and accounted in 1988 for 12.5% of the total market in Copenhagen. Farmers receive a 50% price premium for their milk. Organically produced pork has also been widely available since 1988.

In Sweden, the government is supporting a state advisory service for organic producers to the tune of £400,000 and is providing £8.5 million from 1989 to assist an estimated 1,000 producers during the conversion period. The conversion support will vary between regions depending on soil quality and yield potential, ranging between £65 and £275 per hectare per year for a maximum of three years, but does not cover all the land under conversion and is not payable on horticultural crops. Established organic producers are also to obtain this subsidy. The Norwegian government is following suit more cautiously, by implementing a pilot conversion programme involving 35 farms.

The Swiss cantons of Basel-Landschaft and Berne are paying farmers lump sums of up to £2,000 for education and training and the adaptation and purchase of machinery as well as one-off area payments ranging from £120/ha on hill land to £320/ha on lowland arable land and £1,200/ha on intensively managed horticultural land to help them through the conversion period. In West Germany, the state of Saarland has operated a small scheme since 1987 paying individual producers a total of £3,500 per farm spread over three years, as well as a single payment of £100 for training and £150 to cover the costs of replacement labour while attending a one-week training course.

In the European Community as a whole, recognition has been given to the potential of organic farming in the framing of the new 'extensification' policy. In its earlier form, which emphasised the setting aside of cereal land, the legislation was too restrictive in that it concentrated specifically on combinable crops, but advantage can and has been taken of the set-aside payments to enable arable farmers to convert to an organic system (Beaton, 1989), especially where a combination of permanent fallow (£200/ha) and rotational fallow (£180/ha) are used. Higher rates of payment have been proposed for 'conservation management', such as sowing grass seed mixtures as ground cover with the intention of producing a more diverse habitat for wildlife, and cutting a 15 m strip alongside footpaths at the edge of cereal fields. Such practices are not incompatible with organic farming methods and these higher rates of payment could prove advantageous to organic producers.

The earlier legislation also defined extensification in terms of a minimum 20% reduction of other specific surplus commodities such as beef and sheep without any increase in any other surplus commodity. This again looked likely to prove too restrictive, because organic farming involves a restructuring of the whole farm system and although it might be able to achieve an overall 20% reduction in all surplus commodities, could not necessarily guarantee to do this with individual commodities on each individual farm.

The new legislation published in December 1988 has responded to this problem by including a 'production methods' option. The aim of this option is to achieve a reduction in output based on the adoption of less intensive production methods which may involve, in particular, the use of appropriate farming techniques or varieties and the reduction of intermediate inputs. Member states must demonstrate that the adoption of these methods and the conditions of their application will, under normal circumstances, lead to a reduction in output of at least 20%. It is clear that organic farming meets these conditions and has the advantage that a system of control is already in operation in the form of the Soil Association's Symbol Scheme and the UK Register of Organic Food Standards.

Under the new legislation, maximum payments of 65 ECU/LU (European Currency Units per Livestock Unit) for beef and veal cattle and 55 ECU/LU for sheep and goats are allowed if the production methods approach is adopted. At a stocking rate of 2.0 LU/ha, this is equivalent to about £80/ha. For annual crops, the payments are restricted to 180 ECU/ha, or about £120/ha. These amounts are close to the income reduction during the conversion period estimated above and would go a long way to removing one of the most significant barriers to conversion.

West Germany was the first EC country to introduce in 1989 a specific scheme to support the conversion to organic farming in the context of the extensification of arable production. The actual situation varies between the individual states, but essentially farmers were presented with three options: conversion to organic farming, the production of cereals without sprays or fertilisers (so-called 00 cereals), and the conversion of cereal type from wheat and winter barley to rye, oats, spring barley or spelt wheat. Payments for the organic option are about £155/ha for crops in surplus and £110/ha for other crops. Payments for the non-organic options were set at £110/ha.

The schemes operated by the different West German states have attracted widespread criticism from the organic organisations for a number of reasons. Some states allowed only a very short period for registration, four and six weeks in the case of Rhineland-Palatinate and Hessia respectively, and only two states (Schleswig-Holstein and Saarland) allowed for staged conversions, the majority adhering to the Federal Government's insistence that the whole farm should be converted immediately. Hence a large number of farmers were deterred from taking part. In addition, no support was made available for backup services such as advice, training and marketing and no consideration was given to producers who had already started converting and were therefore deemed ineligible.

Several other EC member states are actively considering an organic option under extensification, including Italy, the Netherlands and Britain. In Britain, emphasis has so far been given to pilot extensification schemes for beef and sheep using the 20% 'quantitative method' option, but separate consideration is being given to an organic option which was formally announced in January 1990, although details were not available at the time of writing. Meanwhile, proposals for an organic farming pilot scheme in West Wales already exist and have been under discussion for some time (Lampkin et al., 1987).

There is a further reason why financial support for organic farming should be considered seriously, which is the impact on public sector spending on agriculture. Conversion to an organic system is likely to be maintained in the long term, even if support payments ceased, because a producer would then be in a position to supply a specific premium market. Neither extensification using the 'quantitative (20% reduction) method' nor set-aside can achieve this. In addition, set-aside applies only to part of the farm with no restriction on intensification on the remainder of the farm. It is probable, therefore, that an organic option under extensification would be more effective at achieving output reduction than any of the other policy instruments currently available.

If the costs of the different policies are compared, as in Midmore and Lampkin (1988), then the public sector expenditure case appears to be even stronger (Table 15.9). The organic set-aside option illustrated involves a green fallow with no financial returns and with payments set at the current rate for set-aside of £200/ha taken out of production. The lower cost per tonne of reduced cereal output arises from the fact that output is reduced not only on the area taken out of production but on the remaining area as well. The organic extensification option assumes a payment of £60 per farm hectare, which is less than the amount allowed under the current extensification legislation. The reason the cost is higher than would be expected from a straight 20% reduction in output is that restructuring of the enterprise mix on the farms has also been taken into account. (On a straight 20% reduction, from 2.0 to 1.6 LU/ha, the cost of a £60/ha payment would be £150 per LU reduced output.) The public expenditure saving effect will also depend on whether a farm is in a Less Favoured Area or not. In the case of beef and sheep, the organic extensification option may have a net cost, but this can also be considered as a measure of the cost of achieving the environmental benefits which organic farming can offer.

Even if schemes to support farmers financially during the conversion period become more widespread, attention will need to be given to the needs of existing organic producers as well. Existing producers have carried the costs of conversion themselves in the past, but are currently able to survive with the help of a premium market. If large numbers of new producers were to convert, the rapid increase in market supply could lead to significant reductions in the premiums obtainable, particularly if the market does not have sufficient time to adapt, and this

Table 15.9 Summary of net costs of current and proposed agricultural policy measures per unit of surplus output.

	Cereals (£/tonne)	Milk (£/hl)	Beef (£/GLU)	Sheep (£/GLU)
Set-aside	69.50	na	na	na
'Organic' set-aside	24.72	na	na	na
Storage	74.81	8.33	268.13	na
Export	49.87	5.25	196.65	na
HLCA	na	na	60.30	45.52
Other premiums	na	na	82.29	187.15
Organic extensification	30.77	8.54	418.18	277.39

Note: Russell & Power (1989) provide the following alternative estimates of the impact on budgetary expenditures of marginal changes in agricultural output (£/tonne) (1 tonne carcass weight ≈ 2 Grazing Livestock Units (GLU) for both sheep and beef).

	Cereals	Beef		Sheep	
		Lowland	Upland	Lowland	Upland
Total support expenditure	102	1085	1308	658	1002
Direct impact on UK Exchequer	10	219	386	−8	165
Total impact on UK Exchequer	77	845	1052	474	770

HLCA = Hill Livestock Compensatory Amount.
hl = hectolitre
Source: Midmore & Lampkin (1988).

could create difficulties in ensuring the financial viability of farms after conversion. This issue has been recognised by the Swedish government in its scheme for supporting organic farmers as well as producers who are converting. In addition, the fact that in many cases existing organic farmers are effectively internalising external costs provides a strong argument for financial support to this group on the same basis as financial support is already given in the UK in the form of management agreements for Sites of Special Scientific Interest (SSSIs) or in Environmentally Sensitive Areas (ESAs).

Ways of solving this particular problem are less easy to identify. In theory, at least, the management agreement and ESA approaches, which specify certain farming practices to be adopted or abandoned, could be one way of achieving this. After all, organic farmers have to agree to stick to a set of practices in order to qualify for recognition and it would therefore not be necessary to negotiate a separate agreement in each individual case. However, the ESAs only cover limited areas of the country; although ESA payments might enable farmers in these areas to think seriously about producing for the organic market. It has been suggested (BOF/OGA/SA, 1989) that an expansion of the ESA concept to include organic farming could involve a lower level of payment (current rates range from £30/ha in more marginal areas up to a maximum of £300/ha in parts of the Breckland ESA), recognising the role that premiums can play in making organic systems economically viable.

The nitrate sensitive area (NSA) pilot schemes introduced in Britain in 1990 could also provide an opportunity to test whether organic farming can in practice make a significant contribution to reducing nitrate leaching. The initial proposals were severely criticised by the organic movement (EFRC et al., 1989) because some of the practices recommended, in particular a prohibition on the use of clover in the most sensitive areas, would make it impossible to implement a viable organic farming system. These proposals have been substantially modified to allow the majority of organic farming practices while emphasising factors such as the use of cover crops on fields left uncropped over the autumn and winter, restricting slurry and manure applications to the spring and summer months only, restricting the conversion of grassland to arable unless it is a short to medium term ley forming part of an arable rotation and restricting the times of year when grassland and cover crops can be ploughed in. Rates of payment vary depending on area from £55–95/ha for the basic rate scheme to £200–380/ha for the premium rate scheme which involves the conversion of arable land to permanent, extensively managed grassland.

There is also a case for capital grants to allow organic producers to re-equip their buildings, fences, water supplies, manure handling systems and other items necessary to help them convert from conventional farming systems and intensive livestock rearing methods. To some extent, these are already available under the MAFF Farm and Conservation Grant Scheme launched in 1989, as well as the Farm Diversification Grant Scheme of 1987 and the various conservation and tree-planting schemes operated by the Nature Conservancy Council, the Countryside Commission and the Forestry Commission. However, farmers within national parks have been eligible for 'top-up' grants on eligible capital investments, and this concept could also be applied to organic producers.

In the long-term, however, a specific policy to encourage organic farming is required. Woodward (1990) argues that organic farming should become the preferred management option in all the nitrate, environment, soil erosion and conservation sensitive areas of Britain and that a policy to bring this about would go a long way towards achieving the Soil Association's objective of 20% of British farmland managed organically by the year 2000.

Marketing and standards
Many more countries have started to support organic farming through the development of marketing initiatives and certification schemes than have given direct financial support. New Zealand, Denmark and Sweden have agreed national standards and are actively promoting the marketing of organic produce. In addition, France, Austria and four states in the USA (California, Minnesota, Texas and Washington State) have legally enforceable certification programmes, and 15 other US states have organic labelling programmes. West Germany, however, has steadfastly refused to get involved in establishing a legislative framework for organically produced food, or even accepting an nationally agreed set of standards, which is in marked contrast to an otherwise relatively positive attitude to organic farming. This situation may change in the near future, because of increasing problems with the 'grey' market for semi or pseudo organic produce and the increasing incidence of fraud.

In Britain, the UK Register of Organic Food Standards (UKROFS), set up by the government under Food From Britain, is responsible for overseeing standards and approving certification organisations, but the standards have as yet no legislative basis. The European Community is due to introduce a regulation on organic farming in 1990/91, with a definition based partly on the IFOAM standards. Each member state will have to have a national administration such as UKROFS to oversee the (private sector) certification organisations.

The introduction of both UKROFS and the EC regulation is likely to create financial difficulties for the private sector certification organisations, many of which, including the Soil Association, are voluntary organisations. In addition, through individual members and corporately, they have provided free of charge major inputs into the government and EC standards. Financial support for these organisations may well be necessary to help them comply with the certification procedures established by the national administrations such as UKROFS, a factor which has already been recognised in Ireland.

Research, advice, education and training
Research conducted within functioning organic systems is essential to overcome some of the technical problems which still need to be resolved and to improve further the environmental sensitivity of organic systems, for example the reduction of nitrate leaching. In Britain, the Ministry of Agriculture (MAFF) is gradually starting to fund research specifically on organic farming, including recent work on nitrogen cycling in organic systems and current work on potash and phosphate availability and the economics of organic farming. Significant research work is also taking place in Scotland (Williamson & Black, 1988) including the launch of Edinburgh University's Organic Farming Centre, funded jointly by the European Community, the Scottish Development Agency and Safeway. Further research is to be funded under an expanded programme in the 1990s.

The efforts of MAFF, however, pale into insignificance when compared with the public funding going into research in West Germany, Switzerland, Sweden, Denmark, the Netherlands and New Zealand. In the United States, millions of dollars each year are going into national low-input sustainable agriculture (LISA) programme, as well as formal low-input sustainable agriculture programmes in eight individual states: California, Iowa, Minnesota, Montana, Nebraska, Ohio, Texas and Wisconsin.

Advisory and extension work is also relatively underfunded in Britain. West Germany, Denmark, Sweden and the Netherlands have full-time state-funded advisers, working both within existing advisory structures and within the organic producer organisations. In Britain, the Agricultural Development and Advisory Service (ADAS) has appointed individuals within each region with responsibility for organic farming, but these are all part-time and their organic work has to compete with other duties. The Scottish Agricultural College is now providing an advisory service for organic producers in Scotland, closely linked with the Edinburgh organic farming centre. The only fully dedicated advisers in Britain are employed by the Organic Advisory Service, based at Elm Farm Research Centre. If an organic farming option were to be established under the extensification legislation, the need for additional, experienced advisers would become critical.

On the education and training side, the agricultural colleges in Britain are taking the initiative and establishing their own organic farming courses, some using funding from EC sources such as the European Social Fund. Courses in commercial organic production are currently offered by the Agricultural Training Board (ATB) at a wide range of locations, and by Worcestershire College of Agriculture, the Welsh Agricultural College, Carmarthenshire College of Technology and Art, Pershore College of Horticulture and the independent Emerson College, which specialises in biodynamic agriculture and horticulture. A full list of available courses, including those which are approved by the Soil Association, can be obtained directly from the Soil Association.

At the University level, however, little is being done, which is in marked contrast to West Germany, with four Chairs in organic agriculture at the Universities of Kassel, Bonn, Giessen and Kiel, the Netherlands with a Chair in Ecological Agriculture at Wageningen, and Canada and the United States where several Chairs in Sustainable Agriculture have been established in the last few years.

Conclusion

Organic farming has attracted increasing recognition, particularly over the last ten years, as a valid and viable method of farming in a way which is potentially environmentally and ecologically sustainable. Its ability to meet at least some of the most important agricultural and environmental policy objectives has been accepted in many countries around the world. But organic farming is not a fixed and static approach to agriculture; it is a dynamic, rapidly developing concept which is certain to evolve and adapt further over the next few years. Organic farming can be helped in this process by well targeted financial support, market and standards development, and the improved availability of new and existing information. It will be essential, however, that the underlying holistic, ecological perspective is not lost sight of due to market or other pressures. If this underlying perspective can be maintained, with the commitment and involvement of individual producers and consumers, then organic agriculture certainly has a future of its own and a significant role in the development of agriculture around the world.

REFERENCES AND FURTHER READING

Quality of organically produced food
Abele, U. (1987) *Produktqualität und Düngung—mineralisch, organisch, biologisch-dynamisch.* Angewandte Wissenschaft (Schriftenreihe des Bundesministers für Ernährung, Landwirtschaft und Forsten) Heft 345

Aehnelt, E. & Hahn, J. (1973) *Fruchtbarkeit der Tiere—eine Möglichkeit zur biologischen Qualitätsprüfung von Futter und Nahrungsmittel?* Tierärztliche Umschau 4: 1–16

Aubert, C. (1987) *Pollution du lait maternel, une enquete de Terre vivante.* Les Quatre Saisons du Jardinage 42: 33–39

Balfour, E. (1943) *The Living Soil.* Faber & Faber

Bulling *et al.* (1987) *Qualitätsvergleich von 'biologisch' und 'konventionell' erzeugten Feldfrüchten.* Regierungspräsidium Stuttgart

Clancy, K. L. (1986) *The role of sustainable agriculture in improving the safety and quality of the food supply.* American Journal of Alternative Agriculture 1: 11–18

Duden, R. (1987) *Image und Qualität von Tomaten.* Gordian 6: 118–119

Dudley, N. (1986) *Nitrates in Food and Water.* London Food Commission

Edelmüller, I. (1984) *Untersuchungen zur Qualitätserfassung von Produkten aus unterschiedlichen Anbausystemen (biologisch-dynamisch bzw. konventionell) mittels Fütterungsversuchen an Kaninchen.* PhD Thesis, University of Vienna

EFRC (1988) *Nitrates in Vegetables.* Research Notes No. 5. Elm Farm Research Centre; Newbury

EFRC (1989) *Quality of Organically Produced Foods.* Proceedings of International Scientific Colloquium, February 1989. Elm Farm Research Centre; Newbury

Elsaidy, S. (1982) *Das Nachernteverhalten von Gemüse, insbesondere Spinat, unter besonderer Berücksichtigung der Nitratanreicherung in Abhängigkeit von den Lagerbedingungen und von der Düngung.* PhD Thesis, University of Giessen

Fischer, A. & Richter, C. (1986) *Influence of organic and mineral fertilisers on yield and quality of potatoes.* In: Vogtmann, H. *et al.* (eds) *The Importance of Biological Agriculture in a World of Diminishing Resources.* Proc. 5th IFOAM Conference, 1984. Verlagsgruppe; Witzenhausen

Gottschewski, G. (1975) *Neue Möglichkeiten zur grösseren Effizienz der toxikologischen Prüfung von Pestiziden, Rückständen und Herbiziden.* Qual. Plantarum 25: 21–42

Hoffmann, M. (1988) *Lebensmittelqualität—Lebensqualität. Eine ganzheitliche Betrachtung.* Lebendige Erde 5/88: 291–300

Kerpen, J. (1988) *Untersuchungen zum Vergleich von Möhren aus ökologischem und konventionellem Landbau.* PhD Thesis. Technical University; Berlin

Knorr, D. (1982) *Use of a circular chromatographic method for the distinction of collard plants grown under different fertilising conditions.* Biological Agriculture and Horticulture 1: 29–38

Knorr, D. (1984) *Feasibility of analytical procedures and unit operations for the distinction between organic, natural or conventional foods.* Biological Agriculture and Horticulture, 2: 183–194

Knorr, D. & Vogtmann, H. (1983) *Quantity and quality determination of ecologically grown foods.* In: Knorr, D. (ed.) *Sustainable Food Systems.* AVI Publishers

Lairon, D. *et al.* (1986) *Effects of organic and mineral fertilisers on the contents in vegetables of minerals, vitamin C and nitrates.* In: Vogtmann, H. *et al* (eds) *The Importance of Biological Agriculture in a World of Diminishing Resources.* Proc. 5th IFOAM Conference, 1984. Verlagsgruppe; Witzenhausen

Lindner, U. (1985) *Alternativer Anbau—Alternative im Erwerbsgemüsebau?* Gemüse 21: 412–418

Lowman, B. (1989) *The eating quality of organic beef.* East of Scotland College of Agriculture; Edinburgh

Maga, J. (1983) *Organically grown foods.* In: Knorr, D. (ed) *Sustainable Food Systems.* AVI Publishers

McCarrison, R. (1926) *The effect of manurial conditions on the nutritive value of millet and wheat.* Indian Journal of Medical Research 14: 351–378

McSheehy, T. (1977) *Nutritive value of wheat grown under organic and chemical systems of farming.* Qual. Plantarum 27: 113–123

Meier-Ploeger, A. & Vogtmann, H. (eds) (1988) *Lebensmittelqualität—ganzheitliche Methoden und Konzepte.* Alternative Konzepte 66. Müller; Karlsruhe

Meier-Ploeger, A. & Vogtmann, H. (1989) *Ökologischer Landbau und Lebensmittelqualität.* Ökologie und Landbau (formerly German IFOAM Bulletin) 69: 9–15

Pearse, I. & Crocker, L. (1943) *The Peckham Experiment.* Allen & Unwin. Reprinted by Scottish Academic Press, 1985.

Plochberger, K. (1984) *Untersuchungen von Auswirkungen verschiedener Bewirtschaftungsmethoden auf die Qualität landwirtschaftlicher Produkte an Hand von Fütterungsversuchen mit Hühnern.* PhD Thesis, University of Vienna

PMB (1989) *Characteristics of organically and conventionally grown potatoes.* Sutton Bridge Experimental Station Annual Report 1988, Chapter 10: 32–37. Potato Marketing Board; Oxford

Popp, F-A. (1987) *Biophotonenanalyse zu Fragen der Lebensmittelqualität und des Umweltschutzes;* Ökologische Konzepte 26: 39–55

Pottinger, F. M. (1946) *Feeding experiments on cats.* American Journal of Orthodontics and Oral Surgery, 32. Reprinted in 1983 as Pottinger's Cats. Price Pottinger Nutrition Foundation: La Mesa, California.

Reinhard, C. & Wolff, I. (1986) *Rückstände an Pflanzenschutzmittel bei alternativ und konventionell angebautem Obst und Gemüse.* Bioland 2/86: 14–17

Samaras, I. (1977) *Nachernteverhalten unterschiedlich gedüngter Gemüsearten mit besonderer Berücksichtigung physiologischer und mikrobiologischer Parameter.* PhD Thesis, University of Giessen

Schüpbach, M. (1986) *Spritzmittelrückstände in Obst und Gemüse.* Deutsche Lebensmittel-Rundschau 82(3): 76–80.

Schuphan, W. (1975) *Yield maximisation versus biological value.* Qual. Plant. 24; 281–310

Schuphan, W. (1976) *Mensch und Nahrungspflanze.* Verlag Eden-Stiftung; Bad Soden

Staiger, D. (1986) *Einfluss konventionell und biologisch-dynamisch angebautem Futters auf Fruchtbarkeit, allgemeinen Gesundheitszustand und Fleischqualität beim Hauskaninchen.* PhD Thesis, University of Bonn. (Summarised in German IFOAM Bulletin 62: 6–9, 1987)

Stopes, C. et al (1988) *The nitrate content of vegetable and salad crops offered to the consumer as from organic or conventional production systems.* Biological Agriculture and Horticulture 5: 215–222

Temperli, A. T. et al (1982) *Einfluss zweier Anbauweisen auf den Nitratgehalt von Kopfsalat.* Schweizerische landwirtschaftliche Forschung 21: 167–196

Teubner, R. (1983) *Zur Qualitätsbestimmung von Nutzpflanzen, insbesondere Medizinalpflanzen, mit Hilfe der ultraschwachen Photonenemission.* PhD Thesis, University of Göttingen

Vogtmann, H. (1981) *The quality of agricultural produce originating from different systems of cultivation.* Soil Association; Bristol

Vogtmann, H. (1986) *Research on organic farming in Europe.* Journal of the Agricultural Society, UCW Aberystwyth, 66: 3–45

Vogtmann, H. et al (1984) *Accumulation of nitrates in leafy vegetables grown under contrasting agricultural systems.* Biological Agriculture and Horticulture 2: 51–68

Vogtmann, H. & Biedermann, R. (1985) *The Nitrate Story—No End in Sight.* Nutrition and Health 3: 217–239

Webb, T. & Lang, T. (1989) *Food Irradiation – the myth and the reality.* London Food Commission.

Wistinghausen, E. v. (1979) *Was ist Qualität? Wie ensteht sie und wie ist sie nachzuweisen?* Schriftenreihe *Lebendige Erde.* Forschungsring für biologisch-dynamische Wirtschaftsweise; Darmstadt

Younie, D. (1990) *Quality of organic and conventional beef.* Paper presented at the British Society of Animal Production winter meeting, Scarborough, Yorkshire.

The environmental impact of organic farming

ADAS (1986) *Hedgerows*. Leaflet P3027

Anon (1989) *Organic farming and birds*. British Trust for Ornithology News 163 (July–August): 4–5. Also: Organic farming good for birds. Habitat 25 (October): 3.

Arden-Clarke, C. (1988a) *The environmental effects of conventional and organic/biological farming systems. I. Impacts on the soil*. Research Report RR–16. Political Ecology Research Group; Oxford

Arden-Clarke, C. (1988b) *The environmental effects of conventional and organic/biological farming systems. II. Impacts on the crop ecosystem, wildlife and its habitats*. Research Report RR–17. Political Ecology Research Group; Oxford

Arden-Clarke, C. & Hodges, R. D. (1987) *The environmental effects of conventional and organic/biological farming systems. I. Soil erosion with special reference to Britain*. Biological Agriculture and Horticulture 4: 309–357

Arden-Clarke, C. & Hodges, R. D. (1988) *The environmental effects of conventional and organic/biological farming systems. II. Soil ecology, soil fertility and nutrient cycles*. Biological Agriculture and Horticulture 5: 223–287

Baldock, D. (1990) *Agriculture and Habitat Loss in Europe*. Worldwide Fund for Nature; London.

Braae, L.; Nøhr, H. & Petersen, B. S. (1988) *Fugelfaunaen på konventionelle og økologiske Landbrug*. Miljøprojekt 102. Miljøministeriet; Copenhagen

Braunewell, R. ; Busse, J. & Marten, S. (1986) *Biologischer Landbau — auch eine Alternative für Flora und Fauna*. Bioland 6/86: 25–26

Croll, B. T. (1986) *The effects of the agricultural use of herbicides on fresh waters*. In: *Effects of Land Use on Fresh Waters: Agriculture, Forestry, Mineral Exploitation, Urbanisation*. Edited by J. F. de L. G. Solbe: 201–209. Ellis Horwood; Chichester

Crouau, M. (1976) *Direkter energieverbrauch in der biol. und klassischen Anbauweise*. Translation from Natur et Progres, 51

Dritschilo, W. & Wanner, D. (1981) *Ground beetle abundance in organic and conventional corn fields*. Environmental Entomology 9: 629–631

Ducey, J. et al (1980) *A Biological Comparison of Organic and Chemical Farming*. University of Nebraska, Lincoln

Elsen, T. van (1989) *Ackerwildkrautbestände unterschiedlich bewirtschafteter Hackfruchtäcker in der Niederrheinischen Bucht*. Ökologie und Landbau (formerly German language IFOAM Bulletin) 71: 7–9

Elsen, T. van (1989) *Ackerwildkrautbestände im Randbereich und im Bestandesinnern unterschiedlich bewirtschafteter Halm- und Hackfruchtäcker*. In: Proceedings of the 3rd IFOAM International Conference on Non-Chemical Weed Control. Linz, 1989. International Federation of Organic Agriculture Movements; Tholey-Theley, West Germany.

Frieben, B. (1988) *Vergleichende Untersuchungen der Ackerbegleitflora auf längerfristig alternativ und konventionell bewirtschafteten Getreideäckern im östl. Westfalen und im norddeutschen Raum— Veränderungen im Vergleich zu den Jahren 1959–1961*. Diplomarbeit, University of Bonn

FWAG (1984) *Conservation and management of old grassland*. Information Leaflet No. 19. FWAG, The Lodge, Sandy, Bedfordshire

Garwood, E. A. & Ryden, J. C. (1986) *Nitrate loss through leaching and surface run-off from grassland: effects of water supply, soil type and management*. In *Nitrogen Flows in Intensive Grassland Systems* H. van der Meer, J. C. Ryden & G. C. Ennick (eds): 99–113. Martinus Nijhoff; Dordrecht

Gordon, I. & Duncan, P. (1988) *Pastures new for conservation*. New Scientist, 17th March

Gremaud, J. K. & Dahlgren, R. B. (1982) *Biological Farming: impacts on wildlife*. In *Workshop on Midwest Agricultural Interfaces with Fish and Wildlife Resources*, R. B. Dahlgren (ed.): 38–39. Iowa State University; Ames

Harwood, R. (1985) *The integration efficiencies of cropping systems.* In: Edens, T. *et al* (eds) *Sustainable Agriculture and Integrated Farming Systems.* Michigan State University Press

Hedging—A practical conservation handbook. British Trust for Conservation Volunteers, 36 St Mary's Street, Wallingford, Oxon

Hooper, M. D. (1987) *Conservation interests in plants of field margins.* In *Field Margins,* (eds) J. M. Way & P. W. Greig-Smith. British Crop Protection Council Monograph 35: 49–52

Ingrisch, S.; Wasner, U. & Glück, E. (1989) *Vergleichende Untersuchung der Ackerfauna auf alternative und konventionell bewirtschafteten Flächen.* In: König, W. *et al. Alternativer und konventioneller Landbau – Vergleichsuntersuchungen von Ackerflächen auf Lössstandorten im Rheinland.* Schriftenreihe der Landesanstalt für Ökologie, Landschaftsentwicklung und Forstplanung Nordrhein-Westfalen, Band 11: 113–271

Jarvis, R. *et al* (1987) *The Boxworth Project.* Cereals '87 Supplement, Farmers Weekly, June 5th, S16–S23

Jenkins, D. (ed.) (1984) *Agriculture and the Environment.* Institute for Terrestrial Ecology and Natural Environment Research Council

Kaffka, S. (1984) *Dairy Farm Management and Energy Use Efficiency.* MSc Thesis, Cornell University

Klepper, R. *et al* (1977) *Economic performance and energy intensiveness on organic and conventional farms in the Corn Belt.* American Journal of Agricultural Economics 59: 1–12

Mercier, J. R. (1978) *Energie et Agriculture.* Edition Debard, Paris

Miljoeministeriet Kopenhagen (1986) *Redegoerelse om miljoemaessige konsekvenser ved overgang till oekologisk jordbrug.* Miljoeministeriet; Copenhagen

Noirfalise, F. (1974) *Ecological consequences of the use of modern production methods in agriculture.* European Community Report on Agriculture No. 137., Brussels

O'Connor, R. J. & Shrubb, M. (1986) *Farming and Birds.* Cambridge University Press

Pimentel, D. *et al* (1983) *Energy efficiency of farming systems—organic and conventional agriculture.* Agriculture, Ecosystems and Environment 9: 359–372

Plakolm, G. (1989) *Unkrauterhebungen in biologisch und konventionell bewirtschaften Getreide-äckern Oberösterreichs.* In: Proceedings of the 3rd IFOAM International Conference on Non-Chemical Weed Control, Linz, 1989. International Federation of Organic Agriculture Movements; Tholey–Theley, West Germany

Pollard, E.; Hooper, M. D. & Moore, N. W. (1974) *Hedges.* Collins; London

Pommer, G. (1989) *Vergleich der agrarökologischen Auswirkungen der Anbausyteme 'Integrierte Pflanzenbau' und 'Alternativer Landbau'.* Lebendige Erde 6/89: 406–416

Potts, C. R. (1986) *The Partridge.* Collins; London

Pouncett, C. L. (1988) *Organic Farming and Wildlife Conservation.* BA Dissertation. University of Bristol

RCEP (1979) *Agriculture and Pollution.*7th Report of the Royal Commission on Environmental Pollution. HMSO; London

Reganold, J. P.; Elliott, L. F. & Unger, Y. L. (1987) *Long-term effects of organic and conventional farming on soil erosion.* Nature 330: 370–372

Ries, C. (1988) *Die Ackerbegleitflora des biologisch-dynamischen und konventionellen Pflanzenbaus in Hüpperdange, Luxemburg.* Diplomarbeit. Universität für Bodenkultur; Vienna

Seymour, J. & Girardet, H. (1986) *Far From Paradise.* BBC Publications

Soltner, D. (1985) *L'arbre et la haie.*7th edition. Collection Sciences et Techniques Agricoles, available from: 'Le Clos Lorelle', Sainte-Gemmes-Sur-Loire, F-49000 Angers

Somerville, L. & Greaves, M. P. (1987) *Pesticide effects on soil flora and fauna.* Taylor and Francis

USDA (1980) *Report and Recommendations on Organic Farming.* United States Department of Agriculture Study Team. US Govt. Printing Office; Washington DC

Vickermann, G. P. (1978) *The arthropod fauna of undersown grass and cereal fields.* Scientific Proceedings of the Royal Dublin Society (A) 6: 273–283

Vine, A. & Bateman, D. (1981) *Organic Farming Systems in England & Wales.* Department of Agricultural Economics, UCW Aberystwyth

Vogtmann, H. (1986) *Research on organic farming in Europe.* Journal of the Agricultural Society, UCW Aberystwyth 66: 33–45

Way, P. (1987) *Farming for the Future Landscape.* BA Dissertation. Gloucestershire College of Arts and Technology

Wolff-Straub, R. (1989) *Vergleich der Ackerwildkraut–Vegetation alternativ und konventionell bewirtschafteter Äcker.* In: König, W. *et al. Alternativer und konventioneller Landbau – Vergleichsuntersuchungen von Ackerflächen auf Lössstandorten im Rheinland.* Schriftenreihe der Landesanstalt für Ökologie, Landschaftsentwicklung und Forstplanung Nordrhein–Westfalen, Band 11: 70–112

Youngberg, E. G.; Parr, J. G. & Papendick, R. I. (1984) *Potential benefits of organic farming practices for wildlife and natural resources.* Transactions of the North American Wildlife and Natural Resources Conference 49: 141–153

Zwingel, W. (1987) *Auswirkungen der Anbauintensität auf die Ackerbegleitflora der Flächen des 'Artenhilfsprogramms Ackerwildkräuter'.* Diplomarbeit. Technical University; Munich

Organic farming and developing countries

Adelhelm, R.; Kotschi, J. (1988) *Environmental protection and sustainable land use: implications for technical co-operation in the rural topics.* English language IFOAM Bulletin 6: 7–11

AGRECOL (1988) *Towards Sustainable Agriculture. Part I: Abstracts, Periodicals, Organisations. Part II: Bibliography.* AGRECOL; Langenbruck, Switzerland

Allen, P. & Dusen, D. van (eds) (1988) *Global Perspectives on Agroecology and Sustainable Agricultural Systems.* Proc.6th IFOAM International Conference. University of California; Santa Cruz. Of particular relevance: Part 6: *Agricultural Development—Case Studies and Perspectives.* (Vol.1: 220–315); Part 7: *Analysis and design of sustainable farming systems.* (Vol 1: 316–431); Part 8: *Traditional farming systems in Latin America.* (Vol 2: 432–473)

Altieri, M. (1987) *Agroecology—the scientific basis of alternative agriculture.* Intermediate Technology Publications; London

Altieri, M. A. & Anderson, M. K. (1986) *An ecological basis for the development of alternative agricultural systems for small farmers in the Third World.* American Journal of Alternative Agriculture 1: 30–38

Augstburger, F. (1983) *Agronomic and economic potential of manure in the Bolivian valleys and highlands.* Agriculture, Ecosystems and Environment 10: 335–345

Breton, R. J. G. Le (1986) *A Tale from the Bunduni Woods: Part 1—The Bunduni National Rural Development Project.* Agricultural Administration 22: 79–87

CET (Centro de Educacion y Technologia) (1983) *La huerta campesina organica.* Inst de Estudios y Publicaciones Juan Ignacio Molina, Santiago, Chile. 45pp

Chambers, R. (1983) *Rural Development: Putting the last first.* Longman; London

Chambers, R. and B. P. Ghildyal (1985) *Agricultural research for resource-poor farmers: the farmer-first-and-last model.* Agricultural Administration 20: 1–30

Chambers, R.; Pacey, A. & Thrupp, L. A. (1989) *Farmer First.* Intermediate Technology Publications; London

Chleq, J. L. & Dupriez, H. (1988) *Vanishing Land and Water: Soil and water conservation in dry lands.* Land and Life Series. Macmillan

Conroy, C. & Litvinoff, M. (eds) (1988) *The Greening of Aid.* Earthscan; London

Conway, G. R. (1985) *Agroecosystem analysis.* Agricultural Administration 20: 31–55

Dalzell, H. W. (1981) *An appropriate technology for Indian agriculture.* In: Stonehouse (ed.) *Biological Husbandry.* Butterworths; London

Dalzell, H. W. *et al* (1987) *Soil Management: Compost Production and Use in Tropical and Sub-Tropical Environments.* FAO Soils Bulletin 56. Food and Agriculture Organisation; Rome.

Dumont, R. (1988) *False Starts in Africa.* Earthscan; London

Dupriez, H. & Leener, P. de (1988) *Agriculture in African Rural Communities: Crops and soils.* Land and Life Series. Macmillan

Durno, J. (1989) *Recipes for fish with rice.* New Scientist, 18th Nov

Egger, K. & Martens, B. (1987) *Theory and methods of ecofarming and their realisation in Ruanda, East Africa.* In: Glaeser, 1987

Egger, K. (1988) *Ecofarming in the tropics.* English language IFOAM Bulletin 6: 3–6

Francis, C.; Harwood, R. & Parr, J. (1986) *The potential of regenerative agriculture in the developing world.* American Journal of Alternative Agriculture 1: 65–74

Glaeser, B. (1984) *Ecodevelopment in Tanzania. An empirical contribution on needs, self-sufficiency and environmentally sound agriculture on peasant farms.* Mouton; Berlin

Glaeser, B. (1984) *Ecodevelopment—Concepts, Projects, Strategies.* Pergamon Press; Oxford

Glaeser, B. (ed.) (1987) *The Green Revolution revisited.* Allen & Unwin; London

Gliessman, S. R. *et al* (1981) *The ecological basis for the application of traditional agricultural technology in the management of tropical agro-ecosystems.* Agroecosystems 7: 173–185

Guepin, M. (1987) *A message from Kenya.* Star and Furrow 69: 21–27

Harrison, P. (1987) *The Greening of Africa.* Paladin; London

Harrison, P. (1987) *A green revolution for Africa.* New Scientist 7th May

Harrison, P. (1987) *Trees for Africa.* New Scientist 14th May

Harwood, R. R. (1979) *Small Farm Development. Understanding and improving farming systems in the humid tropics.* Westview Press; Boulder, Colorado

Howard, A. (1943) *An Agricultural Testament.* Oxford University Press

IFOAM (1990) *Agricultural Alternatives and Nutritional Self-Sufficiency.* Proc. 7th IFOAM International Conference. Burkina Faso, West Africa, January 1989. International Federation of Organic Agriculture Movements; Tholey-Theley, West Germany

Jolly, D. (1988) *Labor-intensive production methods in the Chinese commune system: appropriate or inappropriate technology?* In: Allen & van Dusen, 1988. pp 263–272

Joyce, C. (1988) *The tree that caused a riot.* New Scientist 18th Feb

Kay, R. (1987) *New blood for African farms.* New Scientist 29th Oct

King, E. H. (1926) *Farmers of Forty Centuries.* Cape; London

Kotschi, J. *et al* (1989) *Ecofarming in agricultural development.* Tropical Agroecology 2: 132pp. Verlag Josef Markgraf; Weikersheim, West Germany

MacKay, K. T. *et al* (1988) *Rice-fish culture in northeast Thailand: stability and sustainability.* In: Allen & van Dusen, 1988. pp 355–370

MacKenzie, D. (1987) *Can Ethiopia be saved?* New Scientist 24th Sept

MacKenzie, D. (1987) *Ethiopia's hand to the plough.* New Scientist 1st Oct

Mellow, J. W. (1988) *Sustainable agriculture in developing countries.* Environment 30: 6–7

Millis, F. (director) (1989) *The Fight of the Mossi against the Desert.* A video of agricultural development in Burkina Faso produced on the occasion of the 7th IFOAM International Conference, Ouagadougou, January 1989. Available from: F. Millis, rue d'Albanie 21, 1060 Brussels, Belgium

Morales, H. L. (1984) *Chinampas and integrated farms—learning from the rural traditional experience.* pp 188–195. In: F. di Ostri *et al*, (eds.) *Ecology in Practice.* Tycooly Int. Publ. Ltd; Dublin

Neugebauer, B. (1988) *Agricultural development in central Yucatan and its implications for the promotion of intensive, diversified land-use systems.* In: Allen & van Dusen, 1988. pp 297–306

Nickel, K. J. (1988) *Biologischer Landbau im Süden Brasiliens.* German language IFOAM Bulletin 67: 12–16

Olatunji, A. (1989) *The limits of ecological agriculture in Nigeria.* English language IFOAM Bulletin 8: 7–8

Prakash, B. (1988) *Vigyan Shiksha Kendra.* English language IFOAM Bulletin No. 3, 8–9

Pearce, F. (1987) *A watershed for the Third World's irrigators.* New Scientist 7th May

Rottach, P. (Hrsg.) (1986) *Ökologischer Landbau in den Tropen: Ecofarming in Theorie und Praxis.* Müller; Karlsruhe

Sattaur, O. (1987) *Trees for the people.* New Scientist 10th Sept

Semoka, J. M. R. (ed.) (1983) *Proceedings of workshop on resource-efficient farming methods for Tanzania.* Rodale Press; Emmaus, PA

Smith, A. & Finney, C. (1989) *Hunger in West Africa – a thing of the past?* Food Matters 1: 25–32

Soil Association (1984) *Development plans and the hungry people.* Soil Association Quarterly Review, March

Stoll, G. *Natural Crop Protection—based on Local Farm Resources in the Tropics and Subtropics.* Margraf Tropical Scientific Books, Eichendorffstr. 9, 8074 Gaimersheim, West Germany

Sweeney, S. (1984) *The Soil Association and the task of development.* Soil Association Quarterly Review, June

Sweeney, S. (1986) *An organic approach to rural development in India.* Soil Association Quarterly Review, September

Wilson, G. F. & Kang, B. T. (1981) *Developing stable and productive biological cropping systems for the humid tropics.* In: Stonehouse (ed.) *Biological Husbandry.* Butterworths, London

Wolf, E. C. (1986) *Beyond the Green Revolution: New Approaches for Third World Agriculture.* Worldwatch Paper 73

Zhengfang, L. & Gensheng, Z. (1989) *Studies of the design and construction of the Nanjing Guquan ecological project.* English language IFOAM Bulletin 7: 6–8

Organic agriculture and agricultural policy

Bateman, D. I. & Lampkin, N. (1986). *Economic Implications of a Shift to Organic Agriculture in Britain.* Agricultural Administration 22: 89–104

Beaton, D. (1989) *Set-aside speeds switch to an organic regime.* Farmers Weekly 10th March p. 58

BOF/OGA/SA (1988) *The Case for Organic Agriculture.* British Organic Farmers, Organic Growers Association, Soil Association; Bristol

BOF/OGA/SA (1989) *20% of Britain Organic by the year 2000. A Campaign Policy Document.* British Organic Farmers, Organic Growers Association, Soil Association; Bristol

Brombacher, J. & Hamm, U. (1990) *Ausgaben für eine Ernährung mit Lebensmitteln aus alternativem Landbau.* Lebendige Erde 2/90: 90–101

EFRC et al (1989) *Nitrate Sensitive Areas Scheme—a response to the consultation document of the Agriculture Departments of England and Wales.* Elm Farm Research Centre, Newbury

European Commission (1988a) *Environment and Agriculture.* Commission Communication 338 (8/6/88). Commission of the European Communities; Brussels

European Community (1988b) *Commission Regulation (EEC) No. 4115/88.* Official Journal of the European Community L361: 13–18 (29/12/88)

Harding, D. J. (1987) *Agricultural surpluses? Environmental implications of changes in farming policy and practice in the UK.* Institute of Biology

Henley Centre (1989) *The Market for Organic Food.* Report commissioned by the Development Board for Rural Wales; Newtown

Jenkins, T. N. (1989) *Paying Farmers to Conserve: an environmental approach to agricultural support policy*. Council for the Protection of Rural England

Lampkin, N. (1989) *Organic farming—a policy option for UK agriculture?* Paper presented at UK Agricultural Economics Society conference, Aberystwyth, April 1989

Lampkin, N.; Jones, W.; Mann, M. & Midmore, P. (1987). *Proposal for an Integrated Rural Development Programme Focussing on Organic Production in the Teifi Valley, Dyfed*. University College of Wales; Aberystwyth

Lampkin, N. & Midmore, P. (1988). *Organic Production as an Alternative to Set-Aside. A response to the Agricultural Departments' Consultative Document on Extensification*. University College of Wales; Aberystwyth

Midmore, P. & Lampkin, N. (1988). *Organic Farming as an Alternative to Set-Aside and an Option for Extensification*. Paper presented at the European Association of Agricultural Economists Seminar on Economic Aspects of Environmental Regulation in Agriculture, Copenhagen, November 1988

National Research Council (1989) *Alternative Agriculture*. National Academy Press; Washington DC

Russell, N. P. & Power, A. P. (1989) *UK government expenditure implications of changes in agricultural output under the common agricultural policy*. Journal of Agricultural Economics 40: 32–39

Weinschenk, G. & Dabbert, S. (1988) *Decrease of the Intensity of the Use of Natural Resources as a Way to Reduce Surplus Production*. Paper presented at the European Association of Agricultural Economists Seminar on Economic Aspects of Environmental Regulation in Agriculture, Copenhagen, November 1988

Williamson, C. J. & Black, W. J. M. (1988) *Index of Research and Development Relevant to Organic Systems of Food Production*. Scottish Agricultural Colleges and Scottish Agricultural Research Institutes; Invergowrie

Woodward, L. (1990) *Turning Britain organic*. Living Earth 170 (April/June): 18–20

Recent publications

Environment

Redman, M. (1992). *Organic farming and the countryside*. British Organic Farmers; Bristol.

Greenpeace. (1992). *Green fields—grey future. EC agricultural policy at the crossroads*. Greenpeace; Amsterdam.

Wilson, J. (1993) The BTO birds and organic farming project: 1 year on. *New Farmer and Grower*, 38: 31–33.

Food Quality

Woodward, L.; Stolton, S. and Dudley, N. (eds.) (1992) *Food quality: concepts and methodology*. Elm Farm Research Centre; Newbury.

Policy

EC. (1992). Council Regulation (EEC) No 2078/92 on agricultural production methods compatible with the requirements of the protection of the environment and the maintenance of the countryside. *Official Journal of the European Communities*. **L215**(30/7/92):85–90.

Lampkin, N. H. and S. Padel (eds) (1994). *The economics of organic farming—an international perspective*. CAB International, Wallingford.

Standards for Organic Agriculture

ORGANIC STANDARDS IN BRITAIN

To the outsider, the issue of who decides what is 'organic' and what is not may at times seem extraordinarily complex. This is a pity, because what the consumer and the producer require is clarity and a straightforward, understandable approach.

In Britain, there are three main sets of standards currently in operation. Of these, the Soil Association's Symbol scheme used as the basis for this book is the most widely adopted. Their production standards follow in full in this appendix. These standards have been adopted by the Irish Organic Farmers and Growers Association.

The Bio-Dynamic Agricultural Association operates the Demeter and Biodyn standards for produce from biodynamic systems (see Appendix 2), the Biodyn symbol being used for produce during the conversion period. These standards are in some respects more rigorous than those of the Soil Association, but few commercial producers in Britain adhere to them, although both biodynamic symbols are widely used in mainland Europe.

Finally, the Organic Farmers and Growers Ltd marketing co-operative has established its own standards, including one for produce during the conversion or transition phase.

There are, however, a plethora of other standards which are sometimes used. These include standards operated by commercial interests such as Farm Verified Organic and which apply primarily to imported produce, as well as 'halfway house' standards such as Conservation Grade operated by the Guild of Conservation Food Producers which have little, if anything, to do with organically produced food.

In an attempt to reduce confusion and to unify standards, the British Organic Standards Committee (BOSC) was established in 1981. The committee consists of 12 representatives from producer, consumer, research and related groups within the organic movement. Its role is to develop and update standards for organic agriculture in Britain. These standards are the ones adopted by the Soil Association, but not by Organic Farmers and Growers Ltd although they were also represented on the Committee.

In 1987, the Government established the United Kingdom Register of Organic Food Standards (UKROFS) under the banner of Food From Britain, partly as a response to a forthcoming European Community Regulation on organic food standards and partly as a further attempt to unify standards in

Britain and to overcome continuing disagreements between the Soil Association and Organic Farmers and Growers Ltd. The UKROFS standards were published in May 1989 and efforts are now concentrating on implementing the UKROFS scheme. Both the Soil Association and Organic Farmers and Growers Ltd. have applied for registration with UKROFS and are expected to be approved by UKROFS during 1990.

ORGANIC STANDARDS INTERNATIONALLY

IFOAM Standards

The International Federation of Organic Agriculture Movements (IFOAM) has member organisations in over 50 countries. The IFOAM Technical Committee is responsible for a Standards Document which is used as a basis for national standards operated by member organisations throughout the world. The Soil Association's standards are based on this document, the most recent and wide-ranging revision of which was approved at the IFOAM General Assembly in Burkina Faso in January 1989. At the time of going to press, further changes were being proposed for the IFOAM General Assembly in Budapest in August 1990.

Individual producers cannot use IFOAM Standards, because IFOAM does not operate an inspection or regulatory procedure. They must use the standards operated by a national organisation in the country in which they farm. For similar reasons, produce should not be traded as 'conforming to IFOAM standards' as there is no inspection procedure to ensure that this is the case.

Recognising the increasing international trade in organically produced food, IFOAM has implemented an international evaluation survey of organic standards, policing both procedures and the status of the controlling organisations in member countries. Initially, only the major exporting countries are being assessed. Evaluation of the UK standards took place in May 1987. A directory of their findings is made available to the member organisations which have been evaluated. This enables organisations like the Soil Association to determine whether imported produce meets domestic standards.

The European Community

The European Commission is planning to introduce a regulation which will eventually apply to the production and sale of all organically grown produce in member states. When the regulation becomes law in 1990/91 it will become illegal to sell produce as organically grown unless it carries a quality mark authorised by a member state government. Draft standards were published in December 1989 which attracted widespread criticism, but these have since been modified to accommodate most of the concerns expressed during discussions between the Commission, member state governments and representatives from the organic movement in member states.

Standards for Organic Food and Farming

The Soil Association's Standards for Organic Food and Farming are subdivided into three parts: production (for farmers and growers); food processing, packing and distribution; and industrial products (farm inputs and other non-food products). The production standards (sections 1–6) as at April 1993 are reproduced in full below. The processing and industrial standards, and details of inspection procedures, can be obtained directly from the Soil Association.

THE SOIL ASSOCIATION
SYMBOL SCHEME

86 COLSTON STREET
BRISTOL BS1 5BB
TEL. 0272 290661

SECTION 1

INTRODUCTION

1.1 EUROPEAN & UNITED KINGDOM STANDARDS

THE EUROPEAN COMMISSION

1.101 The European Council Regulation (EEC) No 2092/91 came into force on January 1 1993. It applies to unprocessed agricultural crop products, to products intended for human consumption composed essentially of one or more ingredients of plant origin, and, it introduces specific rules for the production, inspection and labelling of such product.

1.102 The Regulation requires that operators who produce, prepare or import from third country, products specified by the Regulation for the purposes of marketing them, must notify the activity to the competent authority of the Member State in which the activity is carried out (UKROFS) in the case of the UK and they must submit the undertaking to the specified inspection system.

1.103 At this stage, the Regulation does not cover animal production, unprocessed animal products and products intended for human consumption composed essentially of ingredients of animal origin. Proposals were to be submitted by the EC Commission before July 1st 1992, but it is now expected that a Regulation covering these activities will not be published before 1995.

UNITED KINGDOM REGISTER OF ORGANIC FOOD STANDARDS (UKROFS)

1.104   For the purposes of administering the Regulation in the UK, UKROFS has been designated as:
   (a)   The Authority responsible for the reception of notification of Organic activity and for making available to interested parties an updated list containing the names and addresses of operators subjected to the inspection systems;
   (b)   The Inspection Authority responsible for the operation of the inspection system defined in the Regulation (EEC) No 2092/91;
   (c)   The Authority responsible for the approval and supervision of private inspection bodies in accordance with the relevant provisions defined in the regulation.

THE SOIL ASSOCIATION

1.105   The Soil Association is a registered charity which was founded in 1946 by Lady Eve Balfour and others, following the publication of her book 'The Living Soil'.

1.106   The Soil Association exists 'to research, develop and promote sustainable relationships between the soil, plants, animals, people and the biosphere, in order to produce healthy food and other products while protecting and enhancing the environment'.

1.107   The Soil Association's principal activities fall into two areas:
   (a)   Education and Research including detailed research into all aspects of agriculture and related environmental and health issues;
   (b)   Certification of Organic food production and processing through the Symbol Scheme.

1.2   THE SYMBOL SCHEME

1.201   The Symbol Scheme is a Certification Scheme for licensing Organic food production and associated non-food products. The Soil Association launched the Scheme in 1973 to provide an ethical and well regulated basis for establishing the integrity of Organic production systems and food products. It involves the independent inspection and certification of Organic food from its production through the processing and distribution chain to the consumer.

1.202   The Symbol Scheme is registered with the United Kingdom Register of Organic Food Standards (UKROFS) as an Approved Organic Sector Body (Registration No 01051194) and is licensed to certify Organic food production and processing under the European Commission Regulation (EEC) No 2092/91.

1.203   The Symbol Scheme inspection and operating procedures conform to the UKROFS Inspection Requirements and Precautionary Measures and the Symbol Inspectors are registered as UKROFS Inspectors.

1.204   The Symbol Scheme is managed by the Soil Association Organic Marketing Company Limited, a trading company wholly owned by the charity. It is funded by fees levied on Registered Symbol holders.

1.205   The Symbol is a nationally and internationally respected quality mark for Organic food products. It is widely used by all sectors of the Organic industry, and is trusted by consumers, who recognise the leading role played by the Soil Association in promoting Organic farming and the development of Organic standards.

AREAS OF OPERATION

1.206   The Symbol Scheme currently operates in seven areas:-
   (a)   Food Products:
      (i)   Organic Food Production – Symbol Standard food production,
      (ii)   In-Conversion Food Production – production from land in conversion to Symbol Standard,
      (iii) Food Processing, Packing and Distribution – all operations between the primary production and the purchase by the consumer;
      (iv)   International Verification – foreign certification bodies, processors and producers.
   (b)   Non-Food Products (lying outside of the scope of UKROFS but operating

under and complying with the EC Regulation and the UKROFS Standards):

(i)   Industrial Products – farm inputs and other non-food products.
(ii)   Forestry Scheme – sustainable timber production, transport and timber products.
(iii) Education – evaluation of courses in Organic agriculture and horticulture.

1.207   Each of these schemes is based on common production standards as set out in these Standards for Organic Food and Farming and common operating procedures and licensing agreements as set out in the Symbol Scheme Operating Manual. Additional standards have been developed to cover specialist product areas.

1.208   Annual licences are issued to individuals or companies who have been inspected and are approved by the Certification Committee. This licence authorises the Holder to use the Symbol on approved products. Licensees are bound by formal contracts.

1.3   STANDARDS FOR ORGANIC FOOD AND FARMING

1.301   This document defines Organic farming systems, and lays down criteria which must be met and maintained when food products are described as Organic.

1.302   The Standards for Organic Food and Farming are based on guidelines originally established by the International Federation of Organic Agriculture Movements (IFOAM).

1.303   They comply with both the European Community Council Regulation (EEC) No 2092/91 and the United Kingdom Register of Organic Food Standards (UKROFS) – Standards For Organic Food Production as a minimum.

1.304   *The UKROFS Standards For Organic Food Production are indicated by the use of this italic type face.*

1.305   Additional Standards required by the Symbol Scheme and introductions etc. are indicated in this non-italic type face.

1.306   The Production Standards outline the principles and practices of Organic agricultural systems which, within the economic constraints and technology of a particular time, promote:

(a)   The production of high levels of nutritious food;
(b)   The use of management practices which sustain soil health and fertility;
(c)   High standards of animal welfare and contentment;
(d)   The lowest practical levels of environmental pollution;
(e)   Minimal dependence on non-renewable forms of energy and the burning of fossil fuels;
(f)   Enhancement of the landscape, wild life and wild life habitat.

1.307   The Processing Standards outline general criteria for plant and equipment, hygiene, record keeping, labelling and permitted practices and ingredients, and cover the various categories of Organic foods that may be registered and the processing operations that must be registered under the EC Regulation.

1.308   The Standards for Organic Food and Farming were originally formulated and are now kept under constant review by technical sub-committees composed of practising farmers and growers, food processors and manufacturers, input manufacturers, researchers and scientists, ADAS personnel, veterinary surgeons, other professionals with specialist knowledge of Organic agriculture, conservation bodies and representatives of consumer interests. All committee members and contributing technical experts offer their services in an honorary capacity.

1.309   The recent advances in the understanding of Organic systems, the technology of production, and the development of the UKROFS Standards and the EC Regulation (EEC) No 2092/91 mean that it has been necessary to produce this fifth edition of the Standards.

1.310   The Soil Association acknowledges with gratitude the contributions made by the many committee members and corresponding technical experts in the development and drafting of these Standards.

1.4   DEFINITIONS OF TERMS USED IN THE TEXT

AGRICULTURAL PRODUCTS

1.401   *Unprocessed agricultural and horticultural crop products, animals and unprocessed animal products.*

APPLICANT

1.402   *A company, organisation or individual who has applied for but not yet been granted a Certificate of Registration.*

APPROVED BODY/ORGANIC SECTOR BODY

1.403   *A Certification Organisation for Organic production and products holding a valid Certificate of Registration, whose scheme has been approved by the national Certifying Authority.*

APPROVED PRODUCER/OPERATOR

1.404   *A business enterprise or person holding a valid Certificate of Registration from an Approved Organic Sector Body for:*
   (a)   *The production of Organically produced agricultural products; or*
   (b)   *The processing and/or manufacturing and/or preserving and/or packaging of Organically produced agricultural products.*

CERTIFYING AUTHORITY

1.405   The Board of the United Kingdom Register of Organic Food Standards (UKROFS) is the Certifying Authority for the United Kingdom.

CERTIFICATION COMMITTEE

1.406   The committee responsible for the registration and certification process in the Symbol Scheme.

CERTIFICATE OF REGISTRATION

1.407   *A certificate issued under a separate serial number by either:*
   (a)   *The Certifying Authority to an approved Organic Sector Body recognising that their Certification Scheme conforms to the requirements of the UKROFS Certification Scheme; or*
   (b)   *The Certifying Authority or an Approved Organic Sector Body to Approved Producers recognising their operational procedures and practices for a given range of products or land have been assessed and are in accordance with the Standards for Organic Food and Farming.*

CERTIFICATION SCHEME

1.408   *A scheme designed to certify conformity with defined operational procedures and practices that meet the Certifying Authority's Standards.*

HOLDING

1.409   An agricultural holding as defined by the Statutory Body with its own Agricultural Holding Number.

IN-CONVERSION

1.410   Production using permitted techniques and materials as defined in these Standards from land being converted from non-Organic to Symbol Standard and after:-
   (a)   *An Inspection of the holding, approval of the conversion plan and registration with the Conversion Scheme (or equivalent); and*
   (b)   *A conversion period of at least 12 months has elapsed between the last use of materials prohibited in these Standards and the harvest.*

INGREDIENTS

1.411   *Materials of plant or animal origin and substances, including additives used in the processing of Organically produced products that are still present, albeit in a modified form, in the final product.*

INSPECTOR

1.412   A person contracted by the Symbol Scheme to inspect farms and operations and who meets all the criteria for experience and/or qualifications and who holds a valid Certificate of Registration as an UKROFS Approved Inspector.

LABELLING/INDICATIONS

1.413 *Any words, descriptions, trade marks, brand names, pictorial matter or Symbols appearing on any packaging, document, notice, label, board or collar accompanying or referring to a product.*

MARKETING

1.414 *Holding or displaying for sale, offering for sale, selling, delivering or placing on the market in any other form.*

NON-ORGANIC

1.415 All production not registered with an approved Organic Certification Body as In-Conversion or Symbol/Organic Standard.

OPERATING MANUAL

1.416 *The document that contains the details of the specific operating procedures and requirements of the Symbol Scheme.*

ORGANIC/ORGANICALLY PRODUCED

1.417 That food which has been produced in accordance with the EC Regulation No. 2092/91 (and registered with a Certification Scheme such as the Symbol Scheme.

OPERATOR

1.418 *A business enterprise or person who produces, processes or imports from a third country Organically produced products with a view to marketing them or who markets such products.*

PERMITTED

1.419 Practices and materials permitted for use in Symbol and In-conversion Standard production - subject to any qualifications listed.

PROCESSING

1.420 *The operations of processing, manufacturing, preserving and packaging of agricultural products.*

PRODUCER

1.421 *A business enterprise or person managing a holding producing and marketing Organically produced agricultural and horticultural produce.*

PRODUCTION

1.422 *The operations involved in producing agricultural products in a state in which they are normally produced on a farm.*

PROHIBITED

1.423 Practices or materials not permitted for Symbol Standard production under any circumstances.

RECOMMENDED

1.424 Permitted practices or materials fully recommended for Symbol Standard production.

RESTRICTED

1.425 Regulated practices and materials, the need for which must be recognised and approved by the Certification Committee before they can be used.

STANDARDS

1.426 The Standards for Organic Food and Farming as defined in this document.

SYMBOL DEPARTMENT

1.427 The department of the Soil Association Organic Marketing Company Limited responsible for administering the Symbol Scheme.

SYMBOL STANDARD

1.428 Production and/or products that fully comply with the Standards for Organic Food and Farming and are registered with the Symbol Scheme.

THE SYMBOL

1.429 The logo of the Soil Association Symbol Scheme, which may only be used by Registered Producers and Operators holding valid Certificates of Registration.

UNIT

1.430 The components of a holding, including the Organic units as specified in paragraph 2.101 and the non-Organic units as specified in paragraph 2.109, or the premises of a processing operation in which the processing, packing or storage of Organic foodstuffs takes place.

SECTION 2

PRECAUTIONARY MEASURES

2.1 INSPECTION REQUIREMENTS AND DOCUMENTATION

UNITS PRODUCING CROPS AND/OR LIVESTOCK

2.101   *Organic production must take place on clearly defined units of land such that the production and storage areas are clearly separate from those of any other unit not producing in accordance with these Standards.*

ON FARM PROCESSING

2.102   *Processing or packing operations may take place on the holding as part of the licensed production process where the activities are limited to processing or packing their own agricultural products.*

2.103   Where processing or packing operations include bought-in products the operation must be separately registered with the On-Farm section of the Processing and Packing and Distribution Scheme.

STORAGE OF NON-PERMITTED MATERIALS

2.104   *Storage on the unit of input products other than those compatible with Sections 3, 4 & 5 of these Standards is not permitted.*

APPLICATION DOCUMENTS

2.105   Applicants must lodge with the Symbol Department a document setting out:

     (a)   *A full and precise description of the organic unit showing the land areas, the production and storage premises and, where applicable, premises where packaging and/or processing takes place;*

     (b)   *The date of the last application on the land areas concerned of products whose use is not compatible with those listed as Permitted or Restricted in Sections 3, 4 & 5 of these Standards.*

RECORDS

2.106   *Precise and up to date records must be kept to enable the Inspectorate to trace:*

     (a)   *The origin, nature and quantities of all materials bought-in and the use of such materials;*

     (b)   *The nature, quantities and consignees of all agricultural products sold. Quantities sold directly to the final consumer shall be accounted for on a daily basis.*

2.107   *Where the unit itself processes its own agricultural produce the records must contain information regarding:*

     (a)   *The origin, nature and quantities of non-agriculturally produced agricultural products, non-agricultural ingredients and processing aids which have been delivered to the unit;*

     (b)   *The composition of the organically produced products.*

COMPLIANCE

2.108   *When the inspection arrangements are first implemented the producer and the Inspectorate must draw up details of all the practical measures required at the level of the organic unit to ensure compliance with these Standards. The description of these measures must be contained in the inspection report and countersigned by the person responsible for the unit.*

ANNUAL RETURN

2.109 *Each year before the date indicated by the Certification Committee, Approved Producers and those in the process of conversion from conventional to organic production must notify the Symbol Department of their schedule of production of crop products, giving a breakdown by land area and/or, as appropriate, details of their livestock production.*

PRODUCERS WITH ORGANIC, IN-CONVERSION AND NON-ORGANIC PRODUCTION UNITS

2.110 *Where an operator runs several production units in the same area, the unit(s) producing products not covered by these Standards must be subject to regular inspection and the Operator must:*

(a) *Lodge with the Symbol Department a document setting out a full description of the non-organic unit(s) showing the land areas, the production and storage premises and, where applicable, the premises where packaging and/or processing operations take place;*

(b) *Each year before the date indicated by the Certification Committee notify the Symbol Department of the schedule of production of non-organically produced crop products giving a break-down by land area and/or, as appropriate, details of livestock production;*

(c) *Keep written records which enable the Inspector to trace:*

　(i) *The origin, nature and quantities of all materials brought-in and the use of such materials,*

　(ii) *The nature, quantities and consignees of all agricultural products sold.*

　(iii) *Give the Symbol Inspectorate access to the unit(s) for inspection purposes defined in paragraph 2.116 of these Standards.*

　(iv) *That plants of the same variety as those produced on the organic unit are not produced on the non-organic unit(s).*

PROCESSING UNITS

2.111 *Wherever possible, the processing of organically produced products should take place in a unit which is clearly separate from any other unit where non-organically produced products are processed.*

OPERATING PROCEDURES

2.112 *In cases where non-organically produced products are also processed, packaged or stored in the unit concerned:*

(a) *The storage areas used for the organic production, before, during and after processing must be designated for the purpose, clearly identified and must be separate from those used for non-organic production;*

(b) *Operations must be carried out continuously until the complete production run has been dealt with, separated by time and effective cleaning procedures from similar operations performed on non-organic production;*

(c) *If such operations are not carried out frequently they must be announced in advance with a deadline agreed with the Symbol Department;*

(d) *All necessary measures must be taken to ensure identification of lots and to avoid mixtures with non-organically produced products.*

DOCUMENTS

2.113 *Applicants must lodge with the Approved Body a document setting out a full description of the unit showing the facilities used for the processing, packaging and storage of agricultural products before and after the operations concerning them.*

RECORDS

2.114 *Precise and up-to-date records must be kept to enable the authorised Inspector to trace:*

(a) *The origin, nature and quantities of organically produced agricultural products, which have been delivered to the unit;*

(b) *The origin, nature and quantities of non-organically produced agricultural products, non-agricultural ingredients and processing aids which have been delivered to the unit;*

(c) *The composition of the organically produced products;*

(d) *The nature, quantities and consignees of organically produced/processed products which have left the unit.*

2.115 *On receipt of the organically produced products the operator shall check the closing of the packaging or container and the presence of the indications referred to in paragraph 2.121 below. The result of this verification shall be explicitly mentioned in the records referred to in paragraph 2.114 above. Where the check leaves any doubt that the product concerned came from an operator subject to the inspection system provided for in Article 9 of the Regulation (EEC) No. 2092/91 it may only be put into processing or packaging after elimination of that doubt.*

COMPLIANCE

2.116　*When the inspection arrangements are first implemented the producer and the Inspectorate must draw up details of all the practical measures required at the level of the organic unit to ensure compliance with these Standards. The description of these measures must be contained in the inspection report and countersigned by the person responsible for the unit.*

2.117　For the Standards for Processing, Packing and Distribution, see Sections 7, 8 & 9.

GENERAL PRECAUTIONARY MEASURES

2.118　*Applicants, Operators, Approved Producers and those in the process of conversion from conventional to organic production must sign a contract with the Soil Association Organic Marketing Company Limited agreeing to carry out operations in accordance with these Standards and to accept, in the event of infringements, the implementation of measures referred to below:*

　　(a)　*Where an irregularity is found, at the discretion of the Certification Committee, all reference to organic production shall be withdrawn from the crop, animals or production run affected by the irregularity concerned;*

　　(b)　*Where a manifest infringement or infringement with prolonged long term effects is found, the operator concerned shall be prohibited from marketing products under an organic designation for a period of time determined by the Certification Committee and ratified by the Certifying Authority.*

ANNUAL INSPECTIONS

2.119　*Apart from unannounced inspection visits, the Inspector must make a full physical inspection, at least once a year, of the organic unit. Samples may be taken for the detection of substances not authorised under these Standards. However, such samples must be taken where the use of unauthorised products is suspected. An inspection report must be drawn up after each visit and countersigned by the responsible person of the unit.*

2.120　*Producers must give the Inspector, for inspection purposes, access to the unit as well as to the records and relevant supporting documents. The producer must provide the Inspector with any information deemed necessary for the purposes of inspection.*

TRANSPORT

2.121　*Organically produced products which are not in their packaging for the end consumer may be transported to other units only in appropriate packaging or containers closed in a manner which would prevent substitution of the content and provided with a label stating, without prejudice to any other indications required by law:*

　　(a)　*The name and address of the person responsible for the production or preparation of the product;*

　　(b)　*The name of the product;*

　　(c)　*That the product is covered by the inspection arrangements laid down in the Regulation (EEC) No. 2092/91 (ie. by giving the name and address of the Symbol Scheme).*

RESTRICTED PRODUCTS OR PRACTICES

2.122　The need for use must be recognised and approved by the Certification Committee before materials or practices in the Restricted Category are used.

2.2　LABELLING

ORGANIC PRODUCE

2.201　*The labelling and advertising of unprocessed agricultural crop products, animals and unprocessed animal products may only refer to organic production methods where:*

　　(a)　*Such indications show clearly that they relate to a method of agricultural production;*

(b)   *The product was produced in accordance with these Standards by an Approved Producer holding a valid Certificate of Registration or came from another source approved by the Certifying Authority.*

IN CONVERSION PRODUCE

2.202   *During a transitional period ending on 1st July 1994, indications referring to conversion to organic production methods may be given on the labelling and in the labelling of unprocessed agricultural and crop products providing that:*

(a)   *Such indications show clearly that they relate to a method of agricultural production;*

(b)   *The product was produced in accordance with these Standards (or in the case of imported product the equivalent thereof) with the exception of the length of the conversion period;*

(c)   *A conversion period of at least 12 months has elapsed before the harvest has been complied with;*

(d)   *The product was produced or imported by an Operator who is subject to Inspection and Registration under the Conversion Scheme.*

(e)   *The indications concerned do not mislead the purchaser of the product regarding its difference from products which satisfy all the requirements of these Standards.*

(f)   *Compliance with the conditions laid down in c & d has been duly checked by the Inspector and the Certification Committee..*

GENERAL LABELLING REQUIREMENTS

2.203   *All product labels must clearly and accurately describe the product and comply with all relevant legislation.*

2.204   *Where the European logo 'Organic Farming – EEC Control System' is used on the labelling of a product, no claims may be made on the label or advertising material that suggests to the purchaser that the indication of organic production methods constitutes a guarantee of superior organoleptic, nutritional or salubrious quality.*

2.205   *The Soil Association Symbol or name, used direct or implied, indicating that the production system complies with these Standards must only be used on products that have been produced to these Standards by Approved Producers registered with the Soil Association Symbol Scheme.*

2.206   *Where the Symbol is used on a product it may only be used in association with:*

(a)   *The producer's business name as shown on the Approved Producer's Certificate of Registration or the brand mark of the business; or*

(b)   *The purchaser's business name or brand mark providing there is a means by which the producer's name and address can be ascertained by identification marks on the product packaging or label or by means of appropriate documentation.*

2.207   For additional labelling Standards relating to processed food, see Section 7.7.

2.3   RECORD KEEPING

PROCESSORS

2.301   For additional Standards relating to processing, packing and distribution operations, see section 7.5.

PRODUCERS

2.302   *Approved producers and those in the process of conversion to organic production must keep accurate records of their production activities and these must be made available for examination when inspections are carried out by the Symbol Inspectorate.*

2.303   Both physical and financial records of the entire holding, including Symbol, In-Conversion and non-organic sections must be kept as detailed in paragraphs 2.306 to 2.315 below.

2.304   *The records must be sufficiently comprehensive to demonstrate that these Standards have been observed and they must be retained for a period of not less than 3 years.*

2.305   A failure to keep the required records means that the production process cannot be inspected and policed to the satisfaction of the Certification Committee and may, at the discretion of the Committee, result in an applicant not being registered with the Symbol scheme or the licence being withdrawn from Registered Producers.

2.306   Standardised record forms are supplied to all Symbol Registered Producers. Their use is optional but if they are not used, equivalent records must be kept.

PRODUCTION RECORDS

INPUT RECORDS

2.307 Details of the origin, nature and quantities of all materials brought-in and the use of such materials.

OUTPUT RECORDS

2.308 Details of the nature, quantities and consignees of all agricultural products sold. Quantities sold directly to the final customer must be accounted for on a daily basis.

STOCK LEVEL RECORDS

2.309 As appropriate the stock levels for raw materials and finished products.

CROP RECORDS

2.310 Where applicable, the following crop records must be recorded:
  (a) For land In-conversion, the previous treatments over the last three crop years with agro-chemicals, artificial fertilisers and materials not permitted in these standards, by field or area;
  (b) The crop rotational plan or plans;
  (c) The cropping plan by field or area;
  (d) The cropping history of all the fields including crops and yields;
  (e) The source, type, composting treatments and rate of usage of organic materials used for fertilisation and soil conditioning, by field or area;
  (f) The source, type and rate of usage of mineral fertilisers, by field or area;
  (g) The source, type and usage of products used for pest and disease control;
  (h) The source and type of seeds and/or transplants used (including any chemical treatments during propagation).

LIVESTOCK RECORDS

LIVESTOCK MOVEMENTS

2.311 The livestock movement book must be kept up-to-date and complete.

2.312 For brought-in stock:
  (a) Species, source and numbers of brought-in stock;
  (b) Organic status, identification and ages;
  (c) Veterinary history;
  (d) Quarantine measures taken;
  (e) Conversion time by animal or group prior to full organic status.

2.313 For animals sold:
  (a) Species, destinations and numbers of stock sold;
  (b) Organic status, identification and ages.

VETERINARY TREATMENTS

2.314 For any use of a veterinary medicinal product, the following details must be recorded:
  (a) Date of purchase;
  (b) Name of the product and the quantity purchased;
  (c) Supplier of the product;
  (d) Identity of the animals treated;
  (e) Number treated;
  (f) Date treatment started;
  (g) Date treatment finished;
  (h) Total quantity of product used;
  (i) Length of the withdrawal period in number of days;
  (j) Earliest date for sale of the animal or products;
  (h) Name of the person who administered the product.

FEEDSTUFFS

2.315 Details of the following:
  (a) Constituent ingredients and organic status of the feed (Symbol, In-Conversion and non-organic) for each class of stock;

(b)   Proportion of the constituents to the total feed on a dry matter basis;
(c)   Sources of the constituent parts (including brought-in feeds and farm grown feeds).

ACCOUNTS
2.316   The following accounting records where applicable:
    (a)   Sales & purchase Invoices;
    (b)   VAT accounts.

SECTION 3

GENERAL STANDARDS FOR ORGANIC CROP HUSBANDRY

3.1   PRINCIPLES OF ORGANIC PRODUCTION

3.101   Organic (biological) agricultural and horticultural systems are designed to produce food of optimum quality and quantity. The principles of Organic production have been defined by the International Federation of Organic Agricultural Movements (IFOAM).

3.102   The principles and methods employed result in practices which:
- Coexist with, rather than dominate, natural systems;
- Sustain or build soil fertility;
- Minimise pollution and damage to the environment;
- Minimise the use of non-renewable resources;
- Ensure the ethical treatment of animals;
- Protect and enhance the farm environment with particular regard to conservation and wildfire;
- Consider the wider social and ecological impact of agricultural systems.

3.103   The basic characteristics of Organic systems are:
- *The enhancement of biological cycles, involving micro-organisms, soil fauna, plants and animals;*
- Sustainable crop rotations;
- The extensive and rational use of manure and vegetable wastes;
- The use of appropriate cultivation techniques;
- The avoidance of fertilisers in the form of soluble mineral salts;
- The prohibition of agro-chemical pesticides;
- The use of animal husbandry techniques which meet the animal's physiological, behavioural and health needs.

3.104   All food production causes some disruption to the natural environment. However, Organic farming minimises this disruption not only due to the prohibition of synthetic pesticides and soluble fertilisers, but also because the maintenance of ecological diversity within and around cropped land is an essential component of the Organic system. Organic farmers are expected to manage habitats such as banks, hedges, ponds, species-rich pastures, areas of poor drainage and scrubland in accordance with their wildlife value as an integral part of the Symbol Scheme.

3.2 CONVERSION TO ORGANIC PRODUCTION

3.201 Conversion from conventional to Organic production must be effected using permitted materials and practices as defined in these Standards and in accordance with a progressive production plan or conversion plan designed to:

(a) Convert physically separate and identifiable units of land large enough to permit Organic production to be developed and sustained;

(b) Result in a financially separate enterprise with its own accounts and record keeping system complying with the record keeping requirements (see section 2.3).

3.202 Produce may only be sold or classified as In-conversion after:

(a) A production plan or conversion plan has been approved by the Certification Committee.

(b) The land and production has been inspected and registered with the Conversion Scheme;

(c) *A conversion period of at least 12 months from the last use of materials, other than those permitted in these Standards, to the harvest has been complied with.*

3.203 Land and production may be eligible for Registration and products may only be sold under a description that indicates or implies that the product has been produced in accordance with these Standards, after a period of conversion of at least two years. This normally means:

(a) *For arable and horticultural crops – 24 months from the last use of any materials not permitted in these Standards before sowing or planting the Organic crop;*

(b) *For grassland – 24 months from the last use of any materials not permitted in these Standards until the grass is used for Organic grazing or the production of Organic hay or silage;*

(c) *For perennial crops (excluding grassland) – 36 months from the last use of any materials not permitted in Section 3 of these Standards until the harvest of the first Organic crop.*

3.204 The Certification Committee may however, with the approval of the Certifying Authority, decide in certain cases to extend or reduce the conversion period having regard to the previous use of the land area in question.

3.205 Where a producer has a mixture of land designated as Non-organic, In-conversion or Organic and the same crop is to be grown on two or more of these (known as parallel production) a different and identifiable variety must be grown on each differently designated area of land.

3.206 Land contaminated by environmental pollution (e.g. from factories, traffic, sewage sludge) or by residual pesticides may render the holding ineligible for Organic status or require a longer conversion period, at the discretion of the Certification Committee.

3.207 A plan for the reduction of routine use of veterinary medicinal products and an outline of general disease control strategy, including the treatment of 'known farm problems', should be drawn up in consultation with a veterinary surgeon who may seek guidance from a consulting veterinary surgeon recommended by the Symbol Department. Conversion periods for livestock are given in Section 6 of these Standards.

3.208 Where the land was previously under exploitative cropping, the conversion programme must begin with a fertility-building phase.

3.209 Prohibited

(1) The use of materials and practices not authorised in these Standards.

(2) The classification of production (including grazing and forage) as In-conversion which does not comply with paragraph 3.202 of these Standards.

(3) Plants of the same variety produced simultaneously on Organic, In-conversion or Non-organic land

REGISTRATION WITH THE CONVERSION SCHEME

3.210 All land and production being converted to full Organic status or from which the products are intended to be used as or marketed as In-conversion must be registered with the Conversion Scheme. This includes abandoned land, land with no records of inputs or land from which the Certificate of Registration has been withdrawn.

THE PRODUCTION PLAN (CONVERSION PLAN)

3.211 Applicants for the Conversion Scheme must supply a conversion plan with their application which will cover the period of a complete rotation and include the following information where applicable:

(a) The soil management programme (see section 3.3);

(b) Cropping plans and proposed crop rotations (see section 3.4);

(c) The programme for the supply of nutrients to the plants (see section 3.6 & 3.7);

(d) The programme for the control of weeds and pest and diseases (see section 3.8 & 3.9);

(e) Sections allocated to each main enterprise (e.g. cereals, vegetables) (see sections 4.3 to 4.5);

(f) Grazing practices and grassland management that are an integral part of the crop rotation and which seek to eliminate parasites of livestock (see section 4.4);

(g) Environmental conservation measures (see section 4.1);

(h) Practices for livestock conversion, welfare and housing (see sections 5.2 to 5.5);

(i) Feeding regimes for livestock (see section 5.6 & 5.7);

(j) Appropriate disease control programmes drawn up in consultation with a veterinary surgeon (see section 5.7);

(k) The results of a recent soil analysis and any recommendations made;

(l) Field records and histories, including a plan of the holding showing the Ordnance Survey numbers of the fields, a history of the field inputs and crops for the previous four years and a timetable for the conversion stages and completion.

3.212 The Documentation and Record Keeping must comply with section 2.3.

3.213 The Labelling of In-conversion produce must comply with section 2.20.

3.214 Permitted

(1) Production registered with the In-conversion Scheme marketed under the description: SOIL ASSOCIATION APPROVED ORGANIC CONVERSION

3.3 SOIL MANAGEMENT

3.301 The soil must be managed with the aim of developing and protecting an optimum soil structure, biological activity and fertility. The soil management must therefore ensure the following:

(a) A regular input of Organic residues in the form of manures and plant remains to maintain the level of humus, biological activity and plant nutrients;

(b) A level of microbial activity sufficient to initiate the decay of Organic materials and breakdown of non-soluble minerals into simple nutrient salts capable of being absorbed by the plant roots;

(c) Conditions conducive to the continual activity of earthworms and other soil-stabilising agents and the improvement and stabilisation of the soil structure by their production of granular casts, deep burrowing and the incorporation and mixing of Organic matter.

3.302 Recommended

(1) A protective covering of vegetation, e.g. green manure or growing crop, to protect surface living organisms and soil structure from damage by exposure to dry conditions, heavy rain or strong winds.

(2) Appropriate cultivations required for crop production should aim to achieve:

(a) Deep loosening of the sub-soil to break plough or compaction pans.

(b) Minimal disruption of the soil profile;

(c) Timeliness of cultivations or grazing to ensure appropriate tilth and to avoid damage to existing structure.

(3) The monitoring of Organic matter levels, available plant nutrients and nutrient reserves in the soil by means of regular soil analyses.

3.4   ARABLE & HORTICULTURAL CROP ROTATIONS

3.401   The development and implementation of well designed crop rotations is central to Organic production systems. Crop rotations aid in the control of pests and diseases and the maintenance of soil fertility, soil Organic matter levels and structure, whilst ensuring that sufficient nutrients are available and nutrient losses are minimised.

3.402   Whilst there cannot be a definitive rotation, the following requirements must be observed:

(a) A balance must be achieved between fertility building and exploitative cropping;

(b) *Crops with differing root systems must be included;*

(c) *Rotations must include a leguminous crop to provide a balance of nitrogen in the soil for use by subsequent crops;*

(d) *Plants with similar pest and disease susceptibility must be separated by an appropriate time interval.*

3.403   Recommended

(1) Rotations which also:

(a) *Minimise the time that the soil is left uncovered by the maximum use of green manures where appropriate;*

(b) *Maintain or increase the Organic matter levels in the soil;*

(c) Vary weed susceptible crops with weed suppressing crops.

(2) Permanent grassland.

(3) Mixed ley farming with a balance of cropping and grass/clover leys.

3.404   Permitted

(1) Rotations falling short of the above requirements on predominantly horticultural holdings and which rely on the use of external inputs to maintain crop production provided that they are:

(a) Demonstrating that they are moving towards a better balance between fertility building and exploitative management and away from a total reliance upon outside inputs;

(b) Making a maximum use of legumes and green manure catch crops in these systems.

(2) Production systems falling short of the above requirements provided that nutrient supply, weed, pest and disease control is effected by the methods outlined in these Standards, and including the following:

(a) Greenhouse production which includes monocropping or annual cropping of the same genus – excluding alliums, potatoes and brassicas;

(b) Perennial crops such as orchards, vineyards and plantation crops.

(3) The absence of formal rotations on small intensively managed units of less than one hectare where multiple cropping (6 or more crops), intercropping or companion planting are practised.

3.405   Prohibited

(1) Alliums, brassicas and potatoes, as outdoor crops, returning to the same land before a period of 48 months has elapsed from planting date to planting date.

(2) Continuous cropping of alliums, brassicas or potatoes in greenhouses.

(3) Cropping systems not defined above which rely solely on outside inputs for nutrient supply, weed, pest and disease control.

(4) Continuous cereal rotations.

3.5   HEAVY METALS

3.501   *Heavy metals and other metallic elements are naturally present in the soil and some are essential, in trace amounts, to plants and animals. It is however necessary to maintain a correct balance and the concentration in the soil should not be increased beyond acceptable levels by the application of manures, fertilisers and mineral supplements.*

3.502   *Heavy metal levels in manures should not exceed the levels specified for manures. Manures must not be added to the soil where the addition would lead to the heavy metals in the soil exceeding the levels specified.*

3.503   Maximum permitted heavy metal levels in the top soil (on a dry matter basis):

	in the soil		in the manures	
	mg/kg	kg/ha	mg/kg	kg/tonne
zinc	150	336	1000	1.000
chromium	150	336	1000	1.000
copper	50	110	400	0.400
lead	100	220	250	0.250
nickel	50	116	100	0.100
cadmium	2	4.4	10	0.010
mercury	1	2	2	0.002

3.504   At the discretion of the Certification Committee, an analysis of manures, fertilisers and crops may be required before a Certificate of Registration can be granted or renewed.

3.505   In this regard it should be noted that in some circumstances background environmental contamination, residues from previous agricultural practice or levels of naturally occurring substances in the soil may render the land unsuitable for Organic production.

3.6   MANURES AND PLANT WASTES

3.601   Brought-in manures or plant wastes from Non-organic sources must not form the basis of a manurial programme, but should be adjuncts. Exceptions may be made in the case of intensive horticultural systems, where it is recognised that adequate nutrition of the crops is not always possible by the methods given in section 3.5 of these Standards.

3.602   The use of all plant wastes and animal manures from Non-organic sources is Restricted and must be approved by the Certification Committee and receive the treatments specified before use.

3.603   An analysis of the soil and/or manure may be required by the Certification Committee, at the applicant's expense, before approval can be given for a Restricted material.

3.604   Permitted
   (1)   Straw, FYM, stable and poultry manures from Organic sources preferably after being properly composted (see paragraph 3.607).
   (2)   Slurry, urine and dirty water, from Organic sources preferably after being aerated.
   (3)   Plant waste materials and by-products from Organic food processing industries.
   (4)   Manures and composts, registered with the Certified Products Scheme, containing only materials permitted in these Standards.
   (5)   Sawdust, shavings and bark – from untreated timber.
   (6)   Compost activators – microbial and plant extracts.
   (7)   Bio-dynamic preparations.
   (8)   Seaweed.

3.605   Restricted
   (1)   Straw, FYM and stable manure from Non-organic sources – after being properly composted for three months or stacked for six months.
   (2)   Poultry manure and deep litter from the following Non-organic systems – after being properly composted for six months or stacked for twelve months:-
      (a)   Egg Producing (defined by EEC Regulation No. 1274/91):
         (i)   Free Range (max 400 birds/acre),
         (ii)   Semi-intensive (max 1600 birds/acre),
         (iii)   Deep Litter (max 7 birds/sq. m);
      (b)   Deep litter pullet rearing systems (max housing density of birds 17kg/sq m);
      (c)   Meat producing (defined by EEC Regulation No. 1538/91):
         (i)   Free Range,
         (ii)   Traditional Free Range,

          (iii) Extensive Indoor Barn reared (max housing density mature birds 12 hens or 17–25kg/sq m).

(3)   Manures from Non-organic straw-based pig production systems – after being properly composted for six months or stacked for twelve months.

(4)   Plant wastes and by-products from Non-organic food processing industries – after being treated as in paragraph 3.6051 above.

(5)   Mushroom composts made from Non-organic animal manures conforming with paragraphs 3.604 to 3.606 – treated as in 3.6051 above.

(6)   Worm composts made from Non-organic animal manures conforming to paragraphs 3.604 to 3.606.

(7)   Animal slurry from Non-organic sources conforming to paragraphs 3.604 to 3,606 – after aeration.

(8)   Dirty water from Non-organic systems – applied to In-conversion land.

(9)   *Processed animal products from slaughterhouses and the fish industries.*

(10)   *Composts from organic household refuse – treated as in 3.6051 above.*

3.606   Prohibited

(1)   Sewage sludge, effluents and sludge based composts.

(2)   Peat as a soil conditioner.

(3)   *The use of animal residues and manures (other than processed animal products from slaughterhouses and the fish industries) from livestock systems that do not comply with the provisions of sections 5.4 & 5.5 of these standards.*
These include:

(a)   Poultry battery systems and broiler units with stocking rates over 25kg/sq m;

(b)   Indoor tethered sow breeding units;

(c)   Other systems where stock are not freely allowed to turn through 360 degrees, where they are permanently in the dark, or are permanently kept without bedding.

MANURE MANAGEMENT AND APPLICATION

3.607   Composting is defined as a process of aerobic fermentation. A substantial temperature increase can be induced within the heap by means of turning. A temperature of 60°C will facilitate the destruction of most weed seeds, pathogens, chemical residues and antibiotics and the composting process should aim to achieve this. After an initial heating up the compost heap must be turned again, preferably covered and maintained for at least three months.

3.608   The management of the livestock manures and crop residues produced on the holding and materials brought-in should aim to achieve maximum recycling of nutrients with minimum losses.

3.609   Manure treatments, storage systems and applications are expected to conform to the Statutory Code of Good Practice for the Protection of Water under Section 116 of the Water Act 1989 (available from MAFF Publications or the National Rivers Authority).

3.610   *Adequate provision must be made for the storage of manures and slurry prior to application. Manure and slurry stores must be able to:*

(a)   *Cope with the volume of production of manure/slurry on the holding;*

(b)   *Enable flexibility of application timing by provision of adequate storage;*

(c)   *Prevent liquid effluents from manure/slurry from entering water courses and ground water.*

3.611   *Care must be taken when spreading manure/slurry to avoid run-off and the pollution of water courses and ground water. Attention must be paid to the capacity of the ground to absorb the manure/slurry at the time of application. When conditions appear unfavourable and pollution seems likely to occur, application must not take place.*

3.612   In most circumstances the use of animal manures will be assessed in the context of the maintenance of soil fertility through the recycling of nutrients removed by farm livestock consuming feedstuffs produced on the holding.

3.613   The use of approved supplementary manures from outside the holding (as defined in paragraphs 3.601 to 3.606) will normally be restricted to a maximum application rate at the equivalent of 2.5 LSU/ha (approximately 10t/a) per annum.

3.614 Only in the case of exceptional circumstances (including intensive horticultural and glasshouse production) will application rates in excess of this level be permitted, and the Certification Committee reserves the right to further limit the application rate of manures in order to reduce the risk of contamination of water supplies, particularly in high risk areas.

3.615 Recommended
(1) The storage and composting of manures indoors or under plastic sheeting to prevent leaching of nutrients during periods of heavy rainfall.
(2) Steel and concrete slurry tanks and slurry lagoons built to BS 5502, with aeration facilities.
(3) Applications of composted manures and aerated slurries onto fertility building crops, grassland and cultivated land in spring and summer.
(4) Avoiding the spreading of manures within 10 metres of ditches and water-courses and within 50 metres of boreholes.
(5) Avoiding the spreading of manure or slurry on frozen or saturated ground.

3.616 Permitted
(1) The autumn/early winter applications of composted manures to grassland – only whilst nutrient uptake is actively taking place.
(2) Applications of composted manures to greenhouse soils – at any time.
(3) Slurry systems without buffer storage tanks applying slurry over the winter – to grassland only when conditions are suitable.

3.617 Prohibited
(1) Storage systems and practices which result in the pollution of water courses.

3.7 MINERAL FERTILISERS AND SUPPLEMENTARY NUTRIENTS

3.701 Mineral fertilisers must be regarded as a supplement to, and not as replacement for, nutrient recycling within the farm and may be applied only to the extent that adequate nutrition of the crop is not possible by the methods given in Sections 3.4 & 3.6. Production must be planned to minimise the need for brought-in nutrients.

3.702 In the absence of more acceptable materials, restricted use of soluble fertilisers to treat severe trace element or potassium deficiencies may be allowed with specific approval of the Certification Committee. A full soil analysis, including clay fractions, heavy metal content and trace elements must be supplied.

3.703 Permitted
(1) Nitrogen sources:
    (a) Blood meal – restricted to propagating composts and on overwintered crops in spring;
    (b) Hoof & horn meals.
(2) Phosphate sources:
    (a) Natural rock phosphate (e.g. Tunisian rock phosphate);
    (b) Basic slag;
    (c) Calcined aluminium phosphate rock (e.g. Redzlaag);
    (d) Meat & bone meals.
(3) Potassium (potash) sources:
    (a) Natural rock potash – providing it has a relatively low immediate solubility in water and low chlorine content (e.g. Adularian rock potash);
    (b) Wood ash – added to composts and manures;
    (c) Plant extracts (e.g. Kali Vinasse).
(4) Compound fertilisers:
    (a) Fish, Blood & Bone meals – free from non-permitted substances;
    (b) Fish meals;
    (c) Meat and bone meals;
    (d) Fertilisers and liquid feeds registered with the Certified Products Scheme;*
    (e) Liquid feeds made from plants produced on the Organic Unit.

(5) Minor Minerals:
   (a) Calcerious magnesium rock (Dolomitic limestone) – magnesium & lime;
   (b) Gypsum (Calcium sulphate) – calcium;
   (c) Ground chalk & limestone – calcium & lime;
   (d) Calcified seaweed – calcium & lime;
   (e) Epsom salts – for acute magnesium deficiency;
   (f) Magnesium rock (including Kieserite).
(6) Trace Elements:
   (a) Dried seaweed meal;
   (b) Liquid seaweed (free from non-approved ingredients);
   (c) Plant-based foliar sprays and liquid feeds registered with the Certified Products Scheme;*
   (d) Calcified seaweed, limestone & chalk;
   (e) Stone meal (ground basalt).
   * NB Products registered with the Certified Products Scheme contain only materials and substances permitted in these Standards.

3.704   Restricted
(1) Commercial Organic fertilisers and liquid feeds not registered with the Certified Products Scheme – ingredients and nutrient analysis must be supplied for approval.
(2) Sulphate of potash – only where exchangeable K levels are below index 2 and clay content is less than 20%, following soil analysis.
(3) Sulphur.
(4) Trace elements – boron, copper, iron, manganese, molybdenum, cobalt, selenium, zinc – following soil analysis or other evidence of deficiency.
(5) Calcium chloride – for bitter pit in apples.

3.705   Prohibited
(1) The use of fertilisers based on animal slaughterhouse by-products on farms with cattle and sheep (due to the problems with offal possibly being infected with BSE).
(2) Fresh Blood.
(3) Guano.
(4) All other synthetic and natural fertilisers including: Nitrochalk, Chilean Nitrate, Urea, Muriate of Potash, Superphosphates, Kainit.
(5) Slaked lime, Quicklime.

3.8   WEED CONTROL
3.801   Weed control must primarily be approached by adjustments in the management of the system, by giving attention to rotation design, manure management etc.
3.802   Recommended
(1) Balanced rotations.
(2) Varying weed suppressing with weed susceptible crops.
(3) Composting manures and plant wastes.
(4) Slurry aeration.
(5) Hygiene – in the field and on machinery.
3.803   Permitted
(1) Pre-sowing cultivations.
(2) Stale seed bed techniques.
(3) Variety selection for vigour and weed suppression.
(4) Pre-germination, propagation & transplanting.
(5) High seed rates.
(6) Under-sowing.
(7) Utilisation of green manures.
(8) Raised beds and no dig systems.
(9) Mulches.
(10) Mixed stocking & tight grazing.

(11) Re-cleaned seed.
(12) Pre-emergence and post-emergence mechanical operations (e.g. hoeing, harrowing, topping, hand weeding).
(13) Pre-emergence and post-emergence flame weeding.
(14) Plastic mulches.
(15) Steam sterilisation – greenhouse soils only.

3.804 Prohibited
(1) The use of any chemical and hormone herbicides, within the crop, at the edge of fields, within or below hedgerows, headlands and pathways on a registered Holding.

3.9 PLANT PEST AND DISEASE CONTROL

3.901 *Producers are expected to comply with all the specified instructions and withdrawal periods when storing and using natural pesticides approved under the Control of Pesticides Regulations 1986.*

3.902 Purchased equipment that has previously been used to spray prohibited products must be thoroughly cleaned so as to be free from non-approved substances, and must be dedicated thereafter.

3.903 Pest and disease control shall be primarily controlled by a combination of:
(a) An appropriate choice of species and varieties of crops;
(b) A balanced rotational cropping to break the pest and disease cycles;
(c) A proper attention to cleaning routines and hygiene within the holding to minimise the spread of pests and disease.

3.904 *Only in the case of a threat to the crop may recourse be made to the substances referred to in paragraph 3.906 below.*

3.905 Recommended
(1) A balanced supply of plant nutrients.
(2) The creation of a diverse ecosystem within and around the crop to encourage natural predators by:
    (a) Companion planting, under-sowing and mixed cropping;
    (b) Leaving uncultivated field margins, hedges, windbreaks and wildlife corridors.
(3) The use of resistant varieties.
(4) The use of strategic planting dates.
(5) Good husbandry and hygiene practices.
(6) Grafting on to resistant rootstocks.

3.906 Permitted
(1) For controlling insect pests:
    (a) Mechanical controls using barriers;
    (b) *Pheromone traps – for monitoring pest levels only;*
    (c) *Potassium soap (soft soap) and soaps containing plant fatty acids;*
    (d) *Pyrethrum (preparations on the basis of pyrethrins extracted from Chrysanthemum cineraefolium, containing possibly a synergist);*
    (e) *Derris (preparations from Derris elliptica);*
    (f) *Quassia – (preparations from Quassia amara);*
    (g) *Sulphur;*
    (h) *Preparations of Bacillus thuringiensis;*
    (i) Products registered with the Certified Products Scheme;*
    (j) Sticky fly traps – free from insecticides;
    (k) *Biological Pest Control – using licensed naturally occurring predatory organisms.*
    (l) *Granulose virus preparations.*
(2) For controlling fungi:
    (a) *Bordeaux mixture;*
    (b) *Sulphur.*
(3) General pest control:
    (a) Steam – sterilisation of buildings and equipment;
    (b) Steam – sterilisation of greenhouse soils;

(c)   Products registered with the Certified Products Scheme;*
(d)   *Preparations on the basis of Metaldehyde containing a repellent to higher animal species and as far as applied within traps;*
(e)   Mechanical traps, barriers and sound.
(f)   Seeds treated with products not included in paragraph 3.906 above provided that:
    (i)   The treatment is authorised for use in general agriculture in the UK;
    (ii)   The users of such seeds can show to the satisfaction of the Certification Committee that they were unable to obtain on the market non-treated seeds of an appropriate variety of the species in question;
    (iii) Accurate records are kept containing details of the substances used as a seed dressing;
    (iv)   The dressing does not include organo-chlorine dressings (see paragraph 3.907).
(g)   Wetting/sticking agents for sprays – licensed products based on natural plant extracts and oils free from non-permitted additives.

* NB Products registered with the Certified Products Scheme contain only those materials and substances approved for use in these Standards;

3.907   Prohibited
(1)   The storage of prohibited materials on Organic units (see paragraph 2.104).
(2)   All other biocides including:
    (a)   Nicotine;
    (b)   Formaldehyde and phenols for soil sterilisation;
    (c)   Methyl bromide and other chemical soil sterilants;
    (d)   Seed dressings based on mercurial and organo-chlorine compounds (including gamma HCH, Lindane and BHC);
    (e)   Aluminium based slug killers;
    (f)   All other synthetic pesticides.

SECTION 4

STANDARDS FOR CONSERVATION, CROP PRODUCTION AND HANDLING

4.1 CONSERVATION AND ENVIRONMENTAL HUSBANDRY

4.101   Producers are expected to abide by legal and statutory requirements in respect of any aspect of the wider environment at all times.

4.102   Concern for the environment should manifest itself in willingness to consult appropriate conservation bodies and in high standards of conservation management throughout the holding to enhance landscape features, habitat, wild plant and animal species.

ARABLE CROP PRODUCTION

4.103   See section 4.3 for Arable Crop Production Standards.

MEADOW AND GRASSLAND MANAGEMENT

4.104   Species-rich meadows and unimproved pastures are an important habitat for many plants and insects. They enhance the landscape and can make a useful contribution to Organic livestock nutrition due to the diversity of plant species and high mineral and trace element content.

4.105 See section 4.4 for Grassland Management Standards.

TRADITIONAL FIELD BOUNDARIES & HEDGE MANAGEMENT

4.106 Traditional boundaries such as hedges, ditches and stone walls act as important wildlife corridors through agricultural land, help to maintain a diverse ecology, provide a habitat for many beneficial animals and insects and shelter for livestock.

4.107 Recommended
(1) The retention of hedges and stone walling using traditional methods and materials.
(2) Ditch clearance in phased operations maintaining a portion of ditches uncleared each year.
(3) Clearing opposite sides of ditches in successive years.
(4) Hedge trimming, ditch and dyke clearance between January and March.
(5) Leaving at least a metre of undisturbed field margin.

4.108 Restricted
(1) Removal of hedgerows, banks or ditches – following discussion with a conservation advisor; consideration should be given to the need for compensatory environmental work.

4.109 Prohibited
(1) Hedge trimming, ditch and dyke clearance between 1st March and 31st August.
(2) Annual trimming of all hedges unless required by local authorities for road safety.

MOORLAND & HEATHLAND

4.110 Moorland, heathland and other areas of semi-natural vegetation such as scrub are nationally so reduced in extent as to have become of direct conservation interest and approval must be sought from the Certification Committee before any 'improvements' such as gripping or drainage are carried out.

4.111 Recommended
(1) Stocking rates appropriate for maintaining good quality heather and upland vegetation and the avoidance of overgrazing and poaching, as identified by the Statutory Conservation Bodies.

4.112 Permitted
(1) The management of heather by periodic burning as appropriate to keep it in good heart.

WETLANDS AND DRAINAGE

4.113 Prohibited
(1) New or improved drainage affecting areas of significant conservation value.
(2) Exploitation of peat bogs identified by Conservation Bodies to be of conservation value.

TREE AND WOODLAND MANAGEMENT

4.114 Trees and woodland play an important role in maintaining the ecological balance on Organic farms, They provide habitat for wildlife including pest predators. Mature trees and woodland may also have an amenity and recreational value. Individual trees and ancient woodland play a vital part in preserving landscape and species diversity.

4.115 Recommended
(1) The retention and management of trees in accordance with local custom and woodland practice.
(2) Replanting programmes integrated with existing woodland and trees, and using indigenous and local shrubs and trees.
(3) Natural regeneration, coppicing and other traditional management practices.
(4) Fencing of newly planted or regenerating woodland against stock.

4.116   Restricted
        (1)   Clear felling of woodland.
        (2)   Felling of mature specimen trees that are not endangering safety.
4.117   Prohibited
        (1)   New planting on semi-natural or other good wildlife habitat, or other sites of
              particular ecological or archaeological interest.

BUILDINGS AND HERITAGE
4.118   Recommended
        (1)   The siting and construction of new farm buildings should be done sensitively,
              taking account of their environmental and aesthetic impact.
        (2)   Maintaining existing old buildings wherever possible in their original form.
        (3)   Taking advice from the Society for the Protection of Ancient Buildings when
              considering their conversion or demolition.
        (4)   Use of local materials.
        (5)   The provision of roosts and nest sites for bats and barn owls in new buildings
              and conversions.
4.119   Restricted
        (1)   Damage to the nesting and roosting sites of owls, bats and other endangered
              species – without specific approval of the Certification Committee and advice
              and permission from the appropriate Conservation Organisations.
4.120   Prohibited
        (1)   The use of wood preservatives on new or existing buildings which are harmful
              to bats and livestock.
        (2)   Levelling of ridge and furrow fields and cultivation of sites of ancient monu-
              ments, archaeological sites and earthworks.

CREATIVE CONSERVATION
4.121   Recommended
        (1)   Symbol holders are encouraged to undertake creative conservation projects
              wherever possible. Care should be taken to avoid damage to existing sites of
              conservation value.
        (2)   Creating wildlife corridors or linked linear habitats, made up of continuous
              semi-natural habitats incorporating hedgerows, field margins and verges.
        (3)   The maintenance of existing rights of way.

4.2   ENVIRONMENTAL POLLUTION

SPRAY DRIFT CONTAMINATION
4.201   *Where Organic crops are being grown adjacent to non-organically managed crops efforts must
        be made to provide an effective windbreak where there is a risk of spray drift or contamination.*
4.202   Until such a hedge or windbreak is established, the Certification Committee may
        require a 10 metre buffer zone between Organic crops and the source of the
        contamination (20 metres where adjoining sprayed orchards).
4.203   Any known or suspected spray drift contamination must be notified to the Symbol
        Department without delay.

WATER
4.204   Care should be taken to ensure that water used for irrigation is free from con-
        tamination by prohibited materials.
4.205   The washing of Organic/In-conversion quality produce must be done in fresh
        potable water and not in water that has also been used for washing conventional
        produce.

DISPOSAL OF PLASTIC WASTE
4.206   Prohibited
        (1)   Burning of plastic waste

4.3 ARABLE CROP PRODUCTION

4.301 Where an operator has a mixture of land designated as Organic and Non-organic, and the same crop is to be grown on two or more of these (parallel cropping), a different and identifiable variety must be grown on each differently designated area of land and be recorded in the field histories and purchases.

4.302 Recommended
(1) Timing cultivations to minimise disturbance of ground nesting birds.
(2) Leaving at least one metre of undisturbed field margin for wildlife conservation.

4.303 Prohibited
(1) The burning of straw, cereal waste and stubble.
(2) Growth regulators.
(3) Contamination of corn by the combustion products of paraffin or diesel during drying.

4.4 GRASSLAND MANAGEMENT

4.401 Producers are expected to conform to the Statutory Code of Practice for the Protection of Water under Section 116 of the Water Act 1989 when making and storing silage.

4.402 Silage clamps, silos and bags must be constructed and maintained to prevent pollution of water courses and groundwater and either:
(a) Have effluent collection tanks with sufficient storage for unusually wet silage; or
(b) Be protected from water entering the system and causing an overflow.

4.403 Organic, in-conversion and Non-organic forage must be stored separately and fed in the correct proportions (see section 5.6).

4.404 Recommended
(1) Manure applications on unimproved meadows not exceeding an average of 30kgN/a/yr. or equivalent (e.g. approx. 10 tonnes FYM/a/yr.).
(2) Rotating grazing with forage conservation for clean grazing and intestinal worm control.
(3) Rotational stock grazing systems and mixed stocking for intestinal worm control.
(4) Regular soil analyses for hay and silage fields to monitor fertility.
(5) Maintenance of established patterns of cropping for previously defined species-rich (hay) meadows.
(6) Timing of mowing operations to allow grasses and flowers to set seed and to avoid disturbing ground nesting birds, or mowing for hay in a way that permits young birds to escape.
(7) Management of species-rich meadows in keeping with species requirements.

4.405 Permitted
(1) Bacterial and molasses silage additives (in wet seasons when good silage cannot be made by other means).
(2) Set stocking if intestinal worm burdens can be controlled.
(3) Management agreements with Statutory Conservation Bodies (subject to input materials conforming to Sections 3 & 4 of these Standards).

4.406 Restricted
(1) The sale of forage as a cash crop – not more than one year in four from each field unless the soil fertility is monitored by regular soil analyses and can be maintained.
(2) Switching from hay to silage production on unimproved and species-rich meadows.

4.407 Prohibited
(1) Other silage additives.
(2) Pollution of water courses by silage effluent.
(3) Grazing Organic or In-conversion livestock on non-Registered land.
(4) Ploughing of unimproved grassland and species-rich meadows agreed with the Statutory Conservation Bodies to be of conservation interest.

4.5　HORTICULTURAL CROP PRODUCTION

4.501　Where an operator has a mixture of land registered as Organic and Non-organic and the same crop is to be grown on two or more of these (parallel cropping), a different and identifiable variety must be grown on each differently designated area of land and be recorded in the field histories and purchases.

4.502　*If crops are grown from transplants (blocks, modules, root stock and bud materials) the transplants must be grown on a Registered Organic Unit.*

4.503　Recommended
- (1)　Propagation of transplants in Organic propagating composts registered with the Certified Products Scheme.
- (2)　Using Organically grown seed where available.
- (3)　Bare root transplants raised on the Organic Unit.

4.504　Permitted
- *(1)　By way of derogation from paragraph 4.502, until January 1st 1995, transplants may be brought-in from non-registered sources provided that:*
  - (a)　A minimum of 10% of the plants used on the holding are propagated in Organic composts;
  - *(b)　The user can show to the satisfaction of the Certification Committee that they were unable to obtain on the market, organically produced transplants of any appropriate variety of the species in question;*
  - *(c)　Accurate records must be kept of any substances used as a seed dressing and of any non-permitted materials used in, and any chemical treatments applied during the raising of the plants.*
- *(2)　Until January 1st 1996 other vegetative reproductive material such as potato tubers, onion sets, strawberry runners and fruit tree root stock and bud material may be brought from non-registered sources provided that the user can show to the satisfaction of the Certification Committee that they were unable to obtain on the market organically produced materials of any appropriate variety of the species in question;*
- (3)　Propagating media derived from the materials for fertilisation and soil conditioning specified in sections 3.6 & 3.7 of these Standards.
- *(4)　Clay (bentonite and zeolites), vermiculite and perlite which have not undergone chemical treatments with prohibited materials.*

4.505　Restricted
- (1)　Bare-root transplants from non-registered sources.
- (2)　A growing cycle of six weeks or more between the planting and harvesting of transplants from non-registered sources.

4.506　Prohibited
- (1)　Peat as a soil conditioner.

4.6　MUSHROOM PRODUCTION

4.601　Mushroom growing houses must be dedicated to Organic production.

4.602　Recommended
- (1)　Compost containers used for growing the mushrooms dedicated to Organic production.
- (2)　Compost made from Organic manures.
- (3)　Steam sterilisation of buildings and equipment.
- (4)　Physical and barrier methods for fly control.

4.603　Permitted
- (1)　Control of fungal diseases – salt.
- (2)　Control of flies – products for plant pest and disease control specified in section 3.9 of these Standards.

4.604　Restricted
- (1)　Compost made from manures from Non-organic sources – conforming to section 3.6 of these Standards.

4.605　Prohibited
- (1)　Chemical pesticides – either in the compost, sprayed on the crop or as a fog.
- (2)　Chlorinated water for disease control.

(3) Formaldehyde for sterilisation.
(4) Fumigation by methyl bromide.
(5) Bleaching mushrooms.
(6) Post harvest treatment of composts with fungicides.

4.7 HARVESTING AND STORAGE

4.701 *Effective steps must be taken to protect organically grown crops from contamination during harvesting, storage and transportation. In this regard cleaning routines must ensure that:*

    (a) *Harvesting equipment, including vehicles and containers used for transporting the produce are clean, free from Non-organic crop residues and any other materials which may contaminate the produce;*

    (b) *Before use, storage areas used for Organic produce are clean and free from Non-organic crop residues and other materials which may contaminate the produce and are left empty for an appropriate period of time prior to use as a disease and insect break;*

    (c) *Before use, drying equipment, including conveyors and other ancillary equipment, are clean, free from Non-organic crop residues and any other materials which may contaminate the produce;*

    (d) *If sacks are used for storage or delivery of produce to customers, that they are of food grade quality, clean and free from contamination.*

4.702 *Control and operating procedures must be established and maintained to ensure that from harvesting through to dispatch from the Unit, Organically grown produce is clearly and legibly identified.*

4.703 The storage areas and containers used for Organic production must be:

    (a) *Separated from storage areas used for other purposes by a physical barrier in the form of an effective partition;*

    (b) Dedicated to Organic or In-conversion crops only;

    (c) Clearly labelled to prevent mistakes being made between Organic, In-conversion and Non-organic crops;

    (d) Constructed from materials suitable for food use;

    (e) Maintained in a clean and hygienic state;

    (f) Covered to prevent contamination by bird droppings;

    (g) Protected from access and contamination by vermin.

4.704 Any post harvest contamination must be reported to the Symbol Department immediately.

4.705 *As appropriate the crops may be dried by indirect heated air or by other suitable means, including direct fired propane, diesel and paraffin fuelled driers, but they must not be contaminated by the combustion products of the fuel used. A regular maintenance programme must be established to ensure full combustion when in use.*

4.706 Permitted

    (1) For cleaning:

        (a) Vacuum cleaning;

        (b) Steam cleaning;

        (c) High pressure water cleaning;

        (d) Hypochlorite in solution – followed by rinsing with potable water.

    (2) For pest control – products for pest and disease control specified in section 3.9 of these Standards.

    (3) *Static bait traps using licensed poisons for controlling rodents – in locations where there is no risk of product contamination. Substances used must be properly labelled and stored under lock and key away from food.*

4.707 Prohibited

    (1) The use of ionising radiation and synthetic chemicals as an aid to preservation.

    (2) The use of prohibited materials in stores and on premises where Organic or In-conversion crops are stored, including:

        (a) Sprout inhibitors;

    (b)   Fungicidal sprays, dips or powders;

    (c)   Chemical fumigants or pesticides.

(3)   Stores containing wood previously treated with organo-chlorine (gamma HCH & lindane) wood preservatives.

(4)   Contamination by the combustion products of the fuels used for crop drying.

4.8 TRANSPORT

4.801 Organically produced products which are not in their packaging for the end consumer may be transported to other units only in appropriate packaging or containers closed in a manner which would prevent substitution of the content and provided with a label stating, without prejudice to any other indications required by law:

    (a)   The name and address of the person responsible for the production or preparation of the product;

    (b)   The name of the product;

    (c)   That the product is covered by the inspection arrangements laid down in the Regulation (EEC) No. 2092/91 (i.e. by giving the name and address of the Symbol Scheme).

4.802 Products intended for retail sales must be packed and transported to the point of sale in closed packaging. Each consignment must be accompanied by appropriate documentation enabling the origin of the product to be traced.

4.803 All vehicles used for transporting organically produced products should be subjected to a regular cleaning programme to ensure they are maintained in a generally clean state with no build up of Non-organic materials or residues. If they are used for the carriage of other goods or materials, they must be thoroughly cleaned and dried before being used to transport organically produced products.

4.804 Before loading, vehicles and all handling equipment must be inspected to ensure they are clean and free from visible residues and any materials that may contaminate or impair the integrity of the organically produced products to be transported.

4.805 If containers are used they should be of food grade quality, in a state of good repair, clean and free from visible residues or any materials that may contaminate or impair the Organic integrity of the products contained therein.

4.9 PACKAGING MATERIALS

4.901 As far as is reasonably practical, ecologically sound materials should be used for the packaging of organically produced products.

4.902   Non essential packaging should be avoided where possible and consideration should be given to how the end product packaging may be recycled or returned.

4.903 Materials used for packaging must be of food grade quality, clean, unused and sufficiently strong to protect the produce during transport and display.

4.904 The packaging must not affect the organoleptic character of the product or transmit to it any substances in quantities that may be harmful to human health.

4.905 If returnable outer containers are used they must be reserved exclusively for organically produced products, they must be kept in good repair, be clean and free from contamination.

4.906 All packaging materials must be stored off the floor, away from walls and ceilings in clean, dry, hygienic conditions.

4.907   Recommended

    (1)   A full environmental audit for the packaging.

    (2)   Returnable outers and bulk containers.

    (3)   Deposit schemes for cans and bottles.

    (4)   Recycled outer packaging – indicated as such.

    (5)   Single layer, single substance recyclable packaging.

    (6)   Bulk packaging at retail outlets for self selection.

    (7)   Unbleached paper and cardboard.

4.908   Permitted

    (1)   Glass and plastic containers.

(2) Plain and waxed paper and cardboard.

(3) Cellophane, Polyethylene and polypropylene films.

(4) Modified atmosphere packaging films.

(7) Plastic and hessian nets and sacks.

4.909 Restricted

(1) PVC films free from additional plasticisers – non-fat foods only.

(2) Metal foils.

4.910 Prohibited

(1) Expanded polystyrene made with CFCs.

SECTION 5

GENERAL STANDARDS FOR LIVESTOCK HUSBANDRY

5.1 INTRODUCTION

5.101 The Standards for Livestock Husbandry set out the management practices for Organic livestock production. They are designed to encourage positive and dynamic animal health and vitality, thereby promoting greater resistance to disease which should be the aim of good Organic husbandry.

5.102 The MAFF publications 'Codes of Recommendations for the Welfare of Livestock' lay down codes of practice for farm animals covering both welfare and housing considerations. These Livestock Husbandry Standards adhere to the MAFF Codes as a minimum. In some cases more stringent standards are specified.

5.103 The General Livestock Standards applying to brought-in stock, conversion, welfare, housing, bedding, transport and handling, feeding and veterinary aspects of livestock husbandry are outlined in this Section 5.

5.104 Additional specific requirements for each class of livestock are detailed in Section 6.

5.105 The terms Recommended, Permitted, Restricted and Prohibited are defined in section 1.4 of these Standards. A Restricted material or practice must have the approval of the Certification Committee before being used.

5.106 The Standards for Organic livestock must be considered in the context of a whole farm or farming system which is being managed organically. Farmers applying for registration for a livestock enterprise must therefore also comply with Sections 2 to 4 of these Standards.

5.107 Where the text refers to a 'known farm problem', this must first be adequately detailed upon application and/or be identified by the applicant's veterinary surgeon for consideration by the Certification Committee before permission for routine treatment will be granted.

5.2 ORIGIN OF STOCK & BROUGHT-IN STOCK

5.201 *Livestock systems should be planned so that the stock are born and raised on an Organic unit whether or not they are brought-in from other Organic Units.*

5.202 *All animals intended for meat production must be born and raised on an Organic unit whether or not brought-in from other Organic Units.*

5.203 *Subject to the following conditions however, livestock may be brought-in from non-registered* sources for the establishment of herds or flocks, for the expansion of herds or flocks and for replacement purposes:

(a) The purchaser must as far as is reasonably practical ensure that the animal welfare provisions of these Standards are observed at the supplying farm;

(b) Records of all veterinary medicinal treatments administered and all statutory records have been kept;

(c) All brought-in animals should be checked for disease, treated and quarantined for a period as necessary;

(d) Any animals brought-in for breeding or milk purposes must not be culled for Organic meat production;

(e) The animals must undergo a conversion period as indicated in section 5.3 of these Standards before the animals or their products qualify for Organic status and thereafter the animals must be managed in accordance with these Livestock Standards.

5.204 Recommended

(1) Cattle replacements from farms which have no history of using feeds containing meat and bone meal or which have no history of BSE.

(2) Closed herds and flocks to avoid bringing in animals infected with intestinal worms resistant to the permitted Anthelmintics.

5.205 Permitted

(1) Rare breeds and pedigree stock purchased from specialist and pedigree sales even if held at livestock market premises if such animals can be shown to be unavailable from direct sources.

5.206 Prohibited

(1) The purchase of stock through livestock markets.

(2) The culling of animals from Non-organic sources brought-in for breeding, milking or egg laying as Organic meat.

5.3 CONVERSION PERIODS FOR LIVESTOCK

5.301 Conversion periods for livestock may run simultaneously with the conversion of the land with the exception of the feeding requirements for dairy animals which must run after the land is converted.

5.302 Where an existing herd or flock is to be converted, the animals must undergo a conversion period as defined in the relevant paragraph in paragraph 5.303.

5.303 Mature female animals for breeding or milking may be brought-in for the establishment of herds or flocks subject to the following conversion periods, during which time and after which they must be managed in accordance with these Standards for Livestock Husbandry:

(a) Suckler cows must undergo a conversion period of not less than 12 weeks immediately prior to calving;

(b) Dairy cows must undergo a conversion period of not less than 36 weeks except in regard to the feed requirements which must be implemented at least 12 weeks before the end of the conversion period;

(c) Ewes and goats must be mated on an Organic unit;

(d) Sows whose progeny is intended for meat production must be mated on the Organic unit.

5.304 Up to 10% of a herd or flock may be brought-in annually for expansion or replacement purposes, subject to the following conversion requirements, during which time and after which they must be managed in accordance with these Standards for Livestock Husbandry:

(a) Suckler cows must undergo a conversion period of not less than 12 weeks;

(b) Dairy animals must undergo a conversion period of 12 weeks;

(c) Ewes and goats whose progeny are intended for meat production must be mated on an Organic unit;

(d) Sows whose progeny is intended for meat production must be mated on an Organic unit.

5.305 In cases where an expansion of a herd or flock, or where a change of breed requires an intake more than the 10% expansion/replacement allowance, the additional animals must undergo the conversion period as defined in the relevant paragraph in paragraph 5.302.

5.306 Male animals of all species may be brought-in at any time for breeding purposes. No conversion period is required provided the animals are managed in accordance with these Livestock Standards from the time of arrival on the Organic unit.

5.307   *Pullets for egg production may be brought-in up to 10 weeks of age and they must undergo a conversion period of 10 weeks during which time they must be managed in accordance with these Livestock Standards.*

5.308   Poultry for meat production may be brought-in at one day old. *The birds must be managed throughout their lifetime in accordance with these Livestock Standards.*

LIVESTOCK RECORDS

5.309   The livestock movement book must be kept up to date and complete.

5.310   *Accurate and comprehensive records must be kept of all brought-in livestock including species, source and numbers of animals, the veterinary history of the animals, quarantine measures if any and the conversion time, giving dates, by animal or group prior to gaining full Organic status (see section 2.3).*

5.4   GENERAL MANAGEMENT AND WELFARE

5.401   *The techniques employed in livestock management must be directed towards maintaining the animals in good health.*

5.402   The general conduct of animal husbandry should be governed largely by physiological and ethical considerations, having regard to behavioural patterns and the basic needs of animals. The European Convention on Farm Animals requires that they should be kept according to their physiological and ethological needs.

5.403   All stock must have access to pasture during the grazing season unless specifically excepted by the Certification Committee.

5.404   *The livestock plan should normally be an integral part of the crop rotation and must provide sufficient land to:*
   (a)   *Prevent overstocking;*
   (b)   *Allow for rotational or paddock grazing;*
   (c)   *Allow the sward to be reseeded;*
   (d)   *Prevent the build up of parasites.*

5.405   *The plans for livestock systems must allow for the livestock, especially breeding cows and sows, to be kept in reasonably stable groups.*

5.406   Livestock must have access to water at all times. For animals on piped water supplies, the drinking water should be checked regularly.

5.407   The herd or flock size must not adversely affect the individual animal's behaviour patterns.

5.408   *When animals are transported they must be handled with proper care and concern for their welfare and in accordance with section 5.8 of these Livestock Standards.*

5.409   Recommended
   (1)   *Attention to the choice of livestock breeds of the sire and dam in order to produce animals suited to Organic systems, local conditions and to avoid problems at birth.*
   (2)   The maintenance of traditional local or rare breeds of livestock to retain genetic diversity.

5.410   Permitted
   (1)   Outwintering where conditions permit provided that shelter, which is adequate to prevent any welfare problems, is available at all times.
   (2)   *Artificial insemination for breeding.*
   (3)   Tags, ear notching, tattooing and freeze branding animals for identification.
   (4)   Castration and de-horning where it is judged to be necessary for considerations of safety and welfare.

5.411   Prohibited
   (1)   *Animals subjected to any surgical or chemical interference which is not designed to improve the animals' own health or well-being or that of the group, other than those practices permitted in these Livestock Standards.*
   (2)   *Systems of livestock management which involve routine use of prophylactic medication except in accordance with section 5.7 of these Livestock Standards.*
   (3)   Zero grazing systems.
   (4)   Breeding practices which make the livestock systems overreliant on inappropriate technology (e.g. embryo transfer techniques & routine Caesarean sections).

5.5 LIVESTOCK HOUSING

5.501 *Housing and management must be appropriate to the behavioural needs of the animals and birds. All stock must have sufficient room to stand naturally, lie down easily, turn round, groom themselves, assume all natural postures and make all natural movements such as stretching and wing flapping and to walk about freely at least in accordance with MAFF Codes of Recommendations for Animal Welfare.*

5.502 Winter housing must be available for all livestock where severe weather or unhealthy or slippery ground conditions occur.

5.503 Housing must have adequate natural ventilation and lighting.

5.504 *All houses in which livestock of any species are confined, for other than very brief periods or during transit, must be well bedded with straw or other appropriate material and the drainage and other aspects of litter management must ensure that all animals have access to dry lying areas.*

5.505 *No animal should normally be housed out of sight or sound of others of its own species. Where for any reason that is unavoidable, the animal should be housed to allow it the regular sight and sound of human activity or the company of other compatible animals.*

5.506 Building materials treated with paints or preservatives which are toxic to animals must not be in reach of livestock.

5.507 Recommended
(1) Loose housing and straw yards with a minimum area of 1.8 square metres per 100 kg of animal.

5.508 Permitted
(1) Winter housing with a minimum lying area of 1.2 square metres per 100 kg of animal plus extra loafing area.
(2) *Stalls or cubicles in which the animals are confined individually only while feeding provided that:*
    (a) *The animals have free access to them other than at feeding times;*
    (b) *They do not restrict the natural lying behaviour of the animals or prevent them from standing normally.*
(3) *Up to 25% of the floor area may be slatted provided the rest of the floor area is solid and well bedded.*

5.509 Prohibited
(1) Permanent housing of livestock.
(2) Prolonged confining or tethering of animals (shippons, steadings etc.).
(3) Housing of stock without bedding.
(4) Slats on the whole of loafing and lying areas.

BEDDING MATERIALS

5.510 Recommended
(1) Straw from Organic sources.

5.511 Permitted
(1) Straw from Non-organic sources.
(2) Sawdust and wood shavings from untreated wood.
(3) Rubber mats if in use before conversion provided they have an additional layer of bedding on top.
(4) Concrete and sand and soil cubicle bases provided they have a layer of bedding on top.

5.512 Prohibited
(1) Lying areas without bedding.
(2) Sawdust and wood shavings from treated wood.
(3) Peat.

5.6 LIVESTOCK DIETS

5.601 The natural health and vitality of farm livestock is based on sound nutrition from before conception and throughout life. Thus organically grown feedstuffs fed in the form of a balanced ration are the basic requirements of these Standards.

5.602   Particular attention should be paid to the physiological adaptation of livestock to different types of feedstuffs, both in the initial choice of rations and when any changes of diet are contemplated. Sudden changes of diet should be avoided.

5.603   *Accurate and comprehensive records must be kept of all feedstuffs, including the constituent ingredients of the feed, the proportion of the constituents to the total feed on a dry matter basis and the source of the constituent parts.*

5.604   *All livestock systems should be planned to provide 100% of the diet from feedstuffs produced in accordance with these Standards.*

5.605   *However, in cases where this is not immediately possible the following minimum percentage of the dry matter in the diet must have been produced to these Standards (Organic and In-conversion):*

    (a)   *Beef animals*   *at least 90% calculated on a daily basis;*
    (b)   *Breeding ewes*   *at least 90% calculated on a daily basis;*
    (c)   *Dairy stock*   *at least 80% calculated on a daily basis;*
    (d)   *Non-ruminants*   *at least 70% calculated on a daily basis;*
    (e)   *Extensively managed single suckler herds,*
        sheep and goats in Less Favoured Areas   *at least 80% calculated on an annual basis.*

5.606   *At least 50% of the dry matter in the diet must come from feedstuffs produced to full Organic Standard.*

5.607   *The balance of the specified minimum Organic part of the diet, as given in paragraph 5.605 above, may come from Organic Units (or fields) which are:*

    (a)   *In the process of conversion to Organic production and after a conversion period of at least 12 months has been complied with before the harvest;*
    (b)   *The production plan or conversion plan has been accepted and the Holding, Unit (or field) has been inspected and registered by an UKROFS Approved Body.*

5.608   *The balance may be brought-in from non-Registered sources providing:*

    (a)   The Non-organic part of the ration for each class of livestock does not exceed the daily and annual allowances which are specified in Section 6 as follows:

        (i)   Dairy cattle   – paragraph 6.102
        (ii)   Calves   –   ,,   6.113
        (iii) Suckler cows   –   ,,   6.118
        (iv)   Store Cattle   –   ,,   6.124
        (v)   Sheep   –   ,,   6.208
        (vi)   Pigs   –   ,,   6.315
        (vii) Poultry   –   ,,   6.419

    (b)   The ingredients comply with the requirements of paragraphs 5.617 to 5.619.
    (c)   *The sources (and as appropriate the composition) and the conditions under which the feedstuff was manufactured are known to the purchaser.*

5.609   It is anticipated that the Non-organic allowance will be gradually reduced and eventually eliminated when the improved availability of Organic feedstuffs renders supplementation from conventional systems unnecessary.

5.610   *The diet should be balanced and of good quality and should not have levels of protein and energy or other additions associated with intensive production.*

5.611   *At least 60% of the dry matter in all ruminant diets should consist of either fresh green food or un-milled forage produced to these Standards (Organic and In-conversion).*

5.612   *For all breeding and milk producing animals the systems should be planned to make maximum use of grazing. During the winter, when the diet should be based on home produced hay and/or silage and/or straw and/or green food, whenever the weather and soil conditions permit the livestock should have access to forage or grazing.*

5.613   Grazing land must be of Organic or In-Conversion status, and the proportions in which they are used may be implemented on an annual basis rather than on a daily herd/flock basis. The 50/50 rule must therefore be complied with over the year.

5.614   When conserved forage is made from Organic, In-Conversion and Non-organic fields, each must be stored separately, and planning must ensure that their inclusion in the ration is implemented on a daily herd/flock basis.

5.615   Forage from Non-organic sources must be calculated as part of the Non-organic allowance.

5.616   Wherever possible, feed brought in from exploitative or polluted situations should be avoided, e.g. feedstuffs from the Third World or fishmeal from polluted seas or which depletes fish stocks. Sources should therefore be checked and preference should be given to feeds available locally.

ANIMAL FEEDSTUFFS

5.617   Recommended
  (1)   100% own farm produced Organic feedstuffs.
  (2)   Access to green and/or fresh fodder on a daily basis for all livestock.
  (3)   Un-compounded feeds (straights) where possible.
  (4)   Commercial feeds and compounds registered with the Industrial Products Scheme.

5.618   Permitted
  (1)   Non-organic feedstuffs from the following approved list, fed in accordance with the allowances specified in Section 6, (typical % DM in brackets):
    (a)   Fresh cut grass (20%), clover (19%), alsike (15%), lucerne (24%), sainfoin (25%);
    (b)   Conserved forage – hay (85%), silage (clamp 20%);
    (c)   Dried legumes – peas, beans, lupins & vetch (all 86%).
    (d)   Cereals – barley, maize, millet, oats, rye, wheat, quinoa (all 86%);
    (e)   Cereal by-products – wheat germ, wheat feed, oat feed, low grade flour, bran, middlings (all 86%);
    (f)   Maize gluten feed or meal (90%);
    (g)   Brewers grains (22%), distillers grains (25%), malt bran, spent hops (fresh 25%), pot ale syrup, yeast;
    (h)   Sugar beet pulp – (pressed 18%, dried 86%, molassed 90%);
    (j)   Straw (86%);
    (k)   Fodder roots – beet (18%), mangels, swede & turnips (12%), potatoes (21%), carrots (13%);
    (l)   Leaves of roots – (all 10 – 12%);
    (m)   Green crops – rape, kale, cabbage, comfrey (20%);
    (n)   Fruit residues (22%);
    (o)   Expelled oil seed residues – linseed, palm kernel, sunflower, poppy, rape, soya (all 90%);
    (p)   Full fat – soya, linseed, sesame seed, whole soya beans, soya bean screenings (all 90%);
    (q)   Copra (90%);
    (r)   Molasses (70%);
    (s)   Locust Bean (86%);
    (t)   Dairy products – whole milk (12%), whey (6.6%), skim milk (10%);
    (u)   Fishmeal (90%) – for non-ruminants only.
  (2)   Non-organic arable by-products incorporated into Organic silage – calculated as part of the permitted Non-organic allowance.
  (3)   Fats (non-ruminants only), oils and fatty acids provided they are used to balance ingredients of low energy value.

5.619   Restricted
  (1)   Additional straight feedstuffs not listed in paragraph 5.618 above.
  (2)   Commercially produced compounded feeds.

5.620   Prohibited
  (1)   Materials which have been subject to solvent extraction or other processing involving the addition of chemical agents (e.g. Ext. oils & seed residues).
  (2)   Animal by-products (e.g. meat, offal, feather meals).
  (3)   Sawdust, and other non-food ingredients and fillers.
  (4)   Animal manures (e.g. poultry manure).
  (5)   Fishmeal (prohibited for ruminants only).
  (6)   Milk from dairy cows treated with antibiotics – must not be fed to any Organic or In-conversion stock during the manufacturer's stated withdrawal periods.
  (7)   Straw treated with ammonia or caustic soda.

(8) *Fats, oils and fatty acids used to produce high protein diets of high nutrient density designed to achieve very early maturity or high levels of production.*

(9) Urea

MINERAL AND VITAMIN SUPPLEMENTATION

5.621 On well established Organic farms, sound agricultural practices should render mineral supplementation unnecessary. Restricted supplements may be used following approval from the Certification Committee where there is evidence of a suspected dietary deficiency in home grown feeds as a result of soil deficiencies, or there is veterinary evidence for a deficiency within the livestock (e.g. blood analysis).

5.622 Permitted
(1) Seaweed or dried seaweed meal.
(2) Rock salt.
(3) Yeast.
(4) Di-calcium phosphate.
(5) Cod liver oil (including possibly a preservative).
(6) Wheat germ (for Vitamin E).
(7) Limestone/chalk flour.
(8) Calcined magnesite & Vitamin D (see 5.740 for treatments for milk fever and grass staggers).

5.623 Restricted
(1) *Trace element supplements – only where they are necessary to maintain nutrient balance. If such supplements are used, preference should be given to elements in chemically Organic combinations.*
(2) *Concentrated vitamins and pure amino acids – only when necessary to satisfy normal nutrient requirements.*
(3) Proteinaceous copper chelate.
(4) Cobalt chloride or sulphate.
(5) Sodium selenite & potassium iodide.
(6) In-feed straight and/or blended mineral salts.
(7) In-feed synthetic vitamins.
(8) Straight mineral licks free from additives.
(9) Mineral injections & boluses.

5.624 Prohibited
(1) Mineral licks with flavour enhancers, non-mineral additives, preservatives and urea.
(2) *Concentrated vitamins and minerals which produce diets of very high nutrient density designed to achieve very early maturity or high levels of production.*

IN FEED ADDITIVES AND MEDICATIONS

5.625 Prohibited
(1) Use of food additives and in-feed medication for prophylactic reasons without specific approval of the Certification Committee.
(2) *Use of feeds containing non-food ingredients intended to stimulate growth or production by modifying the gut microflora or the endocrine system (e.g. Antibiotics and Probiotics) except where used strictly for therapeutic reasons in accordance with section 5.7 of these Livestock Standards.*

5.7 ANIMAL HEALTH AND VETERINARY TREATMENTS

5.701 The prevention of disease is central to the approach of Organic livestock husbandry. Health in farm animals is not simply the absence of disease, but also the ability to resist infection, parasitic attack and metabolic disorders, as well as the ability to overcome injury by rapid healing.

5.702 *Animals should be sustained in good health by effective management practices, including high standards for animal welfare, appropriate diets and good stockmanship rather than relying on veterinary medication.*

5.703 *The practices employed in the management of livestock must therefore be directed towards maintaining animals in good health and preventing conditions where the use of conventional veterinary medication becomes necessary.*

5.704   If illness does occur, the aim must be to complement the animal's natural powers of recovery and to correct the imbalance which created the disorder, rather than simply to deal with the symptoms of the illness alone.

5.705   Treatment must never be withheld if an animal is seriously ill, suffering or considered by a veterinary surgeon or stockman unlikely to recover fully without treatment. Should treatment be withheld in these circumstances, the Certification Committee reserves the right to withdraw the Registration from that enterprise.

5.706   When any veterinary medicinal product is used, the withdrawal periods specified in paragraphs 5.744 & 5.745 must be observed and treatment records kept as required in paragraph 2.313.

5.707   Recommended
(1)   A plan for the reduction of routine use of veterinary medicinal products and an outline of general disease control strategy, including the treatment of 'known farm problems' compiled in consultation with a veterinary surgeon, who may seek guidance from a consulting veterinary surgeon recommended by the Symbol Department.
(2)   Isolation or hospitalisation facilities for quarantined or sick animals conforming to the MAFF Code of Recommendations for Animal Welfare.

5.708   Restricted
(1)   The routine use of prophylactic veterinary medicinal products – restricted to cases of a known farm problem as indicated by the farm's veterinary surgeon (see paragraph 3.207).

5.709   Prohibited
(1)   Prophylactic use of veterinary medicinal products where no known farm problem exists.

ANTIBIOTICS

5.710   The use of antibiotics and some other conventional products may reduce natural immunity and, although providing rapid initial recovery, can leave an animal more prone to re-infection. Such problems are less likely to occur where veterinary advice is sought on treatment, so that an appropriate product is used at an appropriate dose and for a sufficient period of time.

5.711   Permitted
(1)   The use of antibiotics in clinical cases where no other remedy would be effective or after major trauma as a consequence of surgery or accident.

5.712   Prohibited
(1)   The routine use of antibiotics.
(2)   Dry Cow Therapy.

COMPLEMENTARY THERAPIES

5.713   For conditions requiring treatment and where effective alternative treatments are not available, then conventional veterinary medicinal products should be used, in order to save life, to prevent unnecessary suffering, or to provide the only way to restore the animal to full health.

5.714   Permitted
(1)   Complementary and natural therapies used where:
(a)   These methods have been shown to be effective; or
(b)   Under professional veterinary guidance; and in particular
(c)   When conventional therapies are not available or unsuitable.
(2)   The following complementary treatments where appropriate:
(a)   Homoeopathic nosodes and remedies;
(b)   Naturopathy;
(c)   Acupuncture;
(d)   Herbal – unlicensed herbal preparations should only be used as a tonic or for the treatment of individual animals or a small proportion of the flock or herd on a trial basis.

GROWTH PROMOTERS AND HORMONES
5.715   Permitted
    (1)   Natural prostaglandin or corticosteroid – administered by a veterinary surgeon in the rare case of the need to induce parturition for veterinary reasons.
    (2)   Hormone treatments for specific disorders where no alternative and effective treatment is available to restore the animal to full health.
5.716   Prohibited
    (1)   All growth promoters and hormones for heat synchronization, production stimulation (including Bovine Somatotrapin) and suppression of natural growth controls.

VACCINES
5.717   Vaccination is restricted to cases where there is a known disease risk on a farm or neighbouring land which cannot be controlled by any other means and the diseases present on the farm or in the area must be notified to the Certification Committee.
5.718   Recommended
    (1)   Using vaccines specific to the known farm diseases – replacing multi-vaccines with 4 in 1 or 2 in 1 vaccines if possible.

OTHER VETERINARY TREATMENTS
5.719   Anaesthetics must be used to prevent suffering as advised by a veterinary surgeon and required by law.
5.720   Prohibited
    (1)   The routine use of all other conventional veterinary medicinal products unless:
        (a)   Specifically excepted (see 'Veterinary' section for each class of livestock); or
        (b)   Approval for use of the product has been obtained from the Certification Committee.

SPECIFIC AILMENTS
5.721   *The following control measures may be used in cases where specific diseases or health problems occur and where there is no alternative treatment or effective management practice:-*

INTERNAL PARASITIC WORMS
5.722   The control of intestinal worms must be achieved primarily by good livestock management practices and where appropriate, optimum stocking rates, rotational grazing, clean grazing systems and mixed stocking.
5.723   In the case of a breakdown in the system which requires treatment, a proposal for improving the control of internal parasites by non-veterinary means should be prepared. Adequate monitoring of the efficacy of the control programme, i.e. by the use of worm counts, should be demonstrable.
5.724   Recommended
    (1)   Control of lungworm by allowing suckled calves to develop natural immunity by grazing grass with their dams.
    (2)   Control of internal worms by breeding for, and the use of, breeds of stock with resistance to infection.
    (3)   Control of intestinal worms by grazing management and pasture rotation.
    (4)   Control of Nematodirus by not grazing lambs on the same pasture in consecutive years.
5.725   Permitted
    (1)   The use of Anthelmintics in cases where animals are known to be carrying an unacceptable worm burden but not on a routine basis.
    (2)   The use of Oral Husk Vaccine (Dictol) before turning out weaned calves – in cases of a known farm problem.
5.726   Restricted
    (1)   The routine use of Anthelmintics in cases where:
        (a)   A known farm problem exists;

(b)   *Stocking rates and other factors are acceptable; and where feasible, a programme for controlling the parasite has been supplied to the Certification Committee.*

5.727   Prohibited
        (1)   The use of pour-on worming treatments.
        (2)   Treatments which leave residues in the faeces and which interfere with the normal breakdown of the dung or damage to the flora or fauna (Ivermectin based products).

EXTERNAL PARASITES

5.728   Permitted
        (1)   Where non-chemical controls cannot be used effectively:
                (a)   Flumethrin – (Bayticol or Coopers Gree) for sheep scab;
                (b)   Cyromazin – (Vetrazin) for fly control on sheep;
                (c)   Deltamethrin – (Spot-on) for fly control.
        (2)   Iodiform-based products for fly strike.

5.729   Restricted
        (1)   Ivermectin may be used as a pour-on treatment for warble fly only when enforced by a Statutory Authority to control an outbreak.

5.730   Prohibited
        (1)   Organo-phosphorus and Organo-chlorine (gamma HCH) compounds – as dips, sprays and creams for warble fly, external parasites, sheep scab and fly control.
        (2)   If organo-phosphorous compounds are used in compliance with Statutory requirements then the animals must be permanently marked at the time of treatment and:
                (a)   In the case of meat animals, the animal immediately and irrevocably loses its Organic status;
                (b)   In the case of dairy animals, the animal must undergo the normal conversion period as indicated in paragraph 5.302b before the products again qualify for Organic status. Such animals may not be culled for Organic meat.

FOOT PROBLEMS

5.731   Permitted
        (1)   Zinc sulphate.
        (2)   Copper sulphate.
        (3)   Iodine.

5.732   Prohibited
        (1)   Formaldehyde footbaths.
        (2)   Footrot vaccines.

NAVEL ILL

5.733   Permitted
        (1)   Strong iodine – used at birth to prevent infection.

BLOAT

5.734   Permitted
        (1)   Vegetable oils.
        (2)   Polaxalene.

MASTITIS

5.735   Recommended
        (1)   Frequent stripping of the affected quarter.
        (2)   Cold water treatments.
        (3)   Licensed herbal udder creams.

5.736   Permitted
        (1)   Homoeopathic treatments including nosodes for prevention.
        (2)   Antibiotics – in clinical cases where no other remedy would be effective.

5.737   Prohibited
        (1)   Dry Cow Therapy.

SCOUR IN YOUNG STOCK

5.738   Recommended
        (1)   The prevention of scour by the following practices:

    (a)   Outdoor calving and lambing;
    (b)   Well ventilated housing with clean dry bedding;
    (c)   Colostrum from dam within 6 hours;
    (d)   Sterile utensils;
    (e)   Clean/safe grazing systems.

5.739 Permitted
    (1) Treatment with a glucose/electrolyte solution (oral re-hydration therapy).
    (2) Treatment with veterinary medicinal products for individual cases.

5.740 Prohibited
    (1) Routine use of antibiotics and Anthelmintics.

GRASS STAGGERS AND MILK FEVER
5.741 Permitted
    (1) For the treatment of grass staggers:
        (a) Dusting of pastures with calcined magnesite;
        (b) Dietary/liquid Magnesium supplements.
    (2) For the treatment of milk fever:
        (a) Calcium borogluconate;
        (b) Magnesium & Phosporous salts;
        (c) Vitamin D used judiciously to prevent the condition.

ORF
5.742 Permitted
    (1) Homoeopathic remedies.

BSE AND SCRAPIE

5.743 Due to the long incubation period of BSE and the possibility of compound feeds containing contaminated products in the past, Organic herds are not immune to outbreaks of BSE.

5.744 Recommended
    (1) The following code of practice if an outbreak occurs in an Organic herd:
        (a) A suspected case should be quarantined and reported to the Statutory Authority as soon as possible. Once confirmed the animal must be immediately removed from the herd and the Symbol Department must be informed.
        (b) Farm staff and others should take all precautions to prevent the spread of infective material from the quarantine quarters. Full disinfection procedures must follow the removal of infected animals.
        (c) Records should be kept to ensure that all off-spring of infected animals can be traced and removed from the farm. Compensation permitting, ideally such animals should be slaughtered and incinerated.
        (d) No BSE case should be submitted for surgery of any description (including Caesarean).
        (e) Veterinary surgeons should be asked to use new needles on Organic farms as there is evidence that the infective agent is not killed by sterilisation techniques. Farmers are encouraged to use new needles for each individual animal.

WITHDRAWAL PERIODS

5.744 *All veterinary medicinal products must be used in accordance with their UK product licence. Withdrawal times between administering the veterinary medicinal product and using the products from the animal, shall be at least three times those defined by the licence or the prescribing veterinarian and shall not be less than 14 days or three times the recommended period whichever is the longer.*

5.745 *There are no withdrawal periods for zinc sulphate, copper sulphate, iodine, glucose solution, magnesium supplements, calcined magnesite, calcium borogluconate, magnesium & phosphorous salts and vegetable oil*

VETERINARY TREATMENT RECORDS

5.746 *Precise, accurate and up-to-date records of all treated livestock must be kept, including the details required by Schedule 2 of the Animals, Meat and Meat Products (Examination for Residues and Maximum Residue Limits) Regulations 1991. The animals concerned must be clearly identified. All conventional and other veterinary treatments including the duration, the brand name and manufacturer of any drug used and the actual withdrawal times must be recorded. (See paragraph 2.313 for requirements).*

5.8 HANDLING AND TRANSPORTATION OF LIVE ANIMALS AND BIRDS

5.801 *The principles for the welfare of Organic livestock must also be applied to their handling and transportation. Careful handling of animals in transit will reduce the risk of fatigue, pain, injury and stress-induced physiological changes to the meat at slaughter.*

5.802 *When animals, including birds, are transported they must be handled with proper care and concern for their welfare and in accordance with all relevant legislation and MAFF Codes of Recommendations for Animal Welfare.*

5.803 *During the making up of loads, loading to vehicles, during transit and on unloading, the animals must be handled in conditions which minimise stress and avoid the likelihood of injury. In this regard care should be taken to:*

    (a) *Ensure that the operations are carried out by experienced staff in a relaxed manner;*

    (b) *Avoid the mixing of animals from different social groups;*

    (c) *Avoid the use of unnecessary physical force on the animals;*

    (d) *Ensure that correctly designed and maintained handling facilities are provided at the points of loading and unloading;*

    (e) *Ensure that the vehicles are adequately ventilated throughout the journey.*

5.804 *Liaison between the producer, the haulier and the consignee on the time of collection and arrival should be established and the journey time arranged to ensure that the transit time between the holding and the destination is kept to a minimum.*

5.805 *The vehicles used for transporting animals must be suitable for the purpose, be properly equipped and maintained in a clean and hygienic condition. They must be cleaned and disinfected between loads.*

5.806 *Vehicles must be driven with care, avoiding high speeds, sudden starting or stopping or rapid cornering, in order to avoid damage or injury to the animals.*

CATTLE, SHEEP AND PIGS

5.807 Only fit animals may be transported (unless under veterinary supervision) and they must be presented in a clean and rested condition.

5.808 *Properly designed handling facilities should be provided on farms, and where races and hurdles are required for moving animals, they should be of a solid construction. Driving boards should be used to move pigs in the required direction.*

5.809 *If it is likely that the animals will have to be fed during transit or during a holding period at an abattoir lairage, the producer should provide the requisite amount of Organic feed.*

5.810 *Vehicles used for transporting animals should be properly equipped for the purpose and in particular:*

    (a) *Gates must be used to partition animals from different social groups to ensure that they are kept apart during transport;*

    (b) *When a vehicle is only partly full, gates should be used to restrict the movement of animals during transit.*

5.811 Recommended

    (1) Use of local abattoirs to minimise travelling time.

5.812 Permitted

    (1) A maximum journey time of 8 hours – arrangement should be made to rest and water livestock if journey times exceed this, bearing in mind the additional stress caused by unloading and loading.

5.813 Prohibited

    (1) The use of undue force (e.g. electric goads).

    (2) Overcrowded vehicles.

    (3) The transport of unfit animals – they must receive veterinary treatment as soon as possible.

(4) The mixing of Organic and Non-organic livestock.

(5) The export of Organic livestock for slaughter.

POULTRY

5.814 *Only fit birds may be transported, those which are unfit should be treated without delay or killed as quickly as possible using approved humane slaughter methods.*

5.815 *During loading, unloading and during the period while awaiting slaughter, the birds must be protected from the elements.*

5.816 *During transit, each bird should have sufficient space to rest and stand up without restriction, they should be protected from undue fluctuations in temperature, humidity and air pressure and sheltered from extremes of weather.*

5.817 *The journey time between farm and destination must not exceed 8 hours duration from start to end. The journey time being defined as the time from loading the first bird and the unloading the last bird in the consignment.*

5.818 *Where there is a delay in unloading, vehicles should not be left unattended stationary for a lengthy period unless suitable facilities exist for providing ventilation.*

THE ABATTOIR

5.819 The Standards for slaughtering livestock, meat and meat products (see processing standards), also apply to livestock producers wishing to have animals slaughtered.

5.822 Production establishments (including abattoirs) that process organically produced agricultural products must be registered as Approved Operators and hold a valid Certificate of Registration.

5.821 An abattoir should be licensed either by:

(a) Applying to register directly with the Symbol Scheme; or

(b) Having an application made on their behalf by a producer or operator who treats the abattoir as a sub-contractor (see processing standards).

5.822 Once inspected, approved and registered, a Symbol Meat Stamp will be issued for stamping Symbol Standard carcases, sides and quarters, by the MLC Inspector or other independent personnel.

5.823 All animals must be given access to clean water and comfortable conditions to reduce stress and allow recovery.

5.824 In addition, when animals have to wait for a period before being slaughtered, the following conditions must also be made available:

(a) For an anticipated waiting time of 6 hours or more – bedding must be provided from the beginning and there must be sufficient space for the animals to lie down;

(b) For an anticipated waiting time of 12 hours or more – they must in addition be provided with Organic feed.

5.825 Recommended

(1) Using local abattoirs wherever possible to reduce travel stress.

(2) Delivering animals to an abattoir the previous evening so as to permit them to recover before slaughter where long travel times are involved and slaughtering is to be the first operation of the day.

(*continues overleaf*)

SECTION 6

DETAILED STANDARDS FOR INDIVIDUAL LIVESTOCK CATEGORIES

6.1 CATTLE

6.101 In addition to the General Standards for Livestock Husbandry (Section 5) the following Standards also apply to dairy cattle, calf rearing and beef systems.

DAIRY CATTLE

DIET (NON-ORGANIC ALLOWANCES)

6.102 Permitted
  (1) Brought-in approved feedstuffs (see paragraphs 5.618 to 5.620) from Non-organic sources – fed in accordance with the following daily and annual allowances (kg DM):

Stock	Total Daily DMI	Non-Organic allowance	
Dairy cow		Daily	Annual
(a) Friesian/Holstein	20.0	4.0	650*
(b) Friesian	18.0	3.6	650*
(c) Channel Island	14.0	2.8	650*
(b) In-calf heifer	10.0	2.0	N/A

  * NB 720Kg for Northern England & Scotland where a 200 day winter feeding programme is practised.

DAIRY HYGIENE

6.103 Permitted
  (1) Boiling water or steam sterilisation of milking plant.
  (2) Sterilants and cleansing agents approved by the Statutory Authority for use in milking parlours and dairies.
  (3) Hypochlorite.

CALF REARING

6.104 Ideally, every calf should be reared by its own mother. The natural vigour and resistance to infection that this produces will overcome most, if not all, of the ailments which befall calves reared under artificial conditions.

BROUGHT-IN CALVES

6.105 Restricted
  (1) Beef calves from Non-organic herds bought for multiple suckling on Organic suckler cows (parallel production) are not eligible for Organic status. Different breeds must be used, be identified and managed separately after weaning and not sold as Organic animals.

6.106 Prohibited
  (1) Organic calves purchased from livestock markets.

WELFARE AND HOUSING

6.107 Recommended
  (1) Calving outdoors where conditions permit.
  (2) Group housing in open fronted straw yards.
  (3) Disbudding and castration (where necessary) under three months of age.

6.108 Permitted
  (1) Individual pens up to 14 days, constructed so that each calf can see other calves, and can get up, lie down and turn around without difficulty.

6.109 Prohibited
   (1) Individual pens beyond 14 days.
   (2) Tethering of calves.
   (3) Disbudding and castration when over three months of age unless carried out by a veterinary surgeon.
   (4) The sale of non-pedigree calves for export.

DIET (NON-ORGANIC ALLOWANCES)
6.110 *Cattle systems should be planned to allow calves to be fed on milk produced to these Livestock Standards until they are eating solid food. Where in emergencies this proves impossible, milk replacers without antibiotics or other growth promoters may be used. Preference should be given to the use of reconstituted organically produced milk powder.*
6.111 All un-weaned calves must receive milk at least twice daily for a minimum of the first 9 weeks of life. If housed, they must have access to good quality straw, hay or silage and fresh, clean water.
6.112 Recommended
   (1) Colostrum – preferably suckled within 6 hours of birth.
   (2) Organic whole milk suckled from dam or nurse cow until weaning.
   (3) Continuation of once or twice-daily suckling beyond 9 weeks.
6.113 Permitted
   (1) Bucket or artificial-teat rearing on Organic whole or re-constituted milk.
   (2) Non-organic milk or milk replacer free from antibiotics and additives – emergency use only.
   (3) Brought-in approved feedstuffs (see paragraphs 5.618 to 5.620) from Non-organic sources – fed in accordance with the following maximum daily and annual allowances (kg DM):

Age	Ave. Wt. (kg)	Ave. Daily DMI	Non-organic Allowance	
			Av. Daily	Annual
0–12 mths	168	5.6	0.56	100

6.114 Prohibited
   (1) Non-organic milk except in emergencies.
   (2) Milk replacer containing antibiotics and additives.

WEANING
6.115 Recommended
   (1) Weaning only when a calf is taking adequate solid food to cater for its full nutritional requirements.
   (2) Natural weaning.
6.116 Permitted
   (1) Calves over 6 months of age may be taken straight from the cow and sold off the farm, providing they are eating adequate quantities of dry food.
6.117 Prohibited
   (1) Weaning before 9 weeks of age

BEEF SUCKLER COWS

DIET (NON-ORGANIC ALLOWANCES)
6.118 Permitted
   (1) Brought-in approved feeds (paragraphs 5.619–5.621) from Non-organic sources fed in accordance with the following maximum daily and annual allowances (kg DM):

Stock		Total Daily DMI	Non-organic allowance	
			Daily	Annual
(a)	Suckler cow (Lowland)	14.0	1.4	325
(b)	Suckler cow (LFA)	14.0	N/A	650

STORE & FINISHING CATTLE

BOUGHT IN STOCK

6.119 Restricted
(1) Beef store cattle from non-registered herds bought for finishing on Organic units with an existing Organic beef enterprise – different breeds must be used, the animals must be identified and managed separately and not sold as Organic animals.

WELFARE AND HOUSING

6.120 *As far as possible, systems for producing stores and meat animals must be based on grazing, but animals may be finished in well bedded spacious yards.*

6.121 Where cattle are housed in groups, the size range between the smallest and largest animals must not be so great as to allow bullying.

6.122 Aggressive and horned cattle must be housed separately. If numbers of horned cattle are kept together, additional lying and feeding space must be allowed.

DIET (NON-ORGANIC ALLOWANCES)

6.123 Recommended
(1) Finishing on Organic grass/clover or conserved forage, roots with moderate amounts of cereals where necessary.

6.124 Permitted
(1) Brought-in approved feeds (see paragraphs 5.618 to 5.620) from Non-organic sources fed in accordance with the following maximum daily and annual allowances (kg DM):

Stock		Age	Av.Wt.	Av. Daily DMI	Non Organic allowance Av. Daily	Annual
(a)	Lowland	0–12mths	200kg	5.6	0.56	100*
		12–24mths	500kg	10.0	1.00	180*
(b)	LFA	0–12mths	200kg	5.6	N/A	200*
		12–24mths	500kg	10.0	N/A	360*

* NB For Northern England and Scotland multiply by 1.11 where a 200 day winter feeding programme is practised.

6.125 Prohibited
(1) High energy, low fibre rations and rations containing more than 40% DM concentrates (e.g. barley beef systems).

6.2 SHEEP

6.201 In addition to the General Standards for Livestock Husbandry (Section 5) the following Standards also apply to sheep.

BROUGHT-IN STOCK

6.202 Permitted
(1) The conversion of existing and new flocks and replacements from Non-organic sources (see section 5.3 for conversion periods).
(2) Replacement ewes from Non-Organic sources (see section 5.3 for conversion periods).

6.203 Prohibited
(1) Brought-in ewes with lambs at foot or store lambs from Non-organic sources for finishing as Organic meat.
(2) The annual buying in of flocks from Non-organic sources before tupping.
(3) The grazing of Organic ewes on Non-organic land between weaning and tupping.

WELFARE AND HOUSING

6.204 Stocking rates must be appropriate for the farm conditions and will generally be lower than under conventional management.

6.205 Where housed for lambing or in-wintering, the facilities must ensure that there is:
(a) A minimum of 500mm of trough space per heavy in-lamb ewe for concentrate feeding;

(b) A maximum of 40 ewes per pen if in-wintered;
(c) A maximum of 100 ewes per pen if yarded at lambing;
(d) A sufficient area in the pens for lying and loafing with a minimum straw bedded area of 1 sq. m/ewe, or 2 sq. m/ewe and lambs;
(e) Good ventilation without pockets of stale air or excess draughts.

6.206 Permitted
(1) In-wintering of sheep – subject to the above housing requirements.
(2) Winter shearing – in-wintered ewes only.

6.207 Prohibited
(1) Permanent indoor housing of ewes and lambs.

DIET (NON-ORGANIC ALLOWANCES)
6.208 Permitted
(1) Goat's colostrum for orphan lambs.
(2) Cow's colostrum for orphan lambs (NB may contain anti-sheep antibody and should be tested for this before use).
(3) Milk replacer free from antibiotics and additives for orphan lambs – in emergencies only.
(4) Brought-in approved feeds (see paragraphs 5.618–5.620) from non Organic sources fed in accordance with the following maximum daily and annual allowances (kg DM):

Stock		Total Daily DMI	Non-organic allowance	
			Daily	Annual
(a)	Lowland In-lamb ewe	2.0	0.2	30
	Lambs 0–6 months	0.7	0.07	N/A
(b)	LFA In-lamb ewe	2.0	N/A	60
	Lambs 0–6 months	0.7	N/A	N/A

VETERINARY PRACTICE
6.209 Permitted
(1) Where docking and castration are practised for animal welfare reasons:
(a) Rubber rings used within 72 hours of birth.
(b) A knife and hot iron used between three and six weeks of birth.

6.210 Restricted
(1) Anthelmintics for all ewes at lambing time – in cases where animals are showing signs of carrying an unacceptable worm burden.

6.211 Prohibited
(1) Teeth cutting and grinding.

6.3 PIGS
6.301 In addition to the General Standards for Livestock Husbandry (Section 5) the following Standards also apply to pigs.

WELFARE AND HOUSING FOR BREEDING PIGS
6.302 *Wherever possible pig breeding systems should be planned to allow the sows direct access to the soil and growing green food on free range except when bad weather or unsuitable soil conditions make housing preferable.*
6.303 Housing must allow sows to express their full range of normal behaviour patterns, and must not involve permanent confinement or any housing system which prevents the sows from getting up, lying down and turning round without difficulty.
6.304 The housing facilities must provide:
(a) Ample dry bedding with plentiful natural ventilation and light;
(b) *Access to on outside run for most of the breeding cycle (with dunging, rooting and exercise areas and a rubbing post and implements for play);*
(c) Individual housing for sows with piglets with a minimum lying area of 1 sq. metre/100kg liveweight;
(d) Stable social groups of gilts or sows with a maximum of 10 animals per group.
6.305 There must be ample trough space for all pigs to feed if not fed ad-lib.

6.306 Bullying must be avoided in group-housed dry sows/gilts, particularly at feeding. Simultaneous or individual feeding is preferred.
6.307 Prohibited
    (1) Grazing of sows on Non-organic land or the management of sows other than in accordance with these Livestock Standards between weaning and conception.

FARROWING AND WEANING
6.308 Recommended
    (1) Sows settled into farrowing accommodation well before piglets are due to be born.
    (2) A protective rail, farrowing box or nest.
    (3) Straw bedding (suitable temperatures maintained in the nest by use of straw).
    (4) Weaning at 8 weeks.
6.309 Permitted
    (1) Additional heat in the creep area.
    (2) Sow milk replacer free from antibiotics for orphans.
    (3) Weaning at 6 weeks provided they are taking adequate solid food.
6.310 Prohibited
    (1) The use of Farrowing Crates.
    (2) *Pigs weaned at under six weeks except in emergencies where milk replacers or suitable dry diets without antibiotics may be used.*
    (3) Withholding food and water for drying off sow

WELFARE AND HOUSING FOR FATTENING PIGS
6.311 Recommended
    (1) Fattening pigs on grassland where conditions permit.
    (2) Retaining weaners in family groups.
6.312 Permitted
    (1) The indoor management of fattening pigs provided that:
        (a) The housing must provide ample dry bedding with plentiful natural ventilation and light and outside dunging, rooting and exercise areas with a rubbing post and implements for play;
        (b) The group size does not exceed 30;
        (c) The minimum lying area is:

Kg L wt.	Sq. metres
20	0.20
40	0.35
60	0.50
80	0.65
100	0.75

        (d) A minimum of one drinking point per 10 pigs.
    (2) Grading pigs into size and sex at weaning.
6.313 Prohibited
    (1) Sweat boxes and other housing with artificially controlled environments.

DIET (NON-ORGANIC ALLOWANCES)
6.314 Recommended
    (1) The daily feeding of fibrous roughage and/or green food when access to range is restricted.
6.315 Permitted
    (1) Brought-in approved feeds (see paragraphs 5.618 to 5.620) from Non-organic sources fed in accordance with the following maximum daily allowances (kg DM per pig):

Stock		Total Daily DMI	Non-organic allowance Daily	Annual
(a)	Sow + 6 piglets	4.50	1.35	N/A
	+ each extra piglet	+0.40	+0.12	N/A
(b)	Gilts	2.60	0.78	N/A
(c)	Weaners at 9 weeks	1.00	0.30	N/A
	at 25 weeks	2.65	0.74	80.0

6.316 Restricted
    (1) Higher amounts of Non-organic feed for animals with higher dry matter intakes.
    (2) Swill feeding – ingredients and composition to be supplied.

6.317 Prohibited
    (1) Antibiotics, copper diet supplements and probiotics for growth promotion.

VETERINARY PRACTICE

6.318 Permitted
    (1) Teeth cutting for individual piglets or a litter when necessary to prevent injury to the sow and for breeding boars.
    (2) Ringing of sows, gilts and boars to preserve pasture.
    (3) Ferrous sulphate crystals for anaemia.

6.319 Restricted
    (1) Iron injections for anaemia in the case of iron deficient soils or chronic anaemia in free range systems.

6.320 Prohibited
    (1) Tail docking.
    (2) Routine teeth cutting.
    (3) Castration of all pigs.
    (4) Prophylactic use of iron injections.

6.4 POULTRY

6.401 In addition to the General Standards for Livestock Husbandry (Section 5) the following Standards also apply to poultry.

6.402 These Standards apply primarily to chickens. Those for turkeys, ducks, geese and other poultry will be assessed on the same principles as those for chickens, making appropriate alterations to the specific requirements.

WELFARE

6.403 *All egg production systems must be planned to allow the birds to have continuous and easy daytime access to open air runs, except in adverse weather conditions. The land to which the birds have access must be adequately covered with properly managed and suitable vegetation.*

6.404 Where set stocking is practised, stocking rates should not exceed 250 birds/acre.

6.405 Pasture should be rested for one year in three where set stocking is practised unless stocking densities are low enough to prevent damage to the grassland and avoid disease build-up.

6.406 Birds/colonies/flocks must be provided with protection from predators and have access to shelter. Housing should provide them with sufficient space to move around freely, and there should be litter to scratch in.

6.407 Prohibited
    (1) Permanent housing of poultry.

COLONY SIZES

6.408 Recommended
    (1) Not more than 100 laying birds per housing group
    (2) Not more than 200 fattening birds per housing group.

6.409 Permitted
    (1) Up to 500 birds per housing group

HOUSING

6.410 Housing should be disinfected between batches with iodoform, steam or blow torch, or lime depending on the construction of the house.

6.411 Litter must be regularly replenished and kept in a dry and friable condition.

LAYING BIRDS

6.412 Artificial lighting must not prolong the day length beyond 16 hours.

6.413 Recommended
    (1) Housing designed for a maximum of:
        (a) 15kg (Live wt. of birds)/sq. m floor area;

       (b)   20cm perch space/bird;
       (c)   One nest space/5 birds.

6.414   Permitted
    (1)   Fixed houses may have slatted floor with collection area for droppings and/or solid floor with litter.
    (2)   Houses moved twice per week may have slatted floors with no dropping boards.
    (3)   Housing designed for a maximum of:
       (a)   25kg (Live wt. of birds)/sq. m floor area;
       (b)   15cm perch space/bird;
       (c)   1 nest box/8 birds;
       (d)   They must be supplemented with additional shelter against wind and rain.

6.415   Prohibited
    (1)   Housing densities in excess of 25kg (Live wt. of birds) or 12 birds/sq. m floor area.

FATTENING BIRDS

6.416   Recommended
    (1)   Housing designed for a maximum of:
       (a)   18kg (Live wt. mature birds)/sq. m floor area;
       (b)   75% of floor area littered;
       (c)   25% area as a raised slatted or weld mesh roosting area with enclosed droppings collection area below.

6.417   Permitted
    (1)   Housing designed for a maximum of 25kg (Live wt. mature birds) or 12 birds/sq. m floor area with all the area littered.

6.418   Prohibited
    (1)   Buildings with all wire or slatted floors.
    (2)   Housing densities in excess of 25kg (Live wt. mature birds) or 12 birds/sq. m floor area.

DIET (NON-ORGANIC ALLOWANCES)

6.419   Permitted
    (1)   Brought-in approved feedstuffs (see paragraphs 5.618 to 5.620) from Non-organic sources fed according to the following maximum daily allowances (kg DM):

Bird Type		Total Daily DMI	Non-Organic Allowance	
			Daily	Annual
(a)	Layers	0.118	0.035	12.8
(b)	Table birds	0.077	0.023	N/A
(c)	Turkeys	0.138	0.042	N/A
(d)	Geese	0.150	0.045	N/A
(e)	Ducks	0.150	0.045	N/A

6.420   Restricted
    (1)   Higher amounts of non-organic feed for birds with higher dry matter intakes.

6.421   Prohibited
    (1)   Yolk colourants, in-feed medication or any other feed additives.

VETERINARY PRACTICE

6.422   Permitted
    (1)   Clipping primary flight feathers – individual birds only to prevent escape.
    (2)   Derris powder for parasites in dust baths or directly applied.
    (3)   Coccidiostats in poultry starter rations.
    (4)   Use of Amprolium for Coccidiosis.

6.423   Prohibited
    (1)   Beak clipping and all other mutilations.

6.5 HONEY PRODUCTION

6.501   Honey must be taken from bees that forage only in organically cultivated areas or areas of natural vegetation which are free of herbicides and pesticides, and which have been so for a minimum of two years.

WELFARE

6.502   Hive management and honey extraction methods should be aimed at preserving the colony and sustaining it. The colony must not be destroyed when the honey is harvested.

6.503   The foundation of the comb used in the production of comb honey must be made from organically produced beeswax.

6.504   Permitted
(1)   Selective breeding may be practised.

6.505   Prohibited
(1)   Artificial insemination.

DIET

6.506   Hives must not be placed within 4 miles of conventionally farmed land, private gardens or areas subject to inorganic pollution such as roadside verges and roadways.

6.507   Permitted
(1)   Only organically produced honey and natural pollen supplements.

6.508   Prohibited
(1)   Other materials used for feeding – the hive must be taken out of Organic production for a minimum period of 12 months and all honey residues must be removed from the hive before it is returned to the production of organically produced honey.

VETERINARY PRACTICE

6.509   Permitted
(1)   Wing clipping of queens.
(2)   Chemical medicinal treatment of the bees only when the health of the colony is threatened. After such treatment, the hive must be immediately taken out of Organic production for a minimum of 12 months and all honey residues must be removed from the hive before it is returned to the production of organically produced honey.

TRANSPORT AND PROCESSING

6.510   All bulk containers must comply with the Packaging (see section 8.8), Transport (see section 8.9) and Labelling (see section 7.7) Standards.

6.511   All processing must comply with the Standards for Processing Organic Honey (see section 9.5).

Biodynamic Farming and Gardening

Biodynamic agriculture has been mentioned at several points in this book. Although it has much in common with organic farming, the biodynamic approach also has several unique features which are outlined below. Further information on biodynamic agriculture can be obtained from the Bio-Dynamic Agricultural Association and Emerson College, whose addresses are given in Appendix 4.

SCOPE AND HISTORY

Biodynamic farming and gardening is practised on all continents, in cool and warm climates. A recent survey showed that there are over 80 farmers and smallholders, responsible for 2,500 ha of land, actively practising biodynamic agriculture in Britain; there are more than twice that number practising biodynamic gardening, either on a commercial scale or privately. In 1982, 1,090 commercial farms and gardens, cultivating 17,609 hectares, were reported from some West European countries. Twenty-eight fulltime advisers, 124 processors and 47 wholesalers are connected with them. Larger groups are working in North and South America, New Zealand and Australia. Apart from the larger and family sized farms in other countries, there is an unknown number of smallholdings and private gardens. Top fruit, including citrus, and soft fruit are also produced commercially.

The movement began in 1924, following a series of lectures which Rudolf Steiner, founder of the anthroposophical science of the spirit, gave at the request of farmers. From the outset, a comprehensive approach to farm organisation, husbandry, and the production of quality food was set in motion. Farmers, gardeners, advisers and scientists are today organised into biodynamic associations which are active in many countries. These organisations provide literature, advisory services, and some run their own research facilities.

Another innovative move by the biodynamic movement was the introduction of a certification scheme in 1928 for marketing basic foodstuffs, and today both primary and processed foodstuffs are marketed under the DEMETER and Biodyn trademarks.

PRINCIPLES AND CONCEPTS

In common with other organic farming approaches food quality is defined within the method of production as a whole. Plant and animal husbandry are combined so as to achieve reasonable self-sufficiency with respect to manures and feedstuffs in a manner appropriate to the location of the farm. Other considerations, such as economic viability, marketing potential and human abilities are paid due attention.

The basics of good, sustainable production are also common to organic farming, including the building of soil life and fertility; botanical diversity in leys and crop rotations; careful handling and storage of farm-produced manures; the occasional use of lime and other slowly soluble minerals. In addition, biodynamic producers place special emphasis on the use of seeds that are adapted to the site and farming system, as well as the stimulation of natural processes in soils, composts, manures and plants through the use of biodynamic preparations. Many biodynamic growers also try to observe and study lunar periodicities in plant and animal life. With skilled management, the potential of such systems is considerable and when combined with further inoffensive controls, can reduce diseases, pests and weeds to tolerable levels. Animal husbandry aims to further achieve both production and health through breeding, housing, feeding and conformity with ethical values.

As with organic farming, conservation and environmental protection play an important role, with close observation of plant and animal bio-rhythms helping to integrate the life of the farm into its surroundings in a productive yet sustainable manner.

The farm as an organism

Viewing a biodynamic farm as an organism is the starting point from which the specific steps of fashioning and running it evolve. This idea grows from a constant striving towards a spiritual understanding of the specific forces acting in the mineral, plant and animal kingdoms of nature and their relation to the human being throughout evolution up to the present time. Attention is drawn to the fabric of both obvious and subtle relationship of each living thing to its near and distant surroundings. 'Nothing happens in living nature that is not in relation to the whole', said the German poet Goethe. Steiner's teaching helps the individual in his or her endeavour to develop insight, and supplements and rectifies the narrow, specialised, analytical-technical approach adopted in modern conventional farming.

The essentially sound structure of agriculture, developed over thousands of years, has been torn apart in the second half of this century, resulting in environmental concerns and economic instability. Biodynamic practitioners believe that these problems can be healed by a comprehensive understanding of the ecological, social and spiritual totality of farming.

The biodynamic preparations

Two field sprays known as preparations 500 and 501 are used in biodynamic crop production. These are made in a carefully specified manner from cow manure and quartz and sprayed on the crops in a highly diluted form at rates of 300g and 4g/ha respectively. A further six preparations, made from plants, are added to stimulate compost and manure heaps at a rate of 2–4ppm each. The underlying principle that leads to the use of these substances is new, but it can be rationalised as it is in some detail in the biodynamic literature.

Apart from a broad body of experience gathered from actually working with these substances, their effects have been established in field, pot and laboratory experiments, and positive results have been achieved with most farm and garden crops. These results include morphological and physiological changes in plants, such as ripening rates and dry matter, carbohydrate and protein contents, as well as other factors such as protein balance, enzyme activities, keeping quality and taste of produce. Some of this evidence is indicated in Chapter 15. Influences on the composting process have also been found.

Biodynamic farmers believe that it is the interaction between crop nutrients, water, energy and the formative influences which are provided by light, warmth and the biodynamic preparations which gives biodynamically produced food its particular quality.

Marketing and distribution

Producers, processors and traders can be granted the right to market under the DEMETER and Biodyn trademarks by entering into annually renewable contracts with the Demeter Guild. Binding regulations for management and purchased inputs are issued by this organisation. Usually two persons, the inspector or adviser and a local farmer, are assigned to control adherence to the regulations. The Guild enjoys the confidence of consumer circles.

As a result of the activities of the Demeter Guild around the world, there is a well established market for biodynamic produce, especially processed products including juices, rye crispbreads and babyfoods, which allows producers to make a respectable living (see Chapter 13).

Further information, including details of the Demeter and Biodyn standards, is available from the Bio-Dynamic Agricultural Association in Britain and its related organisations around the world (Appendix 4).

SELECTED READING

Star and Furrow, a twice yearly periodical published by the Bio-Dynamic Agricultural Association.

Bio-Dynamics, the periodical of the North American biodynamic movement.

Lebendige Erde, the bimonthly German language periodical for biodynamic agriculture and horticulture.

Koepf, H. H. *What is Bio-dynamic Agriculture?* Bio-Dynamic Agricultural Association. 44pp.

Koepf, H. H.: Petterson, B. & Schaumann, W. (1976) *Bio-dynamic Agriculture*. Anthroposophic Press; Spring Valley, New York.

Soper, J. (1983) *Bio-dynamic Gardening* Bio-Dynamic Agricultural Association.

Steiner, R. (1924) *Agriculture*. Bio-Dynamic Agricultural Association.

Appendix 3

General Reading

Specific references and further reading have been given at the end of each chapter. This appendix lists selected English language publications.

PERIODICALS

Britain

Elm Farm Research Centre Bulletin provides an up-to-date guide on the latest research results and policy developments

The Living Earth, the Soil Association's quarterly magazine on issues linking agriculture, food, health and the environment.

New Farmer and Grower, a technical quarterly magazine for producers, published by British Organic Farmers and the Organic Growers Association.

HDRA Newsletter, a quarterly newsletter on organic gardening published by the Henry Doubleday Research Association.

Biological Agriculture and Horticulture, a scientific journal on every aspect of organic farming.

International

American Journal of Alternative Agriculture, scientific journal, good on the US situation and development issues, published by the Institute for Alternative Agriculture.

Organic Gardening, the largest circulation organic gardening magazine in North America, published by Rodale Press.

Organic Farmer, a quarterly magazine providing a good overview of developments in organic ideas, certification, and policy issues in the United States.

The New Farm, a magazine primarily aimed at concerned conventional and low-input farmers in North America, published by Rodale Press.

Soil and Health, a quarterly periodical published by the Soil and Health Association of New Zealand.

IFOAM Bulletin/Ecology and Agriculture, international organic movement news and research developments, published by the International Federation of Organic Agriculture Movements.

Abstreco, a quarterly abstract journal published by the Department of Ecological Agriculture, Wageningen, Netherlands.

BOOKS

Altieri, M. (1987) *Agroecology – the scientific basis for alternative agriculture.* Intermediate Technology Publications; London.

Balfour, E. (1975) *The Living Soil and the Haughley Experiment.* Universe Books; New York.

Blake, F. (1987) *The Handbook of Organic Husbandry.* Crowood Press; Marlborough.

669

Boeringa, R. (1980) *Alternative Agriculture*. Elsevier; Amsterdam.

Conford, P. (ed.) (1988) *The Organic Tradition—an anthology of writings on organic farming, 1900–1950*. Green Books; Bideford.

Howard, A. (1948) *An Agricultural Testament*. Oxford University Press.

Knorr, D. (ed.) (1982) *Sustainable Food Systems*. AVI Publishing; Westport, Conn.

Koepf, H. *et al.* (1976) *Bio-dynamic Agriculture*. Anthroposophic Press; Spring Valley, NY.

National Research Council (1989) *Alternative Agriculture*. National Academy Press; Washington, DC.

Neuerburg, W. and S. Padel (1992) *Organisch-biologischer Landbau in der Praxis*. Verlagsunion Agrar, BLV, München.

Oelhaf, R. (1978) *Organic Agriculture*. Wiley; Chichester.

Poincelot, R. (1986) *Towards a More Sustainable Agriculture*. AVI Publishing; Westport, Conn.

USDA (1980) *Report and Recommendations on Organic Farming*.

Vine, A. & Bateman, D. (1981) *Organic Farming Systems in England and Wales*. University College of Wales; Aberystwyth.

Widdowson, R.W. (1987) *Towards Holistic Agriculture*. Pergamon Press.

Wolf, R. (1975) *Organic Farming*. Rodale Press; Emmaus, PA.

Wookey, B. (1987) *Rushall—the Story of an Organic Farm*. Blackwell; Oxford.

CONFERENCE PROCEEDINGS

Allen, P. and Dusen, D. v. (1988) *Global Perspectives on Agroecology and Sustainable Agricultural Systems*. Proceedings of the 6th IFOAM International Scientific Conference, Santa Cruz, 1986. University of California: Santa Cruz.

Besson, J. & Vogtmann, H. (eds.) (1978) *Towards a Sustainable Agriculture*. Proceedings of the 1st IFOAM International Scientific Conference, Sissach, 1977. Verlag Wirz; Aarau, Switzerland.

Besson, J.-M. (ed.) (1990). *Biological Farming in Europe—an expert consultation*. REUR Technical Series, 12. Food and Agriculture Organisation; Rome.

Bezdicek, D. (ed.) (1984) *Organic Farming*. Proceedings of an Agronomy Society of America conference on organic farming. Agronomy Society of America; Madison.

BOF/OGA (1989) *The Case for Organic Agriculture*. Proceedings of the 1989 Cirencester conference. British Organic Farmers; Bristol.

Dlouhy, J. & Nilsson, G. (eds.) (1983) *Comparison of Farming Systems*. Proceedings of IFOAM research colloquium. Swedish University of Agricultural Sciences; Uppsala.

Edens, T.C. *et al.* (eds.) (1985) *Sustainable Agriculture and Integrated Farming Systems*. 1984 Conference Proceedings. Michigan State University; E. Lansing.

Hill, S. & Ott, P. (1982) *Basic Technics of Ecological Farming*. Proceedings of the 2nd and 3rd IFOAM International Scientific Conferences, Montreal, 1978 and Brussels, 1980. Birkhaeuser Verlag; Basel, Switzerland.

IFOAM (1990) *Agricultural Alternatives and Nutritional Self-sufficiency*. Proceedings of 7th IFOAM International Scientific Conference, Burkina Faso, 1989. International Federation of Organic Agriculture Movements; Tholey-Theley, West Germany.

Köpke, U. and Schulz, D. G. (eds) (1993) *Organic agriculture: a key to a sound development and a sustainable environment*. Proceedings of the 9th IFOAM International Scientific Conference, Brazil, 1992. International Federation of Organic Agriculture Movements; Tholey-Theley, Germany.

Lockeretz, W. (1983) *Environmentally Sound Agriculture*. Proceedings of the 4th IFOAM International Scientific Conference, Boston, 1982. Praeger; New York.

Paoletti, M., Stinner, B. R. and Lorenzoni, G. G. (eds) (1989) *Agricultural Ecology and Environment*. Proceedings of International Symposium, Padova, Italy, April 1988. Reprinted from *Agriculture, Ecosystems and Environment* Vol. 27 Nos. 1–4 (1989).

Stonehouse, B. (ed.) (1982) *Biological Husbandry*. Proceedings of 1980 Wye College conference on organic farming. Butterworths; London.

Vogtmann, H. *et al.* (eds) (1986) *The Importance of Biological Agriculture in a World of Diminishing Resources*. Proceedings of the 5th IFOAM International Scientific Conference, Witzenhausen, 1984. Verlagsgruppe; Witzenhausen, West Germany.

Zerger, U. (ed.) (1993) *Forschung im biologischen Landbau*. Stiftung Ökologie und Landbau; Bad Dürckheim.

Addresses

Apart from the section on developing countries, this list is restricted to groups in Britain and Ireland and features particularly organisations mentioned in the text. Addresses of organisations in other countries can be obtained from the:

INTERNATIONAL FEDERATION OF ORGANIC AGRICULTURE MOVEMENTS (IFOAM)
Ökozentrum Imsbach
D-6695 Tholey-Theley
Germany

ORGANIC MOVEMENT ORGANISATIONS

BIO-DYNAMIC AGRICULTURAL
 ASSOCIATION
Woodman Lane
Clent
Stourbridge
West Midlands DY9 9PX
(0562) 884933

BRITISH ORGANIC FARMERS
86 Colston Street
Bristol BS1 5BB
(0272) 299666

CHARLES WACHER TRUST
(*promoting organic growing*)
Aeron Park
Llangeitho
Tregaron
Dyfed
SY25 6TT

ELM FARM RESEARCH CENTRE
Hamstead Marshall
Newbury
Berkshire RG15 0HR
(0488) 58298

HENRY DOUBLEDAY RESEARCH
 ASSOCIATION
National Centre for Organic Gardening
Ryton-on-Dunsmore
Coventry CV8 3LG
(0203) 303517

IRISH ORGANIC FARMERS AND GROWERS
 ASSOCIATION
14 Berkley Road
Dublin 7
Eire
or:
Killegland Farm
Ashbourne
Co. Meath
Eire
(01) 350225

ORGANIC GROWERS ASSOCIATION
86 Colston Street
Bristol BS1 5BB
(0272) 299800

ORGANIC SHEEP SOCIETY
North Wyke Research Station
North Wyke
Oakhampton
Devon EX20 2SB
(0837) 82 558

SOIL ASSOCIATION
86 Colston Street
Bristol BS1 5BB
(0272) 290661

WORKING WEEKENDS ON ORGANIC FARMS
(WWOOF)
19 Bradford Road
Lewes
East Sussex BN7 1RB
(0723) 476286

RESEARCH AND ADVICE

AGRICULTURAL DEVELOPMENT AND
   ADVISORY SERVICE (ADAS)
c/o Roger Unwin
MAFF
Burghill Road
Westbury-on-Trym
Bristol
BS10 6NJ
(0272) 591000

CENTRE FOR ORGANIC HUSBANDRY AND
   AGROECOLOGY
Llandinam Building
University College of Wales
Aberystwyth
Dyfed SY23 3DB
(0970) 622250

ELM FARM RESEARCH CENTRE
(see above)

HENRY DOUBLEDAY RESEARCH
   ASSOCIATION
(see above)

ORGANIC ADVISORY SERVICE
c/o Elm Farm Research Centre
(see above)

ORGANIC FARMING CENTRE
Edinburgh School of Agriculture
West Mains Road
Edinburgh
EH9 3JG
(031) 667 1041

TEAGASC
Johnstown Castle Research Centre
Wexford
Eire
(053) 42888

EDUCATION AND TRAINING

For a complete list of current education and training opportunities in Britain, contact the Soil Association. The Soil Association runs an approval scheme for organic courses. Not all the colleges listed below have had their courses vetted by the Soil Association.

AGRICULTURAL TRAINING BOARD
Management Training Centre
National Agricultural Centre
Kenilworth
Warwickshire
CV8 2LG
(0203) 696511

CARMARTHENSHIRE COLLEGE OF
   TECHNOLOGY AND ART
(Horticultural courses)
Pibwrlwyd Campus
Carmarthen
Dyfed SA31 2NH
(0267) 234151

DERBYSHIRE COLLEGE OF AGRICULTURE
  AND HORTICULTURE
Broomfield
Morley
Derbyshire
DE7 6DN
(0332) 831845

EMERSON COLLEGE
Pixton
Forest Row
Sussex RH18 5JX
(0342) 822238

GREENMOUNT AGRICULTURAL AND
  HORTICULTURAL COLLEGE
Antrim
Northern Ireland BT41 4PU
(0849) 412114

KILDALTON COLLEGE
Piltown
Co. Kilkenny
Eire
(051) 43105

LACKHAM COLLEGE OF AGRICULTURE
Lacock
Chippenham
Wiltshire
SN15 2NY
(0249) 443111

OTLEY COLLEGE OF AGRICULTURE AND
  HORTICULTURE
Otley
Ipswich
Suffolk
IP6 9EY
(0473) 85350

PERSHORE COLLEGE OF HORTICULTURE
Pershore
Worcestershire
WR10 3JP
(0386) 552443

SCOTTISH AGRICULTURAL COLLEGE
Organic Farming Centre, Edinburgh
(*see above*)

WELSH AGRICULTURAL COLLEGE
Llanbadarn Fawr
Aberystwyth
Dyfed
SY23 3AL
(0970) 624471

WORCESTERSHIRE COLLEGE OF
  AGRICULTURE AND HORTICULTURE
Hindlip
Worcester
WR3 8SS
(0905) 51310

WORKING WEEKENDS ON ORGANIC FARMS
  (WWOOF)
(*see above*)

ORGANIC STANDARDS ORGANISATIONS

BIO-DYNAMIC AGRICULTURAL
  ASSOCIATION
(*see above*)

INTERNATIONAL FEDERATION OF ORGANIC
  AGRICULTURE MOVEMENTS (IFOAM)
(*see above*)

IRISH ORGANIC FARMERS AND GROWERS
  ASSOCIATION
(*see above*)

ORGANIC FARMERS AND GROWERS LTD
9 Station Approach
Needham Market
Stowmarket
Suffolk IP6 8AT
(0449) 720838

SOIL ASSOCIATION
(*see above*)

UNITED KINGDOM REGISTER OF ORGANIC
  FOOD STANDARDS (UKROFS)
Food from Britain
301–344 Market Towers
New Convent Garden Market
Nine Elms Lane
London SW8 5NQ
(071) 720 2144

ORGANIC PRODUCER MARKETING CO-OPERATIVES AND ASSOCIATIONS

CLONMEL ORGANIC GROWERS
c/o Honey Pot
Clonmel
Co. Tipperary
Eire

EASTERN COUNTIES ORGANIC PRODUCERS
E.C.O.P.S.
Strawberry Fields
Scarborough Bank
Stickford
Boston
Lincs.
(0205) 480490

GREEN GROWERS ORGANIC PRODUCE
Country Mills
Worcester
WR1 3NU
(08854) 10204

NORTH LEITRIM VEGETABLE GROWERS
  ASSOCIATION
c/o Rod Alton
Rosinver
Manor Hamilton
Co. Leitrim
Eire

ORGANIC FARMERS AND GROWERS LTD
(see above)

ORGANIC FARMERS AND GROWERS
  (SCOTLAND) LTD
Glenside
Plean
Stirling
FK7 8BA
(0786) 818855

ORGANIC FOOD MANUFACTURERS
  FEDERATION
The Tithe House
Peaseland Green
Elsing
East Dereham
Norfolk NR20 3DY
(0362) 83314

ORGANIC GROWERS WEST WALES
Unit 25
Llanbed Industrial Estate
Lampeter
Dyfed SA48 8TL
(0570) 422869

SCOTTISH ORGANIC PRODUCERS
  ASSOCIATION
18/19 Claremont Crescent
Edinburgh
EH7 4JW
(031) 556 6574

SOMERSET ORGANIC PRODUCERS LTD
Wyvern Farms
Seavington St. Michael
Illminster
Somerset TA19 0PZ
(0460) 42149

WELSH ORGANIC LIVESTOCK FARMERS
c/o Matthew Murton
Crynfryn Farm
Penuwch
Dyfed SY25 6RE
(0974) 23206

WEST CORK ORGANIC PRODUCERS
c/o Heiner Miller
Hoolyhill
Ballineen
Co. Cork
Eire

WEXFORD ORGANIC GROWERS
c/o David Storey
Bleachlands
Olygate
Enniscorthy
Co. Wexford
Eire

ORGANIC INPUT SUPPLIERS

The Soil Association has a scheme for agricultural inputs with a list of approved suppliers. Further information can also be obtained from advertisements in New Farmer and Grower and in 'Organic Farming' published by British Organic Farmers and the Organic Growers Association.

ORGANIC PRODUCE OUTLETS

Retail outlets including farm shops and specialist distributors of imported produce (e.g. wines) are comprehensively listed in *Thorson's Organic Consumer Guide* (Thorsons Publishers Ltd, Wellingborough). For this reason, only the major distributors, processors and a few retailers are listed here.

General

ALL ORGANIC COMPANY
39 London Fruit Exchange
Brushfield St.
London E1 6EU
(071) 247 4556/247 4557

THE LAND AND FOOD COMPANY PLC
1 Juniper Cottage
Wick Street
Stroud
Gloucestershire GL6 7QR
(0453) 860844

THE REAL FOODS GROUP
14 Ashley Place
Edinburgh EH6 5PX
(031) 554 4321

REAL FOOD SUPPLIERS
Phil Haughton
36c Gloucester Road
Bristol
(0272) 232015

Vegetables

GEEST PRODUCE MARKETING
West Marsh Road
Spalding
Lincs. PE11 2AL
(0775) 761111 ext. 2291

ORGANIC FARM FOODS (SCOTLAND)
Block 9
Whiteside Industrial Estate
Bathgate
West Lothian EH48 2RX
(0506) 632911

ORGANIC FARM FOODS (WALES)
Unit 25
Llanbed Ind. Estate
Tregaron Rd.
Lampeter
Dyfed SA48 8LY
(0570) 423280

ORGANIC FOODS
c/o Denis Healy
Kiltealy
Co. Whicklow
Eire

OGA PACKAGING
c/o J. Dalby
114 Sundorne Road
Shrewsbury
Shropshire SY1 4RR
(0743) 235638

PRODUCER CO-OPERATIVES
(*see above*)

Cereals
ALLIED MILLS LTD
Beech Flour Mills
Bishop's Stortford
Herts
CM23 3BU
(0279) 655676

BATCHLEY MILL
(Livestock feeds)
John Wakefield-Jones
Batchley
Grendon Bishop
Bromyard
Herefordshire HR7 4TH
(0885) 483377

DOVES FARM FLOUR
Salisbury Road
Hungerford
Berkshire RG17 0RF
(0488) 684880

W M GLEADELL & SONS LTD
Bill Starling
Lindsey House
Hemswell Cliff
Gainsborough DN21 5TH
(0427) 73661 (Work) or
(0673) 83259 (Home)

W JORDANS (CEREALS) LTD
Holme Mills
Biggleswade
Bedfordshire SG18 9JX
(0767) 318222

W H MARRIAGE AND SONS LTD
Chelmer Mills
New Street
Chelmsford
Essex CM1 1PN
(0245) 354455

MORNING FOODS LTD
North Western Mills
Crewe
Cheshire CW2 6HP
(0270) 213261

PIMHILL
R. Mayall & Daughter
Lea Hall
Harmer Hill
Shrewsbury
Shropshire SY4 3DY
(0939) 290342

PRODUCER CO-OPERATIVES
(*see above*)

RUSHALL FARMS
The Manor
Upavon
Pewsey
Wiltshire
(0980) 630264

SHIPTON MILL
Long Newton
Tetbury
Gloucestershire GL8 8RP
(0666) 53620

Meat
GOOD HERDSMEN
Ballybrado
Cahir
Co. Tipperary
Eire

GREENWAY ORGANIC FARMS
Freepost
Edinburgh EH1 0AQ
(031) 557 8111

EVAN OWEN JONES
(Meat wholesaler)
Briwnant
Pumsaint
Llanwrda
Dyfed
SA19 8UT
(05585) 410

KITE'S NEST FARM
Broadway
Worcs. WR12 7JT
(0386) 853320

LLOYD MAUNDER
Willand
Collompton
Devon EX15 2PJ
(0884) 820534

IAN MILLER
Jamesfield Farm
Newburgh
Fife KY14 6EW
(0738) 85498

THE PURE MEAT COMPANY
Coombe Court Farm
Moreton Hamstead
Devon
(0647) 40321

PRODUCER CO-OPERATIVES
(see above)

THE REAL MEAT COMPANY LTD
East Hill Farm
Heytesbury
Warminster
Wiltshire BA12 0HR
(0985) 40436/40060

Milk and Dairy Produce
BUSSES FARM
Harwoods Lane
East Grinstead
West Sussex
RH19 4NL
(0342) 21749

OLD PLAW HATCH FARM
Sharpstone
East Grinstead
Sussex
RH19 4UL
(0342) 810857

RACHEL'S DAIRY
Brynllys
Borth
Dyfed
(0970) 871489

UNIGATE DAIRIES LTD
Station Yard
Totnes
Devon
TQ9 5JP
(0803) 38761

WELSH ORGANIC FOODS
Unit 20/21
Llambed Industrial Estate
Tregaron Road
Lampeter
Dyfed SA48 8TL
(0570) 422772

Beverages
ASPALLS
Aspall Hall
Stowmarket
Suffolk
IP14 6PD
(0728) 860 510

HENRY DOUBLEDAY RESEARCH
  ASSOCIATION
(see above)

PEAKES ORGANIC FOODS AND COPELLA
  FRUIT JUICES
Hill Farm
Boxford
Colchester
Suffolk CO6 5NY
(0787) 210348/210496

ASSOCIATED ORGANISATIONS

A complete list of organic and associated organisations in Britain can be obtained from the Soil Association.

BRITISH ASSOCIATION OF VETERINARY
  HOMOEOPATHY
Chinham House
Stanford-in-the-Vale
Faringdon
Oxon SN7 8NQ

FARM AND FOOD SOCIETY
4 Willifield Way
London
NW11 7XT
(081) 455 0634

THE LONDON FOOD COMMISSION
88 Old Street
London EC1V 9AR
(071) 253 9513

MCCARRISON SOCIETY
24 Paddington Street
London
W1M 4DR
(071) 935 3924

PERMACULTURE ASSOCIATION
8 Hunters Moon
Dartington
Totnes
Devon TQ9 6JT
(0803) 867546

RARE BREEDS SURVIVAL TRUST
4th Street
National Agricultural Centre
Stoneleigh
Kenilworth
Warwickshire
CV8 2LG

DEVELOPING COUNTRIES

More detailed lists of contacts and addresses can be obtained from IFOAM and AGRECOL.

AGRECOL DEVELOPMENT INFORMATION
c/o Ökozentrum
CH-4438 Langenbruck
Switzerland
(062) 601420

APPROPRIATE TECHNOLOGY PROJECT
(Microfiche library containing 900 books)
Volunteers in Asia Inc.
PO Box 4543
Stanford
California 94305
USA

FARMERS THIRD WORLD NETWORK
The Arthur Rank Centre
National Agricultural Centre
Stoneleigh
Kenilworth
Warwickshire CV8 2LZ
(0203) 696969

INFORMATION CENTRE ON LOW
  EXTERNAL INPUT AGRICULTURE
  (ILEIA)
PO Box 64
NL-3830 AB Leusden
Netherlands
(033) 943086

INTERMEDIATE TECHNOLOGY
103–105 Southampton Row
London WC1B 4HH

INTERNATIONAL FEDERATION OF ORGANIC
  AGRICULTURAL MOVEMENTS
(see above)

NORFOLK EDUCATION AND ACTION FOR
  DEVELOPMENT
Development and Environment Centre
38–40 Exchange Street
Norwich
Norfolk NR2 1AX

LITERATURE SOURCES

Books

Most organic farming texts are available from the Soil Association, Henry Doubleday Research Association, Bio-Dynamic Agricultural Association and other organic movement organisations.

Periodicals

BIOLOGICAL AGRICULTURE AND
  HORTICULTURE
AB Academic Publishers
PO Box 97
Berkhamsted
Herts HP4 2PX

HDRA NEWSLETTER
Henry Doubleday Research Association
(*see above*)

THE LIVING EARTH
Soil Association
(*see above*)

NEW FARMER AND GROWER
British Organic Farmers/Organic
  Growers Association
(*see above*)

AMERICAN JOURNAL OF ALTERNATIVE
  AGRICULTURE
Institute for Alternative Agriculture
9200 Edmonston Road, Suite 117
Greenbelt
Maryland 20770
USA

ORGANIC GARDENING
Rodale Press
Emmaus
Pennsylvania 18098
USA

THE NEW FARM
Regenerative Agriculture Association
222 Main Street
Emmaus
Pennsylvania 18098
USA

SOIL AND HEALTH
Soil and Health Association of New
  Zealand
Box 2824
Auckland
New Zealand

IFOAM ENGLISH-LANGUAGE BULLETIN/
  ECOLOGY AND AGRICULTURE
IFOAM
(*see above*)

ABSTRECO
Department of Ecological Agriculture
Haarweg 333
NL-6709 RZ Wageningen
Netherlands

A number of German language periodicals have been extensively referred to in the text—these can be obtained from the addresses below.

GERMAN-LANGUAGE IFOAM BULLETIN/
  ÖKOLOGIE UND LANDBAU
Stiftung Ökologischer Landbau
Postfach 1516/Weinstrasse Süd 51
D-6702 Bad Dürckheim
Germany
(06322) 8666

LEBENDIGE ERDE
Forschungsring für biologisch-dynamische
  Wirtschaftsweise
Baumschulenweg 11
D-6100 Darmstadt
Germany
(06155) 2674

ZUM BEISPIEL
Forschungsinstitut für biologischen
  Landbau
CH-4104 Oberwil
Switzerland
(061) 401 4222

BIOLAND
Bioland e. V.
Barbarossastr. 14
D-7336 Uhingen
Germany
(07161) 31012

Appendix 5

Latin Names of Plants, Pests and Diseases

Abbreviations

spp. – species
ssp. – sub species

form. – formerly
var. – variety

COMMON NAME

LATIN NAME

Non-crop plants

Alison, small	*Alyssum alyssoides*
Amaranth, common (pigweed)	*Amaranthus retroflexus*
Basil thyme	*Acinos arvensis*
Bee's friend	*Phacelia* spp.
Bellflower, creeping	*Campanula rapunculoides*
Bent, black	*Agrostis gigantea*
Bent, common	*Agrostis capillaris* (form. *tenuis*)
Bent, creeping	*Agrostis stolonifera*
Bent, silky	*Agrostis spica-venti*
Bindweed, black	*Bilderdykia* (form. *Polygonum*) *convolvulus*
Bindweed, field	*Convolvulus arvensis*
Bindweed, hedge	*Calystegia sepium*
Bird cherry	*Prunus padus*
Bistort, amphibious	*Polygonum amphibium*
Blackgrass	*Alopecurus myosuroides*
Borage, common	*Borago officinalis*
Bracken	*Pteridium aquilinum*
Bristle grass	*Panicum sanguinale*
Brome, erect	*Bromus erectus*
Brome, field	*Bromus arvensis*
Brome, soft	*Bromus hordeaceus* (form. *mollis*)
Brome, sterile (barren)	*Bromus sterilis*
Bur-parsley, small	*Caucalis platycarpos*
Buttercup, creeping	*Ranunculus repens*
Cabbage, hare's-ear	*Coringia orientalis*
Campion	*Silene noctiflora*
Campion, bladder	*Silene vulgaria*
Campion, white	*Silene alba*
Caraway	*Carum carvi*
Carrot, wild	*Daucus carota*
Cat's ear	*Hypochaeris radicata*
Catchfly, night-flowering	*Silene noctiflora*
Catchfly, small-flowered	*Silene gallica*
Chamomile	*Chamaemelum nobile*
Chamomile, corn	*Anthemis arvensis*
Chamomile, wild	*Chamomilla* (form. *Matricaria*) *recutita*

COMMON NAME	LATIN NAME
Charlock	*Sinapis arvensis*
Chickweed, common	*Stellaria media*
Chickweed, jagged	*Holosteum umbellatum*
Chicory	*Chicorum intybus*
Cinquefoil, creeping	*Potentilla reptans*
Cleavers	*Galium aparine*
Clover, haresfoot	*Trifolium arvense*
Cockspur	*Echinochloa crus-galli*
Coltsfoot	*Tussilago farfara*
Comfrey	*Symphytum officinale*
Corncockle	*Agrostemma githago*
Cornflower	*Centaurea cyanus*
Couch	*Elymus (*form. *Agropyron) repens*
Couch, onion	*Arrhenatherum elatius* var. *bulbosus*
Cress, creeping yellow	*Rorippa sylvestris*
Cress, hoary (pepperwort)	*Cardaria draba*
Cuckooflower (Lady's smock)	*Cardamine pratensis*
Cudweed, Jersey	*Gnaphalium luteoalbum*
Daisy	*Bellis perennis*
Dandelion	*Taraxacum officinale*
Darnel	*Lolium temulentum*
Dead-nettle, red	*Lamium purpureum*
Dock, broad-leaved	*Rumex obtusifolius*
Dock, curled	*Rumex crispus*
Fat-hen	*Chenopodium album*
Fingergrass	*Digitaria* spp.
Forget-me-not	*Myosotis arvensis*
Foxglove	*Digitalis purpurea*
Foxtail, field	*Alopecurus agrestis*
Foxtail, giant	*Setaria faberii*
Foxtail, meadow	*Alopecurus pratensis*
Fumitory, common	*Fumaria officinalis*
Gallant soldier	*Galinsoga parviflora*
Gorse	*Ulex* spp.
Gromwell, corn	*Buglossoides (*form.*Lithospermum) arvensis*
Ground-elder (Bishop's weed)	*Aegopodium podagraria*
Ground-pine	*Ajuga chamaepitys*
Groundsel	*Senecio vulgares*
Heather	*Calluna vulgaris*
Hemlock	*Conium maculatum*
Hemp-nettle, common	*Galeopsis tetrahit*
Holly	*Ilex aquifolium*
Horsetail, field	*Equisetum arvense*
Japanese knotweed	*Reynoutia japonica (*form. *Polygonum cuspidatum)*
Knawel, annual	*Scleranthus annuus*
Knotgrass	*Polygonum aviculare*
Lady's smock, field	*Cardamine pratensis*
Lettuce, prickly	*Lactuca serriola*

COMMON NAME	LATIN NAME
Madder, field	*Sherardia arvensis*
Mallow, dwarf	*Malva rotundifolia*
Mallow, small	*Malva pusilla*
Marigold, corn	*Chrysanthemum segetum*
Mayweed, scented	*Chamomilla (form. Matricaria) recutita*
Mayweed, scentless	*Matricaria perforata* (form. *Tripleurospermium inodorum*)
Mayweed, stinking	*Anthemis cotula*
Meadowgrass, annual	*Poa annua*
Meadowgrass, rough-stalked	*Poa trivialis*
Meadowgrass, smooth-stalked	*Poa pratensis*
Medick, black	*Medicago lupulina*
Mercury, annual	*Mercurialis annua*
Mercury, dog's	*Mercurialis perennis*
Mint, corn	*Mentha arvensis*
Moss	*Bryum, Hypnum & Polytrichum* spp.
Mouse-ear, common/chickweed	*Cerastium fontanum* (form. *holosteoides*)
Mouse-ear, field	*Cerastium arvense*
Mugwort	*Artemisia vulgaris*
Mustard, black	*Brassica nigra*
Mustard, hedge	*Sisymbrium officinale*
Mustard, white	*Sinapis alba*
Neem	*Azadirachta indica*
Nettle, common/stinging	*Urtica dioica*
Nettle, small/annual	*Urtica urens*
Nightshade, black	*Solanum nigrum*
Nightshade, deadly	*Atropa belladonna*
Oat, bristle	*Avena strigosa*
Oat, spring wild	*Avena fatua*
Oat, winter wild	*Avena sterilis* ssp. *ludoviciana*
Onion, wild	*Allium vineale*
Orache, common	*Atriplex patula*
Pansy, field	*Viola arvensis*
Pansy, wild	*Viola tricolor*
Parsley, cow	*Anthriscus sylvestris*
Parsley, fool's	*Aethusa cynapium*
Parsley-piert	*Aphanes arvensis*
Pearlwort, procumbent	*Sagina procumbens*
Pennycress, field	*Thlaspi avense*
Persicaria, pale	*Polygonum lapathifolium* ssp. *pallidum*
Pigweed, prostrate	*Amaranthus blitoides*
Pimpernel	*Anagallis* spp.
Pimpernel, bog	*Anagallis tenella*
Plantain, greater	*Plantago major*
Plantain, ribwort	*Plantago lanceolata*
Poppy, common/corn/field	*Papaver rhoeas*
Radish, wild (runch)	*Raphanus raphanistrum*
Ragwort	*Senecio* spp.
Redshank	*Polygonum persicaria*

Common Name	Latin Name
Runch (wild radish)	*Raphanus raphanistrum*
Rupturewort	*Herniaria* spp.
Rush	*Juncus* spp.
Rush, jointed	*Juncus articulatus*
Salad burnet	*Poterium sanguisorba*
Sand-bur, wild	*Solanum rostratum*
Sheep's bit	*Jasiona montana*
Sheep's parsley	*Petroselinum crispum*
Shepherd's purse	*Capsella bursa-pastoris*
Silverweed	*Potentilla anserina*
Soft-grass, creeping	*Holcus mollis*
Sorrel, common	*Rumex acetosa*
Sorrel, sheep's	*Rumex acetosella*
Sow-thistle, corn/perennial	*Sonchus arvensis*
Sow-thistle, prickly	*Sonchus asper*
Sow-thistle, smooth/milk	*Sonchus oleraceus*
Speedwell, common/field	*Veronica persica*
Speedwell, ivy-leaved	*Veronica hederifolia*
Speedwell, procumbent	*Veronica agrestis*
Speedwell, spring	*Veronica verna*
Spindle tree	*Euonymus europaeus*
Spurge, dwarf	*Euphorbia exigua*
Spurge, sun	*Euphorbia helioscopia*
Spurrey, corn	*Spergula arvensis*
Spurrey, sand	*Spergula rubra*
Stitchwort, lesser	*Stellaria graminea*
Stork's bill	*Erodium cicutarium*
Thistle, creeping	*Cirsium arvense*
Thistle, spear	*Cirsium vulgare*
Thorow-wax	*Bupleurum rotundifolium*
Toadflax, common	*Linaria vulgaris*
Trefoil, common birdsfoot	*Lotus corniculatus*
Venus's looking-glass	*Legousia hybrida*
Vernal-grass	*Anthoxanthum* spp.
Vetch, hairy	*Vicia hirsuta*
Viper's bugloss	*Echium vulgare*
Water pepper	*Polygonum hydropiper*
Weld	*Reseda luteola*
Witchweed	*Striga* spp.
Woundwort, field	*Stachys arvensis*
Yarrow	*Achillea millefolium*

Crop plants

Alfalfa (lucerne)	*Medicago sativa*
Apple	*Pyrus malus*
Artichoke, Jerusalem	*Helianthus tuberosus*
Artichoke, globe	*Cynara scolymus*
Avocado	*Persea gratissima*
Banana	*Musa sapientum*

Common Name	Latin Name
Barley	*Hordeum sativum*
Bean, black	*Castanospermum australe*
Bean, field/broad	*Vicia faba*
Bean, green/dwarf/french	*Phaseolus vulgaris*
Bean, soya	*Glycine max*
Beet, sugar/fodder/root/etc.	*Beta vulgaris*
Broccoli	*Brassica oleracea* var. *botrytis*
Brussels sprouts	*Brassica oleracea* var. *bullata*
Buckwheat	*Fagopyrum esculentum*
Cabbage group	*Brassica oleracea*
Carrot	*Daucus carota* var. *sativa*
Cassava	*Manihot esculenta*
Cauliflower	*Brassica oleracae* var. *botrytis*
Celery	*Apium graveolens*
Chinese leaves	*Brassica rapa* var. *pekinensis*
Clover, alsike	*Trifolium hybridum*
Clover, crimson	*Trifolium incarnatum*
Clover, red	*Trifolium pratense*
Clover, subterranean	*Trifolium subteranneum*
Clover, white	*Trifolium repens*
Cocksfoot	*Dactylis glomerata*
Cocoyam	*Colocasia & Xanthosoma* spp.
Coffee	*Coffea* spp.
Cotton	*Gossypium* spp.
Courgette	*Curcurbita pepo*
Cowpea	*Vigna unguiculata*
Cucumber	*Cucumis sativus*
Currant, black	*Ribes nigrum*
Currant, red	*Ribes sativum*
Fescue, meadow	*Festuca pratensis*
Fescue, red	*Festuca rubra*
Fescue, sheep's	*Festuca ovina*
Fescue, tall	*Festuca arundinacea*
Flax	*Linum usitatissimum*
Garlic	*Allium sativum*
Horseradish	*Armoracia rusticana*
Kale	*Brassica oleracea* var. *acephala*
Kohlrabi	*Brassica oleracea* var. *gongylodes*
Leek	*Allium ampeloprasum*
Lettuce	*Lactuca sativa*
Linseed	*Linum usitatissimum*
Lucerne (alfalfa)	*Medicago sativa*
Lupin, white	*Lupinus albus*
Lupin, yellow	*Lupinus luteus*
Maize	*Zea mays*
Mangel (mangold)	*Beta vulgaris*
Mango	*Mangifera indica*
Millet	*Panicum miliaceum*
Mustard, brown	*Brassica juncea*

COMMON NAME	LATIN NAME
Mustard, white	*Sinapsis/Brassica alba*
Oat	*Avena sativa*
Okra	*Hibiscus esculentus*
Onion	*Allium cepa*
Parsley	*Petroselinum hortense*
Parsnip	*Pastinaca sativa*
Pea, field/threshing	*Pisum sativum*
Pea, fodder	*Pisum arvense*
Peach	*Prunus persica*
Peanut	*Apios tuberosa*
Pepper	*Capsicum* spp.
Potato	*Solanum tuberosum*
Primrose, evening	*Oenothera biennis*
Radish	*Raphanus sativa*
Radish, fodder	*Raphanus sativus* var. *campestris*
Rape, forage	*Brassica napus*
Rape, oilseed	*Brassica campestris* var. *oleifera*
Raspberry	*Rubus idaeus*
Rhubarb	*Rheum rhapontium*
Ribgrass (plantain)	*Plantago lanceolata*
Rye	*Secale cereale*
Ryegrass, Italian	*Lolium multiflorum*
Ryegrass, Westerwold	*Lolium multiflorum* var. *westerwoldicum*
Ryegrass, perennial	*Lolium perenne*
Sainfoin	*Onobrychis viciifolia*
Sesame	*Sesamum indicum*
Sorghum	*Holcus halepensis*
Spinach	*Spinacia* spp.
Squash	*Curcubita* spp.
Strawberry	*Fragaria chiloensis*
Swede/swede-kale (rutabaga)	*Brassica napus*
Sweet potato	*Ipomoea batatas*
Sweetcorn	*Zea mays* var. *saccharata*
Tare – see Vetch	*Vicia* spp.
Timothy	*Phleum pratense*
Tobacco	*Nicotiana tabacum*
Tomato	*Lycopersicou esculentum*
Trefoil	*Medicago lupulina*
Triticale	*Triticosecale*
Turnip group	*Brassica rapa*
Vetch, hairy	*Vicia hirsuta/villosa*
Vetch, kidney	*Anthyllis* spp.
Vetch, summer	*Vicia lathyroides*
Vetch, winter	*Vicia sativa*
Wheat	*Triticum aestivum*

Plant diseases

Blackleg, potato	*Erwinia (carotovora)* var. *atroseptica*
Blight, potato	*Phytophtora infestans*

COMMON NAME	LATIN NAME
Bunt	*Tilletia caries*
Canker, stem oilseed rape	*Phoma lingam*
Chocolate spot	*Botrytis cinerea (B. fabae)*
Clover rot	*Sclerotinia trifoliorum*
Clubroot	*Plasmodiophora brassicae*
Crown gall	*Agrobacterium tumefaciens*
Ergot	*Claviceps purpurea*
Eyespot	*Pseudocercosporella herpotrichoides*
Leaf spot, light	*Septoria tritici*
Loose smut	*Ustilago nuda*
Mildew, downy	*Peronospora* spp.
Mildew, powdery	*Erysiphe* spp.
Scab, common potato	*Streptomyces scabies*
Scurf, black potato	*Corticium (Rhizoctonia) solani*
Stem rust, blackcurrant	*Cronarticum ribicola*
Take-all	*Gaeumannomyces graminis*

Plant pests and beneficial organisms

Aphid, black bean	*Aphis fabae*
Aphid, mealy cabbage	*Brevicoryne brassicae*
Beetle, colorado potato	*Leptinotarsa decemlineata*
Beetle, bruchid	*Bruchus* spp.
Beetle, click	*Agriotes* spp.
Beetle, flea	*Phyllotreta* spp.
Beetle, small ground	*Carabidae*
Borer, European corn	*Ostrinia nubilalis*
Butterfly, cabbage white	*Pieris* spp.
Eelworm – see Nematode	*Heterodera* spp.
Flea, water	*Rivulogammarus pulex*
Fly, cabbage root	*Erioischia brassicae*
Fly, carrot	*Psila rosae*
Fly, crane	*Tripula* spp.
Fly, fruit	*Oscinella frit*
Fly, onion	*Hylemyia antiqua*
Fly, turnip root	*Delia* spp.
Fly, wheat bulb	*Leptohylemyia coarctata*
Fly, white	*Trialeuroides vaporariorum*
Grass grub	*Costelytra zealandia*
Grasshopper	*Orthoptera*
Hoverfly	*Syrhipidae*
Lacewings	*Chrysopidae & Kimminsia* spp.
Ladybird	*Coccinellidae*
Leafminer, cabbage	*Phytomyza rufipes*
Leather jacket	*Tipula* spp.
Locust	*Locusta migratoria*
Mite, red spider	*Tetranychus urticae*
Moth, codling	*Carpocapsa pomonella*
Moth, turnip	*Mamestra brassicae/Agrotis segetum*
Nematode, beet cyst	*Heterodera schachtii*

COMMON NAME	LATIN NAME
Nematode, cereal cyst	*Heterodera avenae*
Nematode, grass cyst	*Heterodera punctata*
Nematode, pea root/cyst	*Heterodera gottingiana*
Nematode, potato cyst	*Heterodera rostochiensis & Globodera* spp.
Nematode, root-knot	*Meloidogyne* spp.
Nematode, stem and bulb	*Ditylenchis dipsaci*
Slugs	*Milax, Arion & Deroceras* spp.
Snail, garden	*Helix aspersa*
Springtail	*Onychiurus* spp.
Thrips	*Limothrips cerealium*
Wasp, chalcid	*Encarsia formosa*
Wasp, cynipid	*Trybliographa rapae*
Weevil, cabbage gall	*Ceuthorryncus pleurostigma*
Weevil, pea and bean	*Sitona* spp.
Wireworms	*Agriotes* spp.

Livestock parasites

Blow fly	*Lucilia sericata*
Coccidia	*Eimeria* spp.
Keds	*Melophagus ovinus*
Lice	*Haematopinus* spp.
Liver fluke	*Faciola hepatica*
Lungworm	*Dictyocaulus viviparus*
Sheep scab	*Psoroptes* spp.
Tapeworm	*Moniezia* spp.
Tick	*Ixodoidea*
Warble fly	*Hypoderma* spp.

Appendix 6

Some Metric Conversion Factors

BRITISH TO METRIC

Length

inches to cm	× 2.54
or mm	× 25.4
feet to m	× 0.305
yards to m	× 0.914
miles to km	× 1.61

Area

sq feet to m^2	× 0.093
sq yards to m^2	× 0.836
acres to ha	× 0.405

Volume (Liquid)

| pints to litres | × 0.568 |
| gallons to litres | × 4.55 |

Weight

ounces to g	× 28.3
pounds to g	× 454
pounds to kg	× 0.454
hundredweights to kg	× 50.8
hundredweights to t	× 0.0508
tons to kg	× 1016
tons to tonnes	× 1.016

Temperature

$(°C × 1.8) + 32 = °F$

METRIC TO BRITISH

centimetres to in	× 0. 394
millimetres to in	× 0.0394
metres to ft	× 3.29
metres to yd	× 1.09
kilometres to miles	× 0.621

sq metres to ft^2	× 10.8
sq metres to yd^2	× 1.20
hectares to ac	× 2.47

| litres to pints | × 1.76 |
| litres to gallons | × 0.22 |

grams to oz	× 0.0353
grams to lb	× 0.0022
kilograms to lb	× 2.20
kilograms to cwt	× 0.020
tonnes to tons	× 0.984

Some double conversions

fertiliser units/acre	× 1.25	= kg/hectare
cwt/acre	× 0.125	= t/ha
lb/acre	× 1.1	= kg/ha
pints/acre	× 1.4	= litres/ha

Index

Page numbers in italics indicate illustrations and tables

691